Remote Sensing

for GIS Managers

Stan Aronoff

ESRI PRESS

REDLANDS, CALIFORNIA

ESRI Press, 380 New York Street, Redlands, California 92373-8100

All rights reserved. First edition 2005
10 09 08 07 06 05   1 2 3 4 5 6 7 8 9 10

Printed in the United States of America

Library of Congress Cataloging-in-Publication Data
Aronoff, Stanley.
    Remote sensing for GIS managers / Stan Aronoff.—1st ed.
      p.   cm.
    Includes bibliographical references and index.
    ISBN 1-58948-081-3 (hardcover : alk. paper)
    1. Remote sensing. 2. Geographic information systems. I. Title.
  G70.4A76 2005
  621.36'78—dc22   2005013391

Ask for ESRI Press titles at your local bookstore or order by calling 1-800-447-9778. You can also shop online at www.esri.com/esripress. Outside the United States, contact your local ESRI distributor.

ESRI Press titles are distributed to the trade by the following:

In North America, South America, Asia, and Australia:
Independent Publishers Group (IPG)
Telephone (United States): 1-800-888-4741
Telephone (international): 312-337-0747
E-mail: frontdesk@ipgbook.com

In the United Kingdom, Europe, and the Middle East:
Transatlantic Publishers Group Ltd.
Telephone: 44 20 8849 8013
Fax: 44 20 8849 5556
E-mail: transatlantic.publishers@regusnet.com

Cover design by Savitri Brant
Book design and production by Savitri Brant
Copyediting by Kandy Lockard
Print production by Cliff Crabbe

To my children, Michael and Laura

Contents

Foreword

At the beginning of GIS some thirty-five years ago, NASA launched and deployed its earliest nonmilitary satellite systems to collect and interpret images using a technology called remote sensing. At that time, remote sensing offered limited value for GIS users, partly because satellite images arrived months later with low spatial and multispectral resolution.

Today, we not only have widely available, high-resolution remote sensing images from earth-orbiting spacecraft like QuickBird, but also a rich selection of data derived from airborne sensors such as lidar and other advanced hyperspectral imaging systems. We're seeing an exponential growth in the coverage and availability of imagery for every place on the planet. Computers and networks are getting faster, and we can store much more imagery and make it available across networks. In ten years, computer speed, storage devices, and networks will increase in ways that will make today's systems seem antiquated by comparison. We'll have more imagery in real time, streaming down from drones and many other types of instruments.

The age of imagery as a component of GIS is becoming quite interesting. GIS software started integrating image data in the early 1980s. This led to the "image integrator," a new environment that let us simultaneously pan, zoom, display, and query raster and vector. This became the foundation for digital photogrammetry and feature extraction. It also served as the framework for many types of spatial analysis, i.e., raster analysis, which led to spatial data modeling using raster grids.

Today, server-based image services can handle tens of thousands of files in continuous maps of image information. We can exploit the analysis of imagery with highly sophisticated classification engines. The emergence of new tools to extract features in mono- and stereoscopic environments results in faster and higher resolution understanding of the earth, done more timely. New tools better integrate this data with standard vector spatial data. This means that imagery has become a major and dominant source of spatially referenced information for the GIS community. GIS technology now integrates remotely acquired image data in more effective ways that make it synergistic with the other data domains. We can leverage remote sensing data with GIS for measurement and observation, and for spatial analysis, visualization, and decision support. Examples of this abound in the text. The accumulation of spatial data will bring better links of imagery to spatial analysis models and more temporal modeling. Improved GIS tools will integrate with multidimensional datasets. The bottom line is that technology-the observation tools and software and networks and computers-will inevitably support more imagery about any place on Earth, from every place on Earth. We'll see the world any time, any place, at any resolution in mediums ranging from black-and-white photography to hyperspectral information.

Stan Aronoff has been around the GIS world for many years. His classic 1989 book, *Geographic Information Systems: A Management Perspective*, clearly communicated the power and effectiveness of GIS as large public and private organizations began to embrace technology. His new book, *Remote Sensing for GIS Managers*, builds a bridge between the image world and the geospatial vector world, explaining the synergy between GIS and remote sensing. This multilayer integration of raster and vector data within the GIS environment is vitally important as we move forward. Remote sensing data is one of the major drivers now accelerating GIS diffusion and effectiveness. I hope and trust this book will provide the information you need to integrate remote sensing information into your unique application of GIS.

Jack Dangermond

Acknowledgments

In writing this book, I am indebted to the many friends, colleagues, and family members who provided me with encouragement and source materials, reviewed drafts of the manuscript, and tended to my well-being through what were sometimes very trying times.

Audrey Kaplan helped to keep the project going when I was running short of patience or energy and gave valuable comments on portions of the text. I also enjoyed the steadfast support of my parents, Allan and Leiba. My father believed this was an important project, and I wish he had lived to see its successful completion.

My children, Laura and Michael, provided an oft needed distraction and perspective on what was really important. When they were younger, they would often say "computer working, Daddy unavailable" when I was busy writing my books. Though they are older now, the sentiment is as true for this book as it was for my first. This writing project took time and energy that made me less available and often less patient with them than I would like to have been.

I would like to thank the authors of the contributed chapters and sections that appear in this book. The field of remote sensing and geographic information systems has broadened and diversified to such an extent that it has become increasingly difficult for a single author to cover all the material. I value their significant contributions to this project as well as their advice and counsel on portions of my text.

I am indebted to the following individuals who gave generously of their time to review and comment on sections of the manuscript: Gerald Bawden, Karl Brown, James Campbell, Michael Cherlet, Richard Dobbins, Tom Eiber, Christopher D. Elvidge, Allan Falconer, Lanny Faleide, Peter Fricker, James L. Galloway, Peter L. Hays, Karl Heidemann, Frank Hissong, Roger Hoffer, Denny Kalensky, Gerald Kinn, Serge Lévesque, Xiaopeng Li, Don Light, Ralph E. Meiggs, Bryan Mercer, Paul Mrstik, Andrew Nadeau, Scott Paterson, Paul Pearl, Mike Renslow, Doug Richman, Vincent Salomonson, Gary Scoffield, Peg Shippert, Walter H. F. Smith, David Swann, Gary Wagner, Stewart Walker, and John P. Young.

I would especially like to thank Dr. Gordon Petrie for his extensive review of several of the chapters and his support and encouragement at critical stages of the project.

My approach to the material discussed in this book is that it is most easily explained when coupled with illustrations. For this reason, *Remote Sensing for GIS Managers* contains an unusually large number of images and diagrams. I appreciate the generosity of the many individuals who made available source materials and gave permission for their use.

This book would not have been published without the support of Jack Dangermond, president of ESRI, and the many individuals at ESRI Press with whom I had the pleasure to work. The project was developed in consultation with Christian Harder. Judy Hawkins managed the project. Tiffany Wilkerson edited the text and steered the project through production. Her numerous suggestions improved the clarity and flow of the text. Savitri Brant designed the cover and the book and redrafted many of the illustrations and tables. Kandy Lockard copyedited and proofread the book. Lesley Downie, Pam Spiva, Tiffany Wilkerson, and Judy Hawkins obtained permission to use the numerous illustrations included in this volume. Cliff Crabbe supervised print production, and Steve Hegle provided administrative support.

I would like to thank some of my colleagues and mentors who guided and encouraged me in my endeavors in the remote sensing and GIS fields: Bob Colwell, Allan Falconer, Val Geist, Allan Levinsohn, Delphine McClellan, Bill Ross, and Grant Ross. The Canada Centre for Remote Sensing has supported my work in many ways throughout my career, and I am indebted to the many individuals who supported me in this book project as they did for my first book, *Geographic Information Systems: A Management Perspective*. I would also like to thank the staff of the Map Library at the University of Ottawa who generously assisted me in locating materials and allowed me access to their facility and resources.

My purpose in writing this book was to collect, in one volume, practical information on remote sensing that would be of value to those who use the data and those who need an appreciation of the capabilities of the technology. I believe that the continued development of geospatial information science requires the intelligent integration of the diverse technologies and analysis methods available. To do so requires the people who use them or manage their use to understand the fundamentals of how the various geospatial technologies (including remote sensing and geographic information systems) work and the types and quality of information these technologies can be used to produce.

Ultimately, I believe it is the technical proficiency and creativity of the people who develop and use these technologies and the vision of those who manage these people that largely determine whether the applications of those technologies will be successful. I hope that this book encourages the integrated use of geospatial technologies in general and, in particular, the wider use of remote sensing products and services.

Stan Aronoff
Ottawa, Canada

Contributors

Zeev Berger
President, Image Interpretation Technologies, Inc.
zeev@iitech.ca

James Campbell
Professor of Geology and Remote Sensing, Virginia Polytechnic Institute and State University
jayhawk@vt.edu

Danny Fortin
Geologist, Image Interpretation Technologies, Inc.
dan@iitech.ca

Steven Franklin
Vice President of Research, University of Saskatchewan
steven.franklin@usask.ca

Kevin Gallo
Physical Scientist, NOAA/NESDIS Center for Satellite Applications and Research
Visiting scientist at USGS National Center for Earth Resources Observation and Science
kgallo@usgs.gov

Timothy Haithcoat
Director, Geographic Resources Center, University of Missouri-Columbia
HaithcoatT@missouri.edu

Ronald Hall
Research Scientist, Natural Resources Canada, Canadian Forest Service, Northern Forestry Centre
rhall@nrcan.gc.ca

James Hipple
United States Department of Agriculture
Risk Management Agency
jhipple@tcq.net

Tom Last
President, ImStrat Corporation
tom_last@imstrat.on.ca

Scott Madry
President, Informatics International, Inc.
Research Associate Professor of Anthropology, University of North Carolina
at Chapel Hill
madry@informatics.org

James Merchant
Professor and Associate Director, Center for Advanced Land Management
Information Technologies, University of Nebraska-Lincoln
jmerchant1@unl.edu

Gordon Petrie
Professor of Topographic Science, Department of Geography and
Geomatics, University of Glasgow
gpetrie@geog.gla.ac.uk

Joseph Piwowar
Assistant Professor of Geography, University of Regina
joe.piwowar@uregina.ca

David Swann
Defense Solutions Manager, ESRI
dswann@esri.com

Michael Wulder
Research Scientist, Natural Resources Canada, Canadian Forest Service,
Pacific Forestry Centre
mike.wulder@nrcan.gc.ca

1 Introduction: Remote sensing and GIS

Stan Aronoff

People have always viewed the world from vantage points high above the landscape to better assess their surroundings. From the tops of rocky cliffs and the branches of the tallest trees, they discerned information vital to their survival, such as the identification of good hunting grounds or the approach of enemies. The value of the "bird's-eye view" increased dramatically when it was captured in the form of drawings and later in maps that could be studied and shared with others. Photography and digital imaging enabled aerial views to be acquired rapidly, making it practical to obtain and analyze imagery for large areas of the earth's surface. Satellite technology enabled the entire globe as well as the planets and many of their orbiting moons to be repeatedly imaged. Today diverse human endeavors depend on a steady flow of images acquired from vantage points high above the earth to inventory, assess, and manage resources from the local to the global scale. Analysis of this imagery is the discipline known as *remote sensing*.

Remote sensing—the science, technology, and art of obtaining information about objects from a distance—takes us well beyond the limits of human capabilities. It allows us to collect information over regions too costly, too dangerous, or too remote for human observers to directly assess. Remotely sensed data takes many forms, including aerial photography, digital satellite imagery, and radar. Vast ocean areas; extensive croplands and forests; active volcanoes; and areas of military conflict, extreme climate, or radioactivity are readily monitored using remote sensing technology.

Remote sensing offers important advantages over other methods of data collection that have led to its use in a wide range of applications. Remote sensing provides an overview that allows us to discern patterns and relationships not apparent from the ground *(figure 1.1)*. The following examples are a sample of the numerous uses and advantages of remote sensing.

Figure 1.1 Remote sensing offers a "bird's-eye view" that allows a person to discern features and patterns difficult to identify from the ground.

Source: Image courtesy of Optech Incorporated.

Aerial photography and digital imagery collected from aircraft, spacecraft, and ships are used to produce maps of the earth's land and seafloor topography, natural resources, and urban infrastructure. The status of earth resources can be assessed and monitored over time using imagery acquired rapidly and repeatedly over large areas and at low cost.

Meteorological satellites monitor the world's cloud patterns and generate global image coverage at least hourly *(figure 1.2)*. Earth observation satellites image the entire globe daily at low spatial resolution with less-frequent coverage provided by higher-resolution sensors. Together, satellite and airborne remote sensing instruments provide data that is essential for resource inventory and monitoring.

Figure 1.2 GOES-12 satellite image of the earth. This color composite image produced from a visible and two infrared spectral bands (channels 1, 3, and 4) shows details not discernible in normal color images.

Source: Image courtesy of NOAA.

Rapid availability of remote sensing imagery has also proved valuable in assessing disaster events and in assessing, planning, and monitoring emergency response activities *(figure 1.3)*. Additionally, archived data serves as a valuable historical record that allows changes in land cover, biophysical processes, and human activities to be monitored and analyzed over time *(figure 1.4 and figure 1.5)*.

Remote sensing technology can record wavelengths the human eye cannot see, such as infrared, to detect phenomena otherwise not visible, such as heat or radioactivity. For example, subtle differences in earth resource features that would normally appear the same or otherwise be obscured can be distinguished on infrared images, making this type of imagery valuable for vegetation analysis, geological mapping, and forest-fire detection *(figure 1.6)*.

Figure 1.3 The Loma Prieta earthquake of October 17, 1989, caused extensive damage in the San Francisco Bay region. Damage assessment made extensive use of remotely sensed imagery such as this aerial photograph of a freeway collapse in Oakland, California.

Source: Image courtesy of Pacific Aerial Surveys and HJW GeoSpatial, Inc.

Figure 1.5 Coal mines in Germany. This simulated natural-color satellite image captured by the ASTER sensor shows a 23 by 14 kilometer region in midwestern Germany. Agricultural fields surround the Hambach opencast coal mine on the right side of the image.

Source: Image courtesy of NASA/GSFC/MITI/ERSDAC/JAROS and the U.S./Japan ASTER Science Team.

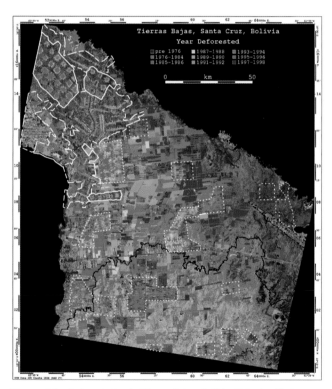

Figure 1.4 Deforestation in Tierras Bajas, Bolivia. This classified Landsat® 7 image shows the gradual conversion of forestlands for agricultural use in the Tierras Bajas region of Santa Cruz, Bolivia. Colors (see key) indicate the first year a site was developed for agriculture. White lines delineate human settlements.

Source: Image courtesy of M. Steininger and NASA/GSFC.

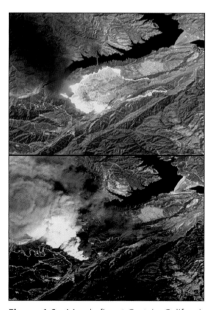

Figure 1.6 Marple fire at Castaic, California, 1996. Thermal infrared imagery flown by an airborne multispectral scanner remote sensing instrument is able to capture visible as well as thermal infrared spectral bands. The natural-color image produced from the visible bands (bottom image) shows what an observer would see. The fire area is obscured by smoke. The image generated from the thermal infrared radiation (top image), not visible to the human eye, clearly shows details of the fire. Hot areas in this images are represented by white, yellow, orange, and red, and the active fire front is white and yellow.

Source: Image courtesy of NASA and the Ames Research Center.

Biophysical measurements of earth resources can be derived from airborne and satellite imagery. For example, sea-surface temperature patterns show the location of currents and upwelling areas *(figure 1.7)*. Satellite-derived water color data is used to estimate near-surface chlorophyll concentrations, an important indicator of ecosystem health and the availability of food for fish stocks. Vegetation indexes calculated from visible and infrared imagery are used to monitor the health of agricultural crops, forests, and wetlands *(figure 1.8)*. Measurements derived from remote sensing data are also used in complex models that predict weather and climate such as El Niño events, which occur when warm water in the Pacific Ocean extends unusually far north, changing global weather patterns *(figure 1.9)*.

Digital elevation models, digital representations of earth surface topography used extensively in geographic information systems (GIS) for analysis and visualization, are produced from aerial photography, airborne digital imagery, and lidar and radar data *(figure 1.10)*. Satellite-based radar can measure sea-state conditions, including wave heights and wind direction. Precise spaceborne radar altimeter measurements of the sea surface are even used in mapping the seabed of the world's oceans, as subtle differences in ocean height are related to water depth.

Figure 1.8 SeaWiFS image of the biosphere. This image showing the Atlantic Ocean and surrounding landmasses uses data from the visible and infrared bands to offer an integrated presentation of plant vigor over land areas as measured by the Normalized Difference Vegetation Index (an indicator of the density of plant growth). Over ocean areas, chlorophyll concentration at the water surface, a measure of the amount of phytoplankton (microscopic plant life) living in the ocean, has been estimated.

Source: Image courtesy of NASA.

Remote sensing images present vegetation, geology, soils, hydrology, transportation network, and settlement patterns together showing their spatial relationship. The imagery used by a geologist to map geologic structures can also be used by wildlife biologists to map habitat and by regional planners to assess development options *(figure 1.11)*. The image data that earth resource satellites have collected for over 20 years now provide a valuable historical archive of multispectral digital imagery with global coverage. This data offers unique opportunities to analyze and monitor regional- and global-scale phenomena and develop models to study and anticipate phenomena such as global warming or depletion of the earth's protective ozone layer.

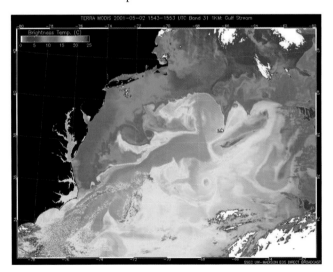

Figure 1.7 Gulf Stream, Eastern North America. This thermal infrared image captured by the NASA MODIS® sensor clearly shows the pattern of the Gulf Stream, the warm water current that flows east toward northern Europe (shown here as yellow, orange, and red, with red being warmest).

Source: Image courtesy of the University of Wisconsin-Madison Space Science and Engineering Center.

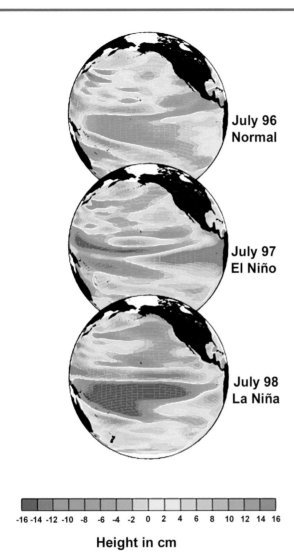

July 96
Normal

July 97
El Niño

July 98
La Niña

-16 -14 -12 -10 -8 -6 -4 -2 0 2 4 6 8 10 12 14 16

Height in cm

Figure 1.9   1997–1999 El Niño /La Niña Events. In 1997, warm water in the Pacific Ocean extended unusually far north. This El Niño condition tends to reduce fish catches. It was followed in the latter half of 1998 by a mild cold event (La Niña). These events affect sea surface height as tracked by a radar altimeter sensor on the NOAA Topex meteorological satellite.

Source: Image courtesy of NOAA.

Figure 1.10   Perspective view with Landsat overlay, Mojave to Ventura area, California. This perspective view was generated from a Landsat satellite image and digital elevation data derived from the satellite radar of the Shuttle Radar Topography Mission (SRTM). It shows the Tehachapi Mountains in the right foreground, the city of Ventura on the coast at the distant left, and the easternmost Santa Ynez Mountains forming the skyline at the distant right.

Source: Image courtesy of NASA, JPL, and NIMA.

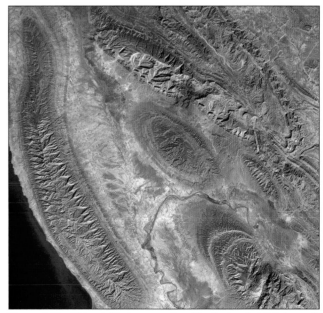

Figure 1.11   This satellite image captured by the Landsat 7 Enhanced Thematic Mapper in February 2000 includes the Mand River and the small town of Konari nestled in the Zagros Mountains of western Iran. This type of image is useful in observing the geology of an area, as well as assessing other resources.

Source: Image provided by the USGS EROS Data Center Satellite Systems Branch.

The commercial availability of satellite imagery with spatial resolutions finer than 1 meter offers the global community a new level of access, providing information on environmental conditions and human activities in regions of the world to which access is denied for military or political reasons. Satellite imagery now provides an independent source of information beyond the control of national governments to monitor activities such as controversial clear-cutting of tropical rain forests, the assessment of natural disasters, or the mobilization of armed forces *(figure 1.12)*.

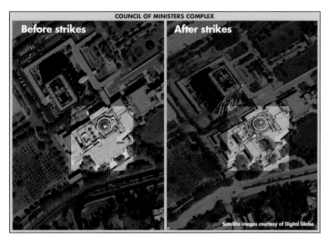

Figure 1.12 Bomb damage to the Council of Ministers Complex, Iraq. During the 2002 Iraq war, selected satellite imagery acquired by commercial operators was released to the media.

Source: Image courtesy of DigitalGlobe.

Remote sensing and GIS

Remote sensing is included within the broader field of *geomatics*—the group of technologies and disciplines that collect, store, and analyze geospatial information (i.e., information that pertains to a geographic location). Geomatics also includes such disciplines as geographic information science, surveying, and cartography.

Many applications implemented using geographic information systems (GIS) depend on datasets derived from remotely sensed imagery or make use of the imagery directly as a background in graphic displays.

Remotely sensed data in digital form can be directly imported into a GIS. Photographic images can also be used if they are scanned and rectified to create GIS-compatible digital imagery. The availability of data in digital form and the development of inexpensive high-speed computers and

image analysis software have made possible the integration of remote sensing information and GIS analyses *(figure 1.13)*. Most GIS software packages can easily import and display remotely sensed data, and many provide basic image manipulation functions.

Obtaining reliable information from remote sensing requires an understanding of sensor technology, systematic analysis procedures, and common approaches used for different applications. There is also an art to the development

Figure 1.13 Investigation of the space shuttle *Columbia* explosion, February 2003. In order to use information from the shuttle debris to guide the search effort, a predictive model was developed. The model calculated the probability of finding debris of a size and type useful in diagnosing the cause of the failure based on the GPS location of debris, its size, type, and quality. A geographic information system (GIS) was used to display areas of significant probability color-coded from, purple (high) through blue, green, and orange to yellow (low) and overlaid on an IKONOS® satellite image with geographic annotations from the vector database of the GIS.

Source: Map courtesy of the Forest Resources Institute, SFASU. Digital elevation model courtesy of the USGS (Nacogdoches 1:100,000 quadrangle). Vector data courtesy of Texas Natural Resources Information System (TNRIS).

of successful applications, that involves making creative use of the many analytical tools available. However, the value of remote sensing lies in the practical information it yields.

Despite the wide use of remotely sensed images, GIS users are often unfamiliar with remote sensing techniques and the capabilities of the technology. *Remote Sensing for GIS Managers* provides an introduction to remote sensing history, technology, and applications tailored to the needs of GIS managers and practitioners.

Chapter 2 presents a brief history of remote sensing. More than a series of technological advancements, the development of remote sensing was shaped by the social, political, and national security issues of the day. From its early beginnings, when aerial photographs were viewed as a mere curiosity, remote sensing has become an information technology of such importance that advanced sensors are closely guarded secrets of the world's most powerful nations.

Remote Sensing for GIS Managers introduces remote sensing with the goal of promoting its use in the production of useful geospatial information. However, to understand how remotely sensed data can be used, some basic information about what the data represents and the methods used to analyze it is needed. Chapters 3 through 9 provide a background to the technology, the fundamental principles on which remote sensing is based, and the instruments developed to detect and record earth observation imagery.

Chapters 10 and 11 present the principles of visual interpretation and computer-based image analysis methods, which are used for extracting information from imagery in conjunction with other data sources. The two approaches are complementary, taking advantage of the unique capabilities of the human interpreter and the power of specialized image analysis systems.

Ultimately, the value of remote sensing lies in the information yielded from the interpretation and analyses. Chapter 12 presents an overview of key remote sensing application areas and examples of their practical use with GIS technology. Chapter 13 discusses approaches to developing a remote sensing strategy suited to an organization's needs and objectives. Effective use of remote sensing does not necessarily require the development of sophisticated in-house expertise. For many organizations, it is more cost-effective to have outside service providers perform some or all of the remote sensing data analysis rather than investing in the equipment and personnel. Unless the need for remote sensing analysis is sufficient to demand the full-time commitment of one or more specialists, the use of outside services is an attractive option.

No single solution suits all situations. Mandates, cultures, and an organization's resources differ widely and change over time. Implementing a strategy for using remote sensing products and services is a process of managing change—matching the use of the technology to the changing needs and objectives of the organization.

The appendix *Rectification and georeferencing of optical imagery* introduces the principles used to transform the geometric properties of imagery to match a chosen map base, principles of particular importance for GIS applications. The appendixes also provide the specifications of satellite imaging sensors widely used for earth resource remote sensing and a list of information and image sources.

The technological barriers to the integrated use of remote sensing and GIS technology have been greatly reduced, offering the capability to develop a wide range of innovative solutions. The objective of this book is to provide information and ideas to stimulate that process.

2 A brief history of remote sensing for earth observation

Stan Aronoff

Remote sensing technology has been used for more than a century. Beginning as photography acquired from tethered balloons and later from aircraft and spacecraft, remote sensing has proven to be a versatile and valuable source of diverse geographic information. A wide range of airborne and spaceborne sensors generate imagery used in a broad range of applications from weather forecasting and resource conservation to land-use planning and defense. Yet GIS users are often unaware that much of the data they use is generated using remote sensing technology or that much more information can be obtained from these sources. This chapter explores the history of remote sensing and some of the political, social, and technological factors that shaped its development.

Evelyn Pruitt, a scientist working for the Office of Naval Research, coined the term *remote sensing* in the late 1950s. Her definition included aerial photography as well as the imagery being acquired by new sensor systems. The term has come to describe the process of observing, measuring, and identifying objects without being in direct contact with them.

The history of remote sensing was influenced by developments in a broad range of interrelated technologies and was shaped by the social goals, political agendas, and national security issues of the day. The early development of aerial photography, a form of remote sensing that evolved from photos taken by cameras carried by balloons and kites, was driven largely by military intelligence needs. First used during the U.S. Civil War, aerial surveillance became a sophisticated intelligence gathering operation during the two World Wars and has continued to develop in scope and sophistication. Information provided by reconnaissance aircraft and later spacecraft over denied territory provided hard evidence of a country's military capabilities. At the height of the cold war, this remote sensing data proved instrumental in preventing incidents such as the Cuban Missile Crisis and U.S. fears of a missile gap with the Soviet Union from reaching open conflict.

While military intelligence has always depended on secrecy to gain advantage, often it is in making remote sensing imagery public that political advantage is gained. Providing open access to independent sources of information such as remote sensing is not only valuable as an independent verification of military capability and activity, but it also changes the political implications of activities such as environmental pollution, deforestation, and destruction of critical habitat by allowing them to be scrutinized by the world community.

Advances in computer and imaging technology made digital imagery practical in the 1970s and led to the development of computer-based image processing systems that enhanced and analyzed new higher-resolution image sources. New sensors were developed, including multispectral scanners, radar (radio detection and ranging), and lidar (light detection and ranging)[1]. In the 1990s, the introduction of sophisticated satellite positioning technology, inertial measurement units, high-speed computers, and advances in photogrammetry

and telecommunications dramatically improved the speed and accuracy with which imagery could be captured, analyzed, and distributed. Multiple civilian satellites now collect images of the earth continuously, providing a wide range of information to all. Unmanned airborne systems can be piloted from and provide imagery in real-time to receiving stations half a world away. With the availability of commercial high-resolution satellite imagery, the use of remote sensing in the media to report on world events has also expanded dramatically, making the internal activities of many nations much more transparent to the world community.

Remote sensing was a major influence in the early development of geographic information systems (GIS), both instrumentally as a major data source and also philosophically as a promoter of an integrated global perspective of the earth's features and resources—a perspective that remote sensing technology could not only show but also measure. The first geographic information systems were designed to support multidisciplinary planning and analysis procedures so that complex environmental trade-offs could be incorporated into the resource management process. Similarly, new civilian applications of remote sensing were used to inventory and monitor the earth's natural resources and to detect natural and human-caused disasters such as flooding and oil spills. Initially, the remote sensing data most widely used in GIS applications was land-cover and digital elevation data. In addition to this data, GIS applications now make greater use of the continuous flow of earth observation data obtained by remote sensing to inventory and monitor resources and human activities from the local to global scale.

1858–1918: Aerial photography—from curiosity to practical use

Modern remote sensing began with the invention of photography and the camera over 150 years ago. Nicéphore Niepce, William Henry Fox Talbot, and Louis Jacques-Mandé Daguerre first demonstrated the new technology in 1839. François Arago, the director of the Paris Observatory, advocated the use of *aerial photography* for topographic surveying as early as 1840. But it was not until 1858 that the first known aerial photograph was produced. The photo was taken by Gaspard Felix Tournachon, also known as Nadar. Nadar used a tethered hot-air balloon to photograph Bièvre, France, from a height of 80 meters. After that, balloon photography flourished as entrepreneurial photographers capitalized on

---

1. Radar and lidar are active remote sensing systems. Radar uses microwaves and lidar uses lasers operating in the visible or near-infrared bands to illuminate the scene and then record the reflected signal. Both radar and lidar are used to produce imagery, digital elevation data, and other information products.

Figure 2.1 Early aerial photograph of Boston, Massachusetts. This view of Boston taken by James Wallace Black in 1860 is the oldest original aerial photograph in existence. A portion of the Boston business district and masts of square-rigged ships in the adjacent harbor can be seen.

Source: *Remote sensing and image interpretation*. 2nd. ed. by T. M. Lillesand and R. W. Kiefer. © 1987 John Wiley and Sons, Inc. Reprinted with permission of John Wiley and Sons, Inc.

the public's enthusiastic demand for images of familiar and exotic landscapes depicted from this novel aerial perspective. The earliest surviving aerial photograph in the United States was taken by James Wallace Black from a captive balloon[2] about 365 meters over Boston in 1860 *(figure 2.1)*.

The first military use of aerial photography occurred during the U.S. Civil War. In June 1862, the Union army used aerial photographs taken from balloons at a height of about

Figure 2.2 Union soldiers inflate an observation balloon at the battle of Fair Oaks during the U.S. Civil War.

Source: Image courtesy of the Library of Congress.

450 meters to assess the Confederate defenses at Richmond *(figure 2.2)*.

The documentary value of aerial photography was recognized in the early 1900s as well by photographers such as George R. Lawrence. A self-taught commercial photographer in Chicago, Lawrence developed his own large-format, panoramic cameras for aerial photography from his Captive Airship. The Captive Airship consisted of up to 17 kites on a piano wire suspending a camera held by a specially designed stabilizing mechanism. The camera was operated from the ground using the current from a battery to activate a solenoid in the camera and trip the spring-operated shutter.

On April 18, 1906, an earthquake and subsequent fires destroyed much of San Francisco. Hearing of the disaster, Lawrence assembled his crew and equipment and within a few weeks documented the devastation of the ruined city in aerial photographs taken 610 meters above San Francisco Bay with his Captive Airship.

Lawrence's images were a sensation around the world *(figure 2.3)*. The images were fascinating not only because of their novelty, but also because they captured an exceptional event in remarkable detail.

2. A free flying balloon drifts with the wind currents. The position of a tethered or captive balloon is maintained by ropes or cables attached to ground points, thereby controlling its observation position and landing site.

Figure 2.3 Captured in May 1906, just weeks after an earthquake and fire ravaged the city, this then novel aerial view of San Francisco Bay drew world attention to the city's plight.

Source: Photograph by G. R. Lawrence.

a

b

Figure 2.4 Pigeons were one of the more novel early platforms for aerial photography. A pigeon fleet (a) operated in Bavaria carried miniature cameras that automatically captured photographs such as this 1903 picture of a Bavarian castle (b). The irregular objects on either side are the bird's flapping wings.

Source: Images courtesy of NASA .

By the early 1900s, photographic technology had improved to the point that smaller cameras and faster lenses and films were available. This enabled aerial photographs to be taken from much smaller platforms. Photographs were successfully taken using kites and even pigeons as platforms *(figures 2.4)*. In 1909, Julius Neubronner published a pamphlet describing carrier-pigeon aerial photography. He also demonstrated the uses of a panoramic and stereoscopic camera and urged its use for strategic information collection.

For aerial photography to become practical, it required a navigable platform that would allow the camera to be placed where it was needed. Wilbur and Orville Wright supplied this platform in 1903 with the invention of the piloted airplane *(figure 2.5)*. But it was not until 1909 when Wilbur Wright took motion pictures over Centocelli, Italy, that the airplane was used as a camera platform. Soon after, German flying students training at English flying schools began taking photographs on advanced training flights.

World War I saw the rapid development of aerial photography and practical photointerpretation methods for military intelligence. Although aircraft were in regular military use by August 1914, during the beginning of the war, aerial photographs were not being taken. Military authorities were initially reluctant to use the new technology. It was only after a British air force officer took aerial photographs of military installations in German-held territory that allied military authorities were convinced of the photographs' intelligence value. By 1915, cameras were developed specifically for aerial

Figure 2.5 On December 17, 1903, Wilbur and Orville Wright carried out the first successful flights of a heavier-than-air flying machine at Kitty Hawk, North Carolina. Their landmark achievement was the result of a sophisticated four-year program of research and development.

Source: National Air and Space Museum, Smithsonian Institution (cropped version: SI-2002-16646)

use *(figures 2.6 and 2.7)*. Photointerpreters became recognized as the "eyes of the armed forces" by all countries in the war, and aerial photographs from biplanes helped both sides make strategic decisions.

The use of aerial photography had a profound effect on military tactics. It was difficult to hide military information from aerial cameras, even with the use of decoys and camouflage *(figure 2.8)*. More important, photointerpreters became skilled at predicting enemy activity from the information in the air photos. By studying the quantities of transportation equipment at railheads and ammunition dumps, the construction of new railroads and medical facilities, and other indicators, they could correctly anticipate military activities in time for counter measures to be planned. By recognizing several independent factors, interpreters can infer information not directly presented. This use of multiple indicators to deduce information is known as *convergence of information* or *data fusion*. It is central to the effective interpretation of all forms of remotely sensed data.

Figure 2.7 Pilot and photographer in AH-14 airplane carrying a Graflex camera, 1915.

Source: Reproduced with permission, the American Society for Photogrammetry and Remote Sensing. R. N. Colwell, ed. *Manual of photographic interpretation*, Washington: ASPRS, 1960. 6.

Figure 2.6 Graflex camera used for aerial photography during World War I.

Source: Reproduced with permission, the American Society for Photogrammetry and Remote Sensing. R. N. Colwell, ed. *Manual of photographic interpretation*, Washington: ASPRS, 1960. 6.

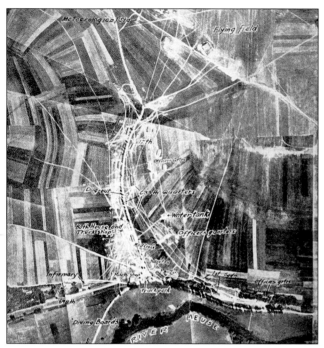

Figure 2.8 Analysis of military facilities in France during World War I.

Source: Reproduced with permission, the American Society for Photogrammetry and Remote Sensing. R. N. Colwell, ed. *Manual of photographic interpretation*, Washington: ASPRS, 1960. 7.

During the middle of the war, detailed photointerpretation studies were carried out. Areas were repeatedly photographed, and detailed comparative studies of terrain and military targets were done. By the end of the war, armies had developed the capacity to produce large quantities of aerial photographs rapidly. French aerial photography units developed and printed 10,000 photographs every night during periods of intense activity. In 1918, during the Allied Meuse-Argonne offensive, 56,000 prints were produced and delivered to the American Expeditionary Force in four days. In all, it is estimated that one million aerial reconnaissance photographs were taken during the course of World War I.

Military photointerpreters could be depended upon to provide accurate and timely information. In 1918, field verification after the armistice showed that U.S. Army interpreters had identified 90% of German military installations opposite the First Army's sector of the front lines. But their experience, gained by trial and error, was not formalized, and much of the hard-won photointerpretation expertise was lost after the war. Training procedures were not well developed, and few aircraft and aerial cameras were available for military training in aerial photography.

1918–1939: Development of nonmilitary applications

In the period between the World Wars, military photointerpretation techniques developed little; however, significant advances were made in commercial and scientific applications. Civilian uses of aerial photography in such fields as forestry, geology, agriculture, and cartography were introduced and led to the development of improved cameras, films, and interpretation equipment. Photogrammetry, the principles and technology of making accurate measurements from photographs, was applied to aerial photography, and instruments were designed specifically for this purpose *(figure 2.9)*. Though the fundamentals of photogrammetry had been developed much earlier, it was during the 1920s with the availability of accurate photogrammetric instruments that the use of aerial photography in government mapping programs became routine. Aerial photography reduced the cost and the time to produce topographic maps. It also made it possible to accurately map areas inaccessible or difficult to reach on the ground. Photogrammetric methods are now used to produce virtually all topographic and natural resource maps.

Initially used for topographic mapping, aerial photography was soon applied to other resource mapping applications, including forest inventory, geologic mapping, soil survey, and collection of agricultural statistics. However, acceptance

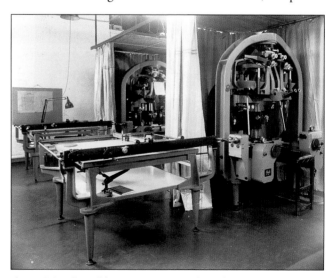

Figure 2.9 The Wild A5 stereo-autograph machine, manufactured in Switzerland, was used extensively for producing topographic maps from stereo aerial photographs.

Source: Image courtesy of R. C. Nesbit.

of aerial photography in science was slow, due in part to the limitations of the technology and techniques and uncertainty about its role in scientific inquiry.

Former military photographers founded aerial survey companies and promoted the use of aerial photography for government mapping programs in the United States, Canada, and Europe. The new commercial survey companies provided air-photo acquisition, interpretation, and mapping services. Government agencies began using the technology in their mapping operations. In the United States, national and regional agencies such as the Forest Service, Agricultural Adjustment Administration, and Tennessee Valley Authority made extensive use of aerial photography as did state, county, and metropolitan planning agencies.

In Canada, regular aerial photography flights for mapping and forest inventory were conducted across the country during the 1920s. Figure 2.10 shows an amphibious bush plane used in Canada for some of the early air-photo missions in the 1920s and 1930s. In 1925, the National Air Photo Library was established to manage all nonmilitary air-photo activities conducted by the Canadian federal government. Regular aerial surveys since that time and subsequent establishment of provincial aerial photo acquisition programs have given Canada one of the world's most comprehensive air-photo archives.

In 1934, the American Society of Photogrammetry (now the American Society for Photogrammetry and Remote Sensing) was founded as a scientific and professional organization to advance the field. By 1940, technical journals in the United States, Germany, and Russia had published hundreds of papers on air photointerpretation. They covered such diverse applications as archaeology, ecology, forestry, geology, engineering, and urban planning.

1939–1945: World War II—the military imperative

In 1938, the German Chief of Staff, General Werner von Fritsch, made what was to become a prophetic statement: "The nation with the best photoreconnaissance will win the next war." Germany foresaw the importance of photointerpretation and was the world leader in the field during the outbreak of World War II in 1939 *(figure 2.11)*. The devastatingly effective German offensive in the spring of 1940 was planned after intensive study of Allied terrain and installations on aerial photography. Complete air-photo coverage had been flown between September 1939 and May 1940. Every important military and transportation facility from Norway to southern France had been photographed—airfields, ports,

Figure 2.11 This air photo of the port of Bremen taken by the German air force in 1939 shows a heavy cruiser under construction in the dry dock at the bottom left and another under construction and moored at center left.

Source: Image courtesy of R. C. Nesbit.

Figure 2.10 This photo, taken in Manitoba in 1924, shows a Vickers Viking IV flying boat, equipped for aerial photography.

Source: Image courtesy of Library and Archives Canada.

canals, and railroads. All the airfields had been found and photographed, the aircraft identified and counted, and the daily working routine identified. This information was used to plan a crippling attack on Allied air power that made use of almost the entire German air force.

Allied forces also took advantage of photoreconnaissance and photointerpretation. After the retreat from Dunkirk in 1940, British intelligence was cut off from its usual sources of military intelligence and had to rely more heavily on its well-developed aerial photography and photointerpretation resources. British photointerpreters detected German barges massed near the coast of France and the Netherlands in the summer of 1940. The British were able to mount an effective air attack that forced Germany to postpone and eventually abandon their planned invasion. Air-photo reconnaissance enabled Allied forces to discover and track German surface ships, hampering German naval power throughout the war.

During World War II, the Allies also developed state-of-the-art photointerpretation techniques. For example, photointerpretation methods were developed to estimate water depths to an accuracy of 0.6 meters for depths up to 10 meters in the clear waters surrounding islands on the Pacific Front. This technology made it possible to plan assaults by amphibious craft and to navigate uncharted waters. As the use of aerial photography for military reconnaissance before, during, and after operations became recognized as essential to modern warfare, the Allies developed mobile processing laboratories to handle thousands of reconnaissance images flown in support of ongoing military operations (*figure 2.12*). The Allies also developed nighttime aerial photography to detect supply convoys operating under the cover of darkness.

As the war progressed, the quality of German photo intelligence declined, due in part to the death of General von Fritsch, while Allied photointerpretation steadily improved

Figure 2.13 Fast fighter aircraft such as the British Spitfire (shown here with aerial cameras) and the American P-51 Mustang were adapted for aerial reconnaissance.

Source: Image courtesy of R. C. Nesbit.

Figure 2.12 An RAF mobile photographic processing unit on its way to Normandy in August 1944.

Source: Image courtesy of R. C. Nesbit.

Figure 2.14 Military photointerpreters processing reconnaissance photography.

Source: Image courtesy of R. C. Nesbit.

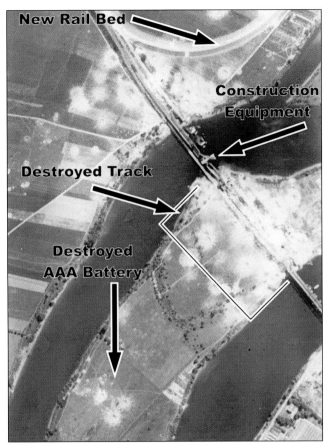

Figure 2.15   World War II bomb damage assessment.

Source: © Imstrat Corporation 2003.

(figure 2.13 and figure 2.14). Allied forces monitored enemy ports, military installations, and transportation facilities. Reconnaissance photography provided assessments of bombing missions *(figure 2.15).* Invasion operations were supported by extensive and repetitive air-photo coverage. For example, V-1 and V-2[3] rocket launching sites were discovered by British photointerpreters in 1944 and targeted for aerial bombardment.

Preparing for the invasion of Normandy, Allied photointerpreters identified military installations, including underground facilities and communication lines, prepared detailed maps of enemy defenses, and constructed air-photo mosaics and terrain models of the area. During the advance across Europe, aerial photographs were enlarged, annotated, divided into grids, and used in place of outdated maps.

World War II also spurred the development of new remote sensing technologies. The introduction of new photographic films and specialized filters led to the development of color infrared film. Used to identify camouflaged military equipment and facilities, the film was sensitive to red, green, and near-infrared wavelengths. Green vegetation appeared red, while objects painted green and cut vegetation used to hide equipment would appear blue-green.

Another remote sensing technology developed during World War II was radar. Great Britain had developed operational aircraft and ship detection radar systems prior to World War II. But it was only during the war, with the development of high-power microwave transmitters operating at short wavelengths, that Britain developed the first successful imaging radar, the *plan position indicator.* A circular display was refreshed with every sweep of the rotating antenna. Objects that strongly reflected the microwave signal appeared as bright spots on the display. First used for aircraft detection from land-based stations, radar gave the British ample warning of German air raids and enabled them to scramble fighters and move civilian populations into shelters. Radar was later adapted for use on aircraft for nighttime and poor-weather bombing and on naval ships protecting supply convoys in the North Atlantic. Despite developments such as these, aerial photography was still the only sensor system in practical use at the end of World War II.

Since World War II, aerial photography and other remote sensing technologies have been considered an essential component of military intelligence. Research and development of military remote sensing applications begun during the war continued apace afterwards. But security concerns restricted the introduction of new sensor technologies for civilian applications.

Before World War II, there was unrestricted access to the most current technology and techniques of aerial photography, photointerpretation, and photogrammetry. After the war proved the military value of the technology, tight restrictions were placed on access to state-of-the-art remote sensing systems such as radar and thermal infrared sensors. Whereas before the war the exchange of information between civilian and military remote sensing specialists could be open and supportive, after the war security restrictions and competition for government funding made collaboration more difficult.

---

3. The V-2 was the first successful long-range missile. Over 2,500 V-2s were launched between September 1944 and March 1945, killing several thousand people. The V-2 was the prototype of all modern ballistic missiles and the first guided missile used in war.

Since that time, the research and development funding and remote sensing technology made available for classified intelligence systems has exceeded by a wide margin that made available for civilian applications.

1946–1971: Continued remote sensing advancement and the introduction of geographic information systems

After the war, the use of remote sensing technology increased exponentially. Thousands of highly trained photointerpreters returned home and applied their skills to civilian professions. The photointerpretation techniques developed in wartime became the established procedures for civilian applications including surveying (topographic mapping), geology, and engineering.

As tensions between the United States and the Soviet Union grew, so did advancements in remote sensing technology. Photos taken from state-of-the-art spy planes were now supplemented with images transmitted from space. Secret technologies developed during World War II, such as multispectral imaging, radar, and infrared sensors, were improved by commercial companies. During the 1970s, geographic information systems changed the way geospatial information (much of it derived from remote sensing) could be organized, analyzed, and displayed. Together these advancements made remote sensing an ever more essential component of military and civilian applications.

New technologies

The period between the late 1950s and the early 1970s saw the continued declassification and development of new remote sensing technologies developed during the war, as well as the refinement of photointerpretation and photogrammetric instruments and techniques. These technologies and techniques included the use of color infrared film and advancements in radar and multispectral imagery.

Color infrared film had been declassified and approved for civilian use after World War II, and it proved particularly useful for vegetation analysis. Crop and forest types were more easily identified and stressed vegetation more easily distinguished from healthy vegetation on color infrared film than on normal color films.

Military research programs in the 1950s developed thermal infrared imaging sensors (*figure 2.16*) in addition to side-looking airborne radar (SLAR) and synthetic aperture radar

(SAR) systems that produced a continuous radar image strip of the terrain to the side of the sensor's flight path (*figure 2.17*). Radar is an active sensor that generates microwave pulses and records the energy reflected from the terrain below. It can collect imagery day or night and penetrate cloud cover. This made the system ideal for imaging cloud-covered regions such as tropical rain forests. It was not until the late 1960s and early 1970s that some of these remote sensing instruments were declassified, enabling civilians to use data produced by thermal infrared and microwave (radar) sensors in their remote sensing applications.

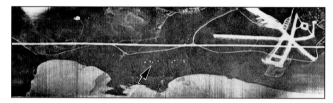

Figure 2.16 Classified American military remote sensing programs such as Tropican developed thermal infrared imaging sensors in the late 1950s. This 1959 image demonstrated the detection of enemy campfires in Puerto Rico (white dots indicated by the arrow).

Source: Courtesy of the Air Force Research Laboratory/Information Directorate (AFRL/IF).

Figure 2.17 Early side-looking airborne radar image of San Francisco. Funded by the U.S. Air Force, Westinghouse developed the first side-looking airborne radar in 1954.

Source: Courtesy of UCSB Geography Department, J. E. Estes (1939–2001).

One of the most ambitious early civilian applications of these new technologies was the RADAM (Radar Amazon) project. Begun in 1971, this project involved the Brazilian government and commercial remote sensing services from the United States. The five-year project for the first time mapped the entire Amazon Basin, an area of almost five million square kilometers, at a spatial resolution of 16 meters using a SLAR. Brazilian scientists interpreted 1:200,000 scale radar imagery to produce maps of geology, geomorphology, hydrology, soils, vegetation, and land-use potential (*figure 2.18*). They also discovered and provided the first complete mapping of a major tributary of the Amazon measuring about 600 kilometers long, as well as 100,000 square kilometers of fertile soil with the potential to support agriculture. The inventory and its maps promoted mineral exploration and facilitated management of the region. It also made possible greater settlement and deforestation. Radar has since become a prime source of information for mineral exploration, forest and range inventory, water supplies, transportation management, and site suitability assessment for remote, cloud-covered regions.

Experiments in multispectral remote sensing, the simultaneous collection of images in multiple spectral bands, were

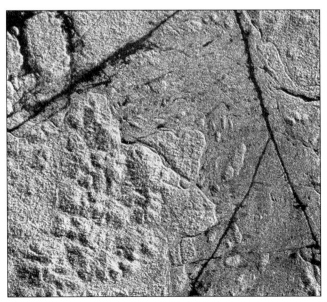

Figure 2.18 Radar image from Project RADAM. Radar image of the Rio Xingu area in the Amazon Basin of Brazil. The distance represented in the image is approximately 14 kilometers across.

Source: Reproduced with permission of the American Society for Photogrammetry and Remote Sensing. R. N. Colwell, ed. *Manual of remote sensing*. Vols. 1 and 2. 2nd ed. Falls Church, Va: ASPRS, 1960.

also first conducted during this time. The experiments used multiple cameras, each with a different film-filter combination mounted in a holder so that they viewed the same target area. Color composite images could be produced by projecting the images from different bands with red, green, and blue light onto a viewing screen or color photographic film. In the mid-1960s an electro-optical scanner that generated digital multispectral imagery was developed and later became the principal remote sensing instrument for earth resource remote sensing from space.

Military airborne reconnaissance

While established air-photo techniques were adapted to civilian use, the beginnings of the cold war between the Western democracies and the Soviet Bloc spurred the development of new reconnaissance techniques. The critical value of the advanced remote sensing technologies developed for intelligence applications made these programs among the most closely guarded military secrets.

The impetus for U.S. development of improved military remote sensing capabilities was the result of several ominous developments in the late 1940s and 1950s. These included the failure of U.S. intelligence to predict the pace of Soviet weapons development programs and also concerns that the Soviet Union might have a substantially larger bomber and missile arsenal than the United States. U.S. fears that the Soviets possessed a substantially larger arsenal of bombers and intercontinental ballistic missiles (ICBMs), the so-called bomber gap and missile gap, made the acquisition of reliable information on Soviet military capability a top priority for the United States.

In 1955, with cold war anxiety rising, U.S. President Eisenhower proposed to his Soviet counterpart, Premier Nikita Khrushchev, that each country conduct reconnaissance flights in the air and from space over the other country and that the imagery be given to the United Nations. Khrushchev rejected Eisenhower's *open skies* policy, and each country proceeded to rapidly develop the world's most sophisticated remote sensing technology in secret.

The United States initiated development of the highly secret Lockheed U-2 aircraft and the WS-117L satellite programs for military reconnaissance. The U-2 spy plane was designed and built by the Advanced Development Projects Division of Lockheed, better known as the Skunk Works. Completed in a remarkably fast nine months, the project was ahead of schedule and under budget. The first flight took place in August 1955, and the first operational U-2 flight over Soviet

Figure 2.19 First flown in August 1955, the original U-2 was a single-seat, single-engine, high-altitude surveillance and reconnaissance aircraft. With glider-like characteristics, the U-2 could lift heavy loads to altitudes above 21,000 meters and achieve flight durations of nine hours or more. Upgraded versions of the U-2 such as the U-2R (developed in 1967) and the U-2S (developed in 1994) continue to fly military reconnaissance as well as peacetime reconnaissance missions in support of disaster relief, search and rescue, drug enforcement, and research.

Source: Image courtesy of the Federation of American Scientists.

Figure 2.20 SR-71 reconnaissance aircraft. Developed for the U.S. Air Force for military reconnaissance, the SR-71 is the world's fastest and highest-flying production aircraft powered by air-breathing engines. It has attained altitudes of over 25,900 m (85,000 feet) and speeds of more than 3,540 kilometers per hour (2,200 miles per hour), more than three times the speed of sound. First flown in 1964, the initial SR-71 design was refined in subsequent versions. It was retired from U.S. Air Force service in 1998 but continued to be used by NASA for high-speed and high-altitude aeronautical research through 1999.

Source: Courtesy of the NASA Dryden Flight Research Center Photo Collection.

airspace took place in July 1956 *(figure 2.19)*. The same group later replaced the U-2 with the high-performance SR-71 Blackbird *(figure 2.20)*.

The U-2 flew above 21,000 meters and produced stereo aerial photography with resolutions as high as 15 centimeters. The aircraft flew too high to be attacked by existing Soviet aircraft and missiles and operated over the Soviet Union and other Eastern Bloc countries unchallenged for four years. But the U-2 was easily tracked by Soviet radar. Soviet protests made U-2 flights politically difficult, and they were severely restricted. The U-2's missions for Soviet reconnaissance ended abruptly on May 1, 1960, when the U-2 piloted by Francis Gary Power was shot down over Soviet territory. Only 24 U-2 missions had been flown over Soviet territory. The imagery was extremely valuable to the United States, but the risks of using the aircraft over Soviet territory had become too great. Faster, higher-flying aircraft were developed, but the ultimate solution was to take intelligence operations into space. Space provided the ideal vantage point, if the world community would accept that satellites could observe any nation without permission.

Military space imaging

Immediately after WW II, in 1946, German rocket technology was used in the development of early satellite launch vehicles by the United States and Soviet Union. The U.S. military used captured German V-2 rockets to photograph the earth at altitudes of 110 to 165 kilometers, which demonstrated the great value of space photography for weather observation *(figure 2.21)*.

But it was the launch of the first satellite, Sputnik-1 *(figure 2.22)*, by the Soviet Union on October 4, 1957, followed four months later by the U.S. Explorer 1 satellite, that ushered in the space age. The rapid development of earth-orbiting satellites quickly led to the development of satellite photography and methods of capturing and transmitting digital imagery from space. It also made possible image acquisition independent of political boundaries. With the cold war at its height, both the United States and the Soviet Union developed extensive satellite image intelligence programs that pushed remote sensing technology to its limits.

In August 1960, just three years after Sputnik, the United States produced its first space photographs from a spy satellite program code-named Corona *(figures 2.23)*. Initiated in 1958, the U.S. Air Force's scientific satellite program Discoverer served as a cover for the Corona missions. The first 12 launches were unsuccessful, but the critical need for satellite reconnaissance imagery provided the impetus for the program to be continued despite the setbacks.

Corona missions were initially short, only two to three days. The film was returned to earth in reentry capsules known as film return buckets, which parachuted to earth over the ocean and were snatched from the air by specially equipped aircraft *(figures 2.24 and 2.25)*. The capsules could also float, which

Figure 2.21   Launch of a V-2 missile. Captured V-2s were used for early research studies of the upper atmosphere and demonstrations of space photography.

Source: Image courtesy of NOAA.

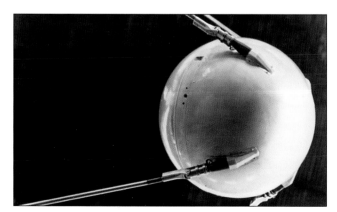

Figure 2.22   Sputnik-1. Launched by the Soviet Union in October 1957, Sputnik-1 was the first satellite to be launched into orbit.

Source: Image courtesy of NASA.

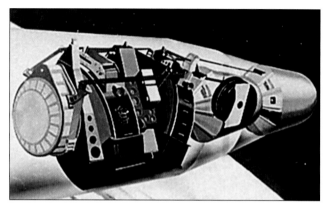

a

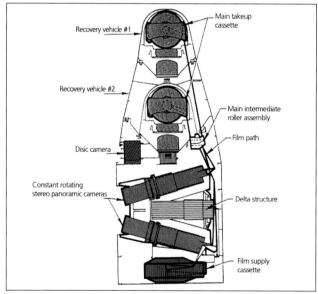

b

Figure 2.23   Corona reconnaissance satellite. Beginning in 1960, Corona spy satellites carried increasingly higher resolution camera and film systems. Early models used a single-camera, single-film return capsule. Later models used two cameras to produce stereo imagery and two capsules that enabled film to be returned twice during a mission.

Source: Courtesy of Central Intelligence Agency Record Group #263.

enabled ocean recovery. The spacecraft carrying the camera system was maneuvered out of orbit and incinerated during reentry. All Corona photography was black-and-white, with the exception of experimental samples of infrared and color film carried on some missions. Early systems carried a single film return bucket, but later systems were equipped with two, thereby extending the useful life of a mission.

The 20 pounds of film from Corona's first mission provided coverage of more than 1.5 million square miles of the Soviet Union, more coverage than all the U-2 missions combined. Although fuzzier than U-2 photographs, the resolution was good enough to identify features about 10 meters in size from photographs collected more than 160 kilometers

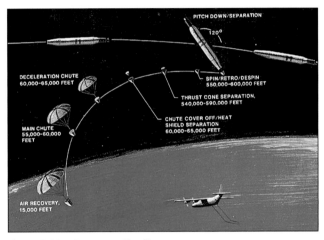

Figure 2.24 Corona satellite film recovery sequence.

Source: Image courtesy of the U.S. National Reconnaissance Office.

Figure 2.25 Recovery aircraft capturing Corona film return capsules as they parachute to earth.

Source: Courtesy of Central Intelligence Agency Record Group #263.

in space. Figure 2.26 shows the first Soviet intercontinental ballistic missile complex (ICBM) identified on a Corona photograph in 1962.

In 1962, a dual-camera system providing stereo coverage was introduced, and by 1967 the Corona system had been improved to resolve objects 2 meters in size with mission durations extended to 15 days. A higher resolution film-recovery system, code-named Gambit, was introduced in 1963. Gambit was capable of resolving objects about half a meter in size.

U. S. President Johnson summed up the value of satellite reconnaissance photography in 1967 when he stated that the $35 billion to $40 billion spent on the space program was worth 10 times that if only for the space photography it produced.

The Soviet Union's first spy satellite, Zenit-2, made its first flight in April 1962. It was a converted Vostok manned spacecraft that carried cameras instead of a cosmonaut. Like the U.S. Corona program, its true purpose was disguised. The Soviets claimed the missions were for scientific exploration under the generic program named Kosmos. Zenit produced color photography resolving objects as small as 5 meters across. Both the Corona and Zenit missions photographed strategically important sites including missile launching and munitions storage and manufacturing facilities, other military installations, and major cities.

However, satellite photography also had important limitations. It could take days or weeks for space images to be recovered, processed, analyzed, and for the resulting information to

Figure 2.26 This 1962 satellite photograph shows the SS-7 missile base at Yurya in Russia, the first Soviet ICBM complex to be identified in Corona images.

Source: Courtesy of Central Intelligence Agency Record Group #263.

be delivered to policy makers and military planners. Satellite availability was also a major problem. In October 1962, U.S. photointerpreters found evidence of the construction of Soviet nuclear missile bases in Cuba *(figure 2.27)*. The ensuing superpower confrontation, known as the Cuban Missile Crisis, illustrated the importance of early detection and superior military intelligence. However, no spy satellite was in orbit or ready for launch during this very sensitive time. To compensate, the U.S. ordered additional surveillance aircraft flights, and two U-2 pilots lost their lives when their planes were shot down. (Similar intelligence gaps occurred during the 1967 Middle East war and the Soviet invasion of Czechoslovakia when satellite images arrived too late to warn of impending events.) Reconnaissance photography from the U-2 and other aircraft alerted the United States to the new Soviet missile installations and later verified that they had been dismantled. Later, U.S. intelligence, gained by surveillance that employed both aerial and satellite imagery, concluded the Soviet Union had on the order of 10 operational ballistic missiles capable of reaching the United States, not the hundreds Khrushchev had claimed. This gave the United States the confidence to call the Soviet bluff and force their withdrawal from Cuba.

Subsequent spy satellite systems offered higher resolution, longer missions, and increased ability to reposition the satellite in orbit. Some systems captured space photography, processed and scanned the film, and then transmitted the image data back to Earth. Satellite reconnaissance systems soon became all-digital and carried a range of sensor systems substantially more advanced than those available for civilian applications. These systems were also capable of greater coverage and faster delivery than civilian systems.

In 1983, the world was given an unexpected glimpse of the capabilities of U.S. classified reconnaissance satellites when U.S. naval analyst Samuel Morison leaked images of a Soviet aircraft carrier under construction on the Black Sea to *Jane's Defence Weekly*. Taken by a U.S. KH-11 photographic intelligence satellite, the image had a spatial resolution of about 0.5 meter to 1 meter, though the satellite was thought to be capable of resolutions as high as 10 centimeters when directly overhead *(figure 2.28)*. Today a multitude of space-based imaging and communication satellites provide the world's intelligence communities with a continuous flow of strategic and tactical military of information.

Figure 2.28   Satellite photograph of a Soviet aircraft carrier under construction. This photo was acquired by an American KH-11 reconnaissance satellite that was thought to be capable of resolutions as high as 10 centimeters.

Source: Courtesy of Central Intelligence Agency Record Group #263.

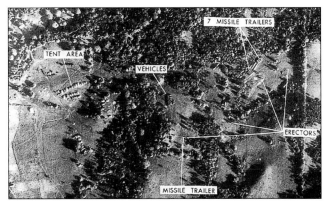

Figure 2.27   This U-2 reconnaissance photo, taken in October 1962, shows a Soviet medium range ballistic missile site in Cuba.

Source: Image courtesy of the Central Intelligence Agency.

Nonmilitary space imaging

The first nonmilitary earth observing satellite was a weather satellite, the Television and Infrared Observation Satellite (TIROS-1), launched by the United States in 1960 *(figure 2.29)*. Early weather satellites offered coarse-resolution images of cloud patterns with little discernible detail of the earth surface. As sensor technology developed, weather satellites developed the ability to collect images of both terrestrial and atmospheric features. Meteorologists could then assess water, snow, and ice features as well as cloud patterns. Though designed for climate and weather applications, TIROS-1 provided the basis for the development of the first earth resource satellite, the Earth Resources Technology Satellite (ERTS-1).

Without the strong support of the United States Geological Survey (USGS) and the U.S. National Aeronautics and Space Administration (NASA), earth resource remote sensing would not have developed as quickly as it did. In the early 1960s, NASA established a research program in remote sensing that continues to support remote sensing research at universities and other institutions throughout the United States.

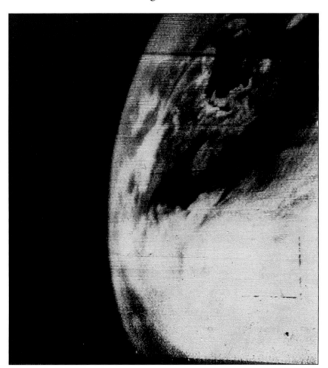

Figure 2.29 On April 1, 1960, NASA's TIROS-1 satellite transmitted this image, the first television picture from space.

Source: Image courtesy of the NASA Earth Observatory.

The development of space-based remote sensing for earth observation by NASA was an offshoot of the effort to land men on the moon. Remote sensors were placed on lunar orbiting satellites to collect data on possible lunar landing sites. These instruments were tested over terrestrial sites similar to the expected lunar terrain, and the interpretations were then checked by visiting the test sites. The value of the information for general geologic studies gathered in these early tests led to a major expansion of the program to include test sites for agriculture, forestry, geography, geology and mineral resources, hydrology, and urban and regional studies.

NASA instituted an aircraft-based remote sensing program to support these studies and the development of hardware, software, and analysis techniques for these new application areas. This program collected intermediate- and high-altitude aerial photography and thermal infrared and radar imagery for large areas of the United States. The data was used to provide training and experience in instrument design, data storage and retrieval, image processing, and image interpretation for scientists and engineers in a wide range of disciplines.

In 1965, the first formal geologic space photography experiment was conducted on the *Gemini 4* earth orbiting manned space flight. The spectacular series of 39 overlapping, near-vertical photographs of the southwestern United States and northern Mexico and some 60 photographs of other selected sites generated intense interest and the support of the geology and geography scientific community *(figure 2.30)*. This led to the development of the USGS Earth Resources Observation Satellite (EROS) program. Recognition of the value of remote sensing as a means of collecting earth resource information as a whole led to the establishment of the NASA Earth Resource Survey Program.

Stewart Udall, U.S. Secretary of the Department of the Interior, and William Pecora, Director of the USGS, initiated the first civil spaceborne Earth imaging project when they announced plans for the EROS program in September 1966. Despite the objections of the U.S. Department of Defense and State Department, the program had the strong support of President Johnson, the U.S. Congress, and the U.S. public.

The EROS program pushed the U.S. space program in a new direction. The United States was in the middle of a cold war with the Soviet Union, a hot war with Vietnam, and a race to reach the moon before the Russians. The fledgling NASA's principal goals had been to land a man on the moon and explore space, goals that embodied national pride and the international prestige of high-profile scientific and engineering endeavors. Earth was considered by many to be

Figure 2.30   This photograph taken by astronauts aboard *Gemini 4* shows the Nile Delta and Sinai Peninsula. The 150-kilometer long Suez Canal can be seen near the center of the image and the Dead Sea can be seen in the upper left-hand corner.

Source: Earth Sciences and Image Analysis Laboratory, NASA Johnson Space Center. eol.jsc.nasa.gov.

sufficiently well known, and therefore an earth observation satellite program was thought to be a costly distraction. Yet it has been earth observation remote sensing satellites and communication satellites that have so far proven to be of greatest economic and social benefit to the world community.

Growing interest in satellite-based earth observation and remote sensing was also supported by major societal changes of the 1960s. The U.S. public was growing increasingly concerned about environmental protection, the ability to renew resources, the rate of depletion of nonrenewable resources such as oil, and pollution. The general public as well as the scientific community was looking for better ways to inventory natural resources, manage development, and minimize environmental impact. The effort to improve the planning process for urban as well as wildland areas led to the development of systematic multiple resource planning methods. The range of data needed for this more ecologically based planning approach was much broader and included soils, vegetation, wildlife capability, agricultural potential, land use, geology, precipitation, hydrography, topography, transportation networks, and cultural features. Almost all of the information needed for this more comprehensive planning

approach made use of remote sensing, either by commissioning photointerpretation specifically for a planning project or by using maps produced through photointerpretation of aerial photography.

The sweeping world changes of the 1960s peaked in 1969. This was the year of the first lunar landing of *Apollo 11 (figure 2.31)* and also the year of the first Woodstock music festival—expressions of a new technological capability and a new cultural philosophy.

Prior to the space age, it had not been possible to repeatedly collect images of large areas of the earth's surface unrestricted by political boundaries. However, with the space age, the world's imagination was sparked by images of the earth rising from behind the moon provided by *Apollo 8* astronauts in lunar orbit *(figure 2.32)*. Spacecraft not only offered a new global view of the earth, but they also provided the technical means to collect and analyze data at a global scale. This ability became more important to a world growing increasingly

Figure 2.31   *Apollo 11* mission, July 1969. Photograph of *Apollo 11* astronaut Buzz Aldrin on the moon taken by Neil Armstrong whose image is reflected in the visor.

Source: Image courtesy of NASA and the *Apollo 11* crew.

more aware of the fragility of "spaceship earth" and the need to conserve and sustain natural resources.

Geographic information systems

Nineteen sixty-nine was also the year *Design with Nature* by Ian McHarg was first published. *Design with Nature* illustrated the technique of using clear plastic map overlays to visually integrate resource datasets the author used for land-use planning *(figure 2.33)*. This manual overlay method was not new. It appeared first in the late nineteenth century when architects and site designers used translucent paper overlays to compare building plans with site surveys. However, the book not only documented the procedure, but more importantly it also presented a compelling case for environmentally sensitive planning that fundamentally changed the land-use planning field. It was the drive to develop computerized implementations of overlay analysis that catalyzed development of practical geographic information systems (GIS) in the early 1970s.

The development of computer mapping and geographic information systems began in university research labs in the 1960s. Hobbled by the limited computer power of the day, early computer mapping was restricted to simple processing of small datasets. At Harvard University, the Harvard graphics laboratory developed the Synagraphic Mapping Program (SYMAP) in 1964. It was the first widely distributed software

Figure 2.32   The earthrise from *Apollo 8*, December 1968. The *Apollo 8* astronauts, the first humans to travel to the moon and back, captured the first photographs of Earth rising above the moon from lunar orbit.

Source: Image courtesy of NASA and the *Apollo 8* crew.

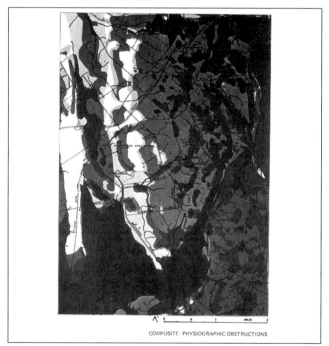

b

Figure 2.33   In the book *Design with Nature*, Ian McHarg illustrated a method that made effective use of land-use plans to reduce environmental impacts. Each environmental factor was remapped onto clear plastic in three gray levels representing suitability (a). By overlaying these maps on a light table, the regions where environmental impacts would be the least were represented by the lightest tones (b).

Source: *Design with nature*. 25th Anniversary Edition by I. L. McHarg. © 1992 John Wiley and Sons, Inc. Reprinted with permission of John Wiley and Sons, Inc.

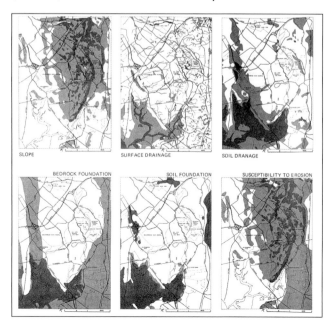

a

for handling geographic data. It was easily implemented using the computer technology commonly available at universities, and maps could be output to line printers. A cell-based program called GRID was later developed that could perform overlay analysis in conjunction with SYMAP. The laboratory subsequently developed ODYSSEY in the mid-1970s, a vector processing system that became a prototype for contemporary vector geographic information systems.

In the early 1970s, software that analyzed digital remote sensing imagery was developed. The Laboratory for Remote Sensing (LARS) at Purdue University developed one of the most widely distributed software packages, LARSYS. The software was designed to run on mainframe computers and generate output onto conventional line printers. With funding from NASA, the designers were able to make the software compatible with a variety of mainframe computers and distribute it at a low cost. As computer technology became more powerful and graphics displays became available, sophisticated remote sensing software was developed to handle the flood of digital imagery that became available in the 1970s.

Early GIS systems were based on either the raster or vector data model with limited capabilities to convert between the two models. The raster data model imposed a regular grid over a geographic area and recorded an attribute value (e.g., the land-cover type such as *forest* or *water*) for each cell in the grid. The vector model encoded the individual geographic features using points, lines, and polygons (areas bounded by straight lines) with attributes assigned to the individual features. The use of the vector data model required handling complex problems of polygon overlay processing and the coordinated management of spatial and attribute data. Raster-based systems employed the same grid array structure as digital images. Though raster-based systems required data files much larger than those required for vector systems to achieve comparable spatial resolution, overlay processing was fast and easy to implement. Also, because they shared a common data model, digital remotely sensed imagery and analysis products such as classifications could be easily imported into raster-based GIS.

Development of what became the first large-scale implementation of a computer-based geographic information system began in Canada during the mid-1960s. In 1966, the Canadian federal government, interested in improving land-use management, funded development of the Canadian Geographical Information System (CGIS), which became operational in 1971. The Canadian Geographic Information System was a means to implement a national-scale, land-inventory program for Canada. Using data generated from air-photo interpretations, CGIS produced 1:50,000 scale maps of shoreline, wildlife, forestry, current land use, and land capability for agriculture. The map products and overlay analyses the system generated became widely used for regional land-use planning.

In the United States, the National Environmental Policy Act of 1970 legislated new land management and environmental protection activities and a wide range of programs that promoted and funded GIS and remote sensing development. Land-use planning and environmental impact assessments became important GIS and remote sensing applications that attracted funding for their development and encouraged the integrated use of the two technologies. The U.S. Department of Defense (DOD) was particularly interested in using the technologies to locate missile sites, plan air-defense strategies, and produce environmental compliance documentation for its extensive land holdings. DOD funding greatly supported early GIS vendors. The formation of the U.S. Environmental Protection Agency and an increasingly complex web of environmental laws motivated state and federal agencies to adopt GIS technology to generate the large-area resource inventory surveys and alternative scenarios demanded. Similar shifts in government policies to land-use management occurred in other countries as well, resulting in the widespread adoption of GIS technology in the 1980s and 1990s.

The demand for GIS technology was met by several companies. In 1969, Environmental Systems Research Institute (ESRI) and Intergraph Corporation (originally called M & S Computing, Inc.) were founded. Both companies became leading GIS manufacturers. Initially ESRI used the raster GIS technology and techniques of the Harvard graphics laboratory and other developing centers of GIS expertise. During the 1970s, ESRI began providing services using both raster- and vector-based GIS, some produced in-house and others produced by outside sources. In the early 1980s, ESRI released ARC/INFO®, the highly successful GIS implementation that stored attribute and location information in separate databases, an idea pioneered during the development of the Canadian Geographical Information System.

The integrated use of GIS and remote sensing

Remote sensing and GIS are complementary technologies. Remotely sensed imagery is a rich source of geospatial data, and GIS is a powerful tool used to manage, analyze, and display geospatial data. Integrating these technologies has been a challenge.

Although data derived from remote sensing analysis, such as land cover and topography, was widely used in geographic information systems, integrated analyses were difficult with the technology available in the 1960s and 1970s, particularly in using the raster-based output of the digital image analysis systems used in remote sensing with vector-based GIS. For example, most land-cover data was visually interpreted from aerial photography or satellite imagery and drawn on paper maps that were then digitized by tracing each feature with a handheld cursor on a digitizing table. When computer-based classification of digital imagery developed in the 1970s, the classification results were output as raster image files. These could be easily used in a raster-based GIS but were difficult to transfer to a vector-based GIS. Vector-based GIS offered an advantage by encoding maps much more efficiently than raster-based systems and could generate high-quality maps with fine line details that raster-based systems could not. But vector-based encoding could not capture the subtle continuous tone and color variations of an image that raster-based encoding could provide.

Many technical complexities made the development of raster-to-vector conversion programs with sufficient spatial accuracy very difficult. For example, a land-cover classification produced from satellite imagery would be encoded as a raster image. In order for software to generate vector elements from the raster file, it would first have to identify groupings of cells that had the same value (i.e., the same land-cover class). The software would then have to decide whether the grouping represented a point, line, or area element in the GIS. Then a suitable boundary around these groupings would have to be defined. Finally, the attribute of the grouping would have to be input to the GIS database and tagged to the element. It wasn't until the 1980s that this capability was successfully demonstrated.

Perhaps more important than the compatibility problems were the differences in the direction of GIS and remote sensing research and development during the early 1970s. Much of the civilian remote sensing community was focused on the development of digital sensors, particularly multispectral scanners, and the software to rectify and classify the digital imagery they generated. The launch of Landsat generated tremendous energy and enthusiasm to find applications for the new data source.

During the same period, GIS development was focused on a very different set of issues. Geographic data was not generally produced in digital format, so considerable effort was devoted to making the time-consuming task of digitizing maps more efficient. Also, the new technology was widely viewed as expensive and difficult to use, so considerable development effort was focused on improving the software's reliability, developing more powerful analysis capabilities, and tailoring GIS technology to meet the needs of specific markets.

While, the different issues and objectives of the remote sensing and GIS communities initially caused them to diverge, the demand by users who needed to work with these technologies together encouraged the development of improved compatibility. Raster-based GIS continued to develop more sophisticated integrated remote sensing and GIS analysis capabilities. However, it was only in the late 1990s that the capability to use large remote sensing images for integrated display, analysis, and output became common among vector-based GIS. Now, photogrammetric workstations produce GIS-compatible output files and in some cases are tightly linked to GIS software. Similarly, GIS software now commonly provide some basic image-analysis tools to handle remotely sensed imagery, and image-analysis software can generate output files compatible with raster- or vector-based GIS software. As remotely sensed imagery becomes more widely used, it is likely there will be further convergence and increased integration of GIS and remote sensing software capabilities.

1972–1986: Introduction of satellite-based earth observation

In 1972, the beginning of a new era for civilian remote sensing began with the launch of Landsat, the first of many civilian satellites dedicated to remote sensing of earth resources launched between 1972 and 1986. Also during this time, several previously restricted military sensors were declassified for use in civilian remote sensing. These sensors included multispectral scanners, thermal infrared systems, and radar imaging systems. For example, airborne thermal infrared imaging systems were declassified and then used for mapping heated effluents from power plant and factory outflows and for monitoring heat-loss from urban housing. Thermal infrared was also used in various experimental applications in geology and soils mapping. Initially flown on aircraft, these instruments were later adapted for use in space.

New platforms also became available for remote sensing in space. Space photography, multispectral scanners, and other experimental sensors were tested on the American *Skylab* experimental space station (1973–74) shown in figure 2.34. Beginning in 1981, the space shuttle program, with its large payload capacity, served as a valuable test bed for advanced sensors as well as for carrying out image acquisition missions *(figure 2.35)*. For example, the Spaceborne Imaging Radar missions SIR-A, B, and C (flown in 1981, 1984, and 1994, respectively) were short-term, highly successful missions that established the value of multifrequency and multiple incidence angle radar imagery. The knowledge gained was used in the development of several operational radar sensors in the 1990s. Development of thermal infrared and radar imaging continued, and new sensors were developed including lidar and hyperspectral scanners, which simultaneously record imagery in over one hundred narrow spectral bands.

Earth observation satellites

Among the many civilian satellites launched from 1972 to 1986, those particularly important in the development of remote sensing for earth resource assessment were the Landsat series, the Advanced Very High Resolution Radiometer (AVHRR) sensor on board the NOAA series of satellites, the Coastal Zone Color Scanner, the Total Ozone Mapping Spectrometer (TOMS), and SeaSat.

Landsat

In July 1972, NASA launched the Earth Resources Technology Satellite (ERTS-1), the first satellite specifically designed to collect information about the earth's natural resources.

Figure 2.34   The first American space station, *Skylab*, was launched May 14, 1973, as part of the Apollo manned space program. During its nine-month mission, three different astronaut crews carried out numerous scientific experiments including the testing of advanced remote sensing instruments.

Source: Image courtesy of NASA.

Figure 2.35   Space shuttle *Endeavour* flares for landing at Edwards Air Force Base in California.

Source: Courtesy of the NASA Dryden Flight Research Center Photo Collection.

The satellite, later renamed Landsat 1 (to distinguish it from the planned SeaSat oceanographic satellite program) operated until January 1978 and was followed by Landsats 2 through 7. Landsat 1 was designed to collect earth resource data at medium spatial resolution in multiple spectral bands. Though the scientific community treated the Landsat program as ongoing, like the weather satellites, Landsat 1 was primarily designed as an experimental system to test the feasibility and value of collecting resource data from unmanned satellites.

The first three Landsat systems carried a return beam vidicon camera or RBV (a variation of analog television camera technology) and a recently developed multispectral scanner system (MSS). The multispectral scanner generated digital images by repeatedly scanning a narrow swath of terrain perpendicular to the flight path. Separate images were simultaneously produced in four wavelength, roughly comparable to the spectral sensitivity of color infrared film. One of the standard image products available used a color scheme similar to that of color infrared film *(figure 2.36)*. However, the digital data had to be processed using expensive computer systems to produce viewable images. The RBV cameras proved to be unreliable, and it was the MSS images that became the standard product. Multispectral scanners became the mainstay of earth resource remote sensing from satellites.

Though Landsat was officially an experimental system, the quality of the imagery and reliable acquisition led users to soon treat the program as operational. Landsat MSS imagery was high quality and showed reasonable detail with minimal distortion. Landsat provided coverage of most of the globe. The imagery was low cost, free of political or security restrictions, and at that time was free of copyright restrictions. Though at first most analyses were done visually using prints or transparencies, the availability of Landsat data in a standard digital format encouraged the development of digital image processing.

Due to the success of Landsat 1, NASA went on to launch Landsats 2 through 7. Landsats 2 and 3 were launched in 1975 and 1978, respectively. The first three Landsat satellites collected multispectral scanner imagery with a ground resolution of 80 meters. A 240-meter resolution thermal infrared band was added for Landsat 3. The entire earth, except portions of the poles, was imaged every 18 days. Instead of the RBV sensor used on previous Landsat missions, Landsat 4 (1982) and Landsat 5 (1984) carried the more advanced Thematic Mapper (TM) multispectral scanner, which had 30-meter spatial resolution in the visible and near-infrared bands and 120-meter resolution in the thermal infrared (figure 2.37). Based on experience gained with Landsat MSS imagery and extensive research, scientists made the spectral bands of the Thematic Mapper optimal for vegetation analysis. They also later added an additional mid-infrared band, which was valuable for geology. The launch of Landsat 6 in October 1993 failed, but data continuity was provided by Landsats 4 and 5 until the launch of Landsat 7 in 1999. Landsat 7 carried an improved TM sensor, the Enhanced

Figure 2.37 Landsat Thematic Mapper image of the Ottawa area (black–water, blue green–urban, red–vegetation).

Source: © Natural Resources Canada.

Thematic Mapper (ETM+), which included a 15-meter panchromatic band.

The Landsat satellite series provided, for the first time, systematic repetitive coverage of the earth's land areas with sufficient detail for practical use in many fields. The dependable availability of multispectral imagery for most of the earth's surface greatly expanded the number of people who developed an interest and expertise in the analysis of this data. As an image archive, Landsat data was used to evaluate contentious environmental issues, such as tropical rainforest deforestation, for regions where access was difficult or denied (figure 2.38). Image analysis techniques were developed to enhance the imagery for visual interpretation and perform automated classifications. Classification proved to be a powerful technique adaptable to a wide range of applications including the generation of land-cover maps, projection of the impacts of development (figure 2.39), change detection, environmental monitoring, and resource inventory. Since classification results were in a raster digital format, area estimates by class could be easily obtained by having the computer program tally the number of pixels assigned to each class. Obtaining these area estimates from a paper map typically involved the time-consuming task of outlining every polygon of a class with a planimeter—a handheld instrument that generates an area estimate from an irregular shaped outline.

Though early Landsat imagery was made available at low cost, the limited capabilities of early image-analysis software

Figure 2.36 The northern Los Angeles metropolitan area (lower right), natural forest and shrub in the mountains to the north (center image), and agricultural areas (light red patches in the upper portion of the image) can be seen in this 80-meter resolution false-color Landsat 1 image acquired June 25, 1974.

Source: Image courtesy of NASA JPL.

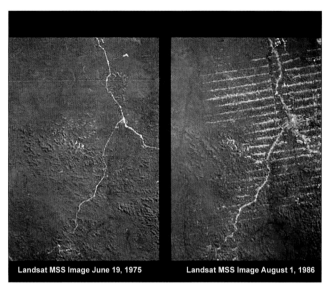

Figure 2.38 Approximately 30% (3.5 million square kilometers) of the world's tropical forests are found in Brazil. These Landsat MSS images acquired in June 1975 and August 1986 illustrate the rapid rate of deforestation in this area, estimated at 15,000 square kilometers per year during this period.

Source: USGS, EROS Data Center, Sioux Falls, S.D.

and the high price of mainframe computers restricted university research and development of remote sensing applications. However, numerous sources provided funding to develop remote sensing software, hardware, and applications. For example, during the 1970s and early 1980s, major government funding from U.S. programs such as the joint NASA-USDA (U.S. Department of Agriculture), LACIE (Large Area Crop Inventory), and AGRISTARS (Agriculture and Resources Inventory Surveys Through Aerospace Remote Sensing) was used to develop practical applications for monitoring agricultural crops and spurred research in other fields and in other countries. The techniques developed for these agriculture applications proved to be valuable not only for assessing national crop production, but also for evaluating crop condition in other countries. Private companies developed crop condition monitoring subscription services used by various commercial interests that traded large quantities of agricultural products from food processing and manufacturing companies to futures traders.

In addition to the United States, other countries such as Australia and Canada, and the European Community provided substantial funding to develop applications for the new space imagery. Landsat spurred the establishment throughout

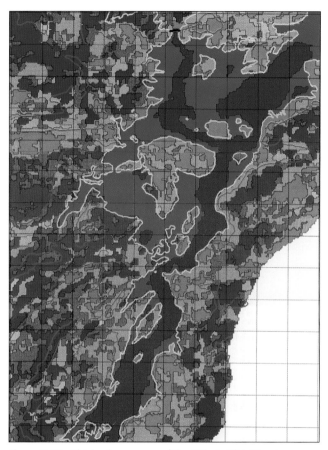

Figure 2.39 This land-cover map of the Little Jackfish River in northern Ontario was generated from a supervised classification of Landsat MSS data to assess habitat loss from flooding if a proposed hydroelectric dam was constructed. Nine land-cover classes were defined: two deciduous forest classes (red shades), three coniferous forest classes (green shades), two wetland classes (orange and yellow), a current water class (dark blue), and a flooded area water class (light blue). The 327-meter contour (yellow line), the proposed water level of the dam, and a higher 350-meter contour, a proposed higher level (light blue line) are also shown.

Source: Image courtesy of S. Pala.

the world of national programs devoted to developing earth resource remote sensing expertise and applications. The costs and benefits of developing these centers may have been hotly debated, but the result was inevitable. Remote sensing expertise had become a required capability for every industrialized nation and arguably for every country. Nations had to know what information foreign interests could gather from remote sensing about their country's natural resources, military assets, national infrastructure, and economic condition.

The Advanced Very High Resolution Radiometer (AVHRR)

The AVHRR was first orbited on the NOAA-6 satellite launched in 1979. The AVHRR sensor was originally designed for meteorological applications—to track clouds and estimate snow cover extent and sea surface temperature. A few years after its launch, its usefulness in monitoring global vegetation was recognized. The instrument's 2,600-kilometer-wide swath width, spatial resolution of 1.1 kilometers, and multispectral imagery with bands in the visible, near-, medium-, and thermal infrared provide daily coverage of most of the globe. AVHRR has the longest service period and widest data distribution and data analysis of any spaceborne instrument in the history of operational meteorology, oceanography, climatology, vegetation monitoring, and land and sea ice observation. AVHRR imagery is used in a great variety of applications including investigation of clouds, land-water boundaries, day and night cloud distribution, and vegetation biomass and condition monitoring. One of the sensor's great advantages is its frequent earth coverage over a long period of time, which provides a large and complete global dataset of earth resource observations that extends from 1978 to the present.

Since the AVHRR sensor provides repetitive coverage with timely and reliable delivery, global users found they could depend on the data and became willing to invest a considerable effort in remote sensing research and development. This in turn accelerated the pace of research and development of practical applications for meteorological data, and particularly AVHRR data for which a broad range of non-weather-related applications from sea ice monitoring to crop condition assessment were developed.

Since 1963, the United States has provided free access to the data from its meteorological satellites. The data is transmitted continuously and can be freely received using relatively simple ground stations anywhere in the world. AVHRR data was made available as part of this free access service, and with a network of thousands of receiving stations around the globe, the imagery has been used worldwide.

The Coastal Zone Color Scanner (CZCS)

The Coastal Zone Color Scanner (CZCS), launched in 1978 on NOAA's Nimbus-7 satellite, was a limited-coverage, proof-of-concept mission[4] that operated until 1986. The mission was designed to find out whether the chosen bands were useful in the assessment of water resource characteristics. The multispectral scanner, designed to analyze water quality, had

four narrow visible bands, a near-infrared band, and thermal-infrared band. The data recorded by the sensor was successfully used to map phytoplankton concentrations, inorganic suspended solids such as silt, and sea surface temperature. Phytoplankton are responsible for approximately 40% of the planet's total annual photosynthetic (i.e., primary) production, and marine phytoplankton represent about 95% of primary production in the oceans. Monitoring marine phytoplankton concentration by means of remote sensing has proven to be of great value in studying global nutrient cycling, ecosystem dynamics, and species distribution (such as fish populations) and in assessing potential impacts of climate change.

The Total Ozone Mapping Spectrometer (TOMS)

Among the many specialized satellites developed and flown by NASA and NOAA for monitoring atmospheric and oceanographic conditions, the TOMS ozone measurement instrument was probably the first to achieve wide public exposure.

In 1985, a British Antarctic survey detected a severe depletion in the ozone layer over the frozen continent. The Total Ozone Mapping Spectrometer (TOMS) instrument, which flew aboard the Nimbus-7 weather satellite, had also recorded low ozone levels over the Antarctic. However, the operations staff assumed the data was incorrect when the ozone values dropped far below the normal range for the season. Prompted by the ground observations, NASA reviewed the TOMS data and found the depletion area or ozone hole was not only large, but it was also increasing in severity from year to year *(figure 2.40)*. Ozone in surface air is a pollutant toxic to humans. About 90% of the earth's ozone is in the stratosphere where it absorbs harmful ultraviolet radiation, preventing it from reaching the earth's surface. Chemical studies concluded that man-made chlorofluorocarbons (CFCs) posed an immediate threat to the ozone layer. As a result, by 1990 an international agreement was reached to end CFC use within ten years. The ozone issue publicized the ability of earth observing systems to monitor global scale changes.

Other satellites, such as Landsat, had revealed rapid deforestation of tropical rainforests in many regions of the world, which sparked concerns about related increases in greenhouse

---

4. A proof-of-concept mission is designed to prove or disprove the value of an instrument or procedure.

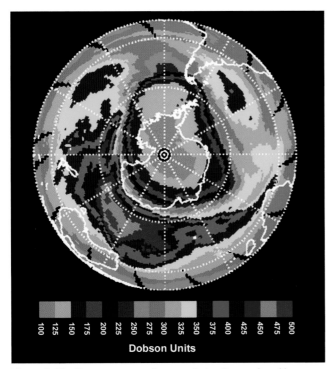

Figure 2.40   Ozone concentration over Antarctica produced by NASA's Total Ozone Mapping Spectrometer (TOMS) instruments, September 25, 1998. This image illustrates the area of low ozone concentration (the ozone hole) that develops during mid-winter over Antarctica. Areas with ozone concentrations lower than 225 Dobson units (the violet colors) delineate the ozone hole. The small black circle in the center is an area where the satellite cannot obtain a reading.

Source: Image courtesy of the Ozone Processing Team, NASA Goddard Space Flight Center, 1998.

gases and global warming. The publicity generated by these issues and others in the mid-1980s not only raised public awareness and concern for global environmental issues but also highlighted the role that remotely sensed data could play in monitoring and researching these issues. In fact, remotely sensed data was the only practical way to collect the global-scale environmental data needed to address them.

SeaSat 1

Launched in 1978, SeaSat provided the first publicly available spaceborne synthetic aperture radar (SAR) images. With a spatial resolution of 25 meters, SeaSat was designed for oceanographic research. Unfortunately, the satellite failed after only 99 days of operation. However, it provided a wealth of imagery that demonstrated an unexpectedly wide range of valuable applications. Originally intended for monitoring the

global surface wave field and polar sea ice conditions, the ocean images in fact proved valuable in analyzing a wide range of atmospheric and ocean phenomena including internal waves, current boundaries, bathymetric features, storms, and rainfall.

Among its instruments, SeaSat carried a radar altimeter that provided measurements of the distance from the satellite to the sea surface below. Using accurate information about SeaSat's orbit, differences in sea height could be derived. Ocean depth is one of several factors that affect sea height (see *Mapping seafloor topography with satellite radar altimetry* in chapter 8). In 1987, scientists published the first map of estimated seafloor topography derived from SeaSat radar altimetry data *(figure 2.41)*. Though it contained major inaccuracies, the map showed anomalies that led to the discovery of major ocean floor features. More sophisticated data analysis and integration with sonar bathymetry (depth measurement using sound waves) has made radar altimetry a valuable data source for seafloor topography mapping. Subsequent radar satellites now provide imagery for monitoring sea state and ice, detecting surface waves and ship wakes, and a range of terrestrial applications as well.

Open skies versus sovereignty

From the 1970s until France launched the SPOT satellite in 1986, the United States was the only country operating a relatively high-resolution satellite-based remote sensing system that systematically collected information of virtually the entire globe: Landsat. The imagery produced by Landsat was and remains publicly available with virtually no restrictions on its use in accordance with the U.S. open skies policy of nondiscriminatory access to imagery collected from space.

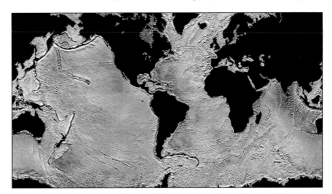

Figure 2.41   The map of ocean floor topography estimated from SeaSat satellite radar altimeter data.

Source: Image courtesy of W. Haxby, Lamont-Doherty Earth Observatory.

With its quasi-monopoly of publicly available earth resource satellite imagery, the United States encountered international pressure to limit access to Landsat data and restrict future development.

The issues of national sovereignty and security in relation to remote sensing are complex. Satellite remote sensing, by providing a new source of unrestricted imagery, could give the user of that imagery a commercial or military advantage beyond the regulatory control of the country being imaged.

Landsat imagery posed little security risk to nations with access to state-of-the-art military surveillance imagery. The Soviet Union and the United States both had access to higher resolution imagery. However, nations without such access were concerned about the security risk in providing unrestricted access to space imagery of their country to potentially unfriendly neighbors. The military value of information is as much related to the alternative information available as it is to the quality of the data.

During United Nations debates it was proposed that

- countries operating satellites obtain the prior approval of other nations before collecting data
- countries should have the right to access all information collected about their country by a foreign nation
- prior consent be obtained from a country before any data collected over its territory is released to a third party

Had these restrictions been applied, only Landsat data for areas of the United States would have been freely available, and only to U.S. citizens.

The Soviet Union suggested limiting the resolving power of civilian remotely sensed imagery to 50 meters, a suggestion supported by India and a number of other countries. However, the debate over limits to the quality and distribution of satellite-based remote sensing images effectively ended in 1986 when France launched the SPOT satellite and, without seeking any international consensus, made 10 meter resolution imagery commercially available to all, with few restrictions.

The debate over imagery for sale

Throughout the 1970s and early 1980s, the Landsat program competed for funding with other NASA programs on the basis of expected commercial and societal benefits. Programs such as the two Voyager satellites launched in 1971 produced startling imagery of Jupiter, Saturn, Uranus, and Neptune; 48 of their moons; and the unique system of

Figure 2.42 Saturn's rings. This image, captured in August 1981 by the Voyager 2 satellite from a distance of 8.9 million kilometers from Saturn, shows variations within the ring system indicating differences in their chemical composition.

Source: Image courtesy of NASA.

rings and magnetic fields those planets possess *(figure 2.42)*. Whereas space exploration and moon landings captured public attention and were considered of value in their own right, remote sensing for earth observation did not achieve the same level of public interest and support and was expected to provide economic and social benefits sufficient to justify its program costs.

Cost-benefit analyses were widely used to justify the Landsat program. This resulted in the inevitable overselling of its capabilities. However, in 1976 the Landsat program was relatively inexpensive for a new technology. The total investment for the first three Landsat satellites was $250 million, less than the cost of one nuclear-powered submarine. In 1977, NASA's entire proposed research budget for remote sensing applications in resource management and environmental monitoring was only $67 million.

Early government efforts to reign in government spending looked at ways to commercialize programs that produced economically valuable products. Earth observing satellite data seemed a likely candidate.

NOAA distributed meteorological data, including meteorological satellite data, internationally at no charge. But an initial study concluded that the value of the data received from other nations was greater than that given away. It would be more costly for NOAA to charge for its data if other countries responded by charging for theirs. So the idea of selling meteorological data was dropped.

However, Landsat data seemed to be a more promising candidate because of the large number of applications of the data being publicized by researchers. As a result, in 1984 the operation of the Landsat program was transferred to a commercial venture, the Earth Observation Satellite Company (EOSAT). Assuming control of the distribution of Landsat 4 and 5 data, EOSAT struggled to make the Landsat program a profitable venture. Even with increased prices, copyright restrictions to prohibit sharing of data among users, large government purchases of imagery, and satellite operation services provided by the government at no charge, there wasn't enough revenue for the enterprise to be profitable. In 1999, with the launch of Landsat 7, operation of the Landsat program reverted to government control.

Until the launch of the commercial satellite IKONOS in 1999, every earth imaging satellite had been subsidized by the government either by funding the program or providing free services such as satellite launches and hardware. The debate over whether the public benefit of remotely sensed data justifies public funding and subsidized pricing continues. Government sales remain a major revenue source for commercial resource satellite ventures that now provide space imagery with spatial resolutions better than 1 meter. However, recognition of the national security benefits of having multiple commercial satellites provide redundant image acquisition and frequent revisit capability has aligned government interests with the commercial success of these private firms.

1986–1999: Global competition, increased commercialization, and the World Wide Web

During the late 1980s and 1990s, concerted efforts were made to commercialize earth observation satellite imagery. New applications of satellite data were developed and the imagery was even used to verify important news stories. Countries besides the United States and Russia increased remote sensing competition by launching earth observation satellites of their own. The implementation of a satellite positioning system by the United States improved the rectification of image products and facilitated the collection of ground reference data. The World Wide Web provided an inexpensive way to inform the public about remote sensing products and distribute them.

Following the Landsat program, other countries initiated earth observation satellite programs. However, the United States controlled civilian space remote sensing because it was the only non-communist country with the means to launch a

satellite. That monopoly ended on February 21, 1986, when France launched its first earth resource satellite, Systeme Probatoire d'Observation de la Terre (SPOT). Producing 10 meter panchromatic imagery and 20 meter multispectral imagery in the visible and near-infrared bands, SPOT® imagery had the highest spatial resolution available. Though the satellite retraced its orbit only every 26 days, the satellite could collect images of an area more frequently using optics that could be pointed to the side (off-nadir viewing) as well as directly beneath the flight path (nadir viewing). As a result, images of areas could be collected one, four, or five days apart depending on the area's latitude. More viewing opportunities are useful for areas with frequent cloud cover and for applications where almost constant monitoring is needed such as the tracking of croplands during the growing season. Another important benefit of off-nadir viewing is that it enables the production of stereo image pairs, two slightly overlapping images of the same area used for photogrammetric measurements such as terrain elevation. Stereo viewing also improves visual image interpretation.

The value of high-resolution satellites with pointable sensors was soon made evident when, two months after SPOT was launched, on April 26, a devastating nuclear accident occurred at Chernobyl, Ukraine. At first, the Soviet Union denied the accident occurred. However, the American Broadcasting Corporation News used imagery from the SPOT satellite to independently confirm the event *(figure 2.43)*. It was the first use of commercial satellite imagery

Figure 2.43   SPOT-1 satellite image of Chernobyl, Ukraine. In 1986, soon after its launch, the SPOT-1 satellite confirmed that a serious accident had occurred at the Chernobyl nuclear reactor site. This combined SPOT and Landsat image reveals the location of the reactor meltdown by its extreme heat as detected by its thermal signature, shown as a red dot.

Source: © CNES 2004/Courtesy of SPOT Image Corporation and courtesy of the USGS.

by a news organization to corroborate a major news story, demonstrating that with satellite imagery events could be monitored in countries where access to the news media was denied. Subsequent thermal infrared imagery from Landsat pinpointed the location of the burning reactor. Fallout from the accident spread far beyond the Soviet Union, and satellite imagery proved invaluable in tracking the progress of emergency management measures.

Other countries soon followed France in launching civilian earth resource satellites. The Russian Resurs satellites were first launched in 1985, and India's first IRS (Indian Remote Sensing) satellite flew in 1988.

The fall of the Berlin Wall in November 1989 was the beginning of the end of the cold war as it affected civilian remote sensing. During the 1990s, with the breakup of the Soviet Union and the increased public availability of satellite imagery, Russia and the United States declassified much of

Figure 2.44 SPIN-2 Russian satellite photograph of New Orleans, Louisiana. SPIN-2 images are satellite photographs taken with the Russian KVR-1000 camera system and scanned to provide 2-meter pixels. The area represented in the image is approximately 2.24 kilometers across.

Source: © 1998 Sovinformsputnik.

their early military reconnaissance images. In 1992, Russia authorized the sale of digitized 2-meter resolution satellite photography from its KVR-1000 camera system on board the SPIN-2 satellite *(figure 2.44)*. As the sole global source of high-resolution satellite imagery, Russia had an early competitive advantage. But internal conflicts between intelligence and commercial interests made for long delays and frequent cancellation of orders. In 1995, the United States followed suit and declassified imagery collected by the Corona intelligence satellite program.

The end of the cold war had another profound impact on the sale of satellite-based remote sensing. It allowed Russia to provide an additional competitive source for satellite launching services. By offering less-expensive launch services and no political interference in the capabilities of the satellites, Russia made it more affordable for countries to design, build, and launch remote sensing satellites in competition with the established players. Having this option available gave independent satellite operators increased bargaining power with all providers of launch services.

Satellites with capabilities that were similar or beyond those of the United States and Russia were launched during the 1990s, including the ERS (European Remote Sensing Satellite), IRS, and SPOT. These satellites provided continuity of service by replacing aging equipment and upgrading sensors. By 1998, the Indian IRS 1C and 1D satellites were providing systematic, near-complete global imagery with a spatial resolution of 5.8 meters. SPOT and Landsat offered lower spatial resolution but additional spectral bands. Continued improvements were made to the sensors of lower spatial resolution satellites such as the NOAA Geostationary Operational Environmental Satellite (GOES) series used for meteorology and the AVHRR sensor widely used for regional crop monitoring, oceanography, meteorology, sea surface temperature mapping, and many other earth resource applications *(figure 2.45)*.

The 1990s also saw the launching of a series of radar satellites. Radar systems provide their own source of illumination and operate at microwave frequencies that penetrate clouds, so these systems provide all-weather day and night image acquisition. In 1991 and 1992, Russia (Almaz-1), the European Space Agency (ERS-1), and Japan (JERS-1) launched SAR satellites with resolutions ranging from 10 to 30 meters. Japan's JERS-1 offered radar imaging (18-meter spatial resolution) as well as an optical multispectral scanner operating in the visible, near-infrared, and mid-infrared bands. JERS-1 was the first civilian satellite to produce along-track stereo

Figure 2.45   Satellite view of hurricane Andrew on August 25, 1992. This composite image was generated with cloud data from the GOES-7 geostationary meteorological satellite. The vegetation cover was derived from the AVHRR instrument on a polar orbiting NOAA satellite.

Source: Image courtesy of NOAA.

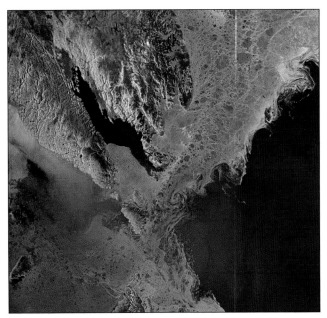

Figure 2.46   Radarsat-1 image of ice in Hudson Strait, Canada. One of the principal applications of Radarsat imagery is mapping ice conditions for navigation in Canada's northern waters.

Source: Radarsat-1 data © Canadian Space Agency 2002. Received by the Canada Centre for Remote Sensing. Processed and distributed by RADARSAT International.

digital imagery by having one duplicate band viewing the area ahead of the satellite. In this way, stereo imagery was generated continuously, avoiding the difficulties of trying to capture a second image of an area from a different orbital track as was done for the first SPOT satellites.

Canada launched Radarsat in 1995 with ground resolutions ranging from 8 to 100 meters *(figure 2.46)*. Developed for monitoring sea ice in Canada's northern coastal waters, the imagery enabled daily monitoring of sea ice conditions and transmission of sea-ice-condition maps to ships operating in those regions.

Radar satellites currently operated by the European community include ERS-2 and Envisat (Environmental Satellite), which, in addition to radar, carry a suite of remote sensing instruments for environmental monitoring including an imaging sensor operating in the visible and infrared bands.

Advances in image processing

While the new remote sensing satellites received much public attention, it was aerial photography that remained the principal remote sensing system used in operational mapping and resource inventory *(figure 2.47)*. The field of

photogrammetry advanced with the introduction of computer-based photogrammetric workstations and then digital photogrammetric workstations, which used digital stereo imagery produced by scanning aerial photographs.

Digital-image processing methods were developed to rectify remote sensing imagery (whether digital images or aerial photographs scanned to produce digital imagery) to match the geometry of a planimetric map[5]. Images rectified in this way are termed *orthophotos* or *orthoimages*. Multiple images with overlapping coverage could be combined by orthorectifying, aligning, and color-balancing the individual images to produce a seamless coverage with the geometry of a chosen map projection—termed an *orthophoto mosaic*, or simply an *orthoimage (figure 2.48)*.

The development of digital methods of orthoimage production significantly improved the quality, reduced the cost, and made the process significantly faster than the largely

---

5. A planimetric map depicts the relative horizontal position of features irrespective of their vertical position.

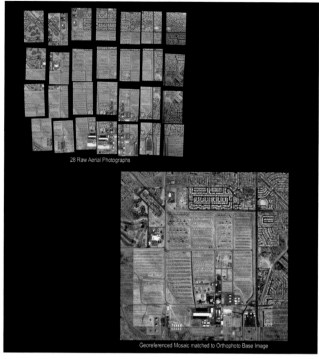

Figure 2.47 Color aerial photography taken by a large-format metric camera. Large-format (230 cm by 230 cm) metric cameras have been the principal remote sensing system for high-accuracy mapping.

Source: Image courtesy of Pacific Aerial Surveys and HJW GeoSpatial, Inc.

Figure 2.48 Orthophoto mosaic. Orthophoto mosaics are produced by geometrically rectifying the images to match the desired map base and then balancing the color and aligning the individual images to form a seamless combined image.

Source: © Positive Systems, Inc.

manual process previously used. What's more, since the digital orthoimages had the geometry of a map, they could be readily used as backgrounds over which vector data could be displayed in a geographic information system *(figure 2.49)*. As geographic information systems become more capable of handling imagery, demand for GIS-compatible digital orthoimage products for use as background image maps suddenly and rapidly increased.

Remote sensing analysis results such as land-cover classifications had been easily imported into raster-based geographic information systems since they were first developed. However, importing such raster data into vector-based geographic information systems to form polygons with class labels stored in the GIS database was a major challenge. Improved database technology and increasing demand from the GIS community led software manufacturers to develop a range of solutions, and the ability to import classifications and other remote sensing analysis results soon became a standard GIS feature. Not only did this increase demand for remote sensing imagery, it also facilitated the use of GIS to store and manage data such

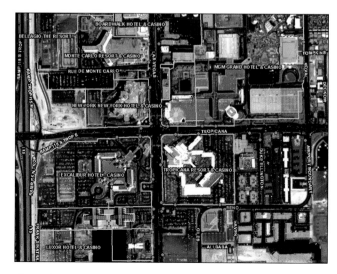

Figure 2.49 Vector data displayed over an orthoimage background in a GIS.

Source: ESRI and DigitalGlobe.

Figure 2.50   Airborne digital camera image of Chesapeake Bay, Washington, D.C. This image was produced from data acquired by an ADAR® 5500 digital camera system. This sensor simultaneously captured images in the near-infrared and visible blue, green, and red spectral bands. The individual bands could be displayed as black-and-white images. By combining the appropriate three bands, normal-color and color-infrared images were produced.

Source: © Positive Systems, Inc.

as training sites used for remote sensing analyses. As a result, the two technologies were increasingly perceived as closely linked and complementary.

Advances in sensor design

Major advances were made in sensor design and reliability, and airborne programs tested the new systems. Airborne digital frame cameras that captured discrete overlapping stereo images like a film camera were introduced for remote sensing. Making use of Global Positioning System (GPS) satellite positioning and aircraft attitude measurement instruments (discussed below), a highly automated process could be used to rectify the images to produce orthoimage maps. These early digital camera systems produced relatively small images (about 1,500 by 1,000 pixels) and were generally used for projects requiring relatively small coverage areas *(figure 2.50)*. Large format photogrammetric film cameras provided much higher resolution and a larger coverage area per frame. However, the trend toward development of all-digital systems to compete with metric film cameras had begun.

Experimental multispectral systems with more bands covering the visible, near-, mid-, and thermal infrared allowed researchers to refine sensor technology and analysis techniques and develop new applications. Hyperspectral imaging,

Figure 2.51   Experimental imagery of southwestern Georgia. This research image was acquired in May by the NASA Advanced Thermal and Land Applications Sensor (ATLAS) for a study of remote sensing applications in precision agriculture. The 15-channel multispectral scanner operates in the visible, near-, mid-, and thermal infrared spectral bands. This 2-meter resolution image was generated from the thermal infrared, near-infrared, and visible red bands. The color variations within fields (e.g., magnified area) indicate differences in soil surface characteristics that can be useful for precision farming.

Source: Image courtesy of the Earth Science Department, NASA Marshall Space Flight Center, Huntsville, AL 35812.

employing a hundred or more narrow contiguous spectral bands, produced imagery from which a detailed spectral curve could be derived for each pixel *(figure 2.51)*. The technology proved especially promising for the identification of minerals in geology over vegetation-free areas.

Radar interferometry, based on a sophisticated computer-intensive method of processing radar image pairs, was found capable of measuring small changes in elevation from two images of the same scene taken from the same position but on different dates. Under favorable conditions, ground uplift or subsidence could be measured with vertical accuracies finer than 1 centimeter from satellite imagery acquired by ERS-1 and 2. The technology was soon adapted for topographic mapping from airborne and later spaceborne (space shuttle) platforms and became a valuable alternative to aerial photography for many applications (see below). Applied to topographic mapping, radar interferometry using two images acquired from different positions with a precisely known offset can provide vertical accuracies on the order of a few meters for airborne systems and tens of meters for spaceborne systems.

Geospatial technologies bring additional precision and a wider audience to remote sensing

During this time, two of the most revolutionary advances in geospatial technology to affect remote sensing were developed—satellite positioning and the World Wide Web.

Satellite positioning

The development of satellite positioning fundamentally changed the way geographic location is determined. The first satellite positioning system, the Global Positioning System (GPS), was implemented by the United States Department of Defense. Beginning with the launch of Navstar-1 in 1978, the program grew to 24 satellites by 1994 *(figure 2.52)*. (As of 2005, the system is typically operated with a constellation of 29 satellites, which provides redundancy and improved performance where the area of viewable sky is restricted.) From the signals broadcast by the satellites, a GPS receiver calculates its three-dimensional geographic position and the precise time of each measurement *(figure 2.53)*.

The cost of GPS receivers fell rapidly as the system became operational and a large global market rapidly developed. One of the first GPS receivers manufactured by Texas Instruments in 1978 cost $158,000, and in 1988 the first hand-held civilian unit was offered by Magellan for $3,500. By

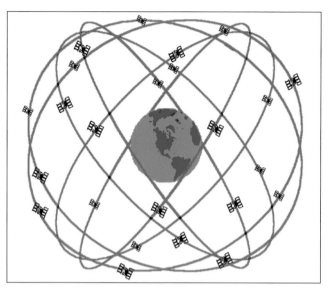

Figure 2.52 The GPS constellation.

Source: Image courtesy of P. H. Dana, PhD.

Figure 2.53 Portable GPS units can be used to accurately record the location of field data collection sites or physical assets.

Source: Image courtesy of Applanix Corporation.

1997, a GPS receiver could be purchased for $100 by anyone anywhere with none of the ongoing subscription or licensing fees to use the service that were previously required.

Before GPS became available, obtaining an accurate geographic position was time consuming. There were three options. First, you could find where you were on a map if you could locate distinctive features for orientation. Or, you could survey your position by taking a series of accurate ground measurements starting from a known position if one existed nearby. Or, you could take accurate sightings of celestial bodies, and with a chronometer (a very precise clock) to provide the time of the sighting and a set of astronomical navigation tables, you could calculate your position. Other options existed for ships and aircraft. Radio-based navigation systems were available for ships to navigate coastal waters, and aircraft navigation used a worldwide network of radio beacons to define flight paths and track position. But these systems were expensive, did not provide worldwide coverage, and could not be built into an inexpensive handheld unit.

GPS for the first time enabled geographic position to be determined with accuracies ranging from about 100 meters to better than 1 centimeter anywhere in the world, depending on the quality of the GPS receiving equipment. (More sophisticated dual frequency GPS receivers can achieve centimeter-level accuracies by collecting data at a location for one to three hours. Accuracies can also be substantially improved using differential GPS, described below). GPS revolutionized navigation as well as the collection of georeferenced data in the field. For remote sensing, the capability to rapidly collect accurate position data enabled the automation of much of the time-consuming and expensive process of image rectification, the process used to transform the geometry of image data to register with maps or other images of the same scene.

The GPS system was designed to provide both a high-accuracy signal for use by the U.S. military and its allies, termed *precision acquisition* or P-code, and a less accurate signal for civilian use, termed *coarse acquisition* or CA-code. A procedure called *selective availability* was used to degrade the CA signal, which produced horizontal accuracies that varied from about 50 to 100 meters. Military receivers could access the P-code signal and achieve horizontal accuracies of about 16 meters worldwide. When the system became fully operational, the precision signal was encrypted and designated P(Y) code.

Civilian users soon found the large errors introduced by selective availability could be overcome by using GPS data simultaneously recorded at a second fixed location with known geographic coordinates. The correction data could be saved and used later to reprocess and correct a set of GPS measurements. Alternatively, GPS measurements could be corrected in real-time by broadcasting the correction signal to a suitable receiver connected to the roaming GPS unit. Since the introduced errors had the same effect over a wide area, the second receiver served as a control, continuously recording the introduced error at each moment in time. These correction methods, termed differential GPS, or DGPS, were so effective that accuracies in the range of 2 to 5 meters were obtained, sufficient for geometric rectification of remotely sensed imagery. Under certain circumstances, accuracies as high as 1 to 2 meters could be obtained. At first, individual users set up their own DGPS reference stations. However, the substantial value of having a network of broadcast reference signals for DGPS soon led government agencies and private firms to implement broadcast networks available free of charge or by subscription.

In May 2000, selective availability was set to zero while retaining the capability to again degrade the signal in times of national emergency. With selective availability removed and the larger 29-satellite constellation currently (as of 2005) in use, handheld 12-channel GPS units can generally provide horizontal accuracies better than 8 meters. Real-time accuracies of 1 to 3 meters can be achieved with equipment as compact as a $1,400 backpack-carried DGPS unit using broadcast correction signals such as the Nationwide Differential GPS Network (NDGN) available free of charge within the contiguous 48 states of the United States.

By integrating DGPS to record position and an inertial measurement unit (IMU) to record precise aircraft orientation within a computerized remote sensing instrument control system, a complete digital sensor position record is produced. This data enables imagery to be rectified with little or no additional ground control to produce orthorectified imagery and mosaics. Use of GPS and IMU technology has quickly become the standard practice for remote sensing image acquisition.

Satellite positioning being essential for both military as well as civilian applications, the former Soviet Union also implemented a satellite positioning system—GLONASS (Global Navigation Satellite System). Similar to the U.S. GPS system, it is based on a constellation of active satellites continuously transmitting coded signals. It is currently managed for the Russian Federation Government by the Russian Space Forces. The first GLONASS satellites were launched into orbit in 1982, but deployment of the full constellation of satellites was completed only in 1996. Funding cutbacks resulted in degradation of the system leaving only eight fully operational satellites in early 2004. The U.S. GPS is the system most widely used and for which the most inexpensive receiving equipment is available. However, dual system receivers that receive both the GPS and GLONASS signals are available. By increasing the number of satellites potentially available, these receivers can provide improved accuracy under some difficult satellite reception conditions.

The European Union and European Space Agency have begun development of a satellite positioning system called GALILEO. It will be similar in design and interoperable with the GPS and GLONASS systems. The fully deployed GALILEO system will consist of 30 satellites and is expected to be operational by 2008.

The World Wide Web

In 1989, Tim Berners-Lee invented the World Wide Web while working at CERN, the European Particle Physics Laboratory in Geneva, Switzerland. In providing a rich graphic user interface implemented through browser software, the World Wide Web provided a means for computers of different designs and using different operating systems to communicate over the Internet.

Within a few years the World Wide Web had become a global communication system. It was a boon to the field of remote sensing and geographic information systems, both graphics intensive disciplines. The rich graphic environment provided a means to search remote databases of imagery and other geographic data, interactively locate areas of interest, preview data products, and transfer the data. In so doing it provided the opportunity to develop networked databases that gave access to large data archives without having them reside at a single physical location. Cooperating organizations could share their data, which gave users access to multiple databases located in widely dispersed geographic locations.

With the development of high-speed access, delivering remote sensing imagery via the Internet became practical, which made the sale and immediate delivery of imagery possible. In addition to satellite operators, independent image vendors began offering remote sensing data online. In June 1998, Terraserver, an online remote sensing image vending enterprise, began operation. The joint venture partnered TerraServer.com with Microsoft, Compaq, and the U.S. Geological Survey to offer access to terabytes of aerial photography and satellite imagery.

Rapid image delivery expanded the possibilities of using remote sensing in time-sensitive applications such as breaking news stories. Inexpensive communication of color graphics allowed educational materials to be made available over the Internet that would be too expensive to distribute in print. Free access to sample images and educational materials provided by vendors and government agencies made remote sensing imagery readily available to anyone with an Internet connection.

1999–The present: Commercial availability of very high resolution imagery

Beginning in 1999 with the successful launch of the IKONOS satellite, 1-meter resolution satellite imagery, formerly available only to the military, was made commercially available to all. Providing detail similar to that of high-altitude photography, it was quickly adopted for a range of applications including military reconnaissance. The high resolution made the imagery valuable to both the civilian and military remote sensing communities. While classified U.S. military reconnaissance satellites generated higher resolution imagery, the commercial imagery could satisfy some reconnaissance needs. Commercial 1-meter resolution imagery was a less expensive source of imagery than launching and operating additional satellites. U.S. government support in the form of volume purchases represented a convergence of the interests of the civilian and military remote sensing communities.

This period also saw the continued development of new sensor technology such as lidar and refinement of others such as radar interferometry. Also, the continued reduction in the cost of computing power and storage and the availability of imagery over the Internet broadened access to and awareness of remotely sensed imagery and applications. Demand for remote sensing image handling led GIS vendors to include basic image analysis capabilities in their software, a trend which continues to the present.

Introduction of 1-meter resolution imagery

The launch of the IKONOS satellite by Space Imaging, Inc. in September 1999 heralded major changes in civilian remote sensing. For the first time, imagery with 1-meter spatial resolution was collected by a private company for sale to both military and nonmilitary customers.

The IKONOS system provides digital panchromatic images[6] with spatial resolutions as high as 1 meter and as high as 4 meters for multispectral imagery. The pointable sensor can acquire stereo imagery in a single pass by collecting a forward viewing and then backward viewing image. Same-pass stereo pairs are preferred because the two images are acquired within minutes, so the scene illumination and content are virtually identical. Across-track stereo pairs are collected from different satellite tracks and may be acquired days or weeks apart depending on a satellite's orbit parameters and cloud conditions that could obscure the scene on the subsequent pass.

IKONOS was followed in 2000 by the launch of EROS A1 by ImageSat—a joint venture between two Israeli companies (Israel Aircraft Industries and Electro-Optics Industries) and an American company, Core Software. EROS A1 provides 1.8-meter resolution panchromatic imagery. DigitalGlobe® QuickBird™ 2 was successfully launched the following year and offers 0.61-meter panchromatic and 2.44-meter multispectral imagery. In 2003, ORBIMAGE successfully launched its OrbView®-3 satellite, which provides 1-meter panchromatic and 4-meter multispectral imagery. Future plans call for the launch of satellites with half-meter resolution (discussed in chapter 7).

As these satellite sensors have obtained higher spatial resolutions, the coverage area captured on each satellite pass has decreased, and the cost per unit area has risen substantially. In general, the cost of remotely sensed imagery per unit area increases with spatial resolution (spatial resolution is discussed in chapter 4). Coarse-resolution imagery, such as AVHRR data with its 1.1-kilometer pixel resolution and 2,400-kilometer swath width, is distributed at minimal or no cost. The Landsat TM sensor images a swath of terrain 185 kilometers wide, whereas higher resolution satellites such as IKONOS and QuickBird produce vertical imagery with swath widths of 11 and 16.5 kilometers respectively. The cost per pixel of IKONOS or QuickBird imagery may be comparable with that of Landsat imagery, but to cover the same area as a single Landsat 30-meter pixel requires 900 (i.e., 30 x 30) 1-meter pixels.

High-resolution satellites with pointable sensors acquire imagery selectively, depending on customer requests, rather than repetitively collecting images of their entire coverage

Figure 2.54 Wildfire in San Bernardino, California. This IKONOS 4-meter resolution false-color image was acquired October 28, 2003. Green living vegetation appears bright red to reddish brown while burned areas are dark gray.

Source: Image courtesy of Space Imaging Corporation.

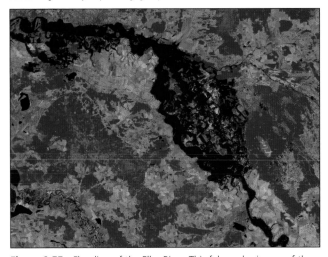

Figure 2.55 Flooding of the Elbe River. This false-color image of the Elbe River and its tributaries (right) was taken on August 20, 2002, by the Enhanced Thematic Mapper sensor on the Landsat 7 satellite. Heavy rains in Central Europe led to heavy flooding that killed more than 100 people in Germany, Russia, Austria, Hungary, and the Czech Republic and caused $20 billion in damages.

Source: Image courtesy of J. Allen, NASA Earth Observatory. Data provided by the USGS EROS Data Center Satellite Systems Branch.

---

6. The term *panchromatic* denotes sensitivity to a broad range of wavelengths, usually including a large portion if not all of the visible spectrum. The IKONOS panchromatic sensor records radiation across almost the entire visible spectrum and a portion of the near-infrared.

Figure 2.57 IKONOS satellite image of captured EP-3 surveillance aircraft. On April 1, 2001, an American EP-3 surveillance aircraft (smaller plane in the image) collided with a Chinese fighter jet and made an emergency landing at the Lingshui airfield on Hainan Island. After 11 days of negotiations, the crew was released and arrangements made for the United States to dismantle and remove the plane on two Russian-made AN-124 cargo aircraft (larger aircraft). The commercial satellite imagery used to illustrate the story in the press and over the Internet added impact and immediacy.

Source: Image courtesy of Space Imaging Corporation.

Figure 2.56 New York World Trade Center disaster, Manhattan. Acquired one day after the September 11th disaster by Space Imaging's IKONOS satellite, this 1-meter resolution satellite image of Manhattan, New York shows an area of white dust and smoke at the location where the 450 meter towers of the World Trade Center once stood. Since all aircraft were grounded at the time, the IKONOS image is the only high-resolution "view from above" of the fire and destruction of the twin towers that day.

Source: Image courtesy of Space Imaging Corporation.

area. However, they are managed to progressively build a complete archive while fulfilling time-critical customer orders. Their high-resolution imagery has proven valuable for mapping and monitoring urban infrastructure, providing information comparable to that obtained from medium-scale aerial photographs. The orthorectified image products have been widely used in GIS applications as backgrounds over which vector information is displayed. The imagery is used for urban and regional planning, earth resource inventory and monitoring, and emergency response to natural disasters such as flooding, hurricanes, wildfires (*figures 2.54 and 2.55*), and human-caused disasters (*figure 2.56*).

The imagery has also been widely used in the media as supporting evidence for news stories and by political factions to influence public opinion. For example, in 2001 a Chinese fighter plane collided with a U.S. reconnaissance aircraft flying over international waters off the coast of China. IKONOS imagery showed the crippled U.S. plane at an airbase on China's Hainan Island shortly after its emergency landing. Repeated imagery monitored the aircraft and its dismantling over a three-month period (*figure 2.57*).

In June 2002, media reports of U.S. security concerns over Iran's Bushehr nuclear reactor were given wide coverage in the media aided by satellite images of the site (*figure 2.58*). Subsequent international pressure led to increased scrutiny of new construction at the reactor site and more stringent agreements on the monitoring of spent fuel.

Web sites devoted to the dissemination of information about the military capabilities and activities of countries throughout the world, such as the Federation of American Scientists and GlobalSecurity.Org, make extensive use of commercial high-resolution satellite imagery, images leaked to the press, or images in public domain. Such initiatives, by increasing transparency (i.e., by exposing and publicizing activities normally hidden from view), can focus public attention on controversial issues related to military, environmental, or other activities with potentially international impacts.

Figure 2.58 Bushehr nuclear reactor site, Iran. Acquired on March 1, 2001, by Space Imaging's IKONOS satellite, this 1-meter resolution satellite image provided an illustration for media stories covering the confrontation.

Source: Image courtesy of Space Imaging Corporation.

Commercial and government satellite imaging: Converging objectives

Before IKONOS, a clear distinction existed between government-owned or government-controlled civilian satellites and classified military image acquisition programs. Military satellites collected imagery with much higher spatial resolution that was rapidly disseminated to intelligence analysts. Civilian earth observation satellites captured imagery with lower spatial resolutions that were disseminated more slowly and used to inventory and monitor a broad range of natural resources and human activities over large regions. The advent of commercial satellites with spatial resolutions of 1 meter or finer and the capability to rapidly deliver image products has begun to blur the distinctions of resolutions and delivery times for civilian and military applications.

Similarly, the distinction has eroded between government-owned civilian satellites and commercial (privately owned but government licensed) observation satellites. Governments increasingly strive to recover program expenses by selling imagery on international markets, and commercial firms have found government agencies to be major customers.

The availability of very high-resolution satellite imagery was not so much a result of technological advances as it was a shift in U.S. policy away from total government control of satellite imaging that originated during the cold war. Civilian remote sensing came to be seen as a dual-use resource, supporting both civilian applications as well as national security needs. Once commercial ventures in other countries could launch and operate very high-resolution satellite imaging sensors (such as SPOT and ImageSat), it was no longer possible for the United States to control access to very high-resolution imagery. Strong incentives existed for the United States to promote a U.S. commercial remote sensing satellite industry that could successfully compete with foreign operators.

A competitive industry would foster innovation that could potentially lead to lower prices and a greater variety and quality of imagery, derived products, and services. Large government purchases of imagery by agencies such as the U.S. National Geospatial-Intelligence Agency (NGA) (formerly the National Imagery and Mapping Agency, or NIMA) not only supported the industry, but it also allowed intelligence services to concentrate their image acquisition resources on more sophisticated high-resolution sensors. Optical sensors on U.S. military reconnaissance satellites are purportedly capable of 10 centimeter spatial resolution imagery and radar sensors of 1-meter spatial resolution[7]. However, the high cost of these sophisticated satellites and their limited number make the use of commercial satellite imagery an important contribution.

In times of national emergency, the image acquisition and processing capacity of commercial firms could also augment that of government agencies to meet surges in image demand. As a major customer, government agencies are well appraised of the capabilities of commercial sensors and can influence the development of products and services to better satisfy their needs.

Also, by licensing commercial satellite operators, their dissemination of sensitive imagery could be controlled. In its licensing policy, the United States reserved the right to limit data collection and dissemination for selected geographic regions in times of national emergency. Other nations

7. Several other countries have or are developing classified reconnaissance satellites producing imagery with similar levels of spatial resolution including China, France, India, Italy, Japan, Russia, and South Korea (Petrie 2004). For many countries, commercial high-resolution satellite imagery is the best reconnaissance imagery they can acquire. Not surprisingly, government purchases are a major source if not *the* major source of revenue for commercial satellite operators.

adopted similar *shutter control* restrictions. In practice, the United States chose to negotiate an exclusive purchase agreement with Space Imaging Inc., operator of the IKONOS satellite, during the war against the Taliban in Afghanistan in November and December 2001, thereby restricting sales of high-resolution imagery of the conflict region to other customers. Vendors made substantial revenues producing high-priority image products in support of national security and intelligence requirements.

However, during the Iraq war in March 2003, the United States chose not to negotiate exclusive rights agreements with commercial satellite operators or impose formal or informal government restrictions on the release of high-resolution satellite imagery of the conflict zone. Satellite images were released during the conflict, widely published in the media, and made available over the Internet *(figure 2.59)*. In part, the presence of journalists in frontline ground units probably made satellite imagery less of a concern to military

Figure 2.59 Satellite view of Baghdad. This IKONOS image, acquired in October 2002, was used to illustrate media stories.

Source: Image courtesy of Space Imaging Corporation.

authorities. As well, the rapid advance of ground forces on Baghdad made the information provided by satellite imagery less representative of current conditions and therefore of less intelligence value. The commercial firms may have also exercised some self-restraint in deciding when and which images of Iraqi locations to release. As a result, an important precedent has been set in that the United States has yet to impose formal shutter control on U.S. commercial satellite operators in times of war, even though the satellite imagery has been used in limited ways to record and report on current military operations.

Technology refinement

The period since 1999 also saw the continued development of new sensor technology. Lidar systems (discussed in chapter 8) became more widely available as instrument costs dropped and accuracies improved. Offering vertical accuracies of ten centimeters or better, lidar can provide elevation data suitable for engineering applications. For areas of subtle relief where perception of stereo is difficult, such as wetland mudflats, lidar can provide more accurate topographic mapping than aerial photography. Continued improvements in radar interferometry have produced airborne systems capable of generating digital elevation models with vertical accuracies of 0.5 meters or better over terrain without vegetation.

In February 2000, during a 10-day mission, the Shuttle Radar Topography Mission (SRTM) onboard the space shuttle *Endeavour* collected data to produce a high-resolution digital topographic database of the earth between 60° N and 56° S latitude (representing about 80% of the earth's land surface). The data is being processed to generate a high-resolution digital elevation dataset with 30-meter spatial resolution for restricted distribution and a lower 90-meter resolution version for public distribution *(figure 2.60)*.

New satellites were launched with improved sensors that provided continuity with established image archives. These included SPOT-5 (SPOT Image Corporation), Envisat (European Space Agency), and the Terra and Aqua satellites (NASA).[8] Terra and Aqua carry the Moderate Resolution Imaging Spectroradiometer (MODIS) sensor, which provides a broad range of imagery with spatial resolutions of 250 meters, 500 meters, and 1 kilometer in 36 spectral bands. As well as

Figure 2.60 Shaded relief image of Southern California generated from the publicly available digital elevation data produced by the Shuttle Radar Topography Mission (SRTM).

Source: Courtesy of B. Zeman at HJW GeoSpatial, Inc.

offering new image products, MODIS is designed to provide continuity with data from the AVHRR sensor.

The trend for airborne remote sensing acquisition to become all-digital has accelerated. Aerial photography, still widely used for high-precision, high-definition, large-scale imaging is being challenged by all-digital metric camera technology. Advances in computer storage technology and processor speeds have made the development of large-format, digital photogrammetric imaging systems technically feasible. Several vendors have introduced multispectral scanner and frame capture systems designed for airborne mapping. With sophisticated differential GPS and aircraft attitude monitoring instruments, imagery can be accurately rectified and processed into mosaics with minimal or no ground control. The large-format images, 12,000 pixels or more across, demand terabytes of onboard computer storage. However, a variety of image composites can be generated from the multispectral dataset, including color infrared, normal color, and panchromatic or infrared black-and-white imagery. The

---

8. The Terra and Aqua satellites are part of NASA'a Earth Observing System, a program designed to collect a comprehensive set of global observations of the earth's land, ocean, and atmosphere in the visible and infrared regions of the spectrum.

all-digital workflow automates much of the rectification and mosaic processing.

Compared with scanned aerial photography, the digital sensors are more sensitive and provide greater shadow detail. The images are free of dust and scratches. High-quality digital images tend to have greater stereoscopic clarity, which enables them to be processed more quickly by automated image matching software. Properly calibrated digital sensors produce imagery that is radiometrically more consistent than film images, which are subject to inconsistencies in processing and handling of the photosensitive film. The color balance of color infrared film is highly sensitive to temperature shifts during handling and processing, making digital solutions particularly attractive. Digital metric cameras have been developed with image capture rates fast enough to produce forward overlaps of 70% or more with spatial resolution as high as three-centimeter pixel resolution *(figure 2.61)*.

Further integration of remote sensing and GIS

During the 1990s, most vector-based geographic information systems incorporated the display of georeferenced raster imagery. This made possible the accurate display of vector GIS data over a background image. With a digital elevation model, a 3D perspective view could be generated as well *(figure 2.62)*. As these capabilities became standard, the demand for orthorectified imagery and digital elevation models from

the GIS community rapidly increased. Integrated raster-vector analysis capabilities continue to be added to GIS software. This in turn has led to the increased use of remotely sensed data, the provision of basic image processing tools within the GIS environment, and improved compatibility with specialized remote sensing analysis software. Remote sensing data vendors responded to the surge in demand by developing products tailored to the GIS environment such as seamless georeferenced image products with accompanying digital elevation models for customer defined areas.

Remote sensing has become one of several key technologies that together constitute the integrated discipline known as *geomatics*. Geomatics encompasses the collection, analysis, and dissemination of all forms of geospatial data and includes geographic information science, surveying, photogrammetry, remote sensing, cartography, and related disciplines.

Earth imagery becomes much more valuable when it is integrated with other data sources using technologies such as geographic information systems and satellite positioning. They provide improved calibration, accuracy verification, and the opportunity to produce a wider range of information products. Particularly valuable is the modeling capability made possible by the fusion of remotely sensed data with other data sources. Remote sensing has become a standard information source in an ever-growing list of application areas. It is within this broader context of geomatics that future developments in remote sensing are being shaped.

Figure 2.61 Comparison of scanned aerial photography and airborne digital imagery. An image from an ULTRACAM™ D large-format metric digital camera acquired with an 11-centimeter ground sample distance (left) is compared with an air-photo film image scanned to give the same resolution (right). In this case, the digital camera provided a sharper image.

Source: Image courtesy of Vexcel Corporation.

a

Figure 2.62   Grand Canyon perspective view. This perspective view looking northward over the Grand Canyon was created from imagery acquired by the Advanced Spaceborne Thermal Emission and Reflection Radiometer (ASTER) on the NASA Terra earth resource satellite. The image data was combined with elevation data to produce this perspective view with no vertical exaggeration. Visible and near-infrared data were combined to form an image that simulates the natural colors of water and vegetation; however, rock colors are not accurate. The blue and black areas on the North Rim represent a forest fire that was smoldering as the data was acquired.

Source: Image courtesy of NASA/GSFC/MITI/ERSDAC/JAROS, and U.S./Japan ASTER Science Team.

b

Satellite imagery with spatial resolutions as high as 0.61 meters is now commercially available to all. Forty years ago such imagery was highly classified intelligence data available to only the most senior officials of the most powerful nations *(figure 2.63)*. Commercial satellites offering even higher resolutions are under development (see chapter 7).

More than just a technological advance, the availability and wide use of high resolution satellite imagery has changed the world's political and social landscape. Remote sensing has become an extension of the public right to information and the freedom of the press. As with the freedom of the press, there remains a need to place some limits on access, but the increased transparency provided by remote sensing offers a valuable and powerful tool to enhance the security and quality of life for the entire global community.

Figure 2.63   The Pentagon, Washington, D.C. Top secret Corona black-and-white satellite photography with resolution of about 2 to 10 meters, such as this image of the Pentagon from the early 1960s (a), would require weeks to be delivered after image capture. Today, publicly available commercial satellite imagery, such as this 0.61-meter pan-sharpened color image captured by a multispectral scanner on the QuickBird satellite in 2001 (b), can be delivered within hours of acquisition.

Source: (a) Image courtesy of Central Intelligence Agency Record Group #263. (b) Image courtesy of DigitalGlobe.

References and further reading

ASPRS. 1983. *Manual of remote sensing.* 2nd ed. 2 vols. Falls Church, Va.: American Society of Photogrametry.

———. 1960. *Manual of photographic interpretation.* Washington, D.C.: American Society of Photogrammetry.

Baker, J. C. 2003. Bringing global transparency into sharper focus. *Imaging Notes* 18(3):6.

Burrows, W. E. 1986. *Deep black: Space espionage and national security.* New York: Random House.

Colwell, R. N. 1997. History and place of photographic interpretation. In *Manual of photographic interpretation.* 2nd ed., ed. W. R. Philipson, 3–47. Bethesda, Md.: American Society for Photogrammetry and Remote Sensing.

Day, D. A., J. M. Logsdon, and B. Latell. 1998. *Eye in the sky: The story of the Corona spy satellites.* Washington, D.C.: Smithsonian Institution Press.

Foresman, T. 1998. *The history of geographic information systems: Perspectives from the pioneers.* Upper Saddle River, N.J.: Prentice Hall.

Gorin, P. 1998. Black Amber: Russian yantar-class optical reconnaissance. *Journal of the British Interplanetary Society* 51: 309–20.

———. 1997. Zenit: The first Soviet photo-reconnaissance satellite. *Journal of the British Interplanetary Society* 50: 441–48.

Gupta, V. 1995. New satellite images for sale. *International Security* 20(1): 94–125.

Hays, P. 2001. Transparency, stability, and deception: Military implications of commercial high-resolution imaging satellites in theory and practice. Presentation at the International Studies Convention, Chicago.

———. 1994. Struggling towards space doctrine: U. S. military space plans, programs, and perspectives during the cold war. PhD diss., Tufts University.

Lillesand T. M., R. W. Kiefer, and J. W. Chipman. 2003. *Remote sensing and image interpretation.* 5th ed. New York: John Wiley and Sons, Inc.

Lowman, P. D. 2002. *Exploring space, exploring earth: New understanding of the earth from space research.* Cambridge: Cambridge University Press.

McDonald, R. A. 1997. Corona between the sun and the earth: The first NRO reconnaissance eye in space. Bethesda, Md.: American Society for Photogrammetry and Remote Sensing.

Nesbit, R. 1996. *Eyes of the RAF: A history of photo-reconnaissance.* Stroud, U.K.: Sutton Publishing.

Peebles, C. 1997. *The Corona project: America's first spy satellites.* Annapolis, Md.: Naval Institute Press.

Petrie, G. 2004. High-resolution imaging from space: A world-wide survey. *GeoInformatics* Part I: North America 7(1): 22–27. Part II: Asia 7(2): 22–27. Part III: Europe 7(3): 38–43.

———. 2003. Iraq: Winning the war, reconstructing the peace. *GI News* (July/August): 34–40.

Richelson, J. T. 1999. U.S. *satellite imagery, 1960-1999.* National Security Archive Briefing Book No.13. www.gwu.edu/~nsarchiv/NSAEBB/NSAEBB13

Steinberg, G. 1998. *Dual use aspects of commercial high-resolution imaging satellites.* Mideast Security and Policy Studies, No. 37. Ramat Gan, Israel: Begin-Sadat Center for Strategic Studies. Bar-Ilan University. www.biu.ac.il/SOC/besa/publications/37pub.html

Taubman, P. 2003. *Secret empire: Eisenhower, the CIA, and the hidden story of America's space espionage.* New York: Simon and Schuster.

Web sites   The Corona Gallery.
www.geog.ucsb.edu/~kclarke/Corona/gallery2.htm

Corona Image Gallery.
www.geog.ucsb.edu/~kclarke/Corona/Corona.html

Estes, J. E., and J. Hemphill. 2003. Some important dates in the chronological history of aerial photography and remote sensing.
www.geog.ucsb.edu/~jeff/115a/remotesensinghistory.html

Galileo, the European Satellite Positioning System.
europa.eu.int/comm/dgs/energy_transport/galileo/index_en.htm

Global positioning system overview.
www.colorado.edu/geography/gcraft/notes/gps/gps.html

GLONASS, the Russian satellite positioning system.
www.glonass-center.ru

GPS Operations, U.S. Naval Observatory.
tycho.usno.navy.mil/gps.html

Hazelton, W. 1999. Early developments in GIS: Something from the past and something for the future.
geomatics.eng.ohio-state.edu/GS_607_Notes/Class_3/Lecture_3.html

NASA History Office.
www.hq.nasa.gov/office/pao/History/index.html

National Air and Space Museum Photo Archives.
www.nasm.si.edu/research/arch/collections/photoarchives.cfm

National Security Archive.
www.gwu.edu/~nsarchiv/NSAEBB

NOAA Photo Library.
www.photolib.noaa.gov/space

The Remote Sensing Tutorial.
rst.gsfc.nasa.gov/Front/overview.html

Smithsonian Institute National Air and Space Museum.
www.nasm.si.edu

Trimble GPS tutorial.
www.trimble.com/gps

3 Remote sensing basics

Stan Aronoff

Information gathered by remote sensing is based on the analysis of energy received from the features being evaluated. Imaging systems, such as an aerial camera and photographic film, systematically record the energy received from an area of the earth's surface and produce a detailed picture of the scene. Imaging systems are the main focus of this chapter. Systems that don't produce images, or nonimaging systems, collect a series of measurements at discrete locations or repeatedly along a single line. Common examples of remote sensing systems that don't produce images are radar and sonar positioning systems used to track aircraft and ships. The geographic positions of the targets are known and can be displayed as bright spots on a screen *(figure 8.3 in chapter 8)*. However, there is not the continuous coverage needed to create a picture-like image of the scene. The handheld radar systems with which traffic police measure vehicle speeds do not even give a readout of the geographic position of the vehicle, just the vehicle speed—enough information to give a speeding ticket.

In order to fully take advantage of remote sensing capabilities, the GIS manager should understand how the remote sensing process works. This chapter explains how remote sensing systems acquire information about the earth features and use that information to produce the data products used in GIS and remote sensing applications.

The remote sensing process

Remote sensing systems derive information about a feature by analyzing the energy reflected or emitted from it *(figure 3.1)*. The source of the energy reflected by the object may be the sun or the remote sensing system itself (a). For example, radar systems illuminate features on the earth by emitting microwave pulses.

The energy illuminating the scene must travel through the atmosphere (b). Atmospheric effects on this energy differ depending on the wavelength of the energy and the atmospheric conditions. Once the energy reaches an object, some of it is *reflected* back, some of it is *transmitted* through the object, and some of it is *absorbed*. Objects reemit absorbed energy as heat.

The energy reflected or emitted from the object (c) then interacts and passes through the atmosphere (d) to be detected by sensor systems (e). Sensor systems detect the energy from the object using film (photography), electronic detectors (scanners, digital cameras, etc.), or antennae (radar) and record it as a physical image or as digital data that can be used to generate an image.

The recorded data is then processed to generate data products for display or further analysis. These products include digital or photographic images, image mosaics (multiple images combined to cover areas larger than a single image), and image maps registered to a standard map base (f). One or more analysis techniques, including visual interpretation, digital image enhancement, and automated classification, may be applied to these products to derive information about the features of interest (g).

Typically, remotely sensed data from one or more sensors is analyzed together with supporting data from other sources, called *ancillary data*, such as measurements taken at selected ground locations or existing information about field conditions like geology maps, soils maps, and statistical summaries. This supporting data, as well as analysis results, are commonly stored within a geographic information system (GIS). While specialized remote sensing analysis systems provide refined tools for sophisticated analyses, a GIS can offer image-processing capabilities as well.

The results of interpretation and analysis are used to make information products suited to the application of a user (h). Information products may include maps, tabular summaries, images, multimedia presentations, and Web sites. They may be in hard copy, or digital format, or provided as data files for input to other information systems such as a GIS.

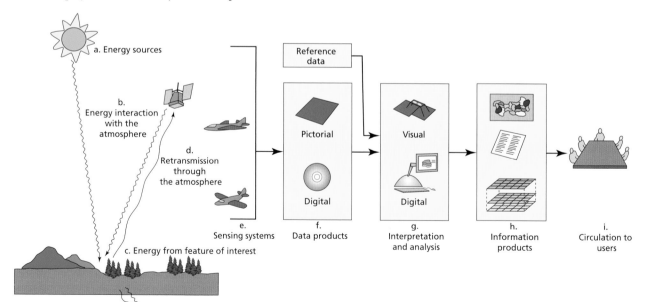

Figure 3.1   The remote sensing process.

Source: *Remote sensing and image interpretation*, 2nd. ed. by T. M. Lillesand and R. W. Kiefer.
© 1987 John Wiley and Sons, Inc. Reprinted with permission of John Wiley and Sons, Inc.

The final step in the remote sensing process is circulation of the information to the users (i). Ultimately, for the information to be of value, it must be used by individuals to make decisions. The decision may be where to construct a road or how much of a forest to harvest, or the remotely sensed data may have been one of several data sources used to make a map or estimate a quantity. Whatever the future use, the justification for collecting and analyzing remotely sensed data, as it is for GIS procedures, is that at some point the information will be of practical use. The recipients may be a select group with access to privileged information, or they may be the public at large. In addition, electronic access over the Internet has made the circulation of information more convenient and inexpensive than the more traditional paper-based report.

Energy sources

The information that can be extracted from remotely sensed data depends on the type of energy the remote sensing system detects. The principle form of energy detected by remote sensing systems is electromagnetic energy. Visible light, x-rays, infrared, microwave, and ultraviolet light are all forms of electromagnetic energy.

Electromagnetic energy behaves in a wave-like fashion[1] that has both electric and magnetic field components (*figure 3.2*). The wavelength (λ) is the distance from one wave peak to the next, and the frequency (f) is the number of peaks that pass a fixed point per unit of time. The wavelength and frequency of electromagnetic energy are related as follows: $c=f\lambda$. The product of the wavelength and frequency of an electromagnetic wave equals the speed of light (c), which is 3×10^8 meters per second. Electromagnetic energy can be specified either by its wavelength or frequency.

Wavelengths range in size from radio waves, which can be several meters long, to visible light, which is measured in micrometers (μm), or millionths of a meter (10^{-6} m). Table 3.1 lists the common units of measure for wavelengths.

The shorter the wavelength of electromagnetic energy, the higher the frequency and the greater the energy level. In general, high-energy electromagnetic radiation has a greater penetrating ability. For example, high-frequency x-rays can pass through the soft tissues of the human body whereas lower

1. The behavior of electromagnetic energy can also be modeled as packets of energy termed photons. The *photon* model is useful in calculations related to the quantity of energy in electromagnetic radiation.

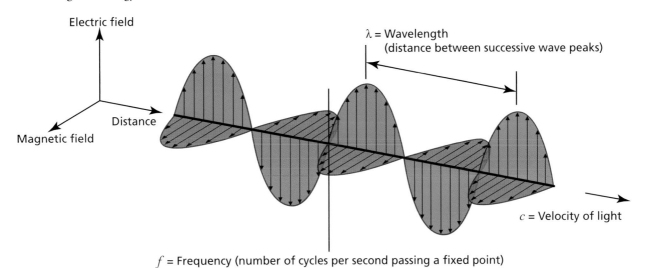

Figure 3.2 Electric and magnetic field components of electromagnetic radiation.

Source: *Remote sensing and image interpretation*, 4th. ed. by T. M. Lillesand, R. W. Kiefer, and J. W. Chipman. © 2000 John Wiley and Sons, Inc. Reprinted with permission of John Wiley and Sons, Inc.

Table 3.1 Units of length used in remote sensing	
Unit	**Length**
Kilometer (km)	1,000 m
Meter	1.0 m
Centimeter (cm)	0.01 m = 10^{-2} m
Millimeter (mm)	0.001 m = 10^{-3} m
Micrometer (μm)[a]	0.000001 m = 10^{-6} m
Nanometer (nm)	10^{-9} m
Ångstrom unit (Å)	10^{-10} m
[a]Formerly called the micron; the term *micrometer* is now used by agreement of the General Conference on Weights and Measures.	

Remote sensing instruments detect energy in the ultraviolet, visible, infrared, and microwave portions of the spectrum covering the wavelengths 0.30 μm to 30 cm. Within this range, only the range from 0.30 μm to 15 μm (a portion of the ultraviolet, visible, and a portion of the infrared region) can be reflected and focused by mirrors and lenses. The longer wavelengths in the microwave region are detected using antennae and sophisticated signal processing systems. Radar systems are one example. The properties of different wave bands determine their value for remote sensing applications.

Not all remote sensing systems depend on electromagnetic energy. Acoustic energy (sound) offers much better penetration of earth and water than electromagnetic energy. Acoustic energy is the variation in pressure produced by the vibration of an object within a medium. The medium may be a gas, a liquid, or a solid. The sound we hear is variations in air pressure. Sonar systems (see chapter 9) measure water depth, detect objects, and image the ocean floor by generating sound pulses that travel through the water and then measuring the reflected returns from features beneath the surface. In seismic surveys, small explosions are produced on the land surface, and a series of acoustic receivers on the ground record the echo reflected from the different rock strata deep within the earth. Complex signal processing procedures convert these acoustic returns into a cross-sectional image from which geologists interpret the depth and shape of rock formations.

frequency visible light cannot. This makes x-rays useful for medical imaging of bone structure.

The range of energy wavelengths and frequencies is termed the *electromagnetic spectrum*. It is divided into regions of wavelengths extending from long-wavelength, low-energy radio waves to short-wavelength, high-energy gamma rays *(figure 3.3)*. These categories of wavelengths conveniently divide the spectrum into frequencies with similar properties. However, there are no sharp breaks at the division points; the properties change gradually across the spectrum. In fact, scientific disciplines differ in the placement of the boundaries for regions in the spectrum. Table 3.2 lists the principle divisions of the electromagnetic spectrum typically used in remote sensing.

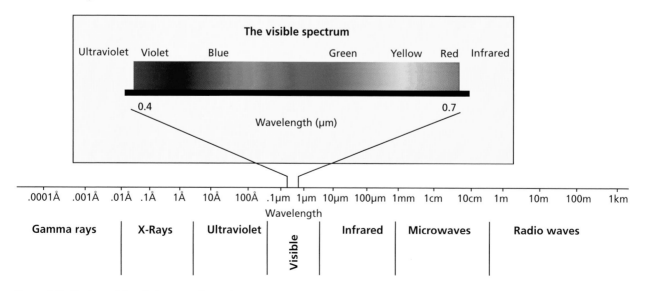

Figure 3.3   Regions of the electromagnetic spectrum.

Table 3.2 Principal division of the electromagnetic spectrum	
Region	**Limit**
Gamma rays	less than 0.03 nm
X-rays	0.03 nm – 240 nm
Ultraviolet	0.24 µm – 0.38 µm
Visible	0.38 µm – 0.70 µm
Infrared radiation	**0.70 µm – 1,000µm (1 mm)**
Near-infrared	0.70 µm – 1.0 µm
Shortwave infrared	1.0 µm – 3.0 µm
Medium wave infrared	3.0 µm – 8.0 µm
Long wave infrared	8.0 µm – 14 µm
Far infrared	14 µm – 1,000 µm
Microwave	1 mm – 100 cm
Radio	greater than 100 cm
Units: 1 millimeter (mm) = 1,000 microns (µm) = 1,000,000 nanometer (nm)	

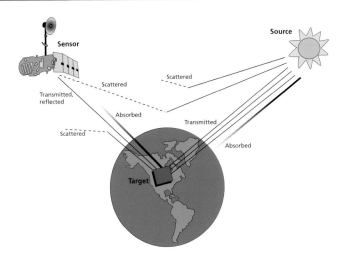

Figure 3.4 Electromagnetic energy is transmitted, absorbed, and scattered by the earth's atmosphere.

Electromagnetic energy and the atmosphere

The atmosphere is a gaseous envelope that surrounds the earth. Electromagnetic energy must pass through the earth's atmosphere when it travels from the sun to the earth's surface and from the earth's surface to the remote sensing instrument. During this process, the atmosphere transmits, absorbs, and scatters electromagnetic energy *(figure 3.4)*. Energy that passes through the atmosphere with little loss of intensity is considered *transmitted*. Atmospheric *absorption* and *scattering* reduce the amount of energy the remote sensor receives from the target.

The blocking characteristics of the atmosphere protect living things from the damaging high-energy radiation from the sun. Without the atmosphere to block most of the ultraviolet radiation, human skin exposed to sunlight would quickly be sunburned and develop skin cancer. For remote sensing, these blocking characteristics are problematic. The atmospheric effects of most concern are absorption and scattering. These processes reduce the amount of energy the remote sensor receives from the target. The longer the path from the target to the sensor, the greater the potential for atmospheric effects to degrade remotely sensed imagery. For this reason, atmospheric effects are more likely to be a problem for imagery acquired from high-altitude aircraft and satellites than from low-altitude airborne platforms. Understanding how the atmosphere changes remotely sensed data is thus important in planning remote sensing missions and analyzing the resulting data.

Atmospheric scattering

Atmospheric scattering is the redirection of electromagnetic energy by particles suspended in the atmosphere, such as dust and smoke, or by large molecules of atmospheric gases, such as water vapor. The more scattering, the more energy is redirected away from its direction of travel and the less energy reaches the sensor from a target. The amount of scattering depends on the size and abundance of these particles, the wavelength of the radiation, and the distance the energy must travel through the atmosphere to reach the sensor.

Atmospheric scattering causes the daytime sky to be blue instead of black because the shorter blue wavelengths of sunlight are scattered more than the longer green and red wavelengths. It is the scattered blue wavelengths that give the sky a blue color as well as an overall brightness or skylight. However, there is still substantial penetration of the shorter blue wavelengths through the atmosphere. We can see detail in shadow areas because it is illuminated by this diffuse skylight. Pictures taken from the moon show a black sky because there is no atmosphere to scatter the incoming solar radiation.

At sunrise and sunset, the sun's rays travel a longer path through the atmosphere to reach an observer on the ground. Only the longer orange and red wavelengths can penetrate the atmosphere without significant scattering, and so the sunset sky appears orange and red.

Scattering has several important consequences for remote sensing. Since atmospheric scattering is high in the visible blue and ultraviolet portions of the spectrum, the brightness of the atmosphere tends to mask variations in the brightness of features in the scene. This is because scattering not only reduces the energy received by a sensor from the target, but it also directs energy from outside the sensor's field of view toward the sensor. This excess energy acts as a mask of uniform brightness that causes dark areas to be lighter and bright areas to be darker than they would otherwise appear. As a result, image contrast is reduced, making it more difficult to distinguish details. Many remote sensing instruments improve image quality by using filters to exclude these shorter wavelengths.

Atmospheric absorption

Atmospheric absorption occurs when energy is lost to constituents of the atmosphere. Energy absorbed by the atmosphere is subsequently reradiated at longer wavelengths. When it is radiated at infrared wavelengths, we sense it as heat.

Three atmospheric gases account for most of the atmospheric absorption of solar radiation: water vapor, carbon dioxide, and ozone. Of the three, water vapor is capable of the most absorption. Water vapor absorbs electromagnetic radiation two to three times more strongly than ozone or carbon dioxide. The concentration of water vapor in the atmosphere changes significantly over time and between regions. Atmospheric effects caused by water vapor are insignificant over deserts but severely limit image acquisition over rainforests.

Atmospheric windows

The transmission characteristics of the earth's atmosphere vary with wavelength. Some wavelengths are transmitted almost perfectly while others are completely blocked. The middle graph in figure 3.5 shows the transmission characteristics of the atmosphere over the range of wavelengths commonly used in remote sensing. Ranges of wavelengths transmitted well by the atmosphere are termed *atmospheric windows*. These regions of the spectrum are used for remote sensing. The bottom panel of figure 3.5 shows the operating ranges of common remote sensing imaging systems.

The most important atmospheric windows for remote sensing of earth resources are the ultraviolet-to-near-infrared bands (0.3 μm–1.2 μm), the mid-infrared bands (3 μm–5 μm and 8 μm–14 μm), and the microwave band. Narrow atmospheric windows exist in the shorter wavelength portion of the microwave region between one millimeter and one centimeter. However, the atmosphere has little blocking effect on longer microwave wavelengths. Passive microwave and active radar remote sensing use these longer wavelengths. It is the relative transparency of the atmosphere to wavelengths longer than one meter that enable television and radio signals to be broadcast long distances.

Electromagnetic energy and earth objects

Electromagnetic radiation interacts with features on the earth's surface the same way as it does with the atmosphere. It can be absorbed, reflected, and transmitted. For example, visible light striking a leaf is partially reflected, creating the image of the leaf we see. The portion not reflected is either absorbed or transmitted (which is why some light can be seen through it). Absorbed energy raises the temperature of the leaf and is reemitted as heat. The leaf's reflectance and absorption characteristics are what give it the color we perceive.

The way an object reflects energy greatly affects the characteristics of the energy detected by remote sensing instruments. The type of reflection that occurs is determined by the surface roughness of the object relative to the wavelength of the incoming energy and by the *incidence angle*, the angle at which the energy strikes the object.

Specular or mirror-like reflection occurs when energy strikes a relatively smooth surface (i.e., when the wavelength is longer than the surface height variation or the particle size of the surface) and is redirected away from the object in a single direction *(figure 3.6)*. Here the angle of reflection is equal to the angle of incidence.

Diffuse reflection occurs when a surface is relatively rough (i.e., when the wavelength is shorter than the surface height variation), and the energy is redirected uniformly in all directions *(figure 3.6)*.

The way energy interacts with an object affects how it is seen. For example, fine-grained sand would appear smooth when illuminated by the microwave energy used in radar, which has wavelengths of several centimeters. However, the same sand would appear rough when illuminated by visible light, which has wavelengths of less than 1/1000th of a millimeter.

Most features are neither perfect diffuse nor perfect specular reflectors but have intermediate reflection characteristics. It is the diffuse reflection that is useful for remote sensing because wavelengths absorbed or transmitted by the object will be reduced in intensity. By analyzing the pattern and

a. Energy sources

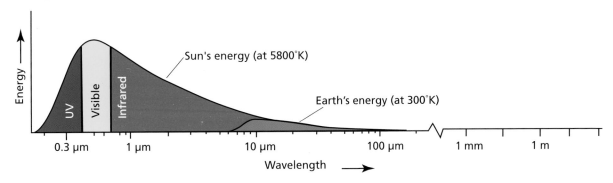

b. Atmospheric transmittance

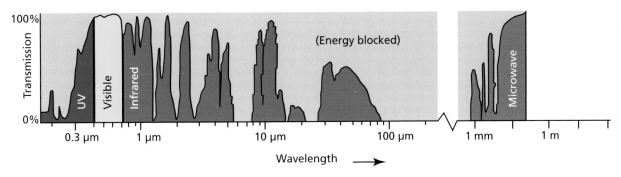

c. Wavelengths detected by common remote sensing systems

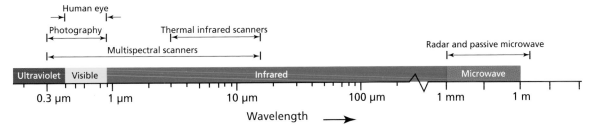

Figure 3.5 Remote sensing in the electronic spectrum. (a) Energy emission. (b) Atmospheric windows important for remote sensing.

(c) Range of wavelengths detected by common remote sensing systems.

Source: *Remote sensing and image interpretation*, 2nd. ed. by T. M. Lillesand and R. W. Kiefer. © 1987 John Wiley and Sons, Inc. Reprinted with permission of John Wiley and Sons, Inc.

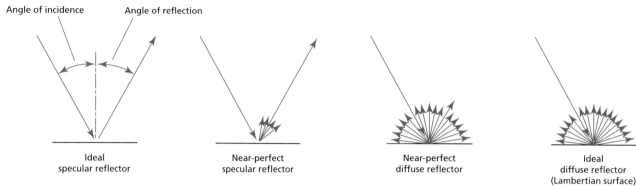

Figure 3.6 Specular and diffuse reflection.

Source: *Remote sensing and image interpretation*, 2nd. ed. by T. M. Lillesand and R. W. Kiefer.
© 1987 John Wiley and Sons, Inc. Reprinted with permission of John Wiley and Sons, Inc.

intensity of the wavelengths in the diffuse reflection, information such as the vegetation type can be derived about the objects from which they were reflected. Specular reflections provide little information about the reflecting object.

Emission of electromagnetic radiation

Some forms of remote sensing depend on the analysis of energy emitted by earth features. The capabilities of these systems are more easily understood with an awareness of the emission process.

All objects with temperatures above absolute zero continuously emit electromagnetic radiation. (Absolute zero is 0° on the Kelvin temperature scale and is equivalent to -273.2° Celsius or -459.7° Fahrenheit.) This radiation is emitted at a broad range of intensities and wavelengths. Warmer objects emit more energy, and at shorter wavelengths, than cooler ones—a phenomenon we experience as radiant heat. This principle is the basis for thermal remote sensing, discussed later in this chapter.

The quantity of energy emitted by an object depends not only on its temperature, but also on its emissivity. The emissivity of an object is a measure of its efficiency in radiating energy. Emissivity varies with a feature's composition and surface characteristics. Measured on a scale from 0 to 1, an object with an emissivity value of 1 would be considered a perfect emitter and absorber of electromagnetic energy, a theoretical object called a *blackbody*. Physicists describing the relationship between temperature and electromagnetic radiation independent of emissivity assume the feature under study to be a blackbody to help model the phenomenon. The emissivity of features in a remote sensing image are generally not known unless samples have been collected and measured—an undertaking generally too costly.

Water is an exception. The emissivity of water is 0.98 (close to the physicist's theoretical blackbody) and is virtually constant, independent of water quality. It is generally the only feature in a scene with a known emissivity. For this reason, airborne and satellite remote sensing imagery can be used to calculate water surface temperature values with accuracies of less than a degree Centigrade. Before the advent of satellite imagery, there was no way to generate an accurate map of sea surface temperature. Today meteorologists depend on daily updates of maps of the surface temperatures of the world's oceans to develop their weather forecasts.

The top portion of figure 3.5 (see page 59) illustrates the theoretical emission characteristics of blackbody sources at different temperatures. The sun's emission curve closely approximates that of a blackbody, having a temperature of 5,800°K. The sun's peak emission is at about 0.5 μm, which is in the visible region of the spectrum—the region to which the eye is sensitive. The first photographic films were developed to reproduce the image the eye sees and so were made sensitive to the visible band.

Naturally occurring earth objects have temperatures on the order of 300°K (equivalent to 27°C or about 80°F), represented by the lowest temperature radiation curve shown in the illustration. At this temperature, most of the emitted energy is at wavelengths of 3 μm to 14 μm. This range of wavelengths is used to estimate the temperature of earth features and is often termed the *thermal infrared band*. Objects with higher temperatures emit progressively larger amounts of energy (represented by the area under each curve) and have peak emissions at progressively shorter wavelengths.

Spectral properties of objects

Remote sensing analysts use spectral properties to help distinguish objects under study. Objects selectively absorb and reflect electromagnetic energy due to differences in the molecular composition of their surfaces. Sunlight has almost equal energy levels at all visible wavelengths. Vegetation appears green because most of the visible light reflected back from the leaves is in the green portion of the spectrum (*figure 3.7*). Similarly a red object reflects more energy in the red portion of the visible spectrum and less in other portions. Although the perceived color of light in the wavelengths from 0.4–0.7 μm progresses from blue to red through all the colors of the rainbow, for convenience, the range is classified as blue (0.4–0.5 μm), green (0.5–0.6 μm), and red (0.6–0.7 μm) bands.

The pattern of spectral response, commonly termed the *spectral signature*, for a particular material can be described in the form of a graph showing the percentage of radiation of different wavelengths reflected from an object, called the *spectral response curve*. By plotting together the spectral response curves of different features, the portions of the spectrum where they differ can be readily identified. In figure 3.8 water, vegetation, and bare soil have substantially different reflectance in the visible portion of the spectrum, and we

would expect to be able to distinguish them easily with our eyes or in a normal color photograph. In viewing an image, we also use other characteristics such as shape and texture to identify objects.

The broadleaf and needle leaf (coniferous) vegetation samples have almost the same spectral response in the visible region (which is why they would both appear green) and would be difficult to discriminate. But they have significantly different reflectance characteristics in the near-infrared band and could be distinguished using data from this band. Vegetation species typically show greater differences in their near-infrared reflectance characteristics than in the visible portion of the spectrum. That is why they are more easily distinguished on photographs taken with color infrared film, which is sensitive to the near-infrared, than with normal color film. Similarly, snow and ice can be distinguished from clouds using data from the mid-infrared band, which is strongly absorbed by water. Since snow and ice have much higher water content than clouds, they have a much lower mid-infrared reflectance (*figure 3.8 and figure 3.11 later in this chapter*).

Ideally, features with different spectral reflectance characteristics would be easily distinguished. In practice, this often is not the case. Features that we classify the same, such as the plants in a field of wheat, will have a range of different spectral signatures. Individual plants will differ in their age and health, which will affect the characteristics of their leaves and

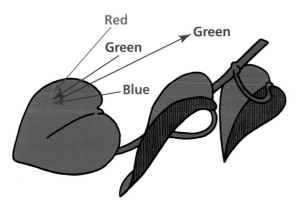

Figure 3.7   Reflection and absorption of electromagnetic energy. Leaves illuminated by sunlight (which has approximately equal energy in the blue, green, and red wavelengths) appear green because more energy in the green portion of the spectrum is reflected than at other visible wavelengths.

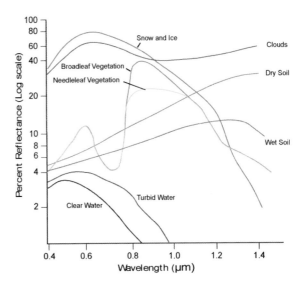

Figure 3.8   Spectral reflectance curves for soil, water, and two vegetation samples.

in turn change their reflectance characteristics. Leaves off the same tree will differ in spectral response according to their age and the amount of light they receive. So the spectral curves in the figure might better be represented as thick lines to represent the range of values that actually occur and the way the curves for different features overlap.

For this reason, a combination of analysis methods is typically used to distinguish features that are confused. Crops that have similar spectral response patterns but differ in the timing of their growth stages can often be distinguished with multiple images acquired on different dates during the growing season. In mountainous regions, elevation data is often used to distinguish vegetation types that appear similar but grow at different elevations. In some cases, a region may be divided into areas amenable to different types of remote sensing analysis and the results later combined.

In multispectral remote sensing, reflectance data for several wavelength ranges are collected simultaneously. The greater the number of bands and the narrower they are, the more closely the data will approximate the spectral reflectance curve of the features detected. A disadvantage in using a large number of bands is that the volume of data becomes very large, making the data more costly to store and analyze. By selecting appropriate bands, features in the landscape can be distinguished by their pattern of spectral reflectance values (*figure 3.9*).

Spectral bands commonly used in remote sensing

More than three decades of research in multispectral remote sensing has identified many valuable applications that have become routine, though new developments provide continuing improvements. With experience, the choice of sensor bands has become more consistent.

The spectral bands typically used for remote sensing of earth resources include visible green, visible red, near-infrared, and one or more mid-infrared bands. Additional bands in the visible blue and thermal infrared are sometimes included. For example, the Thematic Mapper remote sensing system carried on the Landsat satellite series has seven spectral bands representing the principle visible and infrared bands used in earth remote sensing. It has three visible, one near-infrared,

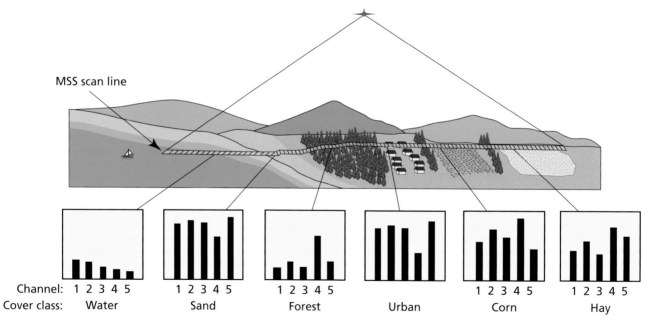

Figure 3.9 Classification by spectral reflectance. The spectral reflectance values in five bands are shown for selected pixels along a scan line: 1= visible blue band, 2 = visible green band, 3 = visible red band, 4 = near-infrared band, and 5 = thermal infrared band.

Source: *Remote sensing and image interpretation*, 4th. ed. by T. M. Lillesand, R.W. Kiefer, and J. W. Chipman. © 2000 John Wiley and Sons, Inc. Reprinted with permission of John Wiley and Sons, Inc.

two mid-infrared, and one thermal infrared band. Radar systems operate in the microwave region and provide a unique all-weather capability, as well as information that cannot be derived from the visible and infrared bands such as wind velocity, wave height, and rainfall intensity.

Remote sensing in the ultraviolet spectrum

Most ultraviolet (UV) radiation is absorbed or scattered by the earth's atmosphere, so it is not widely used in remote sensing. However, some materials, when illuminated by ultraviolet radiation, absorb the UV radiation and reemit it as visible light, a phenomenon known as *fluorescence*. Based on this principle, specialized airborne remote sensing systems have been designed to illuminate a target area with ultraviolet energy and record the visible light emitted. Ultraviolet laser fluorosensors are particularly valuable for oil spill detection. Not only do they provide reliable detection of oil on water, but they are also the only sensors capable of distinguishing oil in floating seaweed and on different types of beaches. Under ideal conditions, it is possible to distinguish different types of oil by their ultraviolet fluorescence signature. A handheld UV light has also been developed for use on ships to detect oil spills at short range at night.

Remote sensing in the visible spectrum

The visible spectrum includes the wavelengths from 0.4 μm to 0.7 μm, perceived by the human eye as colors ranging from purple and blue through green and red *(figure 3.3)*.

The visible spectrum is particularly important not only because we are so familiar with the way objects appear when illuminated by visible light, but also because the amount of energy provided by the sun and available for detection is at a maximum in this range. Remote sensing in the visible portion of the spectrum uses photographic film and cameras, digital cameras, and electronic sensors called scanners, discussed in chapter 6. Compare the normal color image in figure 3.10a to the color infrared image of the same scene in figure 3.10b.

Applications of visible bands

Visible blue band (0.45–0.52 μm)

The visible blue band provides the greatest water penetration but is subject to greater atmospheric scattering and absorption than other bands. It is used for analyses of water characteristics, water depth, and the detection of subsurface features for such applications as water quality assessment and coastal zone mapping. It is also useful for soil and vegetation discrimination, forest-type mapping, geology, and the identification of cultural features.

Visible green band (0.52–0.60 μm)

The visible green band is useful for vegetation discrimination and vigor assessment, and analysis of cultural features and urban infrastructure. This band offers moderate water penetration and is less affected by atmospheric scattering and absorption than the visible blue band, making it useful for

a

b

Figure 3.10 Comparison of normal and color infrared film. Normal color film (a) is sensitive to the entire visible spectrum and is designed to reproduce colors to closely match those seen by the human eye. Color infrared film (b) is sensitive to the visible green, visible red, and near-infrared wavelengths. It is designed to emphasize differences in near-infrared reflectance, which is represented in red and magenta colors. The red truck appears yellow in the color infrared image because reflectance in the visible red band is represented by yellow on color infrared film. Deciduous trees on the hill top are magenta in the infrared image and clearly distinguished from the conifers in the wetland to the right, which appear cyan. Shrubs and grasses are pink. In the normal color image, all the vegetation has a similar dark green color.

water quality studies measuring sediment and chlorophyll concentration (a measure of nutrient load).

Visible red band (0.63–0.69 µm)
The visible red band includes the chlorophyll absorption band of healthy green vegetation. The chlorophyll absorption band is important for discriminating vegetation types, assessing plant condition, delineating soil and geologic boundaries, and identifying cultural features. It is the visible band least affected by atmospheric scattering and absorption, so it generally exhibits the greatest image contrast. The red band provides less water penetration than the visible green and blue, but it provides useful near-surface information on water quality, sediment, and chlorophyll concentrations.

Panchromatic bands (0.50–0.90 µm)
The term *panchromatic* indicates sensitivity to a relatively wide range of wavelengths, usually including most if not all of the visible spectrum. Multispectral scanners on satellites commonly have a panchromatic band sensitive to the visible green, visible red, and a portion of the near-infrared. This single band image data is generally collected at a higher spatial resolution than the other narrower multispectral bands. In a process called image fusion, the panchromatic band can be digitally combined with two or three of the multispectral bands to produce color images with the spatial detail of the panchromatic image and the spectral detail of the multispectral bands (see chapter 4).

Remote sensing in the infrared spectrum
The infrared spectrum includes wavelengths measuring 0.7 µm to 1,000 µm. In remote sensing this range is commonly subdivided into five regions: the near, short-wave, medium-wave, long-wave, and far-infrared bands. The near- and short-wave infrared radiation received from earth features is predominantly *reflected* solar energy, which is energy from the sun that has been reflected back from features on the ground. Earth features *emit* energy at the medium- and long-wave portion of the infrared in relation to their temperature and emissivity (see below). For this reason, the wavelengths in these two regions can be used to measure temperature and together make up the *thermal infrared band*.

Near-infrared band (0.7–1.0 µm)
The near-infrared (NIR) wavelengths, just beyond the visible red, behave in a similar manner to visible wavelengths. The energy can be focused using optical systems and recorded with photographic film as well as with electronic detectors.

The near-infrared band has proven to be particularly useful in distinguishing vegetation types and vegetation condition. Healthy vegetation has a distinctive spectral curve with a low reflectance in the visible red wavelengths and high reflectance in the 0.76–0.90 µm portion of the near-infrared range of wavelengths. Differences in reflectance in this band are useful in distinguishing species. Deterioration of plant condition causes near-infrared reflectance to decrease in this band, often before visible wilting can be seen, thus providing a useful indicator of plant condition. Applications that use near-infrared data include vegetation mapping, crop-condition monitoring, biomass estimation, and soil moisture assessment. Clear water absorbs near-infrared energy strongly, which also makes this band useful for delineating water features. As the turbidity of water increases, the near-infrared reflectance increases, making it useful for assessing water-quality factors that decrease clarity, such as suspended sediment.

Color infrared aerial films sensitive to visible green, visible red, and the near-infrared band were originally designed for vegetation mapping. Comparing the color infrared photograph in figure 3.10b with the normal color image in figure 3.10a above, deciduous trees, conifers, and shrubs are easily distinguished on the infrared image, but not on the normal color photograph. The red truck appears yellow on the color infrared image because it strongly reflects both near-infrared and red wavelengths. Near-infrared radiation received from an object is generally not related to its temperature. However, objects at very high temperature such as molten lava or the flame in a furnace can emit sufficient energy at near-infrared wavelengths to be recorded by infrared photographic film.

Short-wave infrared band (1.0–3.0 µm)
Energy in the short-wave infrared (SWIR) has also proven useful for land-cover classification and vegetation analysis. The 1.55–1.75 µm band is strongly absorbed by water, making it useful for analyzing moisture levels in soil and for monitoring plant vigor and crop condition. When plants are healthy and turgid (stiff), they contain more water than when they are stressed and wilted. This band is also valuable in distinguishing clouds from snow and ice *(figure 3.11)*. Snow and ice, which have high water content, absorb energy in this band strongly, giving them a much lower reflectance than clouds.

Figure 3.11 Use of the shortwave infrared band to discriminate snow and ice from clouds, Great Lakes Region of North America. This color composite image was generated by displaying the visible red, near-infrared, and the 1.58–1.64 μm shortwave infrared bands of the AVHRR satellite sensor as blue, green, and red. In this image the inclusion of the shortwave infrared band makes the snow covered areas light cyan in color, distinguishing them from clouds, which appear white.

Source: © Natural Resources Canada.

The medium-wave (3.0–8 μm) and long-wave infrared (8–14 μm)

The principal wavelengths at which earth objects emit electromagnetic energy are in the medium-wave and long-wave portions of the infrared band, which together make up the *thermal infrared band*. Most thermal imaging sensors operate in the long-wave infrared (8–14 μm) region of the electromagnetic spectrum. The level of energy emitted in this band is closely related to the temperature of the object and its emissivity (the radiating efficiency of the surface), as discussed earlier in this chapter.

The thermal infrared band is useful to measure the temperature of features for which the emissivity is known, particularly water temperature *(figure 3.12)*, and for detecting thermal features that are much warmer or cooler than their surroundings, such as industrial sites, pipelines carrying heated materials such as oil or steam, geothermal sites, and thermal pollution. This band has also been applied to

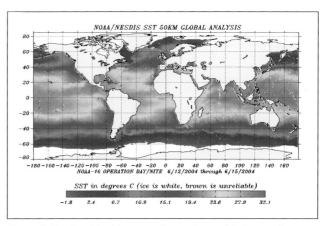

Figure 3.12 Combined image and map shows global sea surface temperature produced from thermal infrared data collected by the AVHRR sensor over a 4-day period. Spatial resolution is 50 kilometers per pixel.

Source: Image courtesy of NOAA/NESDIS.

the analysis of vegetation stress, soil moisture, and geology. Thermal remote sensing is discussed further in chapter 6.

Remote sensing in the microwave region

There is no significant solar radiation in the microwave portion of the electromagnetic spectrum. Therefore, remote sensing systems operating in the microwave region must either detect the weak microwave emissions from earth features (termed *passive microwave remote sensing*) or employ a source of microwave energy to illuminate the scene (radar). Microwave remote sensing systems use antennae and sophisticated electronic signal processing to generate and focus the microwave energy, receive the energy reflected from the target area, and construct an image from the digital data. Passive microwave has proven valuable in the measurement of sea surface temperature. Although microwave temperature measurements are not as accurate as those produced using thermal infrared, measurements are virtually unaffected by cloud cover whereas the thermal infrared imagery must be cloud-free to generate accurate temperature values.

Radar systems, discussed in chapter 8, use one or more microwave energy sources to illuminate the target area.

Within the microwave region, only the relatively short wavelengths are significantly blocked by clouds. Imaging radar systems for earth observation generally use wavelengths that can penetrate clouds, thereby providing all-weather day and night imaging capability *(figure 3.13)*. The longer wavelength radar systems can also penetrate forest canopies and,

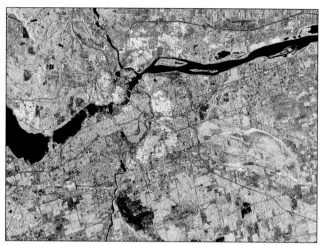

Figure 3.13 Radar image of the Ottawa region. In this radar image, acquired August 29, 1996, by the Radarsat satellite, water areas appear black, roads are dark, the urban area in the center of the image is light, and vegetation is medium to light toned.

Source: RADARSAT-1 data © Canadian Space Agency 1996. Received by the Canada Centre for Remote Sensing. Processed and distributed by RADARSAT International.

to a limited degree, dry soil. The subsurface deposits of riverbed gravel from ancient rivers that flowed in what is now the Sahara desert were discovered using long-wave microwave radar imagery *(figure 8.16 in chapter 8)*.

Sensor systems

A wide range of sensors is used for earth observation remote sensing. They can be broadly grouped into three categories: photographic cameras, passive electronic sensors, and active electronic sensors (radar, lidar, and sonar). The resolution and coverage of remote sensing systems varies with the technology and platform. In selecting image data, four resolution components (spatial, spectral, radiometric, and temporal) should be considered. Aerial photography can provide imagery with the highest spatial resolution while satellite systems offer the broadest coverage. Airborne scanners can produce high-resolution, multispectral imagery in digital form—some systems collecting data in hundreds of spectral bands simultaneously. Airborne and satellite-based radar systems offer day and night all-weather (cloud penetrating) image acquisition. Lidar systems can produce very high-accuracy digital elevation data over land and shallow water areas. Sonar systems are capable of imaging far below

the water surface, providing imagery and digital elevation data of the ocean floor.

In general, the more localized the study area, the higher the spatial resolution that will be appropriate. If the features to be extracted can be distinguished by their spectral response, multispectral scanners should be considered. Where frequent repeat coverage is needed, satellite systems are usually more cost-effective than airborne sensors. For areas of frequent cloud cover or where reliable access to cloud-free imagery is important, such as for emergency response, radar systems may be most appropriate.

Sensor platforms

Remote sensing instruments are operated from a variety of platforms ranging from ground-based vehicles to spacecraft. Ground-based platforms such as truck-mounted booms, tall buildings, water towers, or even a ladder are often used to test remote sensing instruments and to collect detailed spectral measurements from known features in order to test or calibrate imagery.

The term *airborne imagery* refers to data collected from aircraft, helicopters, and other vehicles operating within the earth's atmosphere. Airborne systems offer the flexibility of positioning the instrument wherever and whenever it's needed, weather permitting and so long as there is a suitable staging area and fuel source within range. They can fly at low altitudes to acquire very detailed imagery and with less atmospheric interference than space-based platforms. However, when buffeted by wind, they are a less stable platform than spacecraft.

Remote sensing instruments in space are typically satellite-based, although low-earth orbiting space stations or reusable vehicles, such as the space shuttle, are also used. Space platforms are more stable than aircraft, and their operation is independent of weather conditions on the ground. Although some satellite systems have pointable sensors, the time and frequency of image acquisition is limited by their orbit parameters. A major advantage of satellite-based sensors is consistency. They can offer continuous repeat coverage using the same sensor over long periods of time, providing a valuable source of data for environmental monitoring. The AVHRR sensor aboard the NOAA series of satellites has provided a continuous daily record of multispectral imagery from 1979 until the present.

References

Campbell, J. B. 1996. *Introduction to remote sensing*. 2nd ed. New York: The Guilford Press.

Colwell, R. N., ed. 1983. *Manual of remote sensing*. Falls Church, Va.: American Society of Photogrammetry.

Drury, S. 1998. *Images of the earth: A guide to remote sensing*. Oxford: Oxford University Press.

Fingas, M. F., and C. E. Brown. 2002. Review of oil spill remote sensing. In *Proceedings of the 7th conference on remote sensing for marine and coastal environments*. Ann Arbor, Michigan: Veridien.

Jensen, J. R. *Introductory digital image processing: A remote sensing perspective*. 2nd ed. Upper Saddle River, N.J.: Simon and Schuster.

Lillesand, T. M., and R. W. Kiefer. *Remote sensing and image interpretation*. New York: John Wiley and Sons, Inc.

4 Characteristics of remotely sensed imagery

Stan Aronoff

Geographic information systems provide image analysis capabilities, ranging from basic image handling to sophisticated classification and measurement. Remotely sensed imagery is commonly used as a backdrop over which to display other georeferenced datasets.

All the remote sensing systems discussed in this book produce images. That is, the data collected from these systems is displayed as picture-like representations in which a feature's position is related to its geographic location. If the image has been geometrically corrected to a suitable level of positional accuracy, the image can, in fact, serve as a map and is often referred to as an *image map*.

The information that can be extracted from remotely sensed images depends first on the quality of the data collected and then on the resources available to analyze that data. This chapter discusses important characteristics of remotely sensed imagery that should be considered when assessing its suitability for a particular application—namely positional accuracy and spatial, spectral, radiometric, and temporal resolution. Not every application needs the highest resolution imagery available. For many projects, such as large-area agricultural studies, the information required from imagery can usually be obtained from less-costly low-resolution imagery. The different aspects of resolution affect the information that can be extracted from imagery and are important in choosing data that best suits a specific application.

Photographic and electronic images

Remote sensing systems detect electromagnetic energy photographically or electronically. The term *photograph* is reserved for images that were both *detected* and *recorded* on photographic film. The broader term *image* refers to any picture-like representation, including photographs and data collected by electronic sensors that may later be output on a photographic medium such as a paper print or transparency.

In photography, energy striking the surface of a light-sensitive film causes a chemical reaction that varies according to energy intensity. When the film is developed, variations in brightness are represented as differences in tone or color, which produce black-and-white or color photographs. In photographic systems, the film both detects and records the image.

An image can also be represented digitally. Digital images are subdivided into a grid of equal-sized, usually square cells called *picture elements* or *pixels*. The brightness value for each cell is represented by a numeric value. Each cell's position (or address) within the image is designated by a row and column number.

Figure 4.1 shows a digital image. The enlarged portion of the image shows the individual pixels of which it is composed and the corresponding numeric values that specify their brightness levels. Whether the area represented by a pixel contains a single feature or several light and dark features, only one value, the average brightness, is recorded for the pixel. As a result, there is no detail within a pixel—it is a computer-generated square displayed as a homogeneous gray tone.

A photograph can be enlarged to more easily see detail until further enlargement produces a grainy or fuzzy picture, but no additional information. Similarly in a digital image, once the individual pixels can be seen, no additional detail can be found by further enlargement of the image.

Table 4.1 compares photographic and digital methods of image capture, storage, and manipulation. Aerial and space photographs are often scanned to produce digital images in order to take advantage of digital processing methods that are too costly or not possible using photographic methods. These images have a mixture of characteristics of both photographs and digital images.

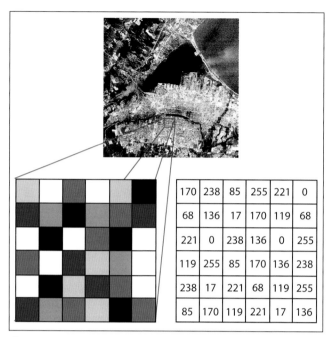

Figure 4.1 Digital image generation from an aerial photograph.

Source: © Natural Resources Canada.

Resolution

Remote sensing systems differ in the level of detail or resolution they can capture. There are four different aspects of resolution important in remote sensing: spatial, spectral, radiometric, and temporal. *Spatial resolution* refers to the smallest feature discernable in an image. *Spectral resolution* refers to the number and width of spectral bands recorded for an image. The number of values available to record the brightness levels in an image is a measure of *radiometric resolution*. The final component is *temporal resolution*, which refers to the frequency of image acquisition. These four sensor characteristics are considered together in selecting the remote sensing data best suited for a specific application.

Spatial resolution

Spatial resolution refers to the fineness of detail visible in an image. A feature is spatially resolved if it can be distinguished from its surroundings. The smaller the area of ground represented by each pixel in a digital image, the finer the details that can be captured and the higher the spatial resolution is said to be. For photographic images, spatial resolution is commonly measured by the number of black and white line pairs per millimeter that can be distinguished (discussed

Table 4.1 Comparison of photographic and digital methods of image capture, storage, and manipulation.

Characteristic	Photographic imaging	Digital imaging
Image structure	Silver grains or dye spots of variable size distributed in a random pattern.	Regular grid of same-sized picture elements termed pixels.
Image capture	Incoming radiation is focused by lenses onto photographic film, which contains photosensitive silver halide crystals. Differences in radiation levels are represented by differences in the degree of chemical alteration of the silver halide crystals.	Incoming radiation is focused by lenses onto a photosensitive solid-state device, typically a CCD or CMOS detector array. Differences in radiation levels produce differences in electrical voltage levels.
Multispectral image recording	Color photographic films have three photosensitive layers that allow image data to be simultaneously recorded in three spectral bands. In normal color film, the layers are sensitive to visible red, green, and blue radiation. In color infrared film, they are sensitive to visible green, visible red, and near-infrared radiation.	Digital cameras use internal filters to determine the spectral band recorded by an individual CCD or CMOS detector site. Digital cameras typically record data in three bands: visible red, green, and blue or visible green, red, and near-infrared. Multispectral scanners use various methods to split the incoming radiation into separate bands and generate a digital image for each band recorded. Scanners can simultaneously record image data in several to hundreds of spectral bands.
Image storage	Photographic film, photographic prints, or conversion and storage as a digital image.	Magnetic tape, CCD, DVD, hard disk, or solid-state media.
Image manipulation	Chemical developing, optical printing.	Digital image processing.
Image transmission	Mail, courier, fax transmission, or conversion to digital image that can be transmitted using digital image transmission modes.	Telemetry, computer networks, telephone line, or physical delivery of images on storage media by mail or courier.
Soft-copy display	Projected slides or movies.	Computer monitors, digital image projection.
Hard-copy display	Photographic prints or transparencies.	Photographic prints or transparencies, dye sublimation, inkjet, laser, or thermal printers.

later in this chapter). Figure 4.2 illustrates the gain in detail obtained when the same image is represented with different spatial resolutions.

Figure 4.3 shows images with three different spatial resolutions commonly used in remote sensing. In the AVHRR (Advanced Very High Resolution Radiometer) image, the width of each pixel in the original image represents a distance on the ground of about 1.1 km *(figure 4.3a)*. Broad land-use and land-cover types such as major water bodies, agricultural land, and forest can be distinguished. The Landsat image *(figure 4.3b)*, which is made up of pixels representing 30 meters of ground distance, shows considerably more detail. A much broader range of vegetation types, cropping patterns, lakes and rivers, major roadways, airports, urban areas, and large building structures can be recognized. In the large-scale aerial photograph *(figure 4.3c)*, in which features smaller than a meter across can be distinguished, individual cars, road markings, and people can be seen.

Images in which the smallest features discernible are large are said to have *coarse* or *low* spatial resolution. When small

objects can be discerned in an image, the image is said to have *fine* or *high* spatial resolution. These terms are relative and depend to a large extent on the application. A 1-km spatial resolution is high for meteorological applications but rather low for vegetation mapping.

For airborne sensors, higher spatial resolutions can generally be obtained by flying the sensor at a lower altitude. Satellite-based sensors normally operate at fixed altitudes, and thus their image resolution is set by the system design.

Factors affecting spatial resolution

The size of the features that can be seen in a remotely sensed image is determined by three factors: the characteristics of the target and its background, the scale of the imagery, and the resolving power of the imaging system. In addition, several external factors can degrade image quality and reduce its effective spatial resolution. These include poor atmospheric and illumination conditions. Smoke, haze, and low light levels reduce contrast and edge sharpness, making it more difficult to distinguish fine details in an image *(figure 4.4)*. An

a

b

c

d

Figure 4.2 Spatial resolution comparison. The same scene is shown as digital images in which the pixel width represents a ground distance of (a) 10 cm, (b) 20 cm, (c) 40 cm, and (d) 80 cm across.

Source: Image courtesy of Terrapoint Canada.

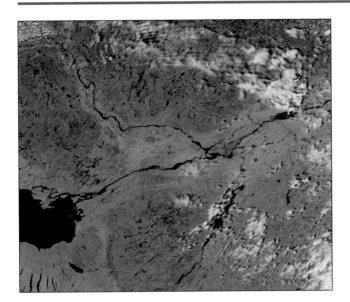

a

b

Figure 4.3   Low-, medium-, and high-spatial resolution imagery of the Ottawa, Ontario, region. (a) This satellite image acquired by the NOAA AVHRR sensor with 1.1 kilometer GSD pixels shows an area 540 kilometers across. (b) This natural-color Landsat 5 Thematic Mapper image shows an area that is 36 km across. (c) This color aerial photograph, acquired at an original scale of 1:6,200, covers an area 650 m across.

Source: © Natural Resources Canada.

c

image can also be blurred by vibration of the sensor system or poor band registration.

Target and background characteristics

The characteristics of a feature and its surroundings affect the ease with which it can be detected by a remote sensing system. A feature can be more easily detected by a remote sensing system if it contrasts moderately with its surroundings: too high a contrast and detail is lost in the highlight and shadow areas, too low a contrast and individual features may be difficult to distinguish. A feature is also more easily detected if it has a distinctive shape with sharp edges or is associated with easily identified features. For example, the image from the QuickBird satellite shown in figure 4.5 has

a

b

Figure 4.4 Poor atmospheric and light conditions can dramatically reduce the effective spatial resolution of a remote sensing system.

These images of the same scene were taken by the same sensor system under good (a) and poor (b) atmospheric conditions.

Source: Courtesy of Imagery Resolution Assessments and Reporting Standards Committee.

a spatial resolution of 61 cm. However, the white lines in the parking lot and tennis court, which are only 15 to 20 cm wide, are readily discernable because they contrast so strongly with the black background.

In general, earth features have moderate contrast with their surroundings. However, tall buildings, trees, mountains, and other tall features cast dark shadows under certain lighting conditions. Insufficient detail and lack of spectral information (i.e., color) make interpretation of features in shadow difficult. Remotely sensed images can be acquired and processed to help overcome problems with shadowing. For example, a well-exposed film system can capture only about 180 brightness levels per emulsion layer (black-and-white films have a single photosensitive layer, color films have three layers), whereas a digital sensor with 12-bit quantization can capture 4,096 levels per band (see *Radiometric resolution* later in this chapter). To take full advantage of the greater sensitivity and range of digital image systems, image processing software can be used to enhance shadow detail while maintaining detail in the brighter areas of the image. This makes the image easier to interpret.

Image scale

The larger a feature appears on an image, the easier it is to identify. The relationship between a feature's size and how it

appears in an image is determined by the *image scale*. Image scale (like map scale) is the ratio of a distance on an image to the corresponding distance on the ground. It is generally represented as a fraction. For example, an image where a 1 cm distance on the image represents 1 km (100,000 cm) on the ground would have a scale of 1/100,000 (also written 1:100,000). The *scale factor* is the inverse of the fraction. In this case, the scale factor would be 100,000. The scale factor is sometimes more convenient to use in calculations as illustrated later in this chapter.

A scale of 1:50,000 is a *larger scale* than 1:100,000 (i.e., the fraction is a larger number, and an object would be larger in size at the larger scale). A distance of 1,000 m on the ground would be 1 cm at a scale of 1:100,000 and 2 cm at a scale of 1:50,000.

The scale of an aerial photograph can be calculated using the height above ground (H) and the focal length (f) of the camera lens[1] as follows:

Scale = f / H

Where f and H are in the same units.

---

1. The focal length of a camera lens is the distance between the optical center of the lens and the position of the plane at which the image it projects is in focus.

Figure 4.5   QuickBird image of tennis courts. The lines on the tennis court are less than 20 cm wide, smaller than the 61 cm pixel resolution of this QuickBird image, but their high contrast with the black asphalt surface makes them visible.

Source: Image courtesy of DigitalGlobe.

For example, an aerial photograph taken with a 150 mm lens at an altitude of 3,000 m above the ground would have a scale of 0.150 m/3,000 m or 1:20,000 and a scale factor of 20,000. That would be the scale of the image on the film, often termed the *contact scale*. A print made the same size as the film image (e.g., by contact printing, where the negative is placed in contact with the printing paper) would be at the contact scale. If the image is enlarged, the scale of the enlarged image, termed the *enlargement scale*, is the contact scale multiplied by the degree of enlargement. So a four-time enlargement of an image with a contact scale of 1:20,000 would result in an enlargement scale of [4 x (1:20,000)] or 1:5,000.

Enlarging an image increases the size of the features in the image, but it does not increase the spatial resolution of the image. The detail in the enlargement can only show what was captured in the original image. Making the image larger only makes it easier to see. In fact, there is some loss of detail any time an image is reproduced. Similarly, when an image

is displayed on a computer monitor, enlarging the image by zooming in increases the scale of the *display*, but not the spatial resolution of the image.

The value of scale in assessing resolution for photographic and digital images

A photographic image has a true scale because it is a physical image on a piece of film or paper. The relationship of the size of features in the image and their size on the ground is fixed. Also, the photographic systems used to collect aerial photography for mapping have similar characteristics. For this reason, larger scale aerial photography can generally be assumed to have a higher spatial resolution than smaller scale photography.

A digital image does not have a fixed physical image form. It can be displayed at any scale. In the case of a digital image, it is not the size of the physical image but the characteristics of the instrument that recorded it that is fixed. A useful indicator of spatial resolution in digital images is the ground area represented by a single pixel, termed the *ground sample distance* (discussed below) in the *original digital image*. Digital images are commonly *resampled*, meaning they are modified to change the image size. During this process, the ground area represented by each pixel is changed to suit different display or printing requirements. As a result, the ground area represented by a pixel in the display image may be quite different from that of the original. For this reason, the scale of a particular display of a digital image, be it on a computer or as a physical print, is *not* in itself a useful indicator of the spatial resolution of the original image.

Application of different photographic image scales

It is more useful to categorize the spatial resolution of imagery by the smallest-sized feature that can be seen in an image rather than by image scale. This applies to both photographic and digital images. However, when all imagery was photographic and the equipment and processing were fairly standard, image scale was a useful and quick basis for comparison. It is still used in comparing photographic image characteristics.

Aerial photographs are commonly categorized as small, medium, or large scale. The scales included in these categories depend on the discipline. For example, a 1:12,000 scale image would be considered large scale for resource management but very small scale for engineering surveys for road construction projects.

For resource management applications, a scale of about 1:50,000 or smaller is considered small scale, 1:12,000 to 1:50,000 is medium scale, and scales of 1:12,000 or larger are large scale. In general, small-scale imagery is used for reconnaissance level mapping, large-area resource assessments, and regional resource planning. Geological mapping, land-use planning, agricultural monitoring, and forest monitoring also commonly use small-scale imagery. Medium-scale imagery is used for identification, classification, and mapping of such features as forest types, agricultural crop types, vegetation communities, soil types, topographic mapping, surface materials mapping, and geology. Large-scale imagery is used for intensive monitoring and detailed measurements, such as surveys of damage caused by natural disasters, detection of diseased vegetation within a field, and identifying buildings, vehicles, and aircraft.

For many applications, the mapping or inventory of a large area is done using multiple scales of imagery together with field verification. To study large areas, small-scale imagery of the entire area is commonly used together with large-scale imagery of selected subareas, and within the subareas sites are selected for field sampling. This process is known as *multistage sampling*[2]. When incorporated into a statistical sample design, this approach can be used to derive reliable quantitative resource estimates at a lower cost than by using a single scale of imagery for the entire area. Multistage sampling is a means of optimizing the trade-off between cost and accuracy. The method provides estimates that are more accurate than would be achieved using small-scale imagery alone and less costly than sampling the entire area at the finest level of detail.

In figure 4.6, satellite imagery, high- and low-altitude aerial photography, and field sampling were used for vegetation mapping and environmental monitoring.

Crop area and timber volume estimates are commonly done using one or more scales of imagery and field sampling in a multistage sampling design. The volume of timber for an area can be estimated by measuring the trees at selected field sites that represent the different forest types that can be identified on aerial photography. Using the area of each forest type as interpreted from the aerial photography and the volume estimates obtained from field data, an estimate of total timber volume can be calculated together with an estimate of the

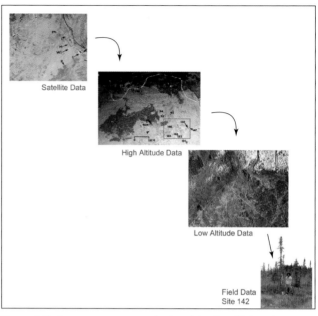

Figure 4.6 Large-area assessment using data at multiple scales. Large areas are commonly monitored, mapped, or inventoried using complete coverage at a low spatial resolution. Additional data at progressively higher detail is collected for successively smaller areas. In this example, data at three spatial resolutions were used: a Landsat MSS satellite image printed at a scale of 1:70,000; 1:10,000 scale color-infrared aerial photography; and field data collected by visiting selected sites.

Source: © Natural Resources Canada and S. Aronoff.

accuracy of the measurement. Similarly, volume estimates produced for selected areas of airphoto coverage can be used to develop estimates for a larger area imaged with small-scale aerial photography or satellite imagery.

Measuring spatial resolution

The resolving power of a remote sensing system refers to its ability to record spatial detail. Many factors affect resolving power and they differ for photographic film-camera systems and for digital imaging systems that use lens systems such as digital cameras and electro-optical scanners (see chapter 6). Imaging radar is a digital system that does not use lenses and is discussed separately in chapter 8.

It is important to distinguish between measurements of the resolving power of individual components as measured under laboratory testing conditions and measurements of a remote sensing system consisting of multiple components under normal operating conditions. While measurements of individual components are useful in designing and calibrating

---

2. Multistage sampling is a particular application of survey sampling—a statistical method used to obtain reliable estimates of group characteristics by measuring a sample of its members. Opinion polls are a familiar example.

systems, it is the overall quality achieved by the system with all of its components operating together that determines the spatial resolution of the imagery produced. For example, in aerial photography, the quality of the lens, the choice of film, and camera shake caused by aircraft vibration all affect the fineness of details visible in the imagery. The chain of components is only as strong as its weakest link. A high-quality camera and film system will produce poor-quality imagery if the camera mount is substandard causing excessive camera vibration.

Assessing film-camera systems

Photographic film can resolve very fine details. Using precision instruments, measurements as fine as about 0.01 mm can be reliably made from aerial photographs. For visual interpretation, features need to be about 0.1 mm in size on the photo for detection. A distance of 0.1 mm on a 1:50,000 scale photograph represents a ground distance of 5 m and on a 1:6,000 photograph it represents 0.6 m.

In photography, spatial resolution is commonly measured by determining how fine a grid of alternating black and white lines can be resolved. The measurement used is *line pairs per millimeter* (lp/mm). Photographic films, as well as aerial photography systems, are evaluated in this way using standard bar targets consisting of progressively smaller squares containing alternating black and white bars *(figure 4.7)*. The lp/mm is determined by finding the block with the finest lines that can be reliably distinguished.

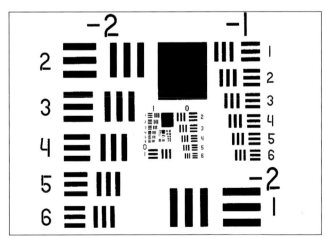

Figure 4.7   Bar target designed by the U.S. Air Force to measure the spatial resolution of remote sensing systems.

Source: Courtesy of the U.S. Air Force.

Films are tested by printing the test target directly onto the film material and analyzing the image. However, a more useful measure is the resolution of the film-camera system under normal operation, which is assessed by flying the system in the aircraft in which it is normally used and photographing a large bar target on the ground. This test imagery incorporates the image degradation from image motion (due to camera vibration), camera optics, atmospheric effects, and other factors that occur under operational conditions. This type of test is thus more representative of the resolution that will actually be obtained.

A typical film used for aerial photography will have a resolution of 100 to 150 lp/mm. Special films can have resolutions as high as 450 lp/mm. However, camera optics and camera motion can reduce the effective resolution of the film-camera aircraft system by half or more, and it is this reduced resolution that should be used in judging the expected resolution of aerial photography.

The combined effects of scale and resolution on image quality can be expressed as the *ground resolving distance* (GRD), a measure of the smallest object expected to be detected. For aerial photography, the theoretical GRD can be calculated from the lp/mm as follows:

GRD = scale factor / R

Where

GRD is the ground resolving distance in millimeters.

R is the system resolution in line pairs per millimeters.

Scale factor is the inverse of the scale fraction.

For example, a system that produces images with a scale factor of 20,000 (a scale of 1:20,000) with a resolution of 40 lp/mm will have a GRD of 500 mm (20,000/40), or about 0.5 m. This result assumes the original film is being examined. A contact print or enlargement from the negative would have a lower spatial resolution because there is always some loss of image definition in the printing process.

This theoretical GRD, while easily determined, is an optimistic approximation of how small a landscape feature might actually be discerned. Seldom will landscape features have such regular shapes, sharp edges, and high contrast as the resolution target. However, because this calculation of GRD is a standardized measure, it is useful for comparing the spatial resolution of different systems under the same operating conditions and for evaluating the same system under different operating conditions.

Assessing digital imaging systems

Bar targets are not well suited for systems that generate digital images because the pattern of straight lines in the target can interact with the grid pattern of the system's detector array to produce image artifacts not present in the scene, a phenomenon known as *aliasing*. Slight changes in the position and orientation of the sensor relative to the target dramatically affect the way the target is imaged. It is a problem peculiar to imaging the resolution target, since landscape features rarely resemble bar targets. (See *Aliasing in digital images* at the end of this chapter).

Figure 4.8 illustrates this effect. In both cases, the pixel size is equal to the width of a bar in the image. In the top row, the scanning grid is perfectly aligned with the lines of the resolution target and the resulting digital image perfectly reproduces the target. In the bottom row, the scanning grid is misaligned, offset by half a pixel. Each pixel views half a black bar and half a white space. Since the value recorded for a pixel represents the average brightness of the ground area each pixel views, an overall gray tone is produced for these pixels, and the individual bars are not resolved.

As discussed before, a digital image does not have a scale per se because it can be displayed and printed at any scale. However, the *ground sample distance* (GSD) at which an image is *acquired* is a function of scale. GSD is the ground distance represented by the width of a pixel. For example, an image in which each pixel represents a 30 m by 30 m area of ground would be said to have a 30 m GSD, and an image where pixels represent a 1 m by 1 m area has a 1 m GSD. The smaller the GSD value, the finer the detail that can be captured and the finer or higher the spatial resolution of the image is said to be. (In cases where the pixel represents a rectangular area, the larger dimension is quoted.)

Just as in a conventional camera, the lens of a digital camera forms an image on a light-sensitive material. Instead of film, a digital camera or scanner focuses the image onto the flat surface of a detector array—a device comprised of thousands of minute light-sensitive detectors arranged in a rectangular grid. The voltage level produced by each detector is proportional to the intensity of the light received, and it is this difference in voltage levels that is used to generate the pixel values. The size of each detector is, in effect, the physical size of a pixel in the sensor, and the ground sample distance of the system is the size of the ground area these pixels represent.

The scale of the image projected on the array is calculated the same way as for a conventional camera:

Scale = f/H

Where

f is the focal length.

H is the height above ground level.

f and H are in the same units.

For example, a camera with a 28 mm focal length lens flown at a height above ground of 1800 m will form an image *on the detector array* at a scale of about 0.028:18000 = 1:65,000 and have a scale factor (the inverse of the scale) of 65,000.

The GSD of the collected image can be calculated as:

Collection GSD = (detector element size) X scale factor

If each detector element in the array is 0.009 mm (9 x 10^{-6} m), then the collection GSD in this case would be (9 x 10^{-6} m X 65,000) or about 0.6 m.

Digital imagery is rarely displayed at the camera scale (i.e., the collection GSD) because it would be too small for comfortable viewing. Instead it is printed in hard copy or displayed on a computer monitor at a different scale called the *product scale*. The product scale of an image is the ratio of the size of a feature on the product to its actual size on the ground. It can be determined by measuring a feature in the image and comparing that to its size on the ground.

The product scale is more commonly calculated as the ratio of the GSD in the product to the collection GSD. For example, image products are commonly printed at 300 pixels per inch (0.0000847 m per pixel), which is about the *visual threshold*—the limit of detail that can be seen by the unaided human eye. Using the collection GSD of 0.6 m

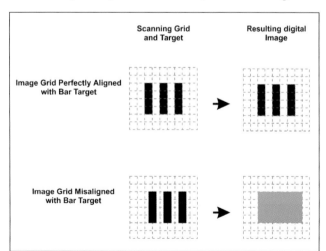

	Scanning Grid and Target	Resulting digital Image
Image Grid Perfectly Aligned with Bar Target	▮▮▮	▮▮▮
Image Grid Misaligned with Bar Target	▮▮▮	▨

Figure 4.8 Digital images produced by scanning a bar target where the pixel size is equal to that of the bars. When perfectly aligned, the target is perfectly reproduced. When offset by half a pixel, the target is represented as a gray square.

calculated previously, the product scale of an image printed at 300 pixels per inch would be calculated as follows:

Product scale = (product pixel size) / (collection GSD)

This would give a scale of 0.0000847 m / 0.6 m or about 1:7,000.

Be aware that the appearance of imagery displayed on a computer is not a good guide to how it will appear when printed. Because the monitor image is emitting light, whereas a printed image is seen by reflected light, the coarser resolution of the monitor display will appear sharp even though it typically has a resolution of only about 72 pixels per inch (about 0.353 mm per pixel), which in the previous example would give a scale on screen of about 1:1,700 (0.000353 m / 0.6 m) if displayed at full resolution (i.e., each image pixel is displayed as a single screen pixel).

The term GSD, unfortunately, is often used to refer to the ground distance represented by the pixel in *image products* as well as in the *original collection* GSD. This causes considerable confusion because while the collection GSD guarantees a certain level of resolution, the GSD of the product *does not*. For example, a satellite image with a collection GSD of 1 m may be resampled to 0.5 m pixels to register with other images. Though in the resampled product each pixel represents a 0.5 m by 0.5 m area, the detail contained in that product is no better than the original 1 m GSD image data from which it was derived. In fact, a poor choice of resampling method can result in a loss of detail. In some cases, an image may be resampled to a coarser GSD than the original image, in which case detail is lost. So the spatial resolution of an image product, be it a paper print or a soft-copy display,

is never better, and can be considerably worse, than the collection GSD of the original image.

Figure 4.9 illustrates this concept. All four images have the same *product* GSD; the width of a pixel in each image product represents a distance of 0.6 m on the ground. However, of the original images from which these products were generated, only image (a) had a 0.6 m GSD; that is, it had a collection *and* a product GSD of 0.6 m. The other images show the effect of using coarser resolution images of the same scene and resampling them to produce a 0.6 m GSD product. The images captured with a coarser resolution have less detail; it doesn't matter that they are displayed at the same GSD as the image captured at the finer resolution.

Thus, the product GSD is not a reliable guide to the level of detail captured in the original image *unless* it was produced directly from the original data *and* equals or exceeds the collection GSD. In using a digital image product, its effective GSD should be considered no better than the coarser of the collection GSD or the product GSD.

How is the GSD of a digital image related to the GRD of a photographic image? As illustrated in figure 4.8, as few as two pixels can resolve a line pair, but only if they are perfectly aligned. Experience has shown that for a digital image to fully capture the visually perceptible detail of a photograph requires two to three pixels per line pair. Values from 2.2 to 2.8 are commonly used. So a 1:50,000 photographic image with a spatial resolution of 50 lp/mm would have a GRD of 50,000/50 mm or 1 m. For a digital image to provide about the same resolution, it would need to have a GSD of 0.45m (1m/2.2) to 0.36m (1m/2.8).

a

b

c

d

Figure 4.9   Collection and product GSD. These four images of the Montreal Olympic Stadium all have the same product GSD of 0.61 m. Only image (a) has an original collection GSD of 0.61 m as well. The other images have coarser collection GSDs and were resampled to generate products with finer GSD values. However, the coarser spatial resolution of the original images is apparent. The collection GSDs of the other images were (b) 1.20 m, (c) 2.40 m, and (d) 4.80 m. (Coarser resolution images were simulated from the original high resolution image.)

Source: Images courtesy of DigitalGlobe.

Aliasing in digital images

Aliasing is the creation of unexpected artifacts (i.e., false patterns not present in the feature being imaged) as a result of the interaction of patterns in the image with the grid pattern of detectors used to record it. It occurs in digital images when distinct features have a repeating pattern oriented parallel to the pixel grid with a spacing of about one to two pixels. These artifacts may appear as color fringing along high contrast edges, swirling patterns termed *moiré* effects *(figure 4.10)*, or distortions of linear features such as exaggerated width or an intermittently appearing and disappearing pattern.

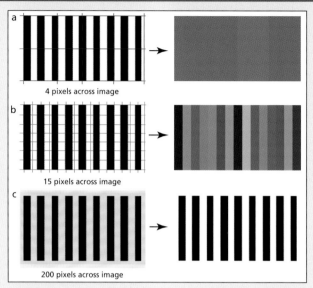

Figure 4.11   Effect of pixel size on aliasing. The digital image of the bar target is affected by the number of pixels used to represent it. When the pixels are large relative to the bars (a, 4 pixels across target), the target is represented as a uniform gray tone. Small pixels (c, 200 pixels across) reproduce the bar pattern of the target. However, a pixel spacing approximately that of the spacing of bars in the target (b, 15 pixels across) creates a false pattern termed *aliasing*.

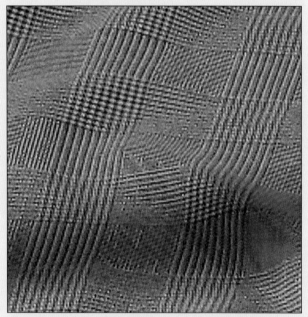

Figure 4.10   Moiré patterns form in digital images when a pattern in the scene such as the texture of fabrics interacts with the pattern of pixels. This artifact can be particulary troublesome in images formed by color interpolation.

Source: © 2002–2004 Foveon, Inc. Illustration used by permission from Foveon, Inc.

Figure 4.11 illustrates this phenomenon by showing how a target of black and white bars is reproduced by varying numbers of pixels. When the distance across the area represented by a pixel is large relative to the pattern, the pattern is not captured, and the target is represented as a uniform gray area. When the distance represented by a pixel is small relative to the pattern, the target is accurately reproduced. But when the distance represented by a pixel is approximately the same as the spacing of bars in the target, an interaction effect occurs, and the representation of the target is unexpected.

In practice, aliasing can easily occur in images of crop rows if the rows are oriented approximately parallel to the pixel grid and the row spacing is about one to two pixels in the image. To eliminate the alias effect, a coarser resolution can be used that does not resolve the individual rows, or a finer resolution can be used so that the individual rows (i.e., a light/dark transition) are imaged with more than two pixels; about 2.2 to 2.8 pixels is generally sufficient[1]. Alternatively, the imagery could be flown with a different flight line so the rows are not aligned with the pixel grid. However, since the row orientation of different fields in a region will vary, but the cropping practices are fairly consistent, in practice it is better to choose a different resolution.

For crop rows with 1 m spacing, artifacts may occur if the crop rows are imaged with a spatial resolution of about

Aliasing in digital images (cont.)

1 m GSD to 0.5 m GSD and oriented approximately parallel to the pixel grid. Using a finer resolution, about 2.2 to 2.8 times that of the pattern, i.e., 0.45 to 0.36 m (1/2.2 = 0.45 to 1/2.8 = 0.36) or finer, would capture the details in the rows without generating artifacts even if the patterns were aligned. Alternatively, using a spatial resolution greater than 1 m GSD would not produce aliasing; however, a lower spatial resolution would not resolve the individual rows.

Aliasing can be subtle enough that an image "looks right." Figure 4.12 is a 2 m GSD digital image produced by an airborne rotating mirror scanner system. The visible green, visible red, and near-infrared bands are displayed as blue, green, and red, which produce a color rendition similar to that of color infrared film. At first glance **A** appears to be a typical image of an agricultural field with crops planted in rows. The vegetation coverage appears sparse, similar to the appearance of crops at an early stage of development. In fact, the apparent details in the field are false—they are artifacts caused by aliasing.

The area at **A** is a cornfield with an even 40% to 50% vegetation cover. What appear to be rows are about the same width as the road at **D**, about 10 m wide, which is far larger than the width of rows of corn. Aliasing has caused the field to appear mottled, similar to the way variations in soil moisture or texture would appear. Yet these patterns are probably not related to the soil, but to variations in the interaction between the rows of crops and the pixel grid of the sensor. Where a pixel fell directly over a patch of vegetation, it would appear pink on this image. A pixel showing only soil would appear mid-to-dark gray. However, where the pixel included soil and vegetation, the reflectance values would be in between that of a pure soil and a pure vegetation pixel. Such mixed pixels make computer-based classification difficult

because the reflectance values are significantly different from that of pixels representing either of the mixture classes alone.

The feature at **B**, which was not bare soil but had vegetation coverage similar to the rest of the field, might have been caused by a slight tilt of the aircraft, which changes the degree of aliasing between the crop rows and pixel pattern. Note the pattern of light and dark bands running in the same direction as the apparent rows but twice as wide. They also are aliasing artifacts. The road at **C** is straight. The slightly sinusoidal shape in this image is caused by the slight change in scale moving outward from the flight line **E**.

In this case, the 2 m pixel resolution caused severe aliasing artifacts that misrepresented the field conditions. Choosing a coarser or finer spatial resolution would have avoided the aliasing problem. Thus, the possibility of alias effects must be anticipated in planning the image acquisition mission and in later quality-assurance inspections of the imagery.

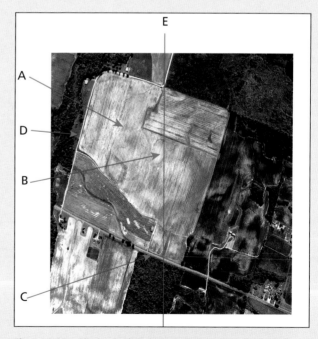

Figure 4.12   Aliasing in digital imagery. The within-field details captured in this 2 m GSD digital scanner image appear to be normal rows and soil variations but are actually aliasing artifacts caused by interaction of the pixel grid with the crop rows. To an observer, the field would appear to be a fully developed continuous field of corn.

Source: Image courtesy of NASA.

---

1. According to the Nyquist Theorem, to accurately reproduce an analog signal (such as the continuous change in brightness across an image), the sampling rate must be equal to or greater than twice the highest frequency in the analog signal. In the case of digital imaging, it means that to accurately reproduce a distinct pattern, such as a pattern of repeating white and black bars, the distance across a white and black bar pair must be at least two pixels. In practice, since the placement of the pixels will not be perfectly aligned to the pattern of the target, the pixel size must be small enough to have somewhat more than two pixels per white and black bar pair, typically between 2.2 and 2.8. When there are between one and two pixels for each bar pair in the target, i.e., the spatial frequency of the pixels approximates that of the bars, a false pattern, or aliasing, develops.

Interpretability ratings for assessing spatial resolution

Measurements of the physical characteristics of images or imaging systems can only be a guide to the smallest-sized feature discernable because they do not take into account other factors that can affect the overall ability for someone to extract specific information from the image, referred to as *interpretability*. An alternative approach is to directly rate the interpretability of an image using standardized targets and determining whether specific features can or cannot be identified.

Our interest in spatial resolution goes beyond detecting the presence of an object. Usually it is important to be able to classify that object (e.g., *Is it vegetation or water?* or *What is the vegetation type?*). Three commonly used categories of information extraction are 1. *detection*, 2. *recognition* or *distinguish between*, and 3. *identification*[3]. At the detection level, the presence or absence of a feature in the landscape can be deduced by its general shape and context within the scene well enough to determine its general class (e.g., a pond, building, stand of trees, agricultural field, aircraft). At the *recognition* or *distinguish between level*, it can be determined that two features within a category are of different types, although the types cannot be identified (e.g., two vehicles are different but the type of vehicle cannot be determined). At the *identification level*, a more detailed characterization is possible such as distinguishing trucks from cars, types of buildings, species of tree, crop species, or types of aircraft.

To some extent, these three levels of information are overlapping categories, yet they are useful in evaluating the spatial resolution needed for a specific application. As a rough guide, it takes about a three-fold improvement in spatial resolution to move from the detection level to the recognition level and about a 10-fold improvement to move from the recognition level to the identification level.

The U.S. military developed both military and civilian versions of the National Image Interpretability Rating Scale (NIIRS), a 10-point, task-based scale from zero (obscured image) to nine (highest level of detail). Using image examples and text descriptions, the analyst examines a set of imagery, looking for examples of progressively smaller features. A rating is assigned to the image based on the smallest test features that can be reliably identified. This type of interpretability rating of image resolution is an effective means of measuring the utility of imagery for specific applications, specifying imagery requirements, and measuring the performance of a remote sensing capability. Table 4.2 lists representative criteria for the civilian NIIRS rating scale.

Because satellite-based sensors can safely acquire imagery of any country, there has long been concern that making higher spatial resolution imagery publicly available provides valuable military information to a country's adversaries. In Table 4.3 a five-level interpretability rating scale was used to illustrate this concern by showing the value of different commercially available satellite image sources for evaluating military targets.

Choosing an appropriate spatial resolution

Complete repetitive global coverage digital satellite imagery is available with spatial resolutions as high as 2.5 m (SPOT 5). Commercially available satellite imagery is collected selectively at resolutions as high as 1 m (IKONOS and OrbView-3) and 0.61 m (QuickBird). Airborne imagery can be acquired at as high a resolution as needed, limited only by cost.

When choosing an appropriate level of spatial resolution, several important trade-offs should be considered, including suitability to the task being performed; the cost of acquiring, processing, and storing the imagery; compatibility with other datasets with which the imagery may be used; and the features that need to be detected in the imagery.

One should also be aware of aliasing, an image artifact produced when a pattern in the image is aligned with the pixel grid and contains distinct features that repeat within a space of about one or two pixels. Crop rows are a common example (see *Aliasing in digital images* earlier in this chapter). The problem can be avoided by knowing the characteristics of the area being imaged and choosing a spatial resolution unlikely to produce alias effects.

Suitability to task

To be considered suitable for a specific task, the spatial resolution of an imaging system should be high enough to produce the required information using the chosen analysis method. In urban infrastructure mapping, the spatial resolution is typically determined by the required positional accuracy of the final map products. For applications that involve classification of imagery, such as maps of land-cover types, the resolution should be high enough to resolve the smallest unit to be mapped, called the *minimum mapping unit*.

However, it is not always necessary or desirable to distinguish individuals of a class in order to assess it. To identify

---

3. This division of interpretability levels is discussed in Colwell (1983), p. 23.

Table 4.2   Examples of civilian NIIRS criteria
Rating Level 0 Interpretability of the imagery is precluded by obscuration, degradation, or very poor resolution.
Rating Level 1 Distinguish between major land-use classes (e.g., urban, agricultural, forest, water, barren). Distinguish between runways and taxiways at a large airfield. Identify large area drainage patterns by type (e.g., dendritic, trellis, radial).
Rating Level 2 Identify large (i.e., greater than 160 acre) center-pivot irrigated fields during the growing season. Detect large buildings (e.g., hospitals, factories). Identify road patterns, like clover leafs, on major highway systems.
Rating Level 3 Detect large area (i.e., larger than 160 acres) contour plowing. Detect individual houses in residential neighborhoods. Detect trains or strings of standard rolling stock on railroad tracks (not individual cars). Distinguish between natural forest stands and orchards.
Rating Level 4 Identify farm buildings as barns, silos, or residences. Detect basketball court, tennis court, volleyball court in urban areas. Identify individual tracks, rail pairs, control towers, switching points in rail yards. Detect jeep trails through grassland.
Rating Level 5 Identify Christmas tree plantations. Identify individual rail cars by type (e.g., gondola, flat, box) and locomotives by type (e.g., steam, diesel). Detect open bay doors of vehicle storage buildings. Identify tents (larger than two person) at established recreational camping areas. Distinguish between stands of coniferous and deciduous trees during leaf-off condition. Detect large animals (e.g., elephants, rhinoceros, giraffes) in grasslands.
Rating Level 6 Detect narcotics intercropping based on texture. Distinguish between row (e.g., corn, soybean) crops and small grain (e.g., wheat, oats) crops. Identify automobiles as sedans or station wagons. Identify individual telephone/electric poles in residential neighborhoods. Detect foot trails through barren areas.
Rating Level 7 Identify individual mature cotton plants in a known cotton field. Identify individual railroad ties. Detect individual steps on a stairway. Detect stumps and rocks in forest clearings and meadows.
Rating Level 8 Identify a USGS benchmark set in a paved surface. Identify grill detailing or the license plate on a passenger truck. Identify individual pine seedlings. Identify individual water lilies on a pond. Identify windshield wipers on a vehicle.
Rating Level 9 Identify individual grain heads on small grain (e.g., wheat, oats, barley). Identify individual barbs on a barbed wire fence. Detect individual spikes in railroad ties. Identify an ear tag on large game animals (e.g., deer, elk, moose).

Source: Table courtesy of the Image Resolution Assessment and Reporting Standards Committee and the Federation of American Scientists.

Table 4.3   Spatial resolution needed to evaluate military targets

Required resolution in meters for the following:

Target	Detection	General ID	Precise ID	Description	Technical analysis
1. Terrain		90.00	4.50	1.50	0.75
2. Urban areas	60.00	30.00	3.00	3.00	0.75
3. Ports and harbors	30.00	15.00	6.00	3.00	0.30
4. Railroad yards and shops	30 to 15	15.00	6.00	1.50	0.40
5. Coasts, landing beaches	30 to 15	4.50	3.00	1.50	0.15
6. Surfaced submarines	30 to 7.5	4.5 to 6	1.50	1.00	0.03
7. Surface ships	15 to 7.5	4.50	0.60	0.30	0.045
8. Roads	9 to 6	6.00	1.80	0.60	0.40
9. Minefields	9 to 3	6.00	1.00	0.03	
10. Airfileds facilities	6.00	4.50	3.00	0.30	0.150
11. Bridges	6.00	4.50	1.50	1.00	0.300
12. Troops units (in bivouac or on road)	6.00	2.00	1.20	0.30	0.150
13. Aircraft	4.50	1.50	1.00	0.15	0.045
14. Command and control headquarters	3.00	1.50	1.00	0.15	0.090
15. Missle sites (SSM/SAM)	3.00	1.50	0.60	0.30	0.045
16. Radio communications	3.00	1.50	0.30	0.15	0.015
17. Radar communications	3.00	1.00	0.30	0.15	0.015
18. Supply dumps	3 to 1.5	0.60	0.30	0.03	0.030
19. Nuclear weapons components	2.50	1.50	0.30	0.03	0.015
20. Vehicles	1.50	0.60	0.30	0.06	0.045
21. Rockets and artillery	1.00	0.60	0.15	0.05	0.045

Maximum resolution of widely used commercial satellite data

10 to 90 meters: Landsat 7, SPOT-4, ERS-1 and 2, Aster, Envisat
2.1 to 9 meters: IRS-1C and 1D, Radarsat-1, SPOT-5
0.6 to 2.0 meters: IKONOS, QuickBird, OrbView 3

Source: Ann M. Florini. The opening skies: Third-party imaging satellites and U.S. security. *International Security* 13(2):91–123. © 1988 The President and Fellows of Harvard College and the Massachusetts Institute of Technology. Reprinted with permission.

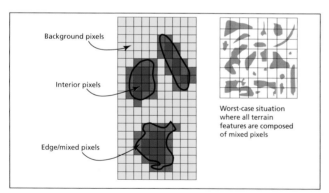

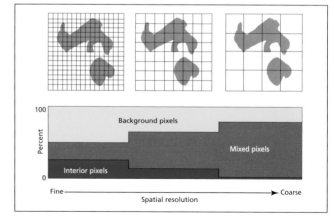

Figure 4.13   Effect of pixel size on the proportion of mixed pixels. In the left panel, smaller and narrower areas increase the proportion of mixed pixels. In the right two panels, coarser spatial resolution can be seen to increase the proportion of mixed pixels as shown in the graph.

Source: *Remote sensing imagery for natural resources* by D. Wilkie and J. Finn. © 1996 Columbia University Press. Image reproduced with permission from Columbia University Press.

an individual feature in a digital image, the spatial resolution should be high enough for several pixels to fall within the feature's boundary. A pixel that falls on a boundary between two different features has a value that is a mixture of reflectance from the two, (i.e., it is a mixed pixel). Mixed pixels contain varying proportions of different classes, so their reflectance values are highly variable and do not identify any single class. The higher the spatial resolution relative to the detail in the information classes being analyzed, the less frequently a pixel will represent more than one feature.

A mixed pixel may cover a boundary between two cover types, called *edge mixed pixels*, or completely include a feature smaller than the pixel. Figure 4.13 illustrates how the pixel size affects the proportion of mixed pixels.

However, distinguishing individuals of a class does not always produce the most accurate results. If the objective is to obtain accurate area estimates of different vegetation classes using computer-based classification of digital imagery, a very high resolution image would show very fine tonal gradations such as the shadowing on a tree. Such detail would tend to make the shaded and unshaded areas of a tree different classes, which effectively reduces the classification accuracy.

Thus, more important than the question of how small an object can be detected in an image is the question of what needs to be detected to generate the information required and how the data will be processed. Although high-resolution images are visually appealing, using the lowest resolution image sufficient for the task is the most cost-effective approach. Often a surprisingly low resolution will suffice. It is not always necessary to distinguish features to collect valuable information about them. Every time we look out over a landscape and distinguish agricultural fields from the surrounding forest we do so without being able to resolve individual trees or crop plants. We might even be able to assess what types of crops are planted—sometimes before the crop has even sprouted—just by knowing the growing season and agricultural practices of the area. What we are using to generate this crop information are *secondary indicators*, indicators other than direct observations of the objects being assessed. For example, rice fields must be flooded at a certain period in the growing season. An assessment of flooded fields at that time can be used to predict the area of cropland planted with rice even though no rice plants are present.

A striking example of the use of secondary indicators took place in the early 1970s when analysts predicted the failure of a Russian winter wheat crop. Though the crop is harvested in the fall, this prediction was made during the preceding winter. Much of the wheat crop produced in northern countries like Canada and Russia is winter wheat. This type of wheat is planted in the fall, germinates and begins to grow underground, and then must survive the winter and continue growing the following spring. For the plant to survive the winter, it must not be exposed to extreme cold. Normally a snow cover during the coldest period of the winter provides sufficient insulation. *Low-resolution satellite data* in January showed the ground was snow free and the temperatures were below normal for several days—enough to kill much of the crop. Analysts correctly predicted the crop failure without ever detecting a single wheat plant! By evaluating the information needed for the application, the scientists realized they didn't need costly high-resolution data to predict the crop failure. It is an example of how a little forethought can go a long way in identifying the cost-effective remote sensing data for an application.

Cost

The cost of remotely sensed imagery depends on several factors, including the degree of government subsidization, the value of the imagery to the user (i.e., what price the market will bear), and competition. The growing number of image sources with overlapping characteristics, particularly of satellite imagery, has increased competition. For many applications, less-expensive imagery from a different sensor can be substituted without compromising results, so users often may have the flexibility to negotiate better prices.

Imagery is generally more costly *per unit area* the higher the resolution. The costs associated with storing and handling the larger datasets are also greater. Data volume increases by the square of the increase in resolution. For example, doubling the resolution would replace a pixel representing a 10 m by 10 m ground area with four pixels, each representing a 5 m by 5 m area, thereby quadrupling the number of pixels and quadrupling the size of the data file. Tripling the resolution would increase the number of pixels nine fold.

Imagery with a higher spatial resolution typically has a smaller coverage area per frame. To create an image of an area from multiple images, they must be combined, a procedure termed *mosaicking*, which entails added costs. There is also a time factor. It might require several overpasses to obtain high-resolution satellite imagery for a large area, and it may take days or weeks to obtain the necessary cloud-free image acquisitions. This may not be acceptable for time-critical applications such as large-area crop monitoring, and other applications such as change detection and automated classification that benefit from a consistent depiction of features throughout the coverage area. Obtaining a larger coverage area within a single frame and more frequent acquisition may be more important than higher spatial resolution. Figure 4.14 and table 4.4 compare the spatial resolution and the coverage of a single image for different airborne and satellite sensors.

Higher spatial resolution will not necessarily improve results, and using imagery with greater detail than required can substantially increase processing, storage, and analysis costs for no benefit. For example, the Crop Condition Assessment Program tracks crop condition and estimates yield of the wheat crop for western Canada (an application described in chapter 13) using 1 km GSD satellite data. Higher resolution imagery would not only be prohibitively expensive to purchase, store, and process, but it would also distinguish details such as shadow effects and the transition zones between vegetation types not resolved on the coarse imagery. This would make automated analyses of crop condition and crop areas less accurate.

Compatibility

In some cases, an important factor in selecting an appropriate spatial resolution is the scale and accuracy of map data with which the imagery is to be used. The scale and accuracy of map products is commonly specified to be compatible with other related geospatial products. For example, in the United States, the U.S. National Map Accuracy Standard

(NMAS), is commonly used. The NMAS requires that for map scales of 1:20,000 or smaller, the geographic position of 90% or more of a set of well-defined features, (e.g., a road intersection, the corner of a building, the end of a pier) as determined from the map, should be within 1/50th of an inch of its correct position and within 1/30th of an inch for larger scales.

Table 4.5 shows, for commonly used map scales, the relationship between the resulting ground sample distance (GSD) assuming 300 pixels per inch (the commonly accepted visual threshold), the corresponding maximum allowable NMAS error, and the one standard deviation error that photogrammetric and other mapping applications are commonly specified to meet. Note that the NMAS error value is larger than the GSD at 300 pixels per inch. It is necessary for the spatial resolution of the product to be at least as fine, and preferably finer, than its level of accuracy to ensure features are displayed with as much accuracy as exists in the image. This does not guarantee that the quality of geometric correction meets the standard, only that features can be discerned at a finer level of detail than the mandated level of accuracy.

This type of table is a useful guide in choosing the spatial resolution for an image so that it can be displayed at the required map scale without appearing pixilated (i.e., so that pixels are displayed at the visual threshold of 300 pixels per inch). For example, to distinguish features as small as 0.85 m in size in an image, the image product scale should be chosen so that a pixel at the visual threshold of 1/300th of an inch, or 8.47×10^{-5} m, is 0.85 m on the product. To find out what scale is needed to produce such a product, refer to table 4.5. The third column refers to the product's GSD at the visual threshold in meters. Locate 0.85. The scale for this product can be found in the first column, Product scale number. The scale of this product would be (8.47×10^{-5} m) /0.85 m or 1:10,000. Moving right to column five, you can then determine what the National Map Accuracy Standard requires for a 1:10,000 scale map. Keeping to this standard will help ensure that the product will be compatible with other geospatial products. For a 1:10,000 scale map, the geographic position as determined from the product for 90% of well-defined test points should fall within 8.5 m of their true geographic position. The last two columns of the table indicate that, for an image with a GSD of 0.85 m, it would be expected that photogrammetric methods would generate maps with position accuracies of 5.1 m or better at the one standard deviation level. This means that if we were to measure a set of test points for which the true ground position

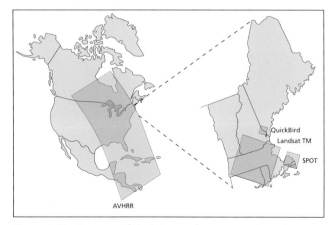

Figure 4.14 Coverage of single images from commonly used sensors.

Source: *Remote sensing imagery for natural resources* by D. Wilkie and J. Finn. © 1996 Columbia University Press. Image reproduced with permission from Columbia University Press.

Table 4.4   Image coverage and spatial resolution of several current sensors

Source of imagery	Spatial resolution	Coverage (swath width)
Aerial photography (230 mm format)	**GRD[1]**	
1:10,000	0.25 m	2 km
1:50,000	1.25 m	11.5 km
1:100,000	2.5 m	23 km
1:150,000	3.75 m	34.5 km
1:250,000	6.25 m	57.5 km
Satellite imagery	**GSD[2]**	
QuickBird	0.61 m	16.5 km
OrbView-3	1 m	8 km
IKONOS	1 m	11 km
SPOT-5 Panchromatic	5 m	60 km
IRS-IC and ID	5.8 m	70 km swath
Radarsat-1 (Fine mode)	8 m	50 km
SPOT-5 HRV	20 m	60 km
Landsat TM	30 m	185 km
Landsat MSS	80 m	185 km
AVHRR	1 km	3,000 km
GOES-9	1 km	>10,000 km
METEOSAT-7	2.4 km	>10,000 km

[1] GRD is ground resolution distance assuming 40 lp/mm system resolution.
[2] GSD is ground sample distance.

Table 4.5   Comparison of visual threshold GSD, U.S. National Map Accuracy Standards, and one standard deviation accuracy levels commonly provided by map and image products generated by photogrammetric processing. Threshold GSD assumes the unaided human eye can distinguish detail as fine as 1/300 inch which is equivalent to 8.47×10^{-5} m or 2.78×10^{-4} ft.

Product scale number	Scale: 1 in. = ft	Visual[1] threshold GSD (m)[1]	Visual[2] threshold GSD (ft)	NMAS[3] horizontal accuracy (90% conf.) (m)	NMAS[4] horizontal accuracy (90% conf.) (ft)	Horizontal accuracy 1 standard deviation (m)	Horizontal accuracy 1 standard deviation (ft)
Standard metric scales							
2,000	166.7	0.17	0.56	1.7	5.6	1.0	3.4
2,500	208.3	0.21	0.69	2.1	6.9	1.3	4.2
5,000	416.7	0.42	1.39	4.2	13.9	2.6	8.4
10,000	833.3	0.85	2.78	8.5	27.8	5.1	16.8
25,000	2,083.3	2.12	6.95	12.7	41.7	7.7	25.3
50,000	4,166.7	4.24	13.89	25.4	83.3	15.4	50.5
Standard English scales							
600	50.0	0.05	0.17	0.5	1.7	0.3	1.0
2,400	200.0	0.20	0.67	2.0	6.7	1.2	4.0
3,600	300.0	0.30	1.00	3.0	10.0	1.8	6.1
4,800	400.0	0.41	1.33	4.1	13.3	2.5	8.1
6,000	500.0	0.51	1.67	5.1	16.7	3.1	10.1
7,200	600.0	0.61	2.00	6.1	20.0	3.7	12.1
7,920	660.0	0.67	2.20	6.7	22.0	4.1	13.3
12,000	1,000.0	1.02	3.33	10.2	33.3	6.2	20.2
24,000	2,000.0	2.03	6.67	12.2	40.0	7.4	24.2
62,500	5,208.3	5.29	17.37	31.8	104.2	19.2	63.1
63,360	5,280.0	5.37	17.61	32.2	105.6	19.5	64.0

[1] Threshold GSD in meters = $(8.47 \times 10^{-5}$ m$) \times$ scale factor
[2] Threshold GSD in feet = $(2.78 \times 10^{-4}$ ft$) \times$ scale factor
[3] NMAS in meters for scales larger than 1:20,000 = $(8.47 \times 10^{-4}) \times$ scale factor; for scales of 1:20,000 or smaller, it is $(5.08 \times 10^{-4}) \times$ scale factor
[4] NMAS in feet for scales larger than 1:20,000 = $(2.778 \times 10^{-3}) \times$ scale factor; for scales of 1:20,000 or smaller, it is $(1.667 \times 10^{-3}) \times$ scale factor

Source: Courtesy of Trimble Navigation Ltd.

were known, the standard deviation of the errors would be expected to be 5.1 m or less, thereby meeting the required accuracy standard of 8.5 m.

While using tables such as 4.5 are useful guides in choosing the minimum spatial resolution to *display* imagery, they do not indicate the level of detail needed to *interpret* the required information from the imagery. That depends on the specific application and analysis procedures to be used.

Spectral resolution

Remote sensing systems are designed to detect energy within selected ranges of energy wavelengths termed *spectral bands* or *channels*. Depending on the application, a band can be set to be relatively narrow or wide. Spectral resolution refers to the number and width of spectral bands detected by a remote sensing instrument. The narrower the wavelength bands the sensor can detect, the higher the sensor's spectral resolution.

Panchromatic imaging systems

Remote sensing systems commonly use a panchromatic band to record the overall brightness of a scene. A camera and normal black and white film is an example of a panchromatic imaging system. The film is comprised of one photosensitive layer that is sensitive to the entire visible spectrum. Earth resource satellites producing digital imagery commonly have a high-spatial-resolution panchromatic band.

Multispectral sensors

Remote sensing systems that simultaneously record multiple wavelength bands are called *multispectral sensors*. A camera with color film is in effect a multispectral sensor in that the film's three dye layers record reflectance from the visible blue, green, and red bands separately. Multiple cameras fitted with different filters have been used to produce multispectral images as well. However, the limitations of film have made scanners the standard for multispectral imaging (discussed in chapter 6).

Airborne and satellite-based electronic multispectral imaging systems are widely used for remote sensing in the ultraviolet, visible, and infrared portions of the spectrum. For example, the Landsat Thematic Mapper collects data in seven narrow spectral bands and a panchromatic band that covers a wider spectral range. Similarly, the IKONOS satellite has a 1 m GSD panchromatic band that detects wavelengths over the visible spectrum and a portion of the near-infrared (0.45–0.90µm). It also has a 4 m GSD multispectral sensor

that records four narrower bands covering portions of the visible blue, visible green, visible red, and near-infrared. This sensor has a lower spatial resolution (larger GSD value) but a higher spectral resolution (narrower spectral bands) than the panchromatic band. The ASTER sensor on the NASA Terra research satellite records 14 narrow bands in the visible, shortwave infrared, and thermal infrared portion of the spectrum (see chapter 7).

Hyperspectral sensors

Systems with very high spectral resolution, called *hyperspectral sensors* or *imaging spectrometers*, can simultaneously record data in hundreds of contiguous, very narrow spectral bands in the ultraviolet (UV), visible, and near- and mid-infrared bands. Hyperspectral sensors are useful in vegetation mapping, detection of plant disease, and geologic mapping. These sensors are discussed further in chapter 6.

A limitation in the design of narrow band imaging systems is that the narrower the band, the less energy is available for detection. For example, a detector that senses a 0.1-µm-wide band will receive on average only a third of the energy as a detector sensing a 0.3-µm-wide band. As the energy available for detection becomes lower, it becomes more difficult to distinguish the desired signal from any undesired electrical noise in the system and from atmospheric effects.

Digital image processing techniques

A color monitor generates an image from a digital file by mixing blue, green, and red light to produce a full range of colors. The monitor can be thought of as having three color channels: red, green, and blue. Each channel displays an image as intensity levels of its color. For example, a monochrome image (i.e., an image that represents brightness levels in a single band such as a black-and-white image) displayed through the red channel would appear as shades of red. Each color channel of the monitor is, in effect, controlled by a monochrome image that governs the brightness with which that color is displayed for each pixel.

Thus, for each pixel location in the image there are three brightness levels, one from each channel. Depending on the relative brightness (the digital number) of each pixel in each channel, the colors combine in different proportions to represent different colors. If the three brightness levels are the same, then the pixel appears as a gray tone. If they differ, then a color is produced. In fact, a black-and-white image is produced on a color monitor by displaying the same image through all three color channels.

If the visible blue, visible green, and visible red spectral bands of an image are displayed through the blue, green, and red channels of a color monitor, a *natural-color* image is produced (i.e., the colors closely resemble what the human eye sees). If the color channels display spectral bands other than the color they reproduce, a *false-color* image is generated. The colors in a false-color image appear significantly different from what an observer would see.

Figure 4.15 shows the seven bands of data for one scene collected by the Landsat Thematic Mapper system, as well as two color composites, a normal-color image, and a false-color composite image using the green and red visible bands and an infrared band. Notice how vegetation types are more easily distinguished in the color-infrared image. Though the colors in the color infrared image are false, in the sense that they are not the colors these objects would appear to the human eye, once an analyst learns how to interpret these images, they can provide a wealth of additional information. The normal-color image looks quite dull due to the significant atmospheric scattering of the short blue wavelengths. By not using the visible blue and visible green bands that are more sensitive to atmospheric effects, and by choosing a color assignment that renders vegetation green, a simulated natural-color image can be generated from the visible red and two infrared bands that has more contrast and appears to be a normal color rendition of the scene *(figure 4.15)*.

Image processing techniques enable a wide range of color images to be generated to allow improved interpretation of the features of interest. In addition, sophisticated classification algorithms have been developed that can simultaneously analyze any number of bands, multiple images, and multiple datasets as discussed in chapter 11.

Image fusion is an image processing procedure for combining information from two different images. It is often used to sharpen a three-band multispectral color image with a higher spatial resolution panchromatic image, or from the opposite perspective, to add color to a panchromatic image with lower resolution multispectral data. The intensity values of the combined image are derived from the higher resolution image and the color values from the color multispectral image *(figure 4.16)*. This type of *pan-sharpened* image not only enhances visual interpretability but in some cases also improves digital classification accuracy. The pan-sharpening process can produce color imagery almost as detailed as multispectral imagery acquired at the higher resolution[4]. Image fusion techniques have also been used to combine images from different sensors, such as Landsat multispectral scanner data with radar images

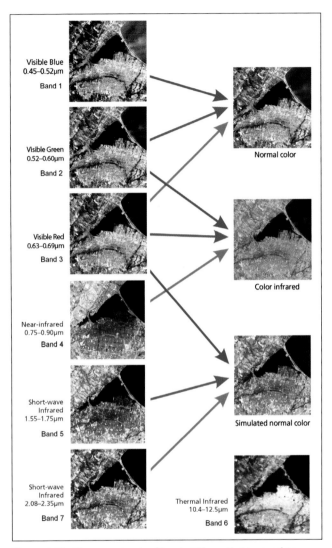

Figure 4.15   The seven bands of Landsat Thematic Mapper data. Normal, color-infrared, and simulated normal color images can be created by combining the images from different bands.

Source: Images © Natural Resources Canada.

(figure 8.13 in chapter 8), or to combine map data with imagery and geophysical data *(figure 4.17)*.

4. For features so small they approach the GSD of the coarser multispectral image but are clearly distinguished in the panchromatic one, there can be noticeable spread of the feature's color to areas beyond its edge. This limitation is well illustrated in Leberl et. al. 2002.

a   b   c

Figure 4.16　Image fusion. The greater detail in the higher spatial resolution 70 cm panchromatic QuickBird image of Tiananmen Square, Beijing (a) has been used to sharpen the 2.8 m resolution color multispectral image (b) to produce a pan-sharpened image containing information from both images (c).

Source: Image courtesy of DigitalGlobe.

Radiometric resolution

Radiometric resolution measures how sensitive a sensor is to differences in the energy, or brightness levels, it detects. The greater the sensitivity, the finer or higher the radiometric resolution is said to be. Increasing a remote sensing system's radiometric resolution can improve the discrimination of features that have similar spectral response patterns but slightly different brightness levels.

A photograph has a continuous range of colors or gray tones. In a digital image, the colors or gray tones are represented by discrete levels that are assigned numeric values. Typically, the lowest brightness level is the value 0, the next highest is 1, and so on. The more brightness levels the sensor can reliably distinguish, the higher its radiometric resolution is said to be.

For film images, radiometric resolution depends on the photochemical characteristics of the emulsion. The measures used are beyond the scope of this text. For digital images, the radiometric resolution is typically quoted as the number of data storage units, or bits, the sensor requires to capture the range of brightness values it can detect (see *Binary encoding*

of digital images later in this chapter). An image may actually be stored using more bits than the radiometric resolution of the data requires because of the standards used for storage and processing of digital data. For example, digital images are typically stored as either 8-bit or 16-bit files. So the data from a sensor that has a radiometric resolution of 7-bits is distributed as an 8-bit image file and 11-bit data would be distributed as a 16-bit file. The greater bit depth of the file doesn't change the radiometric resolution of the image. The file format simply allows more discrete levels to be stored than the data contains.

A single-band image is commonly encoded using eight bits. Using eight-bit encoding, 256 brightness levels ranging from black (0) to white (255) can be recorded. The human eye is able to distinguish about 30 to 40 different gray levels, although an image with more gray levels is perceived as having a smoother gradation of tone. A range of 256 levels are considered sufficient to represent the range of gray tones the human eye perceives. Color images are usually generated from a set of three eight-bit images displayed as blue, green, and red, which offers over 16 million possible combinations

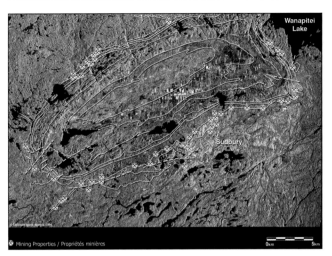

Wanapitei
Lake

Sudbury

Mining Properties / Propriétés minières

0km   5km

Figure 4.17   Fusion of radar and geophysical image data. This combined image has intensity values derived from the Radarsat image and hues derived from the vertical magnetic gradient geophysical data. The radar image shows the surface expressions of structures, while the geophysical data indicates the relative concentration of magnetic minerals. Significant geologic units and linear features have been annotated along with mining sites. The combined image assists in mapping geologic structures.

Source: © Natural Resources Canada.

(256 x 256 x 256). This range of colors and tones more than matches the range of colors and tones the human eye can distinguish.

As noted earlier, digital imaging systems use detectors that generate a voltage that varies with the intensity of the radiation they sense. The range of voltage values is divided into a number of levels, termed *quantization levels*, which are assigned integer values. The finer the divisions, the smaller the differences in radiance levels recorded by the imaging system and the more closely the recorded data will match the variations in the detector output. Thus, the radiometric resolution of the image data depends not only on the sensitivity of the detectors but also on the quantization levels at which the data was recorded. In figure 4.18, the continuously varying output signal from a remote sensing detector is shown recorded with one-bit quantization (producing two brightness levels) and two-bit quantization (producing four levels).

The larger the number of quantization levels, the higher the radiometric resolution of the data recorded. This increases the volume, storage, and processing time of the imagery. However, there is an advantage to recording data with a high radiometric resolution. High-radiometric-resolution data can capture subtle brightness differences that can be digitally enhanced to enable easier image interpretation. Where data is digitally analyzed, the system can take advantage of as large a range of brightness values as are available in the image data.

For example, the high-resolution IKONOS, OrbView-3, and QuickBird sensors use 11-bit encoding. Data products are provided as either 8-bit or 16-bit images. The advantage of using the 16-bit image is that detail is captured in areas that would appear pure white or pure black in the 8-bit image. In a brightly lit scene, features such as mountains and clouds can cast dark shadows. An 8-bit image may not have a large enough range of brightness values to be able to show detail in both the dark shadow areas and the bright areas within the image. So some of the subtle detail in the shadows may all be recorded as the darkest brightness level, whereas the 16-bit image may have enough brightness levels available to depict the dark shadow areas with as many different brightness levels as the detector was able to distinguish. If the additional detail is present in the image file, then enhancement techniques can be used to make that detail visible in a displayed image *(figure 4.19)*.

Earth resources satellites typically use 8-, 10-, or 11-bit encoding (256, 1024, or 2048 levels, respectively), whereas radar systems use 16-bit data. Figure 4.20 shows an image at different radiometric resolutions. As the number of gray levels decreases, the image contrast increases and image detail is lost.

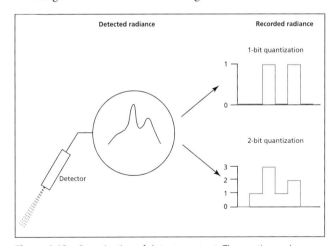

Figure 4.18   Quantization of detector output. The continuously varying output signal from a remote sensing detector is shown recorded with one-bit quantization (producing two brightness levels) and two-bit quantization (producing four levels).

Binary encoding of digital images

The number system we commonly use represents numbers as a string of digits, each of which can have one of ten values from zero through nine. This type of numbering system is referred to as *base ten*. The number of values that can be represented by a series of digits is x^n, where x is the number of values each digit can represent and n is the number of digits used. For example, one digit in a base ten system can represent ten values (10^1); two digits can represent one hundred values (10^2); three digits can represent one thousand values (10^3).

The operation of a digital computer is based on the principal of a switch that can be on or off. This is represented numerically as a one or zero, which is why in a computer numbers are stored as a string of ones and zeros. This type of numbering system is referred to as a *base two* numbering system, also called a *binary system*. For example, the numbers 0, 1, 2, 3 are represented as the binary values 00, 01, 10, 11, respectively.

Each digit in a binary number is termed a *bit*. The more bits, the greater the number of values that can be represented. Table 4.6 shows the number of values that can be represented with different numbers of bits. The greater the number of bits used to encode a digital image, the greater the range of brightness levels that can be recorded; however, this also results in more data transmission, storage, and processing.

Table 4.7 illustrates how a binary value can be converted into the base ten values we commonly use.

Table 4.6 Number of values represented by binary numbers of different bit lengths	
Bits	**Number of values**
1	2
2	4
3	8
4	16
5	32
6	64
7	128
8	256
9	512
10	1,024
11	2,048
12	4,096
13	8,192
14	16,384
15	32,768
16	65,536

Table 4.7 Converting a seven-bit binary value to a base ten value							
7	**6**	**5**	**4**	**3**	**2**	**1**	**Bit position**
1	0	1	0	1	1	1	Binary value
$2^6=64$	$2^5=32$	$2^4=16$	$2^3=8$	$2^2=4$	$2^1=2$	$2^0=1$	Value of a binary 1 value in each bit position
(1 x 64) +	(0 x 32) +	(1 x 16) +	(0 x 8) +	(1 x 4) +	(1 x 2) +	(1 x 1) +	= 87 Base ten value

a

b

Figure 4.19   IKONOS 11-bit data enables detail to be captured in areas within the cloud shadow. In the 8-bit image (a), there is no detail in the black cloud shadow area. Image (b) shows improved shadow detail after enhancement using the full 11 bits of data.

Source: Image courtesy of Space Imaging Corporation.

Data volumes

In creating digital images, the choice of spatial resolution, spectral resolution, and radiometric resolution determines the quantity of information the image records and consequently the size of data file needed to store the image. Increasing resolution can dramatically increase data volumes.

For example, a medium resolution sensor with a 20 m GSD spatial resolution would record a 60 km by 60 km scene as a 3,000 pixel by 3,000 pixel image, a total of 9 million pixels. If each band is recorded as an 8-bit image, 8 bits being 1 byte, the image file would be about 9 megabytes (MB) per band[5]. Doubling the resolution would quadruple the file size. For example, the scene described above recorded with double the spatial resolution, i.e., with 10 m pixels, would generate a 6,000 by 6,000 pixel image with a file size of about 36 MB. A high-resolution image from the IKONOS sensor covers an area of 11 km^2 with 1 m pixels. This produces a single-band image containing 121 million pixels (1,100 x 1,100). The imagery is available with a radiometric resolution of 8-bits, which is 1 byte of data per pixel, or about 121 MB per image. The higher radiometric resolution 11-bit data must be stored as a 16-bit image, which produces 2 bytes of data per pixel. This doubles the file size to over 240 MB.

Data storage is not the only concern in the design and operation of a sensor. In the case of satellite-based systems, the data must also be transmitted to ground stations fast enough to keep up with image acquisition. To optimize system performance, compromises are made. Some satellites, such as IKONOS, offer high spatial resolution with a small number of bands and limited coverage. Others, such as the NOAA AVHRR, offer broad coverage at low spatial resolution with multiple bands and a high radiometric resolution.

Temporal resolution

Temporal resolution is a measure of how frequently a sensor system obtains imagery of the same area, also called the *revisit interval*. The more frequent the coverage, the shorter the revisit time and the higher the temporal resolution.

The analysis of multitemporal imagery, i.e., imagery of the same area acquired at different times, is an important remote sensing analysis technique. Earth observation satellites have been systematically collecting imagery of the earth's surface for decades. By collecting images on a continuous basis, changes that take place on the earth's surface can be monitored over time. These changes include natural processes such as plant development over the growing season and the

5. In the binary numbering system used by computers, a megabyte is actually 2 to the 20th power which equals 1,048,576 bytes. So a million bytes is actually less than a megabyte.

a

b

c

Figure 4.20 Comparison of radiometric resolutions. The same image is displayed with (a) 2, (b) 16, and (c) 256 gray levels, which corresponds to 1-, 4-, and 8-bit radiometric resolutions. Since the human eye does not reliably distinguish more than about 30 gray levels, there is little apparent loss of image quality in the 4-bit image.

Source: © Natural Resources Canada.

flooding of rivers, as well as human-related activity such as urban development and deforestation.

Applications of different temporal resolutions

High temporal resolution is important for monitoring events that change rapidly. Monitoring crop development through the growing season commonly requires weekly imagery, whereas rapid changes in water quality, such as an algal bloom, requires daily imagery or better. Also, for areas with almost constant cloud cover, such as tropical rainforests, the ability to collect frequent imagery increases the likelihood of obtaining cloud-free coverage of an area.

Figure 4.21 illustrates the advantage of using high-temporal-resolution imagery to identify and record the size and timing of an algal bloom event that might have been missed on imagery collected less frequently. In the figure, the rapid decay of an algal bloom over four days is monitored using

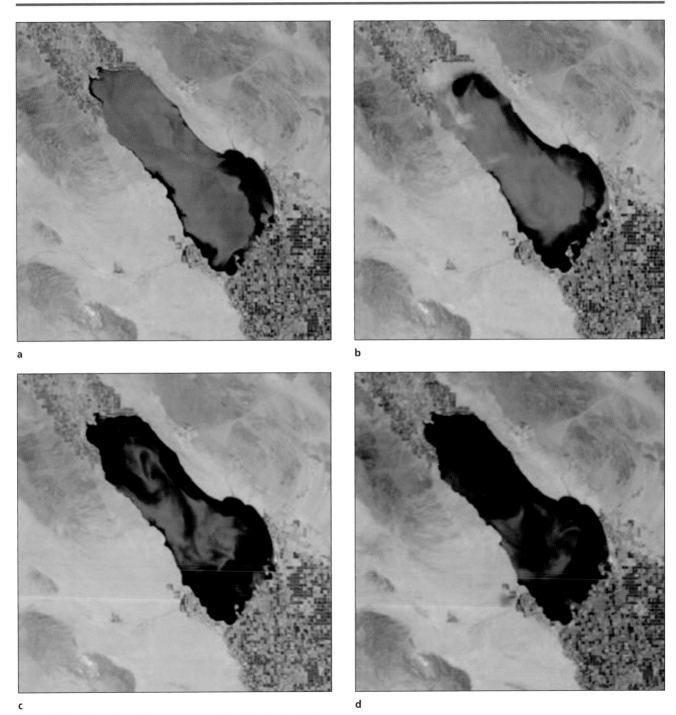

a

b

c

d

Figure 4.21 Salton Sea, California. June 11–16, 2003. Daily satellite imagery from the NASA MODIS sensor monitors the decay of an algal bloom in the Salton Sea.

Source: MODIS Ocean Group NASA/GSFC, SST product by R. Evans et al., U. Miami.

daily satellite imagery. The decay was caused by the agricultural runoff that supplies 90% of the inflow to the Salton Sea, the largest lake in California. This nutrient-rich water regularly triggers rapid growth of algae, which produces large algal blooms that cover the lake and then suddenly die. Their decay consumes so much of the oxygen in the lake that large numbers of fish are killed and water quality declines sharply.

Figure 4.22 shows radar images of the Red River flood in Manitoba during the spring of 1996. Three radar images, acquired on March 23, 1996 (before the flood), April 25, 1996 (when the flood was at its peak), and May 9, 1996 (when water levels had partially subsided), document the flood event. The three images were combined as a multitemporal color composite by displaying them as blue, green, and red, respectively. The color composite was then analyzed to distinguish areas subjected to flooding at different times and of different severity. Areas flooded on both dates (A) appear dark blue. Red features on the left side of the image appear flooded only in the April image (C). The bluish green area north of the town of Morris (D) was flooded only on the May date. The yellow rectangular area (E) is the town of Morris, which was protected by a levee. This imagery was valuable in directing relief efforts and in planning for future flood events.

Multitemporal comparisons from year to year are useful in assessing crop condition relative to previous years or an average of past growing seasons. These analyses can predict shortfalls in crop yield well before the harvest or warn of the spread of plant disease in cropland or forests (see the crop condition assessment examples in chapter 12). The changing appearance of features over time can also be useful in identification. Many crops that appear the same on single-date imagery differ in the times they are planted, when they flower, or when they are harvested. By comparing images collected on multiple dates, differences in the timing of development and cultivation can be used to accurately distinguish similar crop types such as wheat and maize.

Temporal resolution in remote sensing systems

Airborne remote sensing systems can be flown as often as weather permits, so it is possible to collect images of an area frequently, and because aircraft can fly low, the imagery can have high spatial resolution. For example, the spread of an oil spill could be tracked this way if the location is within aircraft range. However, multiple flights are expensive. In addition, the flight path of the aircraft for each mission will not be as consistent as the orbit path of a satellite. So it is

generally more cost-effective to use satellite data for repeated observations if the spatial resolution and repeat period are suitable.

An example of a system that provides high temporal resolution is the AVHRR sensor on the NOAA series of satellites. This system has a revisit interval of 12 hours, which provides daytime and nighttime coverage of the entire globe each day. AVHRR data has a spatial resolution of 1 km, which is lower than the NASA Landsat satellite (30 m). Thus, it takes more than 1,100 Landsat pixels to cover the same ground area as one AVHRR pixel. However, the higher temporal resolution and lower spatial resolution makes AVHRR data valuable for continuous crop condition monitoring over large areas because daily imagery can be obtained during the growing season.

Earth observation satellites imaging in the visible and near-infrared are generally placed in either *geosynchronous* or *sun-synchronous* orbits to optimize imaging of the sunlit portion of the globe. Geosynchronous orbits are about 36,000 km above the earth's equator in a west to east direction. At this altitude they can circle the globe at the same rate and in the same direction as the earth's rotation so as to remain fixed over the same earth location (*figure 4.23*).

Meteorological and communication satellites are commonly placed in geosynchronous orbits to provide constant operation over a region. Because geosynchronous communication satellites remain in a fixed location relative to the earth, antennas at ground stations can be oriented in a fixed position and maintain contact. An international network of meteorological satellites (including the GOES and METEOSAT satellites discussed in chapter 7) positioned at intervals above the equator provides a constant flow of earth images used for weather forecasting. With a pixel resolution of 8 km, the satellites provide images showing the full Earth disc every hour and more frequent observations of smaller areas.

Higher resolution satellite systems, such as AVHRR, Landsat, and SPOT are placed in sun-synchronous, near-polar orbits at lower altitudes (*figure 4.24*). Landsat operates at about 700 km above the earth. The orbits are selected so the satellite passes from north to south at a slight westward inclination to collect images of the terrain below at about the same local time. This ensures the illumination of the imagery is as consistent as possible. However, the sun illumination angle still varies with the seasons.

In order to cross the equator at precisely the same local time during every pass, successive orbital paths must be widely separated (*figure 4.25*). For example, Landsat 5 crosses the equator at 9:45 AM local time with an orbital

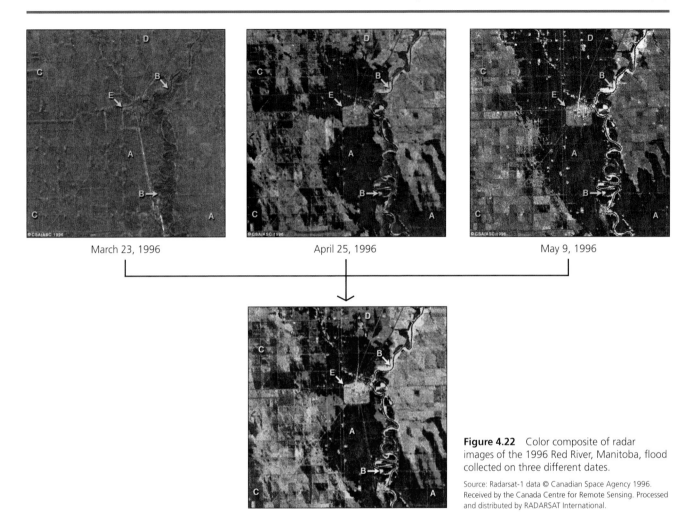

March 23, 1996   April 25, 1996   May 9, 1996

Figure 4.22   Color composite of radar images of the 1996 Red River, Manitoba, flood collected on three different dates.

Source: Radarsat-1 data © Canadian Space Agency 1996. Received by the Canada Centre for Remote Sensing. Processed and distributed by RADARSAT International.

period (the time to complete an orbit) of 99 minutes. The satellite continuously images a 185 km swath of the landscape directly below with successive orbital paths 2,752 km apart. The adjacent swath is imaged seven days later with complete coverage every 16 days. Every location is imaged at least once every 16-day orbital cycle. Areas that lie in the overlap region of adjacent swaths (side-lap areas) are imaged more frequently. Since the orbital paths converge towards the poles, side lap increases with latitude from about 15% at 0° to 85% at 80°. So at high latitudes, some locations will be imaged five times during the 16-day orbital cycle (*figure 4.26*). Thus, the temporal resolution for those areas would be higher than the satellite's revisit interval.

Some satellites, such as SPOT, IKONOS, and QuickBird, have pointable sensors that enable the collection of vertical imagery, as well as imagery viewing to the side, front, or back

of the flight path. In this way, priority targets can be revisited more frequently.

Accuracy of image products

Image resolution characteristics should be chosen to optimize the accuracy of the information that will ultimately be produced. The term *accuracy* refers to the maximum error to be expected in the values of a dataset. In remote sensing, the accuracy characteristics typically of concern are the following:

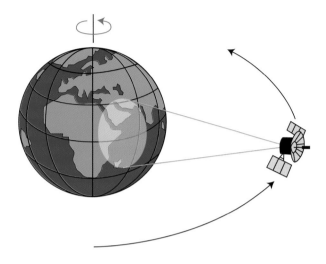

Figure 4.23   Geosynchronous orbit.

Source: © Natural Resources Canada.

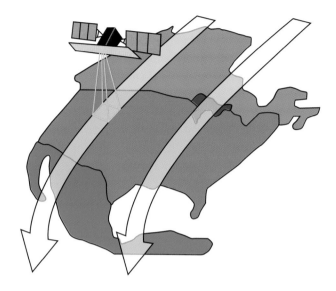

Figure 4.25   Successive orbit paths of sun-synchronous satellites such as Landsat are widely separated.

Source: © Natural Resources Canada.

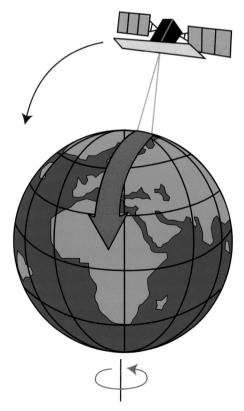

Figure 4.24   Sun-synchronous orbit.

Source: © Natural Resources Canada.

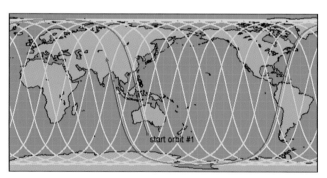

Figure 4.26   The side lap between adjacent swaths is greater near the poles for imaging satellites in sun-synchronous orbits like SPOT and Landsat.

Source: © Natural Resources Canada.

- *positional accuracy*—measurements of the location and size of features identified in an image
- *classification accuracy* (also called *thematic* or *attribute accuracy*)—the accuracy with which a feature is identified as belonging to a specific type or class such as Forest or Water. Classification accuracy is discussed in chapter 10.

The accuracy of a dataset is estimated by comparing a sample of the data with reference information of higher accuracy and considered correct. An accuracy assessment is thus a statistical measure of the maximum amount of error that is expected to occur. It has two components, the magnitude of the error that is calculated from the sample and the probability (referred to as the *confidence level*) that when the data is used it will actually have an accuracy equal to or better than this magnitude. The confidence level is chosen for the assessment. Values of 90% or 95% are commonly used.

For example, the positional accuracy of a dataset might show that measurements of the geographic position of features in an image are within 10 cm of their correct ground position 95% of the time. The classification accuracy of a dataset might be assessed to be that 80% of the features in an image are expected to be correctly classified 95% of the time. It is important to state the confidence level when quoting accuracy because, from the same accuracy assessment, a higher accuracy value can be produced by choosing a lower assessment of confidence level (see *Assessing positional accuracy* later in this chapter).

There is a substantial literature on the measurement of classification accuracy. Classification accuracy is discussed in greater detail in chapter 10, *Visual interpretation*, and Congalton (1991) provides a good review. Positional accuracy is addressed in detail in photogrammetry texts and is discussed below.

Positional accuracy is an important factor in determining an image product's suitability for use. The level of required positional accuracy varies with the application. For some applications, such as meteorology, positional accuracies on the order of kilometers may be tolerated in measuring cloud-cover extent. Imagery for regional-scale mapping may tolerate horizontal accuracies in the tens of meters, whereas the large-scale mapping for urban infrastructure and engineering applications typically require accuracies better than a meter. However, when an image is being used only to show context rather than to obtain accurate measurements, a lower accuracy may be sufficient.

Positional accuracy in aerial photographs

Many factors affect the horizontal and vertical accuracies of the products obtained from the photogrammetric analysis of stereo imagery. These factors include the scale of the imagery (a function of the flying height of the platform and the focal length of the lens), the ground resolution of the imagery (GSD for digital imagery, GRD for film), base-height ratio (the degree of separation between the two images of a stereo pair relative to the flying height), the accuracy with which ground control points are measured, and the performance characteristics of the photogrammetric instruments used.

There is a long tradition of making maps from aerial photography, and photogrammetric procedures for using this type of imagery are well defined. The relationships between the scale and resolution of aerial photography used to produce maps of a given scale and the expected map accuracy are consistent and well understood.

In the United States, the photogrammetric aerial camera with 9-inch film format and a 6- or 12-inch lens has been the standard for mapping applications. Many mapping standards were developed based on this technology and are used worldwide. For example, the U.S. National Map Accuracy Standard (NMAS), last revised in 1947, is widely used in the United States. It specifies that for map scales of 1:20,000 or smaller, position errors must be less than 1/50th of an inch for 90% or more of well-defined test points and 1/30th of an inch for larger scales. Figure 4.27 shows the maximum allowable error for different map scales according to this standard.

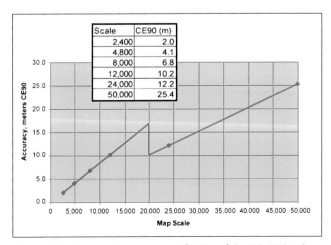

Figure 4.27   Horizontal accuracy specification of the U.S. National Map Accuracy Standard.

Source: Image courtesy of Space Imaging Corporation.

Other standards in common use in North America include those of the U.S. Geological Survey, the American Society of Photogrammetry and Remote Sensing, Natural Resources Canada, Departments of Transportation in the United States and Canada, and various state and provincial government specifications. These standards are very similar to the NMAS. In general they require that the map or orthophoto product generated from imagery have a horizontal positional accuracy such that 90% of well-defined features be within 0.5 mm of their true position on the map. The ground distance will depend on the scale of the map. For example, for a 1:50,000 scale map, 0.5 mm would represent 25 m (0.5 x 50,000) on the ground, i.e., at ground scale.

Similarly, the vertical accuracy requirement is that 90% of all contours be within half the contour interval for open areas and the elevation accuracy for spot elevations be within one fourth of the contour interval. For example, a map with contours at 10 m intervals would be expected to have 90% of the contours within ±5 m of their true elevation and 90% of spot elevations to be within ±2.5 m. A rule of thumb in the aerial survey industry is that to obtain 1 m contours within ±0.5 m of their true position at the 90% confidence level, 1:10,000 scale photography is used for analog photogrammetric instruments or 1:12,000 scale photography for analytical instruments to generate a map at 1:2,000 scale. Other typical scales and contours are given in table 4.8.

Positional accuracy in digital imagery

Digital imagery and satellite photogrammetry have introduced additional variables that affect the final geometric accuracy of orthorectified images and derived map products. As discussed previously, the spatial resolution of a digital image is expressed by the ground sample distance (GSD), the distance across the area on the ground represented by one pixel. However, images with the same GSD may have very different positional accuracies. Positional accuracy depends on instrument calibration, accuracy of the orbital position

(precise location of the satellite within its orbit at a specific time) and the direction the sensor is pointing, the quality and quantity of ground control points (GCPs)[6], and the methods chosen to use this data in rectifying the imagery. For example, the use of more ground control points can correct for errors in satellite position and pointing direction. Imagery can be orthorectified without GCPs; however, doing so may result in relatively large bias errors (i.e., constant error term, see below) that then become the dominant source of position error. Rectification methods are addressed in appendix A, *Rectification and georeferencing of optical imagery*.

Digital images are commonly displayed on a monitor and can be zoomed to any scale. It's difficult to judge what scale of imagery might be suitable for a specific application. A useful measure for comparing the visual interpretability of digital imagery is the scale that results when an image is printed at the visual threshold (i.e., 300 image pixels per inch, as shown in table 4.9). This scale value provides a rough guide to the scale of mapping for which an image might be suited.

In 1998, the U.S. Federal Geographic Data Committee proposed a National Standard for Spatial Data Accuracy (NSSDA) designed to take into account the characteristics of digital geospatial data and the technologies used to produce and analyze them. The NSSDA sets a 95% confidence level for horizontal and vertical positional accuracy instead of the 90% level used in many earlier specifications such as the NMAS. In addition, these standards take into account the characteristics of all types of geospatial data including digital imagery. Where the NMAS specifies the allowable position error as a distance at map scale, the NSSDA leaves the specification of allowable error to the discretion of the agency responsible for generating the data product. In practice, an allowable circular position error of 0.5 mm at map scale is commonly used.

An important distinction in accuracy assessment is whether a product was tested to verify that it met an accuracy standard as opposed to being produced according to established procedures designed to achieve a stated accuracy level. The former involves direct evaluation of the product, whereas the latter does not. Testing typically requires that accurate measurements (e.g., by ground survey) be obtained for about 20 well-distributed points. It is expensive but provides a

| Table 4.8 Contour intervals commonly used for different scales of mapping ||
Scale	Contour intervals (m)
1:1,000	0.5
1:2,000	1
1:10,000	5
1:20,000 to 1:25,000	10
1:50,000	20
1:100,000	50
1:250,000 and smaller	100 is common

6. Ground control points are features for which an accurate ground position is known and that are clearly visible in an image. They are used by software to rectify image geometry.

Table 4.9 Approximate scale factors for images of different ground sample distances (GSD) when printed at 300 image pixels per inch

GSD (m)	Scale factor[1]
0.61	7,320
0.7	8,400
1	12,000
1.5	18,000
2	24,000
2.5	30,000
5	60,000
10	120,000
15	180,000
20	240,000
25	300,000
30	360,000
250	3,000,000
500	6,000,000
1000	12,000,000
5000	60,000,000

[1] Scale factor = 300 pixels/inch x (39.37 in /m) x GSD = 11,811 GSD (in m) or approximately 12,000 x GSD (in m)

temporal resolution available. Faster, less-expensive computers and more efficient image compression and indexing have enabled smooth, continuous roam and zoom to be implemented over multiple images at multiple scales *(figure 4.28)*. With this capability, a GIS analyst can select an area by roaming over a continuous image mosaic of the globe and then zoom to the desired scale, smoothly passing through images of different spatial resolutions obtained from different sensors. Instead of the display being a view of a single image, it becomes a window into a database of images that can be navigated spatially by panning across the area, zooming in or out of the images, and switching to higher or lower resolution images as needed. This concept can be extended to the spectral dimension as well, allowing the analyst to cycle through different band combinations to find a spectral combination optimal for a specific task.

Classification analyses undertaken by remote sensing specialists typically require the use of more specialized image analysis tools than those generally provided in a GIS. Aerial

rigorous measure of the accuracy of the product. Remote sensing products do not necessarily conform to NMAS standards. However, if a product was actually tested and not just produced according to procedures that should achieve the standard, a user can be more confident that the product will indeed meet the standard.

The process of determining the accuracy level required for a given application is perhaps as important as the specification eventually chosen. The choice of an accuracy level is fundamentally a risk-assessment exercise. The accuracy level specified should meet the minimum standards mandated by the organization or client for which the information is being produced. However, it may sometimes be prudent to specify a higher level of accuracy at greater expense if doing so would significantly reduce the risk of potentially costly errors. Risk-assessment techniques can be used to evaluate these trade-offs, but the specification of an appropriate accuracy level requires judgment and experience as well.

Imagery in geographic information systems

A geospatial information system, such as a GIS, may support a wide range of applications that differ in the required level of detail. Consequently, it may be appropriate to have imagery with different levels of spatial, spectral, radiometric, and

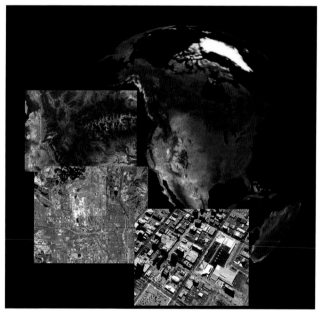

Figure 4.28 Interactive display of multiple image–multiple resolution image database of Salt Lake City, Utah. ArcGlobe™ software uses sophisticated image compression and indexing to offer continuous roam and zoom through very large image mosaics at multiple scales. From the earth globe generated from AVHRR (1 km GSD), the user can continuously zoom through Landsat (30 m multispectral data pan-sharpened to 15 m) and QuickBird (multispectral data pan-sharpened to 0.61 m) satellite imagery.

Source: Image courtesy of DigitalGlobe.

survey firms and mapping agencies generate a range of products for GIS applications from remotely sensed imagery. These products typically make use of high-precision photogrammetric instruments to produce vector line map products as well as orthorectified imagery. The photogrammetric procedures used involve the accurate identification and measurement of point, line, and polygon data and corresponding feature attributes, and significant investment in field verification to minimize errors and omissions. Generating significant quantities of vector data from imagery within a GIS environment is generally less cost-effective than using photogrammetric tools designed for this purpose.

Conclusion

Differences in the spatial, spectral, radiometric, and temporal resolution of remotely sensed imagery affect their suitability for a specific application and the methods used to analyze the data. In addition, accurate georeferencing of the imagery is important to ensure that the information derived from the imagery is accurately registered with any other geospatial data with which it will be used. The growing capability of geographic information systems to make use of remotely sensed imagery has provided GIS users with direct access to this data, both as backgrounds over which to display other geospatial information and as a source of information to update existing data, assess changes at a location over time, and other applications. However, more complex remote sensing analyses are generally done more cost-effectively using photogrammetric or image analysis systems purpose-built for these applications.

Assessment of positional accuracy

Spatial or positional accuracy of maps and rectified image products such as orthophotos are typically assessed by selecting a number of well-defined points (generally 20 or more) and comparing the position coordinates as measured on the product to their true or reference position. These reference positions are obtained from an independent source of higher accuracy (from three to ten times higher accuracy, depending on the practitioner) such as a higher accuracy map or by ground survey measurement.

To test the accuracy of products generated from large-scale aerial photography, ground survey points are typically used. Published maps are commonly used in assessing the accuracy of satellite imagery. It is important to verify whether the map itself is of sufficient accuracy to serve as a suitable reference of higher accuracy. Standard cartographic methods allow certain features to be displaced, which sacrifices positional accuracy for clarity of presentation. For example, roads and railways along a stream are commonly moved to avoid overprinting the stream course. Such adjustments must be identified to ensure the coordinates of the affected features are not used to assess accuracy.

The statistic used to assess positional accuracy is the root mean square error (RMSE) which is the square root of the average of the squared discrepancies between the measured and reference positions. (The RMSE is normally defined in terms of the errors at ground scale, not map scale.)

$$RMSE = \sqrt{\frac{\sum\limits_{i=1}^{n} d_i^2}{n}}$$

Where

d = the distance between the test and true point

i = the ith test point

n = the number of test points

Horizontal accuracy may be measured by the radial distance, or circular error (CE), which is the distance between the test point and its true position. It is calculated from a pair of X- and Y- coordinates as follows:

$CE = sqrt\ [(X' - X)^2 + (Y' - Y)^2]$

Where

X' and Y' are the test point values

X and Y are the reference values considered to be correct

The horizontal accuracy is then stated as a probability or confidence level based on the RMSE value. For example, a circular error of 10 m with a 90% probability, abbreviated as

10 m CE90, indicates that there is a 90% probability that the position of a distinct feature as measured in the image or map will be within 10 meters of its true position. The horizontal accuracy can also be stated as 10 m at the 90% confidence level.

Alternatively, horizontal accuracies of the X and Y components of horizontal position may be calculated separately, in which case the error values of a test point would be calculated as follows:

$X_e = X' - X$ and $Y_e = Y' - Y$

Where

X and Y are the reference values

X' and Y' are the test point values

X_e and Y_e are the errors in X and Y direction, respectively

Since the vertical error (Z_e) is one dimensional, or linear, the discrepancy in elevation is a simple subtraction calculated as follows:

$Z_e = Z' - Z$

Where

Z' is the test point value

Z is the reference value considered to be correct.

Position errors can potentially include the effects of both bias and random errors. Bias errors are systematic—they affect all points in a similar manner. For example, a bias error might cause the coordinates of points in the rectified image to all be shifted 5 meters north of their true position. Bias errors are normally detected and corrected during the image rectification process so that bias error is generally close to zero. Random errors are deviations that differ in magnitude and direction for each point in a random manner. Random errors can be reduced by using more accurate instruments and more rigorous procedures but cannot be eliminated. Random errors contribute to the measured error level.

Calculation of position accuracy at the 90% and 95% confidence levels

Circular error values are calculated as shown below[1]. Note that to meet NSSDA standards, a 95% confidence level is required. The 90% confidence values are given for use in meeting other standards such as the U.S. National Map

---

1. Equations are taken from appendix 3-A of the Geospatial Positioning Accuracy Standards Part 3: National Standards for Spatial Data Accuracy published in 1998 by the Federal Geographic Data Committee.

Assessment of positional accuracy (cont.)

Accuracy Standard.

For horizontal accuracy assessments the circular error values are calculated as follows:

$$CE90 = 1.5175 * RMSE_r$$
$$CE95 = 1.7308 * RMSE_r$$

Where $RMSE_r$ is the root mean square error of the radial distance discrepancies.

Using separate measurements of root mean squared error in the X direction ($RMSE_x$) and in the Y direction ($RMSE_y$), these equations would be:

$$CE90 = 1.5175 * SQRT((RMSE_x)^2 + (RMSE_y)^2)$$
$$CE90 = 1.7308 * SQRT((RMSE_x)^2 + (RMSE_y)^2)$$

If $RMSE_x$ is equal to $RMSE_y$, and if the errors are independent and normally distributed, then the circular error can be calculated from either the $RMSE_x$ or $RMSE_y$ values as follows:

$$CE90 = 2.1460 * RMSE_x = 2.1460 * RMSE_y$$
$$CE95 = 2.4477 * RMSE_x = 2.4477 * RMSE_y$$

Vertical accuracy is assessed in a similar manner except that, since elevation is measured in only one dimension, the discrepancy in elevation is a simple subtraction, $Z_e = Z'–Z$, and the error is a linear error (LE).

Vertical accuracy at the 90% probability level is LE90 = 1.6449 * $RMSE_z$ and vertical accuracy at the 95% probability level is LE95 = 1.96 * $RMSE_z$.

Tables 4.10 and 4.11 illustrate the calculation of horizontal and vertical accuracy from the root mean square error.

From the example, the horizontal accuracy would be 2.92 m CE90, meaning horizontal positions, as determined from the map, would be expected to be within 2.92 m of their true position 90% of the time. Similarly, the vertical accuracy would be 2.70 m LE90, meaning elevation values obtained from the map would be expected to be within ±2.70 m of the true elevation 90% of the time.

Table 4.12 illustrates the $RMSE_r$ values that would be required to provide a horizontal accuracy of 0.5 mm at map scale at both the 90% and 95% confidence levels for different scales of mapping. Note that to achieve the same stated accuracy, the limiting $RMSE_r$ value (i.e., the maximum allowable $RMSE_r$ value) is lower to achieve the more stringent 95% confidence level.

The RMSE includes both random and systematic error components. If there is no obvious bias and a reasonably large set of test points (i.e., more than 30 points), the RMSE will be very close to the standard deviation. However, to convert RMSE values to CE and LE values, the assumption must be made that the bias is zero and the errors are normally distributed.

The use of RMSE as a measure of accuracy assumes that the bias in the measurements, or the sum of the deviations from the reference values, is approximately zero. Photogrammetric procedures are designed to achieve very low bias errors.

Table 4.10 Calculation of root mean square error (RMSE) for horizontal error assessment								
Horizontal accuracy is assessed by circular error (CE) = Sqrt((X' - X)² + (Y' - Y)²) Number of test points (N) = 10								
Horizontal accuracy assessment: Calculated from circular error values						Horizontal accuracy assessment calculated from separate X and Y deviation measures		
Measured value		Reference value		CE	CE²			
Point	X'	Y'	X	Y			X' - X	Y' - Y
1	10	100	11	99	1.414213562	2		-1.00	1.00
2	20	200	19	198	2.236067977	5		1.00	2.00
3	30	300	31	299	1.414213562	2		-1.00	1.00
4	40	400	39	401	1.414213562	2		1.00	-1.00
5	50	500	51	502	2.236067977	5		-1.00	-2.00
6	60	600	58	601	2.236067977	5		2.00	-1.00
7	70	700	69	702	2.236067977	5		1.00	-2.00
8	80	800	81	801	1.414213562	2		-1.00	-1.00
9	90	900	92	900	2	4		-2.00	0.00
10	100	1,000	99	998	2.236067977	5		1.00	2.00
Sum of CE² =						37	Mean X and Mean Y	0.00	-0.10
Mean CE² = Sum of CE²/N						3.7	RMSE_x and RMSE_y	1.26	1.45
RMSEr = Sqrt(Mean CE²)						1.9	RMSE_r = SQRT((RMSE_x)² + (RMSE_y)²)	1.92	
CE90 = RMSEr * 1.5175						2.92	CE90 = RMSE_r * 1.5175	2.92	
Horizontal accuracy is 2.92 m CE90							**Horizontal accuracy is 2.92 m CE90**		

Assessment of positional accuracy (cont.)

Table 4.11   Calculation of vertical accuracy

Vertical accuracy is assessed by the elevation difference $Z_e = Z' - Z$
Number of test points (N) = 10

Vertical accuracy assessment: Linear error

Measured	Reference		
Z'	Z	$Z_e = (Z' - Z)$	$(Z' - Z)^2$
253	254	-1	1
233	231	2	4
195	197	-2	4
220	222	-2	4
216	216	0	0
203	202	1	1
245	244	1	1
233	231	2	4
226	228	-2	4
185	187	-2	4
Sum of LE^2 =		27	
Sum of Z_e / N		2.70	
RMSE$_z$ = sqrt((Sum of Z_e) / N)		1.64	
LE90 = RMSE$_z$ * 1.6449		2.70	

Vertical accuracy is 2.70 m LE90

Table 4.12   Comparison of RMSE$_r$ required to meet 90% and 95% confidence levels for different map scales

Assuming the allowable error at map scale is 0.5 mm

Map scale	Allowable error at map scale (mm)	Allowable error at ground scale	Limiting RMSE$_r$ for CE90 (m)	Limiting RMSE$_r$ for CE95 (m)
1:5,000	0.5	2500	1.65	1.44
1:10,000	0.5	5000	3.29	2.89
1:15,000	0.5	7500	4.94	4.33
1:20,000	0.5	10000	6.59	5.78
1:25,000	0.5	12500	8.24	7.22
1:30,000	0.5	15000	9.88	8.67
1:35,000	0.5	17500	11.53	10.11
1:40,000	0.5	20000	13.18	11.56

Where: RMSE$_r$ is the root mean square error of the radial distance discrepancies
CE90 is the circular error at the 90% confidence level
CE95 is the circular error at the 95% confidence level
Limiting RMSE$_r$ to achieve 90% confidence level = Error at Ground Scale/1.5175
Limiting RMSE$_r$ to achieve 95% confidence level = Error at Ground Scale/1.7308
Allowable error at ground scale = allowable error at map scale / map scale

References and further reading

Campbell, J. B. 2002. *Introduction to remote sensing.* 3rd ed. New York: The Guilford Press.

Colwell, R. N., ed. 1983. *Manual of remote sensing.* Falls Church, Va.: American Society of Photogrammetry.

Comer, R. P., G. Kinn, D. Light, and C. Mondello. Talking digital. *Photogrammetric Engineering and Remote Sensing* (Dec. 1998): 1139–42.

Congalton, R. G. 1991. A review of assessing the accuracy of classifications of remotely sensed data. *Photogrammetric Engineering and Remote Sensing* 37:35–46.

Escobar, D. E., J. H. Everitt, M. R. Davis, R. S. Fletcher, and C. Yang. 2002. Relationship between plant spectral reflectances and their tonal responses on aerial photographs. *Geocarto International* 17(2): 63–74.

Florini, A. M. 1988. The opening skies: Third-party imaging satellites and U.S. security. *International Security* 13(2): 91–123.

Hothem, D., J. M. Irvine, E. Mohr, and K. B. Buckley. 1996. Quantifying image interpretability for civil users. In *Proceedings of the ASPRS/ACSM annual convention and exhibition.* Baltimore, Md., April 22–24. Bethesda, Md.: ASPRS.

Jensen, J. R. 1995. *Introductory digital image processing: A remote sensing perspective.* 2nd ed. Upper Saddle River, N.J.: Simon and Schuster.

Leachtenauer, J. C. 1996. National imagery interpretability rating scales overview and product description. In *Proceedings of the ASPRS/ACSM annual convention and exhibition.* Baltimore, Md., April 22–24. Bethesda, Md.: ASPRS.

Leberl F., R. Perko, and M. Gruber. 2002. Color in photogrammetric remote sensing. In *Proceedings of the ISPRS commission VII symposium.* Hyderabad, India. ISPRS Archives, Vol. 34, Part 7. www.icg.tu-graz.ac.at/~perko/Publications/2002_ISPRS7/leberl02color.pdf

Light, D. L. 2001. An airborne direct digital imaging system. *Photogrammetric Engineering and Remote Sensing* 67(11): 1299–1305.

———. 1993. The national aerial photography program as a geographic information system resource. *Photogrammetric Engineering and Remote Sensing* 59(1): 61–66.

———. 1990. Characteristics of remote sensors for mapping and earth science applications. *Photogrammetric Engineering and Remote Sensing* 56(12): 1613–23.

Lillesand, T. M., and R. W. Kiefer. 2000. *Remote sensing and image interpretation.* New York: John Wiley and Sons, Inc.

Munechika, C. K., J. S. Warnick, C. Salvaggio, and J. R. Schott. 1993. Resolution enhancement of multispectral image data to improve classification accuracy. *Photogrammetric Engineering and Remote Sensing* 59(1): 67–72.

Wilkie, D. S., and J. T. Finn. 1996. *Remote sensing imagery for natural resources monitoring: A guide for first-time users.* New York: Columbia University Press.

Web sites   The Federation of American Scientists.
www.fas.org

Information on image interpretability, including the National Image Interpretability Rating Scale, can be found at the Federation of American Scientists site.
www.fas.org/irp/imint/docs/index.html

The civilian version of the National Image Interpretability Rating Scale.
www.fas.org/irp/imint/niirs_c/guide.htm

GlobalSecurity.org.
www.globalsecurity.org

The JPL Planetary Photojournal.
photojournal.jpl.nasa.gov

NASA's Visible Earth Image Gallery.
visibleearth.nasa.gov

5 Frame-capture sensors: Photographic cameras, digital cameras, and videography

Stan Aronoff and Gordon Petrie

Remote sensing systems that operate in the visible (0.4–0.7 µm), and near-infrared (0.7–0.9 µm) portions of the electromagnetic spectrum use glass lenses to focus the detected energy and capture an image. They can be broadly grouped into frame-capture and scanning systems.

Frame-capture systems, such as photographic cameras, collect images as separate scenes recorded at one instant in time. To record an image of an area, a series of images are captured with overlapping coverage *(figure 5.1a)*. The individual frames can be assembled and stitched together to create one large image of the area, called an *image mosaic*. These images may be recorded on film, as digital images, or as analog or digital video recordings.

Scanning systems progressively build up a continuous image along the flight line by recording a series of narrow contiguous strips perpendicular to the flight path while the instrument moves over the terrain *(figure 5.1b)*. The geometry of a scanner image is different and more complex than that of frame-capture systems because it consists of a large number of image strips recorded sequentially in time to form a continuous strip image. The strip of terrain imaged is called a *swath* and the width of the strip, the *swath width*. While early examples of these scanners recorded their strip images on film, nowadays the images are invariably recorded in digital form.

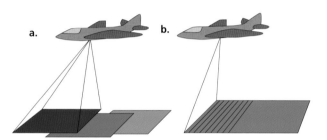

Figure 5.1 Frame-based and scanning imaging sensors. Frame-capture systems (a) record one frame at a time. Scanning systems (b) acquire imagery as a series of narrow strips, producing a long continuous swath of imagery along the flight line.

The two technologies require different processing methods to geometrically rectify the imagery in order to register it to a map base. Digital images have an advantage of being easily enhanced, classified, and rectified using standard digital image analysis techniques, whereas film images often need first to be digitized using a precision film scanner. However, film is still the least-expensive medium to capture and store high-resolution images although, with the continuing decline in digital storage cost, this advantage is diminishing.

This chapter discusses photographic and digital frame-capture systems. Scanning systems are discussed in chapter 6, and geometric rectification for both types of imagery is addressed in appendix A.

Film photographic systems

Photographic cameras have been carried on virtually every type of airborne vehicle devised. Aerial cameras have even been mounted on kites and carrier pigeons (see figures 2.3 and 2.4 in chapter 2). Photographs taken from aircraft remain the most widely used source of imagery for large-

scale mapping. Because aerial photographs are usually high resolution and have a relatively simple geometry, they are easily interpreted with the naked eye or simple instruments. Accurate measurements of ground features can be done manually, and more precise measurements can be made using specialized photogrammetric instruments. Spacecraft such as the NASA *Skylab* and certain space shuttles were also equipped with photographic cameras. Russian mapping agencies have been major users of space photography.

Aerial photographs taken for mapping and resource inventory applications commonly use specialized cameras with low-distortion lenses, referred to as *metric cameras (figure 5.2 and figure 5.3)*. In order to produce accurate maps, *metric cameras* need to be calibrated so that the precise value of the focal length of the lens, the exact position of the principal point (optical center) of the photograph, and the pattern of lens distortion are known. This is in contrast to *reconnaissance cameras*, which are nonmetric cameras that have not been calibrated for mapping purposes. Instead, they are typically optimized to produce high-resolution images for military interpretation purposes.

Standard format metric cameras

Designed to provide extremely high geometric image quality, metric cameras typically use 24 cm wide film in rolls as long as 120 m. They produce 23 x 23 cm (9 x 9 in) film

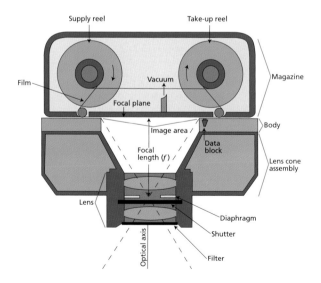

Figure 5.2 Generalized diagram of an aerial mapping camera.

Source: *Remote sensing and image interpretation*. 4th. ed. by Thomas M. Lillesand and Ralph W. Kiefer. © 2000. John Wiley and Sons, Inc. Reprinted with permission of John Wiley and Sons, Inc.

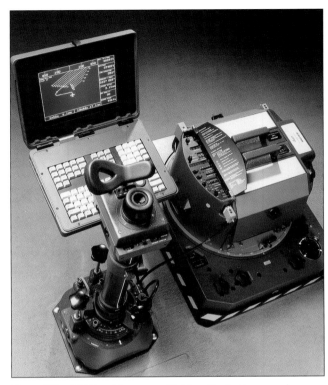

Figure 5.3   RC-30 metric camera. Highly calibrated large- and medium-format cameras called metric cameras are commonly used for large-scale, large-area mapping. The RC-30 produces 230 x 230 mm (9 x 9 inch) images on 240 mm wide roll film. The camera is shown with the navigation sight, which is mounted on the floor of the aircraft to verify the area being photographed, and the camera control system display.

Source: Imagery provided by Leica Geosystems GIS and Mapping, LLC.

images with photo scales typically ranging from 1:1,500 to 1:100,000. A vacuum system holds the film flat against a precision ground metal plate during exposure. The shutter is tripped, and the film is advanced automatically, controlled by a computerized navigation and flight management system that adjusts the photo acquisition rate to the flying height and speed of the aircraft. The navigation and flight system maintains the correct overlap between successive exposures.

During the time the shutter is open, the forward motion of the aircraft could cause a slight blurring of the image. Therefore, metric cameras commonly employ a *forward motion compensation* system that moves the film at exactly the rate of image motion during exposure to negate much of this blurring effect. Gyroscope stabilized camera mounts are typically installed to compensate for some of the aircraft pitch and roll.

When using metric cameras, the following three fundamental elements should be considered:
- the *film format size* of the camera
- the *focal length* of the lens
- the *angular coverage* of the lens and camera

Once the film format size has been defined, the angular coverage is determined by the focal length of the lens being used—which can be interchanged in most modern cameras. The larger the angular coverage, the larger the area viewed. Taking the standard format or image size of 23 x 23 cm, the use of a lens having a focal length (f) of 30 cm (12 in) gives an angular coverage of 60°. This is *normal-angle* coverage. If instead a lens with a focal length of 15 cm (6 in) is used with the same camera, this will give *wide-angle* coverage of 90° over the ground. The use of a lens having a focal length of 8.5 cm (3.5 in) with the same camera will provide *super-wide-angle* coverage of 120° over the terrain.

Not only does the choice of lenses with different focal lengths affect the angular coverage, but it also affects the scale and ground resolution of the resulting images. The actual image scale is defined by the ratio of focal length to flying height (f/H). Thus, for a given flying height of 3,000 m (10,000 ft), the use of these three normal-angle, wide-angle, and super-wide-angle lenses with different focal lengths will result in three quite different photo scales:

(i) $f/H = 0.30 \text{ m}/3,000 \text{ m} = 1{:}10,000$ scale if the normal-angle lens is used

(ii) $f/H = 0.15 \text{ m}/3,000 \text{ m} = 1{:}20,000$ scale if the wide-angle lens is used

(iii) $f/H = 0.085 \text{ m}/3,000 \text{ m} = 1{:}35,000$ scale if the super-wide-angle lens is used

Another important consideration is that the use of short focal length lenses with greater angular coverage means that a greater distance—called the *air base* (b)—will be covered over the ground before the next photo needs to be taken. The ratio of this air base to the flying height (H)—called the *base–height ratio*—has a major impact on the accuracy of measurements that can be made on the resulting photography and, therefore, on the accuracy of the planimetric (X/Y) and elevation (Z) data that can be extracted for mapping and GIS purposes. The following are the base–height ratios for a 60% overlap between photographs taken with different lenses:

(i)   0.3 for normal-angle photos

(ii) 0.6 for wide-angle photos

(iii) 1.0 for super-wide-angle photos

In general, the higher the base–height ratio, the greater the geometric accuracy of the data that can be extracted from the photography. However, this is only achieved with a corresponding reduction in the photographic scale and ground resolution in the resulting imagery.

It is important to understand these characteristics when planning image acquisition. Developing the specifications for an image acquisition mission, called *flight planning*, can be complex because it involves choosing among multiple competing trade-offs, such as the choice of lens focal length affecting ground resolution and geometric accuracy. Besides the characteristics of the camera and available lenses, flight planning also takes into account the type of aircraft available and its performance characteristics (e.g., speed, maximum flying altitude, and flight duration). Very stable, slow-flying aircraft are needed for the acquisition of large-scale photography from low altitudes. However, pressurized aircraft equipped with turbo-charged engines are needed to obtain small-scale photography from high-altitudes. Increasingly, specially modified business jet aircraft are being used to obtain small-scale photography over large areas of terrain.

Small-format cameras

Metric cameras are ideal for large mapping projects with their standard format images, precise film positioning, and highly calibrated lenses. But they are expensive, large, and heavy devices. They require special camera mounts and optical windows be installed in the aircraft and several crew members to operate them. Smaller, lighter, and much less costly 70 mm and 35 mm professional film and digital cameras are used extensively for handheld reconnaissance photography during field investigations. Alternatively, the cameras can be mounted on the aircraft and used to collect vertical aerial photographs. Small-format cameras are particularly valuable for collecting image samples. For example, it may be less expensive to assess scattered forest regeneration sites from aerial photography by photographing individual sites with a small-format camera rather than using a standard-format metric camera that may image far more area than is needed. Crop condition assessment for precision farming also makes extensive use of small-format imagery, commonly digital imagery.

However, when large area coverage is needed, the use of a standard format metric camera greatly reduces the number of images and the flight time required (since fewer flight lines are needed). A 230 mm x 230 mm (9 x 9 in) image (using 240 mm wide film) has about 17 times the area coverage of a 55 mm x 55 mm image (using 70 mm wide film) and 61 times the coverage of a 24 mm x 36 mm image (using 35 mm wide film). What is more, the 30% side lap (overlap between images from adjacent flight lines) further increases the number of flight lines needed. For example, edge to edge, four 55 mm images almost extend across a 230 mm standard format aerial photograph. But to achieve a 30% side lap, six lines, not four, would be needed to equal the coverage of a single flight line of 230 mm wide imagery.

Large-format cameras

In order to increase the angular coverage and improve the base-height ratio of the resulting photography, a number of large-format metric cameras (LFCs) have been built, principally for military mapping purposes. These have been used extensively onboard high-flying aircraft such as the Lockheed U-2 and onboard the NASA ER-2. The National Ocean Survey (NOS) has also made considerable use of this type of camera for the mapping of coastal areas. Large-format cameras have also been operated from space onboard such vehicles as NASA space shuttles and on numerous Russian satellites.

As their name implies, large-format cameras produce larger format images than standard-format metric cameras. For example, Itek large-format cameras use the same 240 mm wide film used in standard-format metric cameras; however, the format size is doubled in the film direction, resulting in a very rectangular 23 x 46 cm format. LFC cameras are normally operated with a 30 cm (12 in) focal length lens. The increased format size of the LFC can be used either to give greater across-track coverage of the terrain or to give an increased base-height ratio with the longer side pointing in the along-track direction. With the Russian TK-350 large-format camera, the format size is even larger than the Itek camera—30 x 45cm—while the focal length of the lens is 35 cm (13.8 in). This type of camera has been used extensively to provide imagery for the compilation of Russian military mapping, especially in the Middle East, North Africa, and Southern Asia.

Panoramic cameras

Panoramic cameras have been used extensively for military reconnaissance purposes, both from the air and from space. It is a scanning type of photographic frame camera in which the lens (or prism) rotates in front of the film to sweep across the terrain in the across-track direction *(figure 5.4)*. Thus the area being imaged is covered at a relatively narrow angle in the along-track direction but with a very wide angle in the across-track direction. While this process results in a frame image being produced, the film rests on a cylindrical surface as it is being exposed. This gives rise to a very distinctive geometry in which the scale of the image changes rapidly in the across-track direction. Besides which, additional displacements in the image result from the rapid forward motion of the airborne or spaceborne platform during the time it takes for the lens to sweep across the terrain (and the film).

Panoramic cameras are well-suited for the acquisition of high-resolution images from very high altitudes. Thus, many different types of panoramic camera have been built using long focal length lenses in the range of f = 60 cm (24 in) to f = 1.20 m (48 in). For example, the Itek KA-80 panoramic camera uses a lens with a focal length of 60 cm (24 in). This is used in conjunction with 12.7 cm (5 in) wide film and produces photos with a format size of 11.5 x 127.8 cm (4.5 x 50.3 in). From space, the Corona reconnaissance photography, which has now been declassified and is available for civilian use, was acquired using panoramic cameras. Much of the Corona imagery was taken using twin Itek panoramic cameras—one tilted by 15° in the forward direction, the other tilted by 15° in the backward direction. This configuration allowed the acquisition of photography with a 30° convergent angle that, in turn, allowed stereo-measurements of the terrain to be carried out. Each individual Corona image is only 5.5 cm (2.15 in) wide, but 74.5 cm (29.32 in) in length! Panoramic cameras were also carried aboard Apollo missions to carry out lunar mapping *(figure 5.5)*.

Russian mapping agencies have also used panoramic photography extensively from space. High-resolution space photography of large parts of the earth taken with the Russian KVR-1000 panoramic camera are available from Russian agencies such as Sovinformsputnik, from various commercial suppliers in the United States such as EastView Cartographic or LandInfo, and from suppliers in Europe such as Eurimage.

Vertical and oblique aerial photography

Aerial photographs can be classified as either vertical or oblique. *Vertical photography* is acquired with the camera axis oriented vertically to the ground below. It is by far the most common type of photography used in remote sensing. However, a perfectly vertical photograph is rarely obtained. Even with gyroscope stabilized camera mounts that adjust for the tip and tilt of the aircraft, vertical aerial photographs can have a tilt of as much as one to three degrees. Spaceborne platforms are more stable and can acquire photography that is nearly truly vertical.

Photographs intentionally taken with the camera axis pointed away from the normal vertical orientation are called *oblique photographs (figure 5.6)*. High-oblique photographs include the horizon while low-oblique photographs do not. Oblique imagery tends to be preferred for certain commercial applications, such as for advertising or recording site conditions. The oblique perspective is more familiar and easier to interpret. Closer features appear larger, visually exaggerating the foreground, which can be useful for illustrations. Accurate measurements and mapping can be done from oblique imagery, but the procedures are much more complex because the scale decreases rapidly from the foreground to the background.

Much of the photography acquired by military air forces for reconnaissance purposes is oblique photography, often using multiple cameras to give increased coverage of the terrain. The use of *twin oblique cameras* is typical. Twin oblique

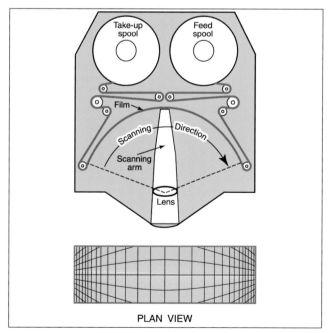

Figure 5.4   Panoramic camera. The upper diagram is a cross section of a panoramic camera, showing (i) the cylindrical image surface on which the negative film rests and (ii) the lens that pivots around its center (nodal point) to sweep the image of the terrain on to the film. The lower diagram shows the effect of the panoramic sweep on a grid of equal squares marked out on the terrain. The image scale changes rapidly toward the edges of the photograph.

Source: Image courtesy of G. Petrie and M. Shand.

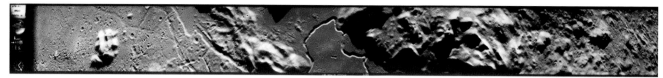

Figure 5.5 *Apollo 15* panoramic camera image of the lunar surface. Panoramic cameras were carried on several Apollo missions to collect imagery for topographic mapping of the lunar surface. This image shows the Hadley-Appenines adjacent to the *Apollo 15* landing site.

Source: Image courtesy of NASA.

cameras provide coverage across-track to the left and right of the flight line respectively. The resulting images are often called *split-verticals*, though they are in fact low-oblique photographs. Fans of four or six photographs may be used to provide still further coverage in the across-track direction.

Tri-metrogon photography is another well-known camera arrangement *(figure 5.7)*. It comprises a single wide-angle vertical camera and two high-oblique wide-angle cameras pointing left and right of the flight line in the across-track direction. This arrangement provides horizon-to-horizon coverage. Tri-metrogon photography was used extensively during and after World War II to provide the image data that was used in the compilation of small-scale maps and air navigation charts of remote areas.

Yet another distinctive type of oblique photography is *long-range oblique photography* (LOROP). This is taken by cameras equipped with very long focal length lenses. LOROP was used extensively by NATO air forces during the cold war, where it allowed high-flying aircraft to acquire images of areas or targets across borders or marine shorelines from long ranges.

For certain applications, small-format handheld oblique images acquired from a helicopter, light plane, or even from atop a tall building are all that is required. For example, oblique aerial photographs can document the progress of building construction to allow the head office to monitor one or more sites or to provide evidence of completion of project milestones to secure financing. Bank loans may be structured to progressively release funds upon completion of certain construction milestones such as pouring the foundation or completing the roof of a building. Imagery generally does not always need to be rectified for such use.

Stereo aerial photography

Most aerial photography is acquired as a series of overlapping parallel flight lines. Airphoto missions are normally flown with an overlap of 60% between successive images (for complete stereoscopic coverage) and a 30% overlap between adjacent flight lines to ensure complete coverage *(figure 5.8)*.

Figure 5.6 July 1979 oblique aerial photograph of Parliament Hill, Ottawa.

Source: © Natural Resources Canada.

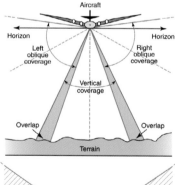

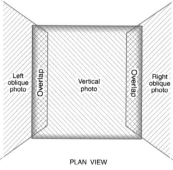

Figure 5.7 Tri-metrogon photography. The upper diagram shows the basic geometry of tri-metrogon photography, which is based on the use of one vertical and two high-oblique aerial cameras to provide horizon-to-horizon photographic coverage of the terrain. The lower diagram shows the corresponding coverage of the terrain.

Source: Image courtesy of G. Petrie and M. Shand.

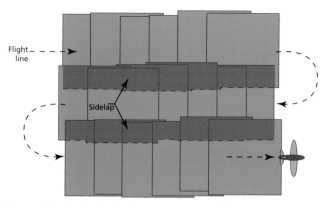

Figure 5.8 Overlap in air-photo missions. Aircraft normally fly over a project area in a series of parallel flight lines spaced to give a 30% side lap. Successive airphotos are normally acquired with a 60% forward overlap.

Source: *Remote sensing and image interpretation*. 2nd. ed. by Thomas M. Lillesand and Ralph W. Kiefer. © 1987. John Wiley and Sons, Inc. Reprinted with permission of John Wiley and Sons, Inc.

Figure 5.9 Stereoscopes. A stereoscope is an optical instrument for viewing pairs of photographs acquired so that when each eye views one of the images, the scene is perceived as three-dimensional. The two smaller instruments are collapsible for compact storage. The larger instrument is a mirror stereoscope with angled mirrors that enable stereopairs to be viewed when placed farther apart than with the smaller instruments—allowing standard format aerial photographs that are 230 mm (9 in) across to be placed side-by-side.

Adjacent pairs of overlapping vertical photographs are called a *stereopair*. In the area of overlap, they show the same portion of the landscape but viewed from different camera positions, giving them a different perspective.

Stereo effects and parallax

The ability of the human eye and brain to interpret differences in the perspective of two images as depth is called the *stereo effect*. The three-dimensional stereo image is called a *stereomodel*. When looking vertically down at the landscape, one can perceive differences in the heights of features, which gives a three-dimensional view. The strength of this stereo effect depends on the magnitude of the difference, called the *parallax*, in the two views. The closer the objects are to the camera (i.e., the taller they are), the more different are their relative positions within the image and the closer we perceive them to be. By measuring the parallax and knowing the specifications of the camera and the flying height, the elevation of features in the image can be calculated.

Stereo viewing of hard-copy images

When viewed using a *stereoscope (figure 5.9)*, the left image is presented to the left eye only and the right image to the right eye only. The difference in perspective is perceived as a three-dimensional view or *stereomodel*. Figure 5.10 is an air-photo stereopair positioned for viewing with a pocket stereoscope. Some people are able to view the image in stereo without a

Figure 5.10 Air-photo stereopair of Parliament Hill in Ottawa. This pair of photographs can be viewed in stereo using a pocket stereoscope. The complex of buildings in the upper portion of the scene is the Canadian Houses of Parliament overlooking the Ottawa River. In the lower portion of the scene, the locks of the Rideau Canal are visible. When viewed in stereo, the relative heights of features appear exaggerated, as explained in the text. The ground distance across each image is 360 m.

Source: © Natural Resources Canada.

stereoscope by relaxing their eyes and viewing the image at a distance so that each eye sees the same feature in the corresponding photograph.

Many applications of aerial photography make extensive use of stereoscopic coverage. The geometry of the stereo model enables precise three-dimensional measurements to be made for topographic mapping and the determination of elevation. Furthermore, stereo viewing dramatically improves the quantity and quality of information that can be extracted from remotely sensed data. For example, the three-dimensional shape of a landform is one of the factors that can be used to identify its composition. Gravel, sand, silt, and clay materials can be accurately interpreted from 1:40,000 scale black-and-white aerial photographs viewed in stereo. As well as providing the three-dimensional perspective, stereo viewing also improves the perceived resolution of the photos. Air photos tend to appear sharper and are more easily analyzed when viewed in stereo than the same photographs viewed singly.

Stereo viewing of digital images

There are several ways to view digital stereo images. Stereo-pairs can be displayed side-by-side using one or two monitors and viewed with a stereoscope *(figure 5.11)*. Many systems use complementary filters to present left and right images to the viewer's corresponding eye. *Anaglyphs* superimpose the two images of the stereopair on a single color monitor, displaying one image in red and the other in blue or green. When viewed with glasses having the appropriate red and

green or red and blue filters, a 3D black-and-white image is perceived.

Polarizing filters can provide full-color 3D viewing. One approach uses two display monitors, one placed in front of the viewer and the other mounted in front pointing vertically upward with a large semi-reflecting mirror placed between the two monitors so the observer sees both screen images superimposed. Each monitor has a polarized filter in front of its screen, one horizontally polarized and the other vertically polarized. Using special glasses, one eye with a horizontally polarized filter and the other with a vertically polarized one, a 3D image is perceived. Probably the most commonly used 3D stereo-viewing system used on digital photogrammetric workstations is one that displays the left and right images alternately at high speed on a single monitor. The image is then viewed with active viewing glasses equipped with alternating electronic shutters synchronized to the display so each eye only sees the display when the corresponding image is shown *(figure 5.12)*. Alternatively, a liquid-crystal filter is placed in front of the display that rapidly alternates between clockwise and anti-clockwise polarization synchronized to the alternating display of left and right images. When viewed with glasses having the corresponding polarized filters, a 3D image is perceived *(appendix A, figure A.38)*.

Vertical exaggeration

When the landscape is viewed from far away, such as from an aircraft, the terrain appears more level with much less relief than when viewed from the ground. This is because

Figure 5.11 Stereosopes, such as the ScreenScope shown here, can be mounted on computer monitors to provide stereo viewing.

Source : Image courtesy of Stereoaids. www.stereoaids.com.au

Figure 5.12 Digital photogrammetric workstation.

Source: Imagery provided by Leica Geosystems GIS and Mapping, LLC.

the distance between our eyes is small, just a few centimeters, relative to the distance to the ground below so there is little difference in the perspective, i.e., both eyes see features in the faraway scene in about the same relative position. In an air-photo stereopair, the positions at which the two photographs were taken were hundreds or thousands of meters apart, much farther apart than the few centimeters between a person's eyes. As a result, the difference in the relative positions of objects is greater in the photographs than would normally be seen, and the heights of features appear to be exaggerated. This vertical exaggeration in aerial photographs improves the accuracy with which the elevation of features can be viewed and measured and improves image interpretation. Vertical exaggeration is also commonly used in generating perspective views for landscape visualization to make small differences in terrain characteristics easier to see. It is important to clearly indicate the degree of exaggeration being used and take measures to ensure the enhanced viewing is not misleading for the intended use.

The stereopair in figure 5.10 illustrates vertical exaggeration. When seen through a stereoscope, the canal in the center of the image looks like a trench and the buildings appear to be tall, thin slabs teetering above the ground.

The degree of vertical exaggeration is determined by the base–height ratio. The larger the base–height ratio, the greater the vertical exaggeration. As noted above, the base–height ratio is the distance between the camera positions from which the two photographs were taken (called the *air base* or *stereo base* of the stereo pair) divided by the height above ground *(figure 5.13)*.

Auxiliary instruments

GPS (global positioning system) satellite navigation receivers are commonly integrated into the camera control system to record flight-line direction, flight-line spacing, and control photo exposure intervals. The GPS data provides an accurate geographic position and elevation of the camera for every image frame. Some camera systems also incorporate an *inertial measurement unit* (IMU) to record the precise three-dimensional orientation of the camera at each exposure. The IMU and GPS data for each image frame is then recorded as a digital file. This data is later used to calculate the camera position and attitude for image rectification and to plot flight lines and image centers on index maps as a record of the mission.

For very high accuracy positioning and height determination, differential GPS (DGPS) or Kinematic GPS (KGPS)

may be used. In *differential* GPS, the accuracy of the GPS data collected by the aircraft is mathematically enhanced with GPS data simultaneously collected from base stations operating at known locations on the ground at the same time and in the same area as the imagery was being collected. *Kinematic* GPS is a similar procedure, but it uses the carrier phase of the signal to further refine errors introduced by the atmosphere. Base stations may be needed within 30 to 50 km of the flight line, depending on the accuracy specifications of the mission. However, the range to a base station may be extended to several hundred kilometers, depending on the relative position of the GPS satellites in view and the final accuracy requirements. Typically, these base stations are installed specially for the overflight.

In some cases, the data from government-operated GPS base stations can be used if they are close enough to the study area *(figure 5.14)*. In Canada, the United States, and other countries, a network of continuously operating reference stations (the CORS system in the United States) have been implemented for this purpose. Conventional aerial surveys also use well-defined ground features and set out targets, or panel points, at surveyed locations before

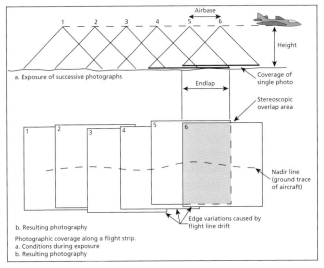

Figure 5.13   Photographic coverage along a flight line. Successive exposures along a flight line produce a series of overlapping photographs. The ratio of the air base (the distance between successive camera positions) and the flying height determines the degree of vertical exaggeration perceived when viewing successive image pairs in stereo.

Source: *Remote sensing and image interpretation*. 2nd. ed. by T. M. Lillesand and R. W. Kiefer. © 1987 John Wiley and Sons, Inc. Reprinted with permission of John Wiley and Sons, Inc.

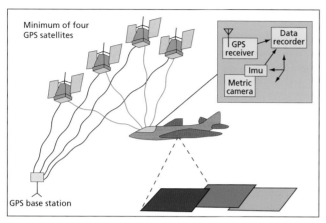

Figure 5.14 Illustration of supporting technology for airborne remote sensing. Airborne remote sensing systems are typically operated with an onboard GPS and record the aircraft location for each frame of imagery. An inertial measurement unit, rigidly affixed to the sensor mounting, records the three-dimensional movement of the sensor as the aircraft pitches and rolls during flight. GPS data collected at the time of overflight at base stations placed in the area or government-operated GPS stations nearby may later be used to improve the accuracy of the onboard GPS data.

an air-photo mission. Later, these points in the image are used to set up and georeference stereomodels and rectify the photography during map production. With IMU and GPS technology providing attitude and geographic position data for each frame of photography, fewer, and in some cases no, ground reference points may be needed, substantially reducing the time and cost to produce accurate mapping. This process, using GPS and IMU technology without ground control points for geometric rectification, is referred to as *direct positioning*.

However, the present cost of high-precision integrated DGPS/IMU units is extremely high. The major cost lies in the provision of the IMU unit. The result is that many smaller aerial survey companies and mapping organizations cannot afford to make the investment that is required. Thus, a great deal of aerial photography is still flown without these expensive, high-accuracy units. In which case, only the GPS required for air navigation and camera operations will be used. In this situation, the required ground control points (GCPs) will be provided by field survey in the traditional way, usually supplemented by the minor control points (MCPs) provided by aerial triangulation.

Photogrammetry

Photogrammetry is the engineering field concerned with obtaining accurate measurements from aerial photographs. The more general term *stereogrammetry* is applied to the analysis of all types of stereo imagery, including aerial photographs, radar, and electronic scanner data.

Photogrammetric instruments are used to measure parallax very precisely to provide accurate elevation measurements. The operator views the stereo image through the eyepieces of the instrument. Older instruments used hand wheels and foot wheels to control the horizontal and vertical movement of a measuring mark in the stereomodel *(figure 5.15 opposite page)*. Current instruments, such as the analytical *stereoplotter* shown in figure 5.16, use handheld cursor control buttons. (Some models can be fitted with hand- and foot-wheel controls as used in older mechanical instruments.) The operator adjusts the height of the measuring mark to rest on the surface of the terrain and traces the objects to be mapped or draws contour lines. The instrument's computer records the movement of the cursor and calculates the horizontal and vertical position of the points. Similarly, spot elevations can be determined, such as building roof heights.

The analytical plotters are, in turn, being replaced by all-digital instruments called *digital photogrammetric workstations* (DPWs). DPWs use digital images as well as produce digital output, which eliminates the requirement for a high-precision optical system *(figure 5.12)*. These all-digital

Figure 5.16 Zeiss P3 analytical stereoplotter. The stereo images are mounted in the housing above the eyepieces, and the operator controls all functions from a single control unit operated with two hands. The output in digital format is linked to a high-speed computer graphics workstation to produce basemaps, topographic maps, and other high-accuracy products.

Source: Photo courtesy of M. Kitaif, Cardinal Systems, LCC.

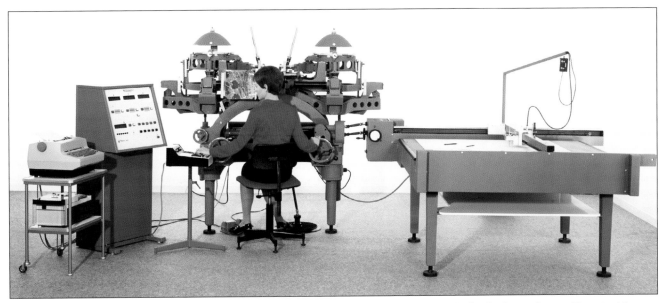

Figure 5.15   Mechanical stereoplotters use hand wheels and foot wheels to physically move images in the viewing equipment. The output is generated on the plotter to the right of the instrument. Though eclipsed by current digital systems, some older technology such as this Wild A8 and the later Wild A10 are still in use.

Source: Imagery provided by Leica Geosystems GIS and Mapping, LLC.

systems are designed to use the ever-growing variety of digital remotely sensed imagery as well as scanned aerial photographs. Digital photogrammetric systems with their greater versatility and unique capabilities have been rapidly replacing optical systems as the cost of high-speed computers and large-capacity disk storage continues to fall.

Digital elevation data from aerial photographs

To produce *contour maps*, the measuring mark in an analytical plotter or DPW is set to the elevation of each contour line and moved across the stereo model so the mark appears to remain in continuous contact with the terrain surface. The accuracy of the resulting (continuously plotted) contour lines is normally assessed to lie within 1/1,000 to 1/2,000 of the flying height at which the aerial photography was acquired.

Alternatively the measuring mark in the analytical plotter or DPW can also be moved in a series of parallel lines step-by-step across the stereomodel to generate a grid of elevation data points, often called *spot heights*. This grid of elevation points would then form a digital elevation model (DEM)—a set of elevation measurements representing the terrain surface. The more closely spaced the data points, or *postings*, in a DEM, the more accurately the surface is represented, and the more

costly it is to generate. The photogrammetric instruments used to produce digital elevation models provide elevation accuracies in the order of a 10,000th of the flying height.

Organizations differ in their terminology for elevation datasets. For example, the USGS uses the term DEM, the National Geospatial-Intelligence Agency (NGA) uses the term digital terrain elevation data (DTED), while the term digital terrain model (DTM) is common in Europe. To add

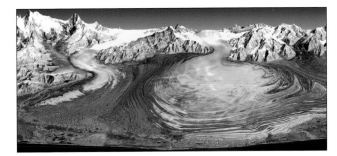

Figure 5.17   Perspective view of the Malaspina Glacier, Alaska. This perspective view was generated from Landsat 30 m multispectral scanner data and 30 m digital elevation data obtained by radar interferometry from the space shuttle *Endeavor* in 1994. The area shown is 55 km by 55 km.

Source: Image courtesy NASA, JPL, and NIMA.

to the confusion, some organizations use the terms DEM and DTM to refer to different products. The DEM as previously defined and the term DTM refer to a broader dataset consisting of elevation data and other terrain information, such as slope and aspect, stream centerlines, and elevation break lines that show the local height of land.

Digital elevation models are used in a broad range of GIS analyses. They are also used to display spatial data such as maps and remotely sensed image data as *perspective views* of the terrain with options of choosing the viewing position, illumination direction, and degree of vertical exaggeration (*figure 5.17*).

Figure 5.18 illustrates *shaded relief images* produced from four different DEM datasets for the same area. The DEMs were generated by different agencies or with different spatial resolutions (postings). Notice that the finer resolution datasets

reveal greater detail. Additionally, the two images generated from 30 m datasets from aerial photography and from lidar technology (laser direction and ranging, see chapter 8) differ in the detail they have captured. In addition to resolution and capture technology, the same data points processed using different algorithms can produce somewhat different results.

Digital vector data from aerial photographs

Except at the very largest scales, where field survey methods can still be used, virtually all topographic base maps are produced from aerial photographs using stereo-photogrammetric procedures. GIS data can then be derived from these hardcopy maps by digitizing the point, line, and polygon data contained in them. However, nowadays, the map and GIS

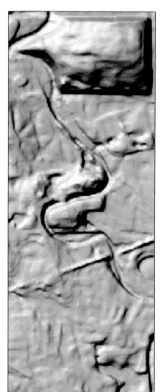

Figure 5.18 Comparison of digital elevation models for Arlington Heights, Illinois. The four shaded relief images illustrate differences in the level of detail captured in digital elevation datasets with different spatial resolutions (postings) and produced using different methods. The image on the left is a shaded relief display of data collected using photogrammetric methods and stereo aerial photography. The other three images were generated from datasets collected using lidar technology, at progessively finer spatial resolution. Lidar captures greater detail but is generally more costly to acquire and process.

Source: Image courtesy of the USGS.

data is almost always extracted directly from the aerial photographs using an analytical plotter or a digital photogrammetric workstation (DPW). The procedure involves the use of 3D stereomodels. These models are set up in the instrument or workstation, employing a special orientation procedure with a suitable pair of overlapping aerial photographs. This operation also fits the stereomodel to a set of ground control points (GCPs), thus ensuring that the 3D model is correctly georeferenced to the terrestrial coordinate reference system.

Once this preliminary orientation procedure has been completed, the data required for the compilation of the topographic map or for use in the GIS system is then extracted from the stereo model by a trained operator. Each feature that falls into a particular type or class of data (e.g., roads, railways, field and forest boundaries, streams, built-up areas, individual buildings) that is required for the production of the map or for inclusion in the GIS is identified and measured very accurately in the stereomodel using a special 3D cursor or measuring mark. As these measurements take place, the precise x,y,z coordinates of each measured point are recorded continuously in digital form. During this systematic measuring operation, the appropriate attributes associated with the features being extracted are also being entered into the computer by the operator. After editing and field completion, the resulting vector data is then ready for entry into the GIS.

Digital image maps from aerial photography

Remotely sensed images generally require some form of geometric correction in order to use them directly with other geographic information. The degree of correction required varies with the application. Low-level corrections that, in effect, stretch the image so that major features align to their map locations might be sufficient for small-scale imagery.

Orthophotos produced from a single aerial photo

An orthorectified image is precision corrected to remove distortions so that the image has a consistent scale throughout. Three types of distortions are removed:

- Camera calibration information is used to remove *lens distortion* effects.
- The camera tilt is determined from ground control points or the data derived from an inertial measurement unit. If the camera was not exactly vertical, the resulting *tilt displacements* on the image are calculated and removed.

- Variations in terrain relief cause displacement due to scale variations, termed *relief displacement*. A digital elevation model is used to calculate and remove the effects of relief displacement *(figure 5.19)*. See figure A.39 in appendix A.

Orthophotos produced from stereo pairs of aerial photographs

When large blocks of aerial photos have to be orthorectified, as distinct from individual photographs, they are usually produced photogrammetrically from stereopairs using highly automated image matching or correlation techniques. Each stereomodel is set up and georeferenced using the same orientation procedure that has been described above for the compilation of map and GIS data. However, once the 3D stereomodel has been set up, it is then scanned in a systematic fashion, usually on a patch-by-patch basis. The image data contained in each patch on the one photo of the stereopair is then matched precisely to the corresponding image data on the second photo of the pair. This allows the x-parallaxes resulting from the relief displacements to be determined and, in turn, the ground elevation values to be calculated. In this way, the digital elevation model of the area covered by the stereopair of photographs is created directly using an automated procedure. Thereafter, the orthorectified image is produced from the stereomodel by orthogonal projection also using an automated procedure *(figure A.37 in appendix A)*.

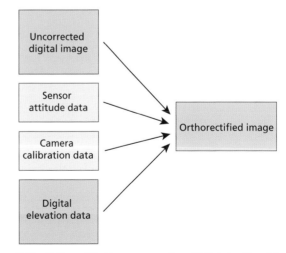

Figure 5.19 Orthorectified image generation. Digital elevation data, camera viewing direction information, and calibration data are used to generate an orthorectified image.

Characteristics of orthophotos

Ideally, the resulting orthorectified image appears as if every point is viewed from directly above, termed an *orthographic projection*, and is corrected to fit the selected map base so that other data layers displayed with the same map projection and scale are in correct registration (see the sidebar, *Relief displacement and orthographic projections*).

Orthorectified images differ in their level of correction. Standard orthorectification corrects the relief displacement that is present in ground level features. Above ground features, such as the tops of buildings and bridges, aren't normally corrected because their elevations are not recorded in the digital elevation model. As a result, these features appear noticeably distorted, particularly at large scales: buildings lean and bridges are warped. Higher precision orthorectification removes the relief displacement of above ground features. For example, building lean is corrected so the tops of buildings appear directly over their base and the image data for the ground around the building is visible *(figure 5.20)*.

Correcting displacement effects for tall features with a relatively small base, such as a building, requires large-scale high-precision digital terrain data (typically produced from aerial photographs, lidar, or radar) and more sophisticated and more expensive processing. It would be too expensive to individually rectify small tall features, such as lamp standards and power line towers, so they are left to appear leaning in the orthocorrected image, as in the original aerial photograph (see the sidebar, *Relief displacement and orthographic projections*).

Orthophoto mosaics

Multiple images can be orthorectified and combined to produce seamless *digital image mosaics* that contain all the information of the original images positioned in their correct map location, in effect having the geometric integrity of a map. In addition to orthorectification, mosaicking requires additional processing to match the brightness, color, and contrast of the individual images so as to produce seamless image coverage *(figure 5.21)*.

Called *image maps*, these precision-corrected orthophoto mosaics are widely used in geographic information systems

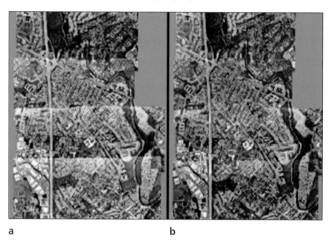

a   b

Figure 5.21 Image mosaics. Image mosaics are comprised of multiple images geometrically corrected to the same map base (a) and then radiometrically corrected to the same brightness color balance and contrast (b).

Source: © The Sanborn Map Company.

a   b   c   d

Figure 5.20 Orthorectification of above ground features. The elevation of buildings and bridges are not included in standard digital elevation models, so buildings lean (a) and bridges appear wavy (c) in

standard orthorectified images. By using detailed elevation data for these features, these artifacts can be removed (b and d).

Source: © The Sanborn Map Company.

Relief displacement and orthographic projection

Features on a map are portrayed as if every point was viewed from directly above, called an *orthographic projection*. Like a map, an aerial photograph represents a three-dimensional scene (the landscape) as a two-dimensional image (the photograph). But even in vertical aerial photographs, features are not all portrayed as if viewed from directly overhead. Except for the point directly below the camera position at the center of the photograph, features are viewed at an angle. For features that rise above the terrain, such as a building or a tree, some or most of the side of the feature is imaged. This is because objects that are closer appear larger (i.e., are represented at a larger scale) and are displaced radially outward

from the center of view. For this reason, tall buildings appear to lean outward in figure 5.22. Features that are farther away are displaced inward toward the center of view, the way the edges of a road appear to converge into the distance. This is referred to as *perspective projection*, the same projection produced by the human eye, and is thus very familiar. The apparent displacement of a feature's position relative to its observation distance is called *relief displacement (figure 5.23)*

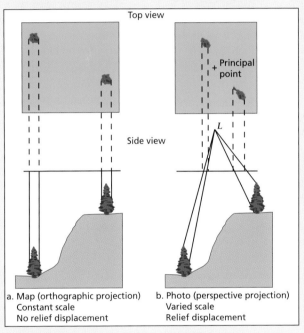

Figure 5.23   Comparison of the orthographic projection used in maps and the perspective projection of an aerial photograph. Note that the tree farther from the perspective center of the photograph is portrayed larger and more of the side of the tree is imaged.

Source: *Remote sensing and image interpretation.* 2nd. ed. by T. M. Lillesand and R. W. Kiefer. © 1987 John Wiley and Sons, Inc. Reprinted with permission of John Wiley and Sons, Inc.

Figure 5.22   Radial displacement. This vertical aerial photograph of Parliament Hill in Ottawa Canada illustrates the way radial displacement causes buildings to appear to lean outward from the perspective center of the image. Tall buildings also appear to taper downward because their tops, being closer to the camera, are imaged at a larger scale than their bases at street level. The distance across the image is 400 m.

Source: © Natural Resources Canada.

as a background image over which other data layers can be overlaid and viewed together. The background image mosaic is not only valuable to orient the viewer for presentations or for analysis, but it can also be used to update the geospatial database or to serve as the source of information that would be too costly to digitize.

Orthophotos in a GIS

One of the major advantages of digital orthorectified imagery is that it can be readily used in a GIS environment. Vector map data can be displayed as overlays or processed with digital elevation data to produce perspective views. Information can be interpreted directly from the screen image and selectively digitized to produce new data layers. Point, line, and polygon features can be collected and then integrated with existing tabular and spatial datasets. Orthorectification makes it easier for multiple users with diverse interests to view, analyze, and extract information for their application.

Large-scale orthorectified image mosaics have been particularly valuable within geographic information systems for urban areas to monitor changes in land use, property improvements, and to plan maintenance activities. For example, ownership boundaries and building outlines can be overlaid on orthophoto coverage of a city. Improvements such as new roads, building additions, and other improvements not in the GIS database are readily identified on the image and can be outlined and added to the GIS. The orthophoto mosaic can also be maintained as an additional data layer in the GIS as a record of features that are too detailed or too costly to digitize, such as the condition of vacant lots.

Until recently, most of the image maps used in municipal GIS have been derived from aerial photography. When orthorectification is being carried out by computer-based image processing, the aerial photographs must first be scanned in a photogrammetric film scanner to generate suitable high-resolution digital images (see sidebar, *Scanning aerial photographs*). As digital imaging technology has become more reliable and cost-effective, the use of high-resolution airborne digital camera and scanner data has steadily increased. The commercial availability of very high-resolution orthorectified satellite imagery has also led to it being rapidly adopted for GIS use.

Photographic film

Four types of film are commonly used in remote sensing—panchromatic (standard black-and-white film), black-and-white infrared, natural color (standard color film), and color infrared *(figure 5.24, opposite page)*. Black-and-white films are sensitive to a single range of wavelengths. The intensity of light is recorded as shades of gray ranging from black to white—the reason why these films are commonly referred to as black-and-white films. Color films are comprised of three layers, each sensitive to a different range of wavelengths. In a normal or natural color film, these layers are made sensitive to the visible blue, green, and red wavelength ranges, and the film is designed to approximate as closely as possible the colors perceived by the human eye. In color infrared films, the three layers are made sensitive to the green, red, and near-infrared wavelengths *(figure 5.25)*, and the color representation is false—near-infrared appears as red, visible red as green, and visible green as blue.

Most aerial photography is acquired using natural color film or the less expensive panchromatic film. More subtle distinctions can be seen in a color image than in the gray tones of black-and-white. Expert photointerpreters can distinguish fifteen to thirty shades of gray on panchromatic images but can distinguish thousands of hues on color photographs. However, greater differentiation of ground features is not always a benefit. Analysts mapping surficial materials such as sand and gravel often find black-and-white photos more useful because moisture characteristics, important in recognizing these deposits, are expressed as variations in tone. Color tends to obscure the subtle tonal shift. Normal color photography is used extensively for topographic mapping, surface water quality assessment and the mapping of soils,

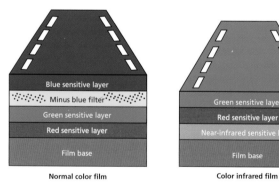

Figure 5.25 Color films contain three layers. In standard color films these layers are made sensitive to blue, green, and red wavelengths and produce images with natural colors—similar to their appearance in nature. In color infrared film, the layers are sensitive to green, red, and near-infrared and produce a color representation quite different from the way the color of the features appear in nature.

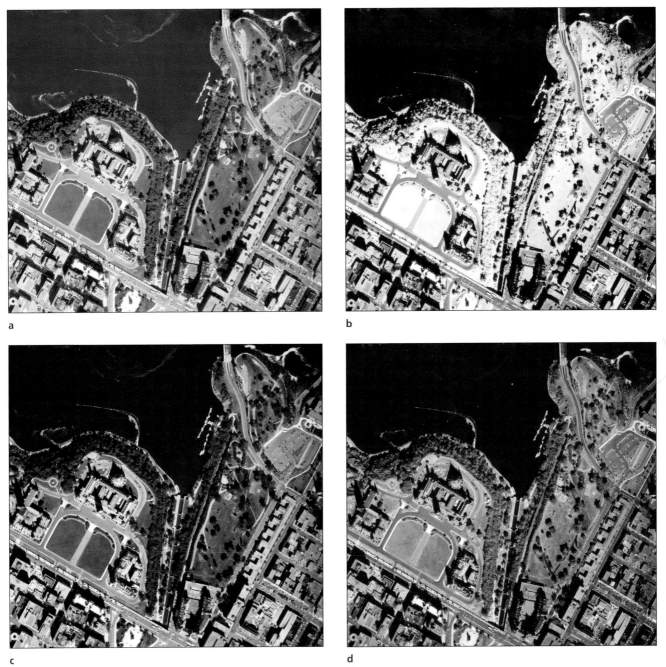

a

b

c

d

Figure 5.24 Aerial photography of Parliament Hill, Ottawa, Canada. The same scene is shown as it would appear on four types of films used for aerial photography: (a) panchromatic black-and-white film, (b) black-and-white infrared, (c) natural color, and (d) color infrared. The imagery was acquired June 24, 1978, at a flying height of 1,524 m (5,000 ft.) using a 305.3 mm (12 in) lens, giving a photo scale of 1:5,000.

Source: © Natural Resources Canada.

Scanning aerial photographs

Aerial photographs are scanned in a high-precision film scanner to produce high-resolution digital images for such application as the following:

- digital photogrammetric instruments
- the automated production of orthorectified images and digital elevation models
- the production of the digital image maps now widely used in geographic information systems

The geometric accuracy of these photogrammetric film scanners is very high—typically on the order of ±2 to 3 μm. The cost of these devices is correspondingly high—in the range of $25,000 to $100,000. While scanner spot sizes as small as 7 μm (i.e., 7/1000th of a mm) are available, a much coarser spot size of 25 μm is commonly used for producing digital orthorectified images for use as background image maps in a GIS. However, for high-accuracy engineering or GIS work, smaller pixel sizes from 10 to 15 μm will be used. The practical resolution limit for much aerial film photography is about 12 μm, at which point, the spot size approaches the effective resolution of the image.[1] Table 5.1 illustrates the relation of spot size and air-photo scale to ground resolution.

Table 5.1 Relation of digitizer spot size and airphoto scale to pixel ground resolution

Digitizer spot size		Pixel ground resolution in meters for different scales of photography							
Dots per Inch	um	1,200	2,400	4,800	10,000	20,000	40,000	50,000	80,000
100	254.0	0.30	0.61	1.22	2.54	5.08	10.16	12.70	20.32
200	127.0	0.15	0.30	0.61	1.27	2.54	5.08	6.35	10.16
300	84.7	0.10	0.20	0.41	0.85	1.69	3.39	4.23	6.77
600	42.3	0.05	0.10	0.20	0.42	0.85	1.69	2.12	3.39
1200	21.2	0.03	0.05	0.10	0.21	0.42	0.85	1.06	1.69
2000	12.7	0.02	0.03	0.06	0.13	0.25	0.51	0.64	1.02

Useful scanning conversions:

DPI= dots per inch	DPI to microns	(2.54 X 10,000)/DPI
um = micrometers or microns	Microns to DPI	(2.54 X 10,000)/um
m = meters = 106 microns	Inches to meters	0.0254 X in
in = inches	Meters to inches	39.37 X m

Calculation of pixel ground resolution:
S = photo scale

	Pixel size in ft	Pixel size in meters
Using DPI	(S/DPI)/39.37	(S/DPI)/12
Using m	S X um X .000001	S X um X .00000328

For example, a 1:5,000 scale aerial photograph scanned at 500 DPI would give a pixel resolution of (5000/500)/39.37= 0.254 m per pixel or (5000/500)/12 = 0.833 ft per pixel.

Source: Adapted from Jensen 1996.

1. The spatial resolution typically achieved by aerial film camera systems is about 40 line pairs per millimeter (lp/mm) or 25 μm per line pair. (Film achieves higher lp/mm values under laboratory conditions, but there is a loss of resolution when it is subjected to the atmospheric effects, vibration, and other factors in actual use.) To capture this resolution in a digital image, 2.2 pixels per line pair is often used, giving a pixel size of 25 μm / 2.2 or about 12 μm.

geology, vegetation, and urban and industrial sites. Color infrared film is generally used for specialized vegetation mapping applications.

Panchromatic film is sensitive to radiation in the range of wavelengths from 0.35 μm to 0.7 μm (the whole visible spectrum and the edges of the ultraviolet and near-infrared bands). Atmospheric scattering at the shorter wavelengths (blue and ultraviolet) causes haze, which reduces image contrast and makes features appear less distinct. To reduce this effect, a minus-blue haze-reduction filter is used to block these shorter wavelengths.

When processed, the film produces a *negative image* in which the tones are reversed. This means that the negative image has to be copied on to paper or film in a contact printer so that a *positive image* can be achieved. The opportunity is often taken during this stage to alter the contrast in the image through the use of an *electronic dodging printer*. Thus, the hard-copy images that reach the user often have been dodged to increase their interpretability. The film scanners used to convert the image on the negative film to digital form also have this dodging capability.

Black-and-white infrared film is sensitive to radiation in the range of wavelengths from 0.3 μm to 0.9 μm (the edge of the ultraviolet band, the visible bands, and the near-infrared band). The film is normally used with filters that pass only the 0.7–0.9 μm near-infrared band or pass the visible green, red, and near-infrared (0.5–0.9 μm). Since the shorter wavelengths are filtered out, these film-filter combinations offer good haze penetration and typically provide images with greater contrast than panchromatic film. Infrared energy is strongly absorbed by water, making open water appear black or dark gray, whereas vegetation reflects near-infrared energy and appears in tones ranging from gray to white on the image. The heightened contrast is an advantage in mapping the boundaries of water bodies such as lakes and rivers. Also, vegetation types are more easily distinguished on black-and-white infrared film than panchromatic film, making it more suitable for vegetation mapping.

Color film was developed in the 1930s. However, its high cost, slow speed, and low resolution made it unsuitable for mapping. After World War II, when higher resolution and higher speed color films became available, normal color film became more widely used. Both negative and positive color films are available, the basis for the familiar retail print and slide film respectively. Color aerial photography generally uses positive transparency film because the original film image can be viewed directly, and the color to be viewed is determined when the film is developed. The process of printing from a negative introduces variation in the color rendition (there is color variation in the printing process) and a loss of sharpness in the print produced from the original film. Color negative is the most widely used film for producing orthophoto image products for use in geographic information systems. Again, these color negative images need to be converted to digital form in a suitable color-film scanner if they are to be used in a digital image processing system or a digital photogrammetric workstation. As with black-and-white panchromatic film, the blue sensitivity of color film causes image quality to be strongly degraded by haze. Missions flown at altitudes above 3,000 m or in poor atmospheric conditions may produce low-contrast images with poor quality color that are difficult to interpret. However, the use of a suitable color electronic dodging printer again helps to improve the quality of the images.

Color infrared film was developed during World War II to identify camouflaged military equipment and facilities. Hence, it was often referred to as *camouflage detection film*. Sensitive to near-infrared wavelengths, as well as the visible red and green bands, it produced a very useful image with nonnatural or false colors (hence the film is often called *false-color infrared* film). Green paint and green vegetation both reflect green and absorb red and blue wavelengths, which is why they appear green. However, healthy green vegetation that is chlorophyll rich has a high reflectance in the near-infrared band, whereas the camouflage paint of the time did not. Branches cut to camouflage army vehicles were also easily identified because, when the leaves wilted, their near-infrared reflectance decreased and changed their appearance on color infrared film. Color infrared film was engineered to depict healthy, green vegetation in reddish colors while objects painted green appeared blue or brown and could be easily detected. Green paints have since been developed that mimic the infrared response of vegetation. However, the film is still used to distinguish healthy from unhealthy vegetation and for differentiating plant species—an unintended, but beneficial, use of the film. The film was later optimized for vegetation analysis.

Color infrared film has much better haze penetration characteristics than normal color film because the shorter wavelengths are not recorded. It is used for mapping forest and agricultural areas and is particularly effective in the detection of vegetation stress caused by insect damage, disease, flooding, or other factors. Water, which absorbs near-infrared radiation strongly, appears blue to black, which clearly delineates the water's edge and makes it easier to map. Even streams partially obscured by vegetation are much more easily distinguished on color infrared film than on normal color imagery.

The increasing use of digital infrared imaging sensors has reduced the demand for color infrared film. Kodak is the only remaining supplier. Color infrared film is difficult to handle, the film needs to be stored frozen, and there can be considerable variation in the color rendition between production runs. It is subject to *flight line syndrome* whereby the images gain a blue cast caused by the low humidity at altitude, especially if the camera system needs to be heated because of cold flying weather. Also, one of the best stabilizers for processing the film was formaldehyde, which can no longer be used because of environmental considerations. Radiometric qualities of color infrared photography are not consistent, which is a difficulty for studies such as crop condition assessment for agriculture. Digital color infrared imaging provides more consistent color rendition, a wider radiometric sensitivity, and avoids the difficulties of cold storage to handle the film. For these reasons, digital color infrared imaging is rapidly replacing film.

The near-infrared reflectance of earth features is generally determined by their reflectance of the infrared component of sunlight. So in most cases, it is this reflected infrared energy that color infrared film detects. The infrared energy emitted by earth features is generally at much longer wavelengths than those recorded by film. The exceptions are very high temperature features that emit sufficient energy in the near-infrared to be imaged by false-color film, such as molten lava with a temperature of 1100°C (see *Energy sources* in chapter 3).

Obtaining aerial photography

Aerial photography is obtained either by commissioning an aerial survey firm to fly imagery for an area of interest or by using existing imagery available from government or private sources. Commercial air survey firms offer diverse image acquisition, processing, and mapping services and may also be a source of existing imagery.

Aerial photography is regularly acquired by government agencies and, at least in North America, it is generally made available to the public at low cost. The prices charged generally cover little more than the cost of reproduction, not the substantial cost of flying the photography and processing and storing the film. Many municipalities acquire air photos that can be purchased directly from the aerial survey firm that produced them. In addition, many utilities and resource companies acquire imagery that may be made available to outside users.

In Canada, the National Air Photo Library in Ottawa is the repository for aerial photography collected for the national topographic mapping program. There is complete repeated coverage for the entire country at large to small scales. Provincial mapping agencies and municipalities also produce complete air photo coverage of their jurisdictions.

In the United States, the U.S. Geological Survey (USGS) coordinated the National High Altitude Photography (NHAP) program from 1980 to 1987. The program was then changed to the National Aerial Photography Program (NAPP), which acquires larger scale imagery. The objective of these programs is to build a uniform aerial photography archive for multiple uses and eliminate duplicate efforts in various government agencies. NHAP photography missions simultaneously acquired 1:80,000 scale black-and-white and 1:58,000 scale color infrared photography. NAPP missions are flown at a scale of 1:40,000 in black-and-white or color infrared, depending on state or federal requirements, with the objective of providing complete coverage of the United States on a seven-year cycle *(figure 5.26)*. The NHAP/NAPP program is supported by the U.S. Department of Agriculture, National Resources Conservation Service (formerly the Soil Conservation Service), the Bureau of Land Management, the USGS, and the Tennessee Valley Authority. These agencies as well as the U.S. Forest Service and many state-level agencies and municipalities maintain archives of aerial photography.

In the United Kingdom, systematic aerial photographic coverage of most of the country at 1:10,000 scale has been produced by two competing commercial organizations. The first of these is U.K. Perspectives, which is a partnership between two large air survey and mapping companies—Simmons Aerofilms and Infoterra. The other competing organization is GetMapping, which obtains its photography from a subcontractor, Cooper Aerial Surveys. The coverage from both organizations includes the whole of England and Wales and the central part of Scotland. The photos are available for sale either in hard-copy or digital form and can be purchased either as unrectified images or in orthorectified form. Besides these commercial offerings, the national mapping organization, the Ordnance Survey of Great Britain, also has extensive aerial photographic coverage taken at different times and at different scales according to its requirements for mapping.

Other countries in Western Europe have similar national coverage. In the case of Belgium, full national coverage has been carried out by commercial companies on different occasions. However, for most countries in North Africa, the Middle East, and South Asia, national mapping organizations do not release aerial photography for reasons of national security.

Russia and the United States have commercialized some of their intelligence satellite camera imagery. Aside from their historical interest, declassified images from cold-war archives may be useful in documenting historical land use for sites that may have been exposed to persistent chemical contaminants and may have been off-limits for commercial aerial survey flights.

Electronic frame-capture systems

Digital and film-based cameras work on essentially the same principle. They consist of a lens system precision mounted to focus an image onto a light-sensitive material. In conventional cameras, the light-sensitive material is photographic film; in digital cameras, it is a gridded array of light-sensitive detectors. Each detector generates a voltage proportional to the intensity of light striking its surface, and the voltage is converted to a digital pixel value. One image pixel is generated for each detector.

Figure 5.26 National Aerial Photography Program's seven-year acquisition plan.

Source: Data available from USGS, EROS Data Center, Sioux Falls, S.D.

Detector technologies

Two detector technologies are used in digital cameras—CCD (charge coupled device) and CMOS (complementary metal oxide). They are sensitive to wavelengths from about 0.4 μm to 1.1 μm covering the visible and near-infrared region of the spectrum and offering a spectral sensitivity that extends further into the near-infrared than film. Until recently, high-end digital cameras used CCDs exclusively (*figure 5.27*). They had a greater dynamic range (i.e., they could distinguish more brightness levels from light to dark), were more light sensitive (could operate under lower light conditions), and had less noise (introduce fewer random pixel values). Developments in CMOS technology are overcoming many of these limitations, resulting in their more frequent use in high-end systems.

CMOS uses the same silicon chip manufacturing techniques as most semiconductors whereas CCDs require a unique process. This makes CCDs more costly to manufacture. CMOS chips also consume less power, as little as 1/100th of that needed by a CCD chip. When a detector cell on a CCD is exposed to excess light, it tends to increase the voltage of the surrounding cells, increasing their pixel values. This effect, called *blooming*, is not present in CMOS sensors.

A new type of CMOS detector captures three bands of image data at each detector location. The depth of light penetration into a silicon chip depends on the wavelength. By embedding three layers of detectors within a silicon chip, this property can be used to generate three values for each pixel, representing the brightness in the red, green, and blue wavelength bands, respectively (*figure 5.28*). So for every pixel on the sensor array, there is a stack of three detectors thereby producing a full-color image without interpolation. However, compared with CCDs, there is more overlap between the three bands, lower sensitivity, and a lack of near-infrared capability compared with CCDs. These characteristics limit the use of this new CMOS chip for remote sensing applications.

Digital versus film images

While film and digital images may appear similar, there are fundamental differences in their structure.

Film-based images are formed by varying sizes of randomly arranged grains of silver halide that, during development, are converted to grains of silver in black-and-white films or, in

Figure 5.27   CCD array. This Kodak CCD sensor, used in digital cameras, measures about 3.4 cm across. The CCD array in the center consists of almost 4.2 million detectors, each 9 μm x 9 μm in size, and captures a 2042 x 2042 pixel image.

Source: Image courtesy of © Eastman Kodak Company.

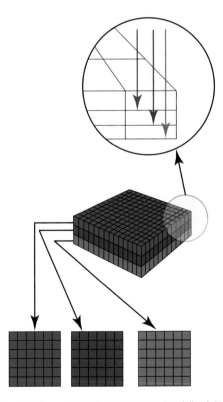

Figure 5.28   Multilayer CMOS detector arrays. Specialized CMOS chips have been designed with three layers of detectors to simultaneously measure the brightness values for the blue, green, and red bands at each detector site. Separate digital images can then be generated for each band.

the case of color films, are replaced by colored dyes. Thus, the smallest element in a film-based image is the silver grain or spot of dye; i.e., there is no detail to be seen within these elements.

Digital images consist of a regular array of colored square picture elements (pixels). The color displayed for each pixel is determined by one or more digital values assigned to that pixel location. A pixel is the smallest element for which a value is recorded in a digital image file; thus, there is no detail within a pixel—the entire pixel is represented as a single color. In images displayed with large numbers of pixels (i.e., at high resolution), the pixels can be small enough to be imperceptible to the eye.

Although film and digital images may look the same at the normal viewing magnifications, at extreme enlargements they look very different. A film image appears grainy, but the digital image looks like a patchwork of colored squares—i.e., it is pixilated *(figure 5.29)*. These differences in structure can produce characteristics such as aliasing in digital images (discussed in chapter 4) that must be taken into account in their use.

Digital cameras

The format size of digital cameras is given by the number of pixels in the sensor array. A 3,000 x 2,000 pixel array has 6 million pixels and is referred to as a 6 megapixel camera. The size of the individual detectors in the sensor arrays used in

the digital cameras employed in airborne remote sensing lie in the range 8 to 15 μm.

Advantages that digital cameras have over their film counterparts include the ability to capture imagery at lower light levels and a greater dynamic range, i.e., the ability to reproduce a wider range of light levels.

It is important to note that the digital sensor arrays employed in the digital camera systems used in remote sensing are fundamentally single-band detectors[1]. Filters are used to make the image generated by an array represent a specific band, or in the case of a mosaic filter, a pattern of filters over the detector array generates a single image with individual pixels representing the brightness level in one of three different bands. Color films contain three emulsion layers, each made sensitive to different spectral bands. As a result, a color film simultaneously captures a three-band image.

Digital frame cameras used in remote sensing can be divided into four general categories: large-format cameras, single cameras using mosaic filters, multiple camera designs using a single detector per camera, and multiband cameras using beam splitting.

1. As previously discussed, the CMOS sensor manufactured by Foveon, Inc. simultaneously captures brightness values for three bands for each pixel. However, other characteristics of the sensor have so far precluded its use for remote sensing applications.

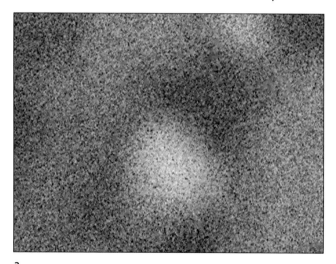

a

b

Figure 5.29 Comparison of image structure in film and digital images. A film image consists of varying sizes of randomly distributed grains or spots of colored dye (a). A digital image is comprised of a regular distribution of same-sized, usually square, picture elements (b).

Large-format digital cameras

Large-format digital cameras capture one brightness value for each pixel in a single wavelength band. This approach has been used extensively in all the highest resolution and largest format digital cameras that have so far been developed. These include the Z/I Imaging® DMC® (digital mapping camera), which comprises four individual cameras housed within a single unit, each fitted with a 7,000 x 4,000 pixel (28 megapixel) CCD area array. The four cameras are integrated within a single box, arranged in a star-type arrangement with each camera pointing outward to generate a tilted (low-oblique) panchromatic image. The images from the four tilted cameras overlap slightly and are exposed simultaneously. Later the images are processed (i.e., rectified) to form a single vertical perspective pan image that is 13,500 x 8,000 pixels (108 megapixels) in size. The Z/I Imaging DMC includes an additional multiple camera system comprising four small-format cameras, each with a 2,000 x 3,000 pixel (6 megapixel) array that produces multispectral imagery in the near-infrared and visible blue, green, and red bands. Natural color and color infrared composite images can be generated from these multispectral images.

The Vexcel ULTRACAM D camera also consists of multiple digital cameras housed within a single unit, but in this case, they all point in the vertical (nadir) direction with parallel optical axes *(figure 5.30)*. Four of the cameras are aligned linearly to view adjacent areas. Their shutters are opened sequentially, with a tiny time delay between them, so that each image is acquired from a single exposure station in the air. When stitched together, they form a single complete pan image, 11,500 x 7,500 pixels (86 megapixels) in size. The ULTRACAM D has four additional cameras within the same instrument aligned to view the same area to acquire smaller multispectral images (4,000 x 2,700 or 10.8 megapixels in size) in the near-infrared and visible red, green, and blue portions of the spectrum. Normal color and false-color images can be generated from this data.

In addition to these commercial products built in Europe, other large-format digital cameras have been produced in the United States for military reconnaissance purposes. These include a series of cameras built by the Recon/Optical company with format sizes up to 10,000 x 5,000 pixels (50 megapixels). Several models of digital camera have also been built by BAE Systems that use its 9,000 x 9,000 pixel (85 megapixel) CCD area array. Both of these American companies use single CCD area arrays in their cameras, rather than the multiple arrays used in the Z/I Imaging and Vexcel cameras.

Figure 5.30 Vexcel ULTRACAM D large-format digital frame camera. This camera uses eight lenses for its eight CCD arrays. Four of these are located in-line to produce adjacent segments that are combined to form one 86 megapixel panchromatic large-format frame image. The remaining four acquire four images of the same area to produce a four-band multispectral image 10.8 megapixels in size.

Source: Image courtesy of Vexcel Corporation.

Single digital cameras with mosaic filters

This is the technology used in digital cameras built for the consumer and professional photography market. These cameras record three bands to produce color images: the red, green, and blue visible wavelengths. Color infrared cameras record the green, red, and near-infrared bands. These systems need to generate three values for each pixel to represent the brightness in each of the three bands. Since there is only one detector for each pixel, the missing values are estimated from the surrounding pixels using a process called *Bayer interpolation* (see *Interpolated color* later in this chapter).

Multiple camera design using a single detector per camera

This design, developed extensively in the United States, uses multiple small-format CCD cameras. These cameras are coupled together with their optical axes set parallel to one another and their shutters set to operate simultaneously to cover the same piece of terrain. Each camera is equipped with a different spectral filter. In most systems, four cameras are used, typically with filters that allow the simultaneous acquisition of images in the blue, green, red and near-infrared portions of the spectrum. From these component images, true-color or false-color images can be created. With this approach, no interpolation is required, and each pixel contained in the final image will have the brightness value that was actually measured in the air during its exposure. However, while the individual cameras are less complex and less expensive than a multidetector system, the use of multiple cameras is obviously more expensive than using a single camera with a Bayer array and interpolation. Moreover, there is the need to accurately register the separate images so that pixels representing the same location on different images are precisely overlaid. Multiple cameras housed in the same unit, discussed previously, are more accurately calibrated and considerably more expensive.

Multiband camera using beam splitting

In this design, the incoming light is split by a prism into (usually) three channels (red, green, and blue), each of which is focused on a different detector array *(figure 5.31)*. The three arrays produce images of exactly the same scene in different bands. Filters can be inserted in the channel path to further narrow the wavelength bands. The use of filters greatly improves the radiometric accuracy of the imagery, which is important for remote sensing where spectral analysis is involved. However, the high-quality prism and complex alignment required make this approach more costly than using a single detector array. Separate sensors for each band also necessitate precise registration of the separate band images that comprise each scene. Beam splitting necessarily reduces the brightness of the individual images, which gives a poorer

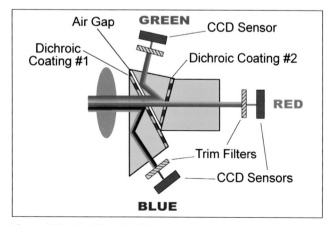

Figure 5.31 Multiband/multidetector digital cameras use a prism to split visible light into separate wavelength components. These bands may be further narrowed by the use of filters.

Source: Image courtesy of Redlake MASD, LLC. www.redlake.com

signal-to-noise ratio, whereas with large-format and single digital cameras (described previously) the separate images are captured with separate lenses, avoiding this light loss.

The red, green, and blue bands typically used in color imaging are based on a convenient equal division of the visible light spectrum into blue (0.4–0.5 µm), green (0.5–0.6 µm), and red (0.6–0.7 µm). However, for remote sensing, results are improved if the wavelength bands detected are optimized for the application; e.g., for a vegetation mapping project, the bands will be chosen that will best distinguish the expected vegetation types. Narrow bands are not necessarily better; the narrower the band, the less energy available to be detected and the poorer the signal-to-noise ratio.

Manufacturing large-format sensor arrays is costly. As noted above, some system designs use multiple sensor arrays to capture one very large image frame. Each design has its trade-offs, making it useful for certain applications and less effective for others. Multidetector systems can offer greater flexibility and control of band selection but increase the risk of reduced sharpness caused by slight band-to-band misregistration. In addition, the quality of the implementation is an important factor in the performance of the instrument, whatever the design.

Remote sensing application of digital cameras

Digital camera technology is advancing rapidly. Each year, cameras with larger pixel arrays producing sharper imagery are becoming available in the consumer market. Some of this technology is readily adapted for remote sensing either by using off-the-shelf components to build specialized remote sensing systems or by using existing digital cameras, with or without modifications.

Aircraft-mounted digital cameras are being adopted for many applications formerly served by aerial photography, such as natural resource management, precision farming, mining, military surveillance, urban planning, and environmental monitoring. By providing overlapping stereo imagery in digital form and precise aircraft position and attitude data from GPS and inertial measurement units integrated with the camera system, the process of generating orthorectified imagery and digital elevation models can be highly automated. Some airborne systems are capable of delivering orthorectified imagery upon completion of the acquisition flight, with a higher precision product available within a few days.

The application of digital camera technology in remote sensing can be broadly divided into four categories: hand-

Interpolated color

A digital camera produces color images by measuring the brightness level in three roughly equal contiguous ranges within the visible wavelength band. These three ranges are designated the red, green, and blue bands for convenience; in fact, they contain wavelengths the eye perceives as a range of distinct colors. The wavelength perceived as violet, for example, would be included in the visible blue band.

By simultaneously measuring the intensity level in these three bands, three brightness values are obtained for each pixel. The data is typically stored as red, green, and blue raster images, commonly called red, green, and blue *channels*, which together comprise an *RGB image*. From the three values stored for each pixel location, a full range of colors can be generated.

The digital cameras developed for amateur and professional photography typically use a single detector array to capture a full color image, i.e., there is only one detector per pixel. These detectors are sensitive to the full visible spectrum and usually the near-infrared as well.

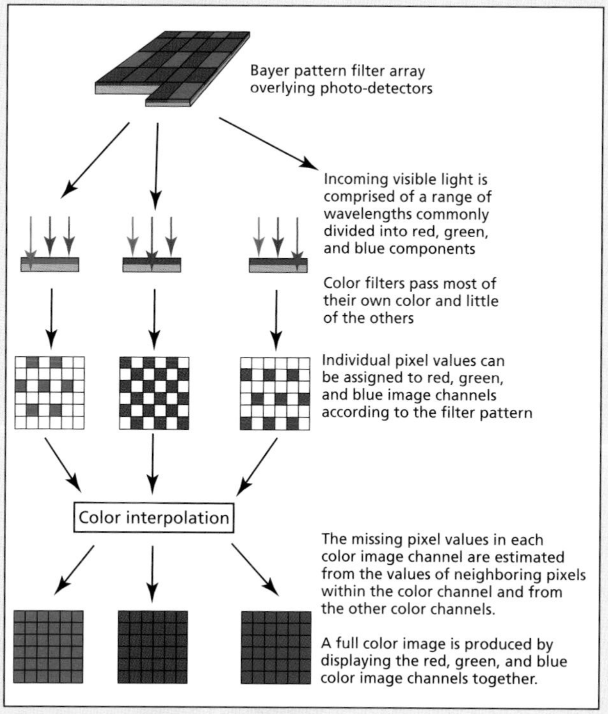

To produce the three pixel values of an RGB image, the detectors are covered with a layer of filters that pass the red, green, or blue wavelength ranges *(figure 5.32)*. The filters are registered to the underlying photo-detectors so that each detector records the energy from one wavelength band (red, green, or blue). The pattern of filters most commonly used is the Bayer pattern, consisting of four-pixel groupings with one red, one blue, and two green pixels. (The extra green is present because the human eye is most sensitive to that wave band.) From the image data, three sparsely populated RGB image components are generated (i.e., 25% of the pixels to the red channel, 25% to the blue channel, and 50% to the green channel). The missing pixel values in each channel are then estimated or interpolated from the surrounding pixels to produce fully populated red, green, and blue raster images for each channel.

The interpolation algorithms are very effective. The imagery produced from Bayer arrays can be as sharp, or sharper, and radiometrically as accurate as imagery produced by other designs that use multiple sensor arrays. This is because multiple sensor arrays must contend with other factors that degrade image quality. For example, misregistration of the images captured by the different arrays can degrade image sharpness and reduce radiometric accuracy. Blur filters are sometimes used to reduce alias effects caused by the interaction between the size and pattern of the pixel array and that of features in the image. When used, blur filters generally have a positive effect on image quality, and the blurring is not generally detectable. Where blurring is perceived, it is more likely caused by camera motion or haze than by the blur filter.

Bayer pattern filter array overlying photo-detectors

Incoming visible light is comprised of a range of wavelengths commonly divided into red, green, and blue components

Color filters pass most of their own color and little of the others

Individual pixel values can be assigned to red, green, and blue image channels according to the filter pattern

Color interpolation

The missing pixel values in each color image channel are estimated from the values of neighboring pixels within the color channel and from the other color channels.

A full color image is produced by displaying the red, green, and blue color image channels together.

Figure 5.32   Interpolated color.

held small-format, aircraft-mounted small-format, aircraft-mounted medium-format, and aircraft-mounted large-format metric cameras.

Handheld small-format digital cameras

Handheld small-format digital cameras are generally used to collect oblique aerial images and ground photos of selected targets. For example, the images may be used to document field data collection or as a record of status at a point in time, such as the progress of a construction project or land development. GPS data is often collected simultaneously and some cameras enable the data to be included in the image file.

Rangefinder designs, where the image is viewed through a separate window, are lighter in weight than single lens reflex (SLR) cameras where the image is viewed through the same lens used to acquire the image. However, SLR cameras are usually preferred for their interchangeable lenses, faster cycle times between shots, and other advanced features *(figure 5.33)*. Imagery is stored in the camera's flash memory card. Some models can output imagery directly to an external storage device such as a hard disk. The pixel resolution of these cameras has increased rapidly. The resolution of current amateur and professional digital SLRs range from about 6 to 16 megapixels. The higher resolution cameras use CMOS arrays.

Aircraft-mounted small-format cameras

Aircraft-mounted small-format cameras are typically used to collect imagery over small areas, such as for crop condition monitoring of tens of square kilometers, and to record high-resolution, low-altitude imagery to document linear features such as wetlands along streams and rivers or utilities such as power lines, seismic lines, and other corridor assessment applications. They are often flown to produce an image record as a reference for data collected with other sensors such as lidar or digital video. These systems may use multiple cameras to capture narrow band multispectral imagery or a single color or color infrared digital camera. The sensor arrays in these cameras have increased in size with 14 and 16 megapixel sensors available from such manufacturers such as Kodak and Canon. With larger sensors being used in small-format bodies, the image size of these cameras overlaps that of medium-format digital cameras.

For remote sensing applications, models that can be computer controlled are preferred as they enable flight management software to control the shutter. In addition, external image storage is important for practical handling of the large volume of imagery typically collected. Figure 5.34 shows a Canon EOS-1D on its mounting bracket for use in light aircraft or helicopters, together with a sample image. The upgraded Canon EOS-1Ds Mark II offers a larger 16 megapixel image resolution.

a   b

Figure 5.33 Canon EOS-D60 digital single lens reflex camera. Digital SLRs typically have a small color display on the rear panel for viewing the current and stored images.

Source: © Canon, Inc.

a

b

Figure 5.34   The Canon EOS-1D small-format digital camera is shown mounted next to a lidar sensor on a bracket designed for use in small aircraft and helicopters (a). This digital camera produces 2,464 by 1,648 pixel images (b). This image shows a ground distance 250 m across, with a spatial resolution of 10 cm GSD, a resolution typical for imagery acquired during a lidar survey.

Source: Image courtesy of Terrapoint Canada Inc.

infrared imagery with 12-bit quantization. Typical ground resolutions range from 0.3 m to 1 m. The German company, IGI produces both small- and medium-format integrated systems. With the rapid development of sensor technology, these systems are regularly upgraded to provide higher pixel resolutions.

Aircraft-mounted medium-format cameras

Aircraft mounted medium-format cameras are typically used for multispectral imaging of small areas, up to a few hundred square kilometers. With image sizes from 16 to 22 megapixels, they offer considerably greater coverage than the smaller cameras but are much smaller and less expensive to operate than a standard-format metric camera. These systems, available from such manufacturers as Applanix, Spectrum, and Rollei, are suited to small, single-engine aircraft with little or no modification. Off-the-shelf professional medium-format cameras from Contax, Hasselblad, and Rollei, fitted with digital imaging camera backs and control systems, are being used for remote sensing *(figure 5.35)*.

The Applanix DSS *(figure 5.36)* is an integrated airborne digital camera system designed to be remotely operated by the pilot in a single engine aircraft. The flight-management system provides direct geopositioning and software for mission planning and execution. The unit shown provided 4,000 x 4,000 pixel interpolated natural color or color

Aircraft-mounted large-format digital metric cameras

Aircraft-mounted large-format digital metric cameras produce images on the order of 60 to more than 100 megapixels. The DMC large-format metric digital camera manufactured by Z/I Imaging is an example of this type of digital metric camera. This camera system produces 7,680 x 13,824 pixel panchromatic images (i.e., 106 megapixels at 12-bit quantization per band), and four-band multispectral images with 3,000 x 2,000 pixel resolution. It incorporates a digital forward motion compensation system and a gyro-stabilized mount, which enables ground sample distances as small as 5 cm to be captured. The camera unit weighs 80 kg and the flight-management system computer, another 20 kg.

These large-format digital cameras are designed to replace conventional metric film cameras. The economics of these systems versus film cameras is hotly debated. For example, the DMC costs about $1.4 million, which is about three times the cost of a standard format metric film camera. Though the

Figure 5.35 Medium-format digital camera system built on a conventional film camera body installed in its mounting bracket with inertial measurement unit connected. Here a Hasselblad 553ELX is fitted with a BigShot™ 4,000 x 4,000 pixel panchromatic digital camera back. The camera system has the electronic control functionality to design custom software to integrate an inertial measurement unit and GPS to provide a fully digital computer controlled remote sensing system.

Source: Image courtesy C. K. Toth, Center for Mapping, Ohio State University.

Figure 5.36 Medium-format digital metric camera system. This digital camera system, the DSS from Applanix Corporation, produces 4,092 x 4,079 pixel interpolated-color images using 12-bit quantization. The georeferencing data collected during image acquisition enables the rectification of the imagery to be highly automated for rapid product generation.

Source: Image courtesy of Applanix Corporation. © Trimble Navigation, Ltd.

imagery is generated in digital form, DMC panchromatic imagery requires extensive preprocessing of its tilted images to generate the single large-format image suitable for photogrammetric use. (Similarly, the Vexcel ULTRACAM large-format digital camera generates four subimages that need to be stitched together before the single pan image is produced.)

However, the limiting factor in the application of large-format metric cameras is the fact that even the DMC 108 megapixel image is still much smaller that that of a standard-format film image digitized at a pixel size of 10 μm, which will generate 23,000 x 23,000 pixels (over 529 megapixels).

Another potential limitation for some users is the volume of data produced. At a radiometric resolution of 12 bits stored as 16-bit values, the DMC captures each frame as a panchromatic image of about 220 MB and 48 MB for the four multispectral images at a maximum rate of one frame every two seconds—i.e., about one gigabyte of image data must be stored every eight seconds if all sensors are used simultaneously. In practice, these systems are operated to collect about 500 GB of imagery during a flight. Other digital metric camera systems produce smaller images and therefore require a lower rate of data storage. But ultimately, high-resolution imagery of large areas requires that terabytes of data be stored and processed. The cost effectiveness of these systems is dependent on the availability of greater storage capacities and higher speed computers at lower cost, together with the availability of suitable software to handle these very large datasets and make available the required image data rapidly.

Trade-offs between film and digital image acquisition

Though digital image capture may seem to have all the advantages for producing digital imagery for photogrammetric applications and the production of orthorectified imagery, there are significant trade-offs.

Comparisons between film and digital image capture are not straightforward. The characteristics of the media are quite different. Film is an amorphous medium; the image is formed of randomly distributed silver halide crystals that produce a more or less continuous gradation of tone and color values. A digital image consists of a regular grid of uniform sized pixels, each of which can be assigned a number of values predetermined by the system design, albeit the range of values may be very large. The economics of acquiring and maintaining the equipment differ substantially, and the procedures of handling and storing film and digital data demand different skill sets. Also, film, being the older technology, has a large installed base and well-developed procedures. Some of the major trade-offs to consider are listed below:

- Film and digital capture require different preprocessing. Film has to be developed and scanned to produce digital imagery, whereas imagery captured with digital camera technology is already in digital form. However, as previously noted, large-format digital images are captured in multiple image segments that must be rectified and stitched together.

- Some medium- and small-format digital camera systems have integrated data from onboard GPS and inertial measurement units and automated image rectification to the point that, using a preexisting DEM, an orthorectified image can be available within hours of capture. Usually, the geometric accuracy is less than if additional processing is done to meet standard map accuracy specifications. However, the accuracy satisfies the requirements for many applications, especially when rapid response is essential.

- For some applications, the capture of very large detailed imagery is important. Large-format film cameras with image sizes of 23 x 23 cm and larger produce imagery with more detail in a single frame than one large-format digital camera frame can capture. Film formats are much larger than those of digital cameras. Indeed, digital cameras with comparable formats have still to be developed.

- High-quality imagery from a standard-format aerial metric camera produces imagery with a spatial resolution of between about 35 to 55 line pairs per mm (lp/mm). To capture a digital image with the spatial resolution of 40 lp/mm (25 μm per line pair), there should be somewhat more than 2 pixels per line pair. A value of 2.2 pixels per line pair or about 12 μm per pixel is commonly used. A standard metric camera image is 230 mm x 230 mm (9 in x 9 in) in size. At 12 μm per pixel, the equivalent digital image would be about 19,000 x 19,000 pixels. The radiometric resolution of aerial photography is generally no better than about 8 bits (256 gray levels) for black-and-white film and 24-bits for color films (8-bits for each of the three separate layers). So a liberal estimate of the equivalent information content for film is about 360 MB for a black-and-white image and three times that, or about 1 GB, for a color image. A single 600-

image roll of color aerial film would thus be equivalent to about 600 GB of digital image data.

- Electronic detectors have wider exposure latitude than film (higher radiometric resolution). That is, they can record a wider range of brightness values. As previously noted, film can typically record a brightness range of about 8 bits (256 levels) per emulsion layer, whereas digital sensors offer 11 bits (2048 levels) or 12 bits (4096 levels). Digital capture technology offers a higher light sensitivity and better signal-to-noise ratio than film (typically about 40:1 for film versus 200:1 for a CCD sensor). This allows faster shutter speeds to be used, which reduces motion effects, allowing more detail to be retained in shadow and highlight areas and imagery to be collected at lower light levels.

- Though digital imagery may not be immediately available in a fully rectified form, imagery can be immediately viewed to ensure proper capture. There can be substantial cost savings and quality improvement if missed or poorly imaged areas can be immediately reflown. Not only may the cost to schedule a subsequent mission be saved, the image consistency maintained by reflying imagery on the same day and with approximately the same lighting conditions can be important for many applications.

- Digital cameras can acquire multispectral imagery in wavelength bands customized for the application as required for automated classification and other applications dependant on accurate measurement of intensity in multiple spectral bands. However, digital scanning technology (discussed in chapter 6) is the preferred sensor for multispectral imaging. Film cannot provide multispectral imagery suitable for these applications.

- Digital capture offers more consistent image characteristics than film. Differences in film lots and processing chemistry cause variations in photographic images. Calibration steps in the process of digital camera operation and image generation provides higher radiometric accuracy and therefore more consistent imaging than film.

- Film is currently a much less costly method to store large volumes of high resolution image data but, as the cost of digital storage continues to drop, this advantage will diminish.

- There is a large, well-established aerial survey industry to fly and process standard-format aerial photography.

These firms commonly have photographic equipment that is fully paid for and can be used for many years at little additional expense. Conventional photographic systems are rugged, have been used for decades, and are less complex than digital systems. They are extremely cost effective. But they are also less capable. Digital camera technology is new and developing rapidly.

A firm can develop a small-format digital camera imaging capability at relatively modest cost for the camera and supporting technology and personnel to produce marketable products. However, medium- and large-format metric digital cameras become substantial investments in technology, personnel, and specialized training that must be justified by a profitable product and significant demand.

In summary, while digital capture offers a number of advantages over film, including higher radiometric resolution, multispectral image capture, more consistent image characteristics, and elimination of film scanning, there can also be significant disadvantages, such as higher equipment and data storage costs and current limitations on the maximum size of image that can be captured in a single digital image frame. The choice of technology will largely depend on the requirements of a specific application, the skill sets of the personnel, and the technology available to complete the work. However, the trend is moving toward an ever-increasing use of digital imaging technology.

Aerial and underwater videography

Aerial videography uses standard analog or digital video recording and playback equipment. The equipment can be handheld or mounted and can be used from virtually any platform from light aircraft and helicopters to land- or water-based vehicles. The system is usually integrated with a GPS, which provides continuous recording of geographic position and elevation along with the image data. Voice recordings of observations made during acquisition can also be stored in the video file.

The data can be viewed as continuous moving images or one image at a time using freeze-frame playback. Analog video can also be converted to individual digital images using a special analog-to-digital converter called a *frame grabber*—available as an add-on hardware board for computer systems.

Videography systems have been integrated with GPS to provide rapid, low-cost, spatially-referenced image acquisition for

such applications as environmental sampling, shoreline mapping, surveillance, and asset management, e.g., continuous recording of road surface conditions from a truck-mounted video system.

Underwater video systems

Underwater video imagery has long been used to document seabed habitats during diving surveys. More recently, GPS and depth measurement have been integrated with the underwater video system for mapping applications. There are three approaches to underwater videography: the equipment is either carried by a diver, towed, or mounted on a remotely operated vehicle (ROV).

Video imagery is very effective in communicating complex information about marine habitats and the operation of marine equipment such as fishing gear. Video systems collect large quantities of imagery quickly, demanding efficient archiving and retrieval systems. In fact, processing and interpreting the imagery and encoding the information as digital data can be extremely labor intensive and a major bottleneck. Underwater video systems can be operated at all depths. With the use of low-light, red and near-infrared lighting below the visual sensitivity of target organisms, records of undisturbed behavior and abundance can be obtained. Advances in imaging and positioning technologies have made it possible to generate accurate 2D and 3D measurements of size and position from video imagery.

Among its limitations, video is limited to areas of reasonable water clarity, and the footprint of video is much smaller than the side-scan sonar and airborne multispectral scanner technologies used for marine habitat mapping.

The Seabed Imaging and Mapping System (SIMS) is an example of an inexpensive towed underwater video system developed in Canada by Coastal and Ocean Resources Inc. and Archipelago Marine Research Ltd. to map shallow, near-shore habitats *(figure 5.37)*. Operating in shallow water depths up to 25 m, SIMS collects underwater video imagery of the near-shore subtidal seabed, georeferenced using differential GPS (DGPS)[2]. The video imagery is reviewed by a geologist and a biologist, and their interpretations for each

---

2. DGPS satellite positioning is used to continuously record the position of the surface vessel towing the underwater instrument from which its velocity and travel direction can be determined. In addition, the length of cable spooled out is recorded, and an instrument on the towfish provides data on its depth. From this data, the position of the towfish can be calculated to within ±4 m at the 95% confidence level.

second of video imagery are recorded in database format using a standardized classification system for marine substrate, flora, and fauna *(figure 5.38)*. Using the position and depth data for the towfish, the classifications are integrated within a GIS to produce maps of bottom habitats as well as precise mapping of point-specific features *(figure 5.39)*. The system has also been used to identify storm water and sewage outfalls, map cable and pipeline corridors, monitor compliance, and inventory regional resources (e.g., eel grass).

Airborne video systems

Real-time videography systems that transmit live video with synchronized GPS information to a receiving unit at a ground command post are available. There the signal is automatically plotted onto a digitized moving map display. The airborne platform is tracked on a mapping system which displays both the live position of the aircraft together with the live video picture stream. Standard navigation data is displayed, including position, time, date, course over the ground, altitude, and velocity. Industry standard map formats are used to facilitate integration with commonly used GIS, and the operator can zoom in, zoom out, pan, and even perform live data queries. Some systems also incorporate camera position data to provide target positioning of ground objects. Recently much public attention has been focused on the successful use of airborne video cameras mounted

on unmanned airborne vehicles (UAVs) during the recent conflicts in Afghanistan and Iraq.

The principal disadvantage of the video camera is its relatively poor spatial resolution compared with that of film or digital cameras. The effective shutter speed of video systems is relatively slow, 1/30th to 1/60th of a second for the NTSC RS-170 standard used in North America and Japan. A 700 x 570 pixel image output is typical. To obtain good quality images, an automatic exposure adjustment can be used (the

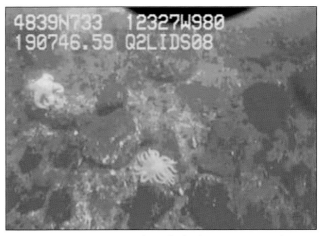

Figure 5.38   Screen capture of a seabed video image. Image shows boulders lying over bedrock at a depth of about 10 m. The yellow star (Pycnopodia) is approximately 50 cm across.

Source: Courtesy of J. R. Harper, President, Coastal and Ocean Resources, Inc.

Figure 5.37   The Seabed Imaging and Mapping System (SIMS) consists of a towed video camera that images the seabed and is precisely positioned with real-time differential GPS (DGPS). A specialized seabed classification system allows precision mapping of the near-shore substrate and flora and fauna, as well as identification of specific seabed targets. The system is operated to depths of about 30 m.

Source: Courtesy of J. R. Harper, President, Coastal and Ocean Resources, Inc.

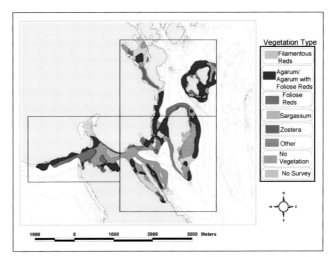

Figure 5.39   Vegetation type map generated from interpretation of underwater video mapping system (SIMS) imagery.

Source: Courtesy of J. R. Harper, President, Coastal and Ocean Resources, Inc.

white balance adjustment). However, this automated correction will make similar land-cover features look different, so color alone cannot be used to classify cover types.

Despite these limitations, aerial videography offers a number of advantages:

- Video equipment is relatively low cost and light weight. It can be flown on small aircraft that are relatively inexpensive to operate, or it can be built into small underwater housings. Video images can be viewed as they are acquired and used immediately, whereas film must first be developed. Video tapes are less expensive than the cost of film and processing, and the tapes can be reused. Video equipment also provides an audio track to record observations during acquisition.

- Videography has been used in applications where rapid acquisition is of critical importance, and the limitations of image quality can be tolerated, such as for crop condition monitoring, disease detection, and emergency response. For example, a videography system integrated with a GIS was used for emergency response at the Salt Lake City, Utah, Olympic Games in 2002. The video feed from helicopter-based video equipment was transmitted to the Operations Control Center and displayed in real time alongside a GIS-based map showing the helicopter position (*figure 5.40*).

Other applications include reconnaissance-level agricultural, rangeland, and natural resource management; detection of hazardous waste sites; monitoring of utility right-of-ways; detection of forest disease and insect damage; and irrigation mapping. The Coastal Alaska Web site illustrates an innovative use of airborne videography. The Web site, sponsored by several government agencies and local government, provides data on multiple resources of the Alaska coast. In addition to online interactive map creation and free downloading of existing datasets in GIS-compatible format, georeferenced video imagery of the coast collected by helicopter during low tide to show the intertidal zone can be viewed. This imagery is being used by scientists, resource managers, and recreational visitors to select sample sites, document existing habitat conditions, identify essential fish habitat, such as eel grass and kelp beds, and plan emergency response to events such as oil spills.

Small-format digital cameras operated with a GPS to record the location of each image frame have replaced aerial videography for many applications. Digital cameras offer higher resolution, relatively low cost, and better image geometry than video. Images can easily be accessed from a GIS as

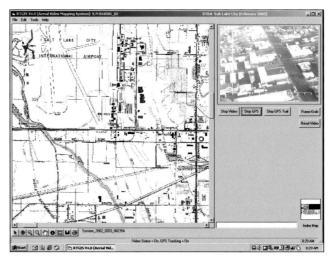

Figure 5.40 Real-time videography integrated with a GIS. As part of the emergency response capability for the Salt Lake City Winter Olympic Games in 2002, a GIS-based videography application provided simultaneous real-time display of the video imagery from a helicopter-based camera and the helicopter position overlaid on a detailed area map.

Source: Image courtesy of BlueGlen Technology, Ltd.

separate views, or vertical imagery can be mosaicked to map small areas. However, aerial videography still offers unique advantages for applications that require real-time continuous image capture.

Conclusion

Much of the information stored in a GIS is generated from remotely sensed imagery. In most cases, photogrammetric methods are used to extract vector point, line, and polygon data and assign attributes to generate accurately georeferenced data products suitable for input to a GIS. Orthorectified imagery is commonly used as a background for orientation over which vector-format maps and other geospatial information are displayed or in the generation of perspective views.

Imagery can also serve as a cost-effective means of storing information that is required only selectively or unpredictably. When information can be readily interpreted directly from the image and most of the data will rarely if ever be needed, it may be more cost effective to store the image than digitize and store the derived data.

Considerable savings can be realized by avoiding the cost of analyzing and digitizing features for large areas when little of the data will ever be needed and can be readily interpreted directly from the image. Features that need to be recorded can be digitized directly on the screen image as needed. For example, the presence of structures on a municipal lot could be checked by viewing the imagery with an overlay of property lines. Analysis of drought damage might only be required for those areas for which assistance payments were being offered. An additional benefit of analyzing imagery as needed is that the most recent imagery can be used as soon as it is delivered.

Ultimately, to the GIS manager contracting for airborne digital image products, it is the quality of the product and timeliness of delivery that is the deciding factor in choosing a contractor. However, because the technology is changing rapidly, the suite of technology and personnel skills available to a contractor vary widely and directly affect the cost at which a given product can be produced. Small differences in the time to delivery, resolution, and other specifications can greatly impact the cost for one contractor while having little impact on the cost for another to deliver the same product. Thus, considerable savings can be realized by a judicious

requirements assessment to determine minimum product requirements for the intended applications, e.g., the coarsest resolution and the lowest level of geometric accuracy. Service providers can then be assessed and selected by their ability to reliably meet the minimum requirements most cost effectively. Further negotiations with the chosen provider can identify additional products and services that might be added at little or no extra cost.

Airborne film and digital frame-capture systems are the major sources of imagery used to generate the georeferenced data products and orthorectified image mosaics used in geographic information systems. While digital cameras are being rapidly adopted for remote sensing applications, there remain many situations, particularly large-area, high-spatial-resolution projects when standard-format film cameras may be the most cost-effective solution. The rapid progress of digital imaging technology promises not only improvements in the quality of imagery but also in the reliability of these complex systems. Together, these developments should steadily reduce the cost of operation and barriers to entry for digital frame-capture systems so the technology and expertise are more widely available and the products less costly.

References and further reading

Ang, T. 2002. *Digital photographer's handbook.* New York: DK Publishing.

Campbell, J. B. 2002. *Introduction to remote sensing.* 3rd ed. New York: The Guilford Press.

Escobar, D. E., J. H. Everitt, M. R. Davis, R. S. Fletcher, and C. Yang. 2002. Relationship between plant spectral reflectances and their tonal responses on aerial photographs. *Geocarto International* 17(2): 63–74.

Harvey, E., and M. Cappo. Video sensing of the size and abundance of target and non-target fauna in Australian fisheries. National Workshop 4-7 September 2000, Rottnest Island, Western Australia. Fisheries Research Development Corporation. Report 2000/ 187 Fisheries Research and Development Corporation. Deakin, Australia. www.aims.gov.au/pages/research/video-sensing/index.html

Jensen, J. R. 1996. *Introductory digital image processing: A remote sensing perspective.* 2nd ed. Upper Saddle River, N.J.: Simon and Schuster.

Lillesand, T. M., and R. W. Kiefer. 2000. *Remote sensing and image interpretation.* New York: John Wiley and Sons, Inc.

Petrie, G. 2003. Airborne digital frame cameras: The technology is really improving. *GeoInformatics* 6(7): 18–27.

———. 2001. 3D stereo-viewing of digital imagery: Is auto-stereoscopy the future for 3D? *Geoformatics* 4(10): 24–29.

U.S. Geological Survey. National aerial photography program (NAPP) information. Sioux Falls, S.D.: EROS Data Center. U.S. Geological Survey. edc.usgs.gov/products/aerial/napp.html

Web sites

Digital frame camera systems

Airborne data systems.
www.airbornedatasystems.com

Applanix Corporation.
www.applanix.com

DIMAC Systems.
www.dimac-camera.com

IGI mbH.
www.igi-ccns.com

Redlake MASD, LLC.
www.redlake.com

Spectrum Mapping.
www.spectrummapping.com

Vexcel Corporation.
www.vexcel.com

Visual Intelligence.
www.visidata.com

Z/I Imaging USA, Inc.
www.ziimaging.com

Video mapping systems

Archipelago Marine Research Ltd.
www.archipelago.bc.ca

Asashi.
www.aeroasashi.co.jp

BlueGlen Technology Limited
www.blueglen.com

Coastal Alaska.
www.coastalaska.net

Coastal and Ocean Resources Inc.
www.coastalandoceans.com

Observera.
www.observera.com/video.html

Paravion Technology, Inc.
www.paravion.com

Red Hen Systems, Inc.
www.redhensystems.com

USGS Video Mapping of Coral Reefs.
coralreefs.wr.usgs.gov

Victoria and Esquimalt Harbours Ecological Inventory and Rating Project (HEIR) Subtidal.
www.veheap.crd.bc.ca

VTT Information Technology.
www.vtt.fi

6 Line scanners

Stan Aronoff

Film and digital camera systems discussed in chapter 5 operate in the visible and near-infrared bands with wavelengths from about 0.4 to 0.9 µm. The line scanners discussed in this chapter operate over a much wider wavelength range, from 0.3 to 14 µm, and can simultaneously collect data in a greater number of wavelength bands *(figure 6.1)*.

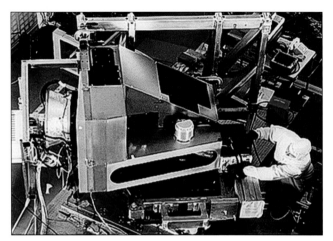

Figure 6.1 This photo shows the final testing of a multispectral scanner for the MODIS sensor.

Source: Image courtesy of NASA.

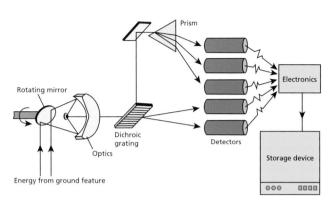

Figure 6.2 Diagram of a multispectral scanner.

Source: *Remote sensing and image interpretation.* 2nd ed. by T. M. Lillesand and R. W. Kiefer. © 1987 John Wiley and Sons, Inc. Reprinted with permission of John Wiley and Sons, Inc.

Whereas frame capture systems collect the image of a scene at one instant in time, a multispectral scanner records narrow swaths perpendicular to the flight path to build up an image of the terrain below. The scanning rate is adjusted to the ground speed so that successive scans view adjacent swaths.

Line scanners can be designed to detect energy in the visible, near-ultraviolet, and infrared bands. They contain precision optics like that of a telescope. High-precision large-format metric line scanners typically capture a single panchromatic band. Line scanners that simultaneously collect imagery in multiple bands are termed multispectral scanners. In older multispectral systems, the incoming electromagnetic energy passed through one or more beam-splitting optics, separating the energy into its constituent wavelengths. Figure 6.2 illustrates a multispectral scanner that uses a rotating mirror to scan across the image and a prism and dichroic grating to split the incoming energy. With the availability of large linear and area array sensors, multispectral scanners were developed without beam-splitters. Instead, multiple sensor arrays with different filters to detect different wavelengths are positioned side-by-side within the focal plane of the instrument (see, for example, figure 24 in appendix A.)"

Four types of scanning systems are commonly used for remote sensing: across-track (whiskbroom), along-track (pushbroom), spin scanners, and conical scanners.

Across-track scanners

Across-track scanners (also known as *whiskbroom scanners*) record brightness values along a scan line, one pixel at a time. A rotating or an oscillating mirror directs the sensor's field of view to repeatedly sweep across the terrain *(figure 6.3)*. For each scan line, a single detector per band continuously outputs a voltage proportional to the incoming energy. The voltage level is repeatedly measured (sampled) during each scan to generate the individual pixel values *(figure 6.4)*. The forward motion of the aircraft or satellite brings the next image segment into view, progressively building a continuous image from the narrow image segments. Since the sensor is moving forward while a line of pixels is being scanned, the trace of the scan line on the terrain is slightly skewed—an effect that is removed when the image is rectified.

Practical across-track scanners were developed before the along-track designs in part because they only required a small detector array, which could be produced using the early detector fabrication methods. Later, when large detector arrays could be economically produced, along-track scanners were developed.

Consequently, early airborne and spaceborne multispectral scanners, such as those used in the Landsat satellites, were across-track designs. When along-track scanners became available, their advantages made them the preferred design.

An across-track scanner collects the energy within the system's *instantaneous field of view* (IFOV), the cone-shaped angle of view sensed by the detector *(figure 6.5)*. The patch of ground sensed at any instant is called the *ground resolution element* or

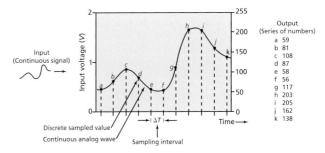

Figure 6.4 Across-track scanner image generation. The detector in an across-track scanner generates a continuously varying voltage as it records the change in brightness levels within a scan line. This analog signal is repeatedly sampled, and the voltage level at each sampling time is used to generate the pixel values in the digital image.

Source: *Remote sensing and image interpretation.* 4th ed. by T. M. Lillesand and R. W. Kiefer. © 2000 John Wiley and Sons, Inc. Reprinted with permission of John Wiley and Sons, Inc.

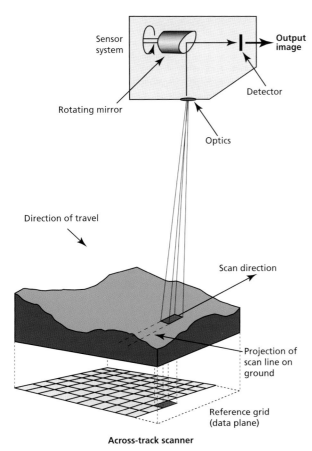

Across-track scanner

Figure 6.3 Across-track or whiskbroom scanner. An across-track scanner builds up an image one pixel at a time as it scans the terrain perpendicular to the direction of travel.

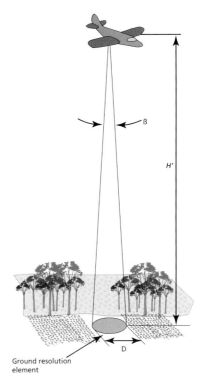

Figure 6.5 This diagram illustrates the instantaneous field of view (IFOV) and resulting ground area detected by a multispectral scanner.

Source: *Remote sensing and image interpretation.* 2nd ed. by T. M. Lillesand and R. W. Kiefer. © 1987 John Wiley and Sons, Inc. Reprinted with permission of John Wiley and Sons, Inc.

ground sample distance. The diameter of the ground resolution cell is loosely referred to as the system's spatial resolution and is calculated as follows:

$$D = H\text{ß}$$

Where

  D = diameter of the ground area viewed

  H = flying height above the terrain

  ß = IFOV of the system expressed in radians

For example, a system with an IFOV of 2.5 milliradians (mrad) operating at an altitude of 2,000 meters would view a ground area diameter of about 5 meters (calculated as 2,000 meters x 2.5 x 10^{-3} radians) directly beneath the sensor.

A scanner collects data from the point directly below the sensor (called the *nadir*) and to each side of nadir to form an image. The distance between the sensor and the ground increases as the sensor moves along the scan line to each side of nadir. The greater the viewing distance from the sensor to a ground point, the smaller the scale at which that point is going to be represented. This variation in scale along a scan line causes a panoramic distortion that makes the uncorrected image appear compressed at the edge of the scan line *(figures 6.6 and 6.7)*—an effect much more noticeable in imagery flown at low altitudes.

The signal from the detector is recorded, or sampled, at a rapid rate during each scan to produce a series of pixel values. The scanner's sampling rate and forward motion are adjusted so that, ideally, successive sample values and scan lines advance a ground distance exactly equal to the size of the ground resolution cell. In this way, the scanner generates an image as a contiguous grid of cells for which reflectance values in one or more energy bands have been recorded.

The *dwell time* is the length of time during which a sample measurement is collected over the area of a single ground resolution cell. The smaller the ground resolution cell, the more frequently a scan line needs to be generated and the faster the scan rate needs to be to image successive pixels within a scan line, both of which shorten the dwell time. A

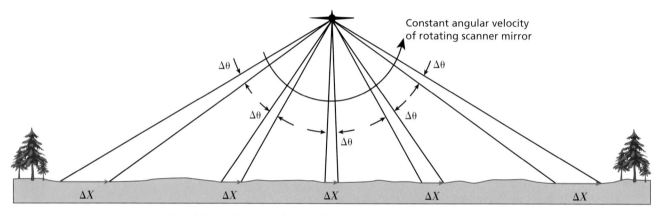

Resulting variations in linear velocity of ground resolution element

Figure 6.6   Variation of the ground coverage of a pixel across a scan line.

Source: *Remote sensing and image interpretation.* 2nd ed. by T. M. Lillesand and R. W. Kiefer. © 1987 John Wiley and Sons, Inc. Reprinted with permission of John Wiley and Sons, Inc.

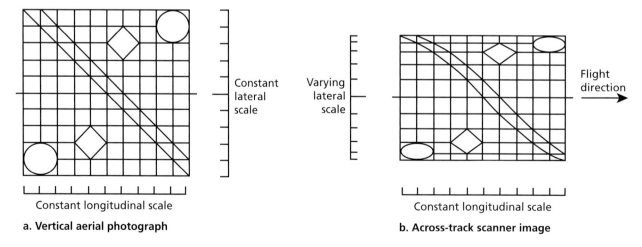

Figure 6.7   Lateral compression of scale in unrectified across-track scanner imagery.

Source: *Remote sensing and image interpretation.* 2nd ed. by T. M. Lillesand and R. W. Kiefer. © 1987 John Wiley and Sons, Inc. Reprinted with permission of John Wiley and Sons, Inc.

shorter dwell time reduces the total energy detected for each image pixel, increasing the noise in the data. Electronic circuits generate spurious low-level electrical background signals called *noise*. When the level of the signal being measured is relatively high, the much lower background noise signal is easily identified and can be removed, or it may not be noticeable and can just be ignored. However, when the signal level is low, it approaches that of the background noise, making it difficult to distinguish the noise from the desired signal. The resulting imagery tends to have low contrast and be grainy. Where the reflectance is too low, the signal may not be distinguishable from the background noise, and a usable image cannot be captured.

In across-track scanners, the ground distance between adjacent pixels (*the ground sample distance* or GSD) is determined by the sampling rate and is not necessarily equal to the ground resolution cell. However, in practice it is usually about the same size, and so both the ground resolution cell and GSD are often referred to as the system's spatial resolution.

The IFOV for airborne multispectral scanners typically ranges from an angle of 0.5 to 5 mrad or milliradian (one milliradian is 1/1000th of a radian). The smaller the IFOV, the higher the spatial resolution, or the smaller the details that can be recorded. However, a larger IFOV provides a larger signal because energy is collected from a larger ground area. As previously noted, a larger signal permits more sensitive measurement of scene radiance (higher radiometric resolution) because the signal is much greater than the system's background electronic noise. This characteristic of signal quality is commonly expressed as the signal-to-noise ratio. The greater the *signal-to-noise ratio*, the more the scanner is able to distinguish small differences in brightness levels. The signal-to-noise ratio can also be improved by detecting a wider range of wavelengths (lower spectral resolution), thereby increasing the total energy collected. Thus, there is a trade-off between spatial, spectral, and radiometric resolution in selecting a scanner's operating parameters.

Satellites are very stable platforms. In contrast, airborne platforms tip and tilt with the changing air currents, which causes additional image distortions *(figure 6.8)*. Accurate aircraft attitude data is collected continuously from onboard inertial measurement units during airborne missions for later use in image rectification. All scanner data is processed after acquisition to remove geometric distortion and correct for detector errors. Figure 6.9 shows a scanner image before and after correction for aircraft tip and tilt and lateral compression.

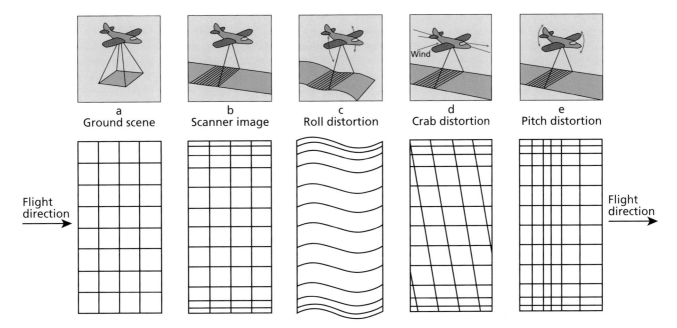

Figure 6.8　Image distortions caused by aircraft attitude variation.

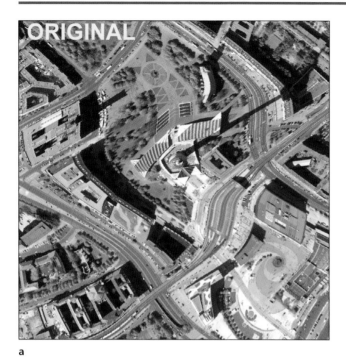

a

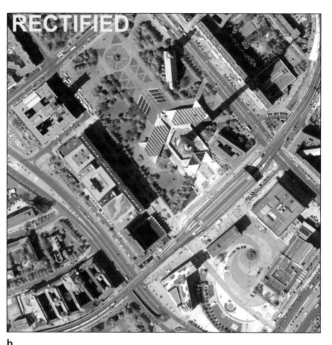

b

Figure 6.9 Imagery from a high-resolution airborne multispectral scanner before (a) and after (b) correction for aircraft tip and tilt during image acquisition. The distortion in the uncorrected image is extreme because the scanner was not mounted on a stabilized platform.

Source: Imagery provided by Leica Geosystems GIS and Mapping, LLC.

The geometry of scanner imagery is different from the perspective geometry of frame capture systems such as cameras. Since scanners image a single across-track plane, relief displacement occurs only along a scan line, so tall features appear to lean at right angles away from the flight path, whereas on an aerial photograph, vertical features are displaced radially out from the center of the image *(figure 6.10)*.

Along-track scanners

Along-track scanners, also known as *pushbroom scanners*, capture data in narrow image strips to build up an image. They do not use rotating or oscillating mirrors. Instead, a linear array sensor consisting of thousands of charge-coupled device (CCD) detectors arranged end-to-end, one per pixel, simultaneously records the brightness values for all the pixels of the swath width at one time *(figure 6.11)*. The forward motion of the aircraft or spacecraft moves the linear array over successive swaths of terrain. Linear arrays may be 10,000 detectors long or more. Multiple banks of linear arrays can be used to capture multiple wavelength bands or multiple viewing directions simultaneously.

Along-track scanners have several advantages over the mirror scanning systems. A scanning mirror introduces errors caused by inconsistencies in the motion of the mirror and the pattern of the scan lines. Because an along-track scanner has no moving parts, it tends to have higher geometric accuracy and a lower overall cost of development and operation. In addition, the larger number of individually calibrated detectors in the detector arrays of an along-track scanner tends to give it a higher radiometric accuracy than an across-track scanner.

Along-track scanners also have a longer dwell time, providing a stronger signal. Since each detector captures data for only a single pixel within one scan line instead of for the entire scan line, it can collect the incoming energy for each pixel over a longer period of time. The longer dwell time improves the detector's signal-to-noise ratio, which improves the detector's ability to produce a usable image under low light levels, i.e., it gives improved radiometric resolution. A higher signal-to-noise ratio allows the sensor to be designed with a smaller IFOV, meaning a pixel represents a smaller

Figure 6.10   Comparison of relief displacement on a frame photograph and on scanner imagery.

Source: *Remote sensing and image interpretation.* 2nd ed. by T. M. Lillesand and R. W. Kiefer. © 1987 John Wiley and Sons, Inc. Reprinted with permission of John Wiley and Sons, Inc.

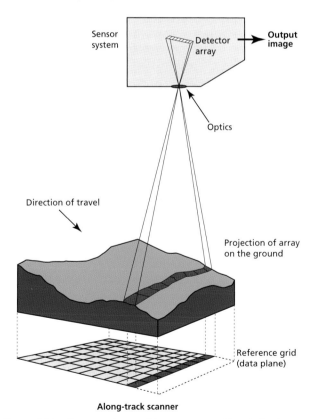

Along-track scanner

Figure 6.11   Along-track or pushbroom scanner. The pixels of an entire swath width are captured at one time, so the image is built up a row at a time in the direction of travel, i.e., along the ground track.

ground area, which gives a higher spatial resolution. A high signal-to-noise level also allows the incoming energy to be split (e.g., by a prism or filters) into multiple spectral bands, which means the system has an increased spectral resolution. Each band has a lower energy level than the incoming energy, but the separate spectral bands provide valuable information that one broader band cannot.

Along-track scanners are generally smaller, lighter, and consume less power than across-track scanners. Since they have no moving parts, along-track scanners have a higher reliability and longer life expectancy, making them the most widely used design for earth resource remote sensing. Most earth resource imaging satellites, including SPOT, IRS, IKONOS, and QuickBird, use along-track scanners. There are now linear and, in most cases, area array sensors available for detection of every wavelength band used in remote sensing, including the shortwave infrared (SWIR), medium wave infrared (MWIR), and long wave or thermal infrared (LWIR or TIR).

The image geometry of along-track scanners is similar to that of across-track scanners in that relief displacement is at a right angle to the scan line. However, the absence of a moving mirror gives along-track scanners greater geometric stability and accuracy.

Spin scanners

Geostationary meteorological satellites, such as GOES, maintain the same position over the earth (see chapter 7).

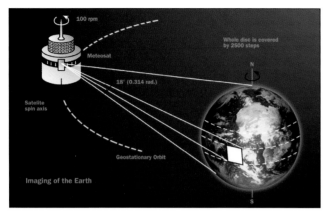

Figure 6.12 Spin scanner. As the satellite spins, the scanner's field of view moves from east to west across the earth's surface. With each rotation, the scanner's viewing angle is stepped lower, from north to south, to scan the adjacent swath.

Source: © Eumestat 2004.

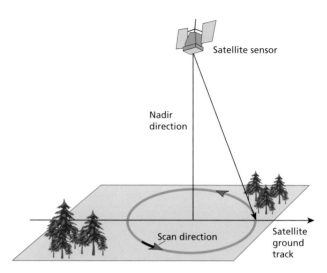

Figure 6.13 Principles of conical scanning.

Since the earth is not moving beneath them, they cannot use the forward motion of the spacecraft to view successive image swaths. One method used in some of the early meteorological satellites was to rotate the satellite itself on its north–south axis to scan the terrain below. With each rotation, the sensor would record a single line of image data in an east–west direction. A mechanical stepping motor changed the angle of the viewing mirror in discrete steps after each rotation to collect adjacent scan lines, building up a full image *(figure 6.12)*.

The mirror on the Visible Infrared Spin-Scan Radiometer (VISSR) spin-scanner instruments, flown on the GOES-7 satellite in 1987 and earlier, stepped through 1,821 positions to provide north-to-south viewing while the 100 rpm rotation of the satellite provided for west-to-east scanning. The instrument used a flexible scan system that produced full-earth disc images every 18 minutes, as well as sector images of smaller areas. Ground pixel sizes ranged from 1 km to 5 km, depending on the image product and satellite.

More recent meteorological satellites do not spin. Instead an across-track scanner operates in the east–west direction, and the field of view is stepped in the north–south direction for successive scans.

Conical scanners

Conical scanners record images of a circular or elliptical track on the terrain below. If the sensor is oriented to scan a circular track about a vertical (nadir) axis, then the IFOV will remain constant, and the angle at which the sensor views the terrain will also be constant *(figure 6.13)*.

Satellite-based conical scanners are commonly used for meteorological applications. These sensors operate at microwave[1] and infrared wavelengths at relatively coarse spatial resolutions of about 1 km GSD in the infrared and about 50 km GSD in the microwave band.

For example, the Tropical Rainfall Monitoring Mission microwave imager collects data in five microwave bands with a spatial resolution of about 50 km. The data products generated include sea surface temperatures, surface wind, atmospheric water vapor, liquid cloud water, and rainfall rates.

The Along-Track Scanning Radiometer (ATSR) carried on the ERS-2 satellite is a conical scanner operating in seven bands in the visible and infrared (see *ATSR and AATSR sensors* in chapter 7). A unique feature of this scanner is that it has an inclined conical scan enabling data of the earth's surface to be collected from two different angles, a vertical (nadir) view and a forward view, within a few minutes of each other. (Only data for the portions of the scan approximately perpendicular to the flight path are used). Having two observations with different atmospheric path lengths for

---

1. The microwave sensors are passive, meaning they detect energy emitted from earth features at microwave wavelengths. Radar is an active sensor that illuminates the terrain with pulses of microwave energy and measures the return signal reflected from the terrain below.

each location allows better correction for atmospheric effects and thus more accurate sea surface temperature measurements than would be obtained with a single measurement.

Hyperspectral scanners

Most scanner systems detect 3 to 11 wavelength bands simultaneously. *Imaging spectrometers* or *hyperspectral scanners* simultaneously record images in tens or hundreds of narrow contiguous spectral bands over a continuous spectral range *(figure 6.14)*. These bands are typically 0.015 μm in width or less. From the image data, a spectral curve can be generated for each pixel in the scene with much more detail than can be produced from data collected from a few broader and widely separated bands *(figure 6.15)*. Under certain conditions, the greater detail in the derived spectral reflectance curves can be used to distinguish ground features that are not distinguishable on conventional multispectral scanner imagery.

Hyperspectral remote sensing concepts

Imaging spectroscopy, the use of multiple contiguous narrow band images to study spectral properties of a signal, was originally developed to obtain geochemical information from inaccessible sites such as planets within the solar system. Several hyperspectral image algorithms have been developed to classify the data. Some algorithms are based on comparing the spectral curves generated from pixels in the image with the spectral curves in a database of known feature types. Other algorithms, such as spectral *unmixing* [2], generate classifications based on the characteristics of the spectral curves derived from the image data and then determine the identity of the classes by comparison with field data. For example, hyperspectral images have been used to distinguish spectrally similar materials such as geologic minerals or vegetation types. In some cases, after studying the full range of image data, a few bands can be identified that will be sufficient for the analysis.

2. Information on spectral unmixing and other algorithms used in processing remotely sensed data can be found in Mather 1999.

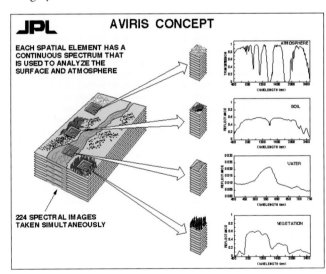

Figure 6.14 Concept of imaging spectrometry. Hyperspectral scanners can collect image data in hundreds of bands, represented here as a stack or cube of images. The series of pixel values obtained by drilling down through the image stack represents a series of reflectance values in different wavelength bands. These values for a single pixel location can be plotted as a reflectance curve and used to identify characteristics of the feature at that location.

Source: Imagery courtesy of NASA/JPL.

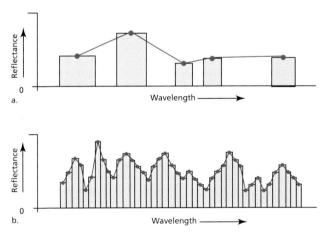

Figure 6.15 Comparison of spectral data captured by conventional multispectral scanners and hyperspectral scanners. The bar charts compare the reflectance data collected from the five wider spectral bands of a conventional multispectral scanner (a) compared with the 42 narrow contiguous bands covering the same spectral range in a hyperspectral scanner (b). The spectral curves generated from the hyperspectral data contain additional peaks and troughs, details that may distinguish features in the hyperspectral imagery that are not distinguishable in the imagery of the conventional scanner.

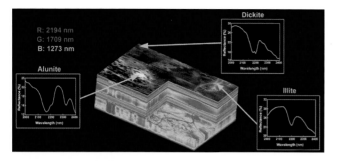

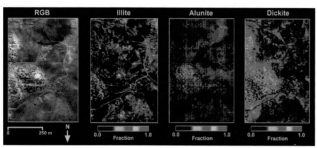

Figure 6.16 Hyperspectral remote sensing for mineral identification. Hyperspectral data for the Virginia City area in Nevada is displayed as an image cube (top) in which reflectance is color-coded from blue (low) to red (high). The lower portion of the figure shows the same color-composite image and abundance maps for each of these minerals generated by classification of the hyperspectral imagery.

Source: Borstad Associates Ltd. and Natural Resources Canada.

Hyperspectral images are often presented for display as an image cube in which each narrow band image is represented as a horizontal layer and relative reflectance is color coded from blue (low) to red (high) *(figure 6.16, top)*. The image cube illustrates that the set of images contain as much information in the spectral dimension (shown vertically) as they do in the spatial dimension (shown horizontally). This image cube shows the Virginia City region in Nevada. The top of the cube is a color composite generated from three short-wave infrared images centered on 2.194 µm (red), 1.709 µm (green), and 1.273 µm (blue). Field verification sites for the minerals alunite, dickite, and illite are shown. The lower portion of the figure shows the same color composite image together with abundance maps generated for each of these minerals by classification of the hyperspectral data.

Hyperspectral scanner systems

A hyperspectral system widely used for research is the Airborne Visible Infrared Imaging Spectrometer (AVIRIS), developed by the Jet Propulsion Laboratory (JPL). It simultaneously records data in 224 spectral channels, 0.0096 µm wide, in contiguous bands between 0.4 µm and 2.45 µm. Designed to support research, the wide use and valuable results obtained with AVIRIS spurred the development of commercial systems. Several hyperspectral sensors are commercially available.

The Compact Airborne Spectrographic Imager (CASI) from ITRES Research Limited in Canada is an along-track

Figure 6.17 Hyperspectral scanner image, Fish Lake, Nevada. This color-composite image was created from three narrow band (0.015 µm wide) images: a mid-infrared (centered on 2.045 µm), a near-infrared (centered on 0.885 µm), and a visible blue (centered on 0.465 µm) image were displayed as red, green, and blue.

Source: Courtesy of HYVista Corporation Pty. Ltd. Baulkham Hills, Australia.

scanner. The 550 model can record a 550 pixel swath width detecting 288 wavelength bands between 0.4 μm and 1.0 μm with individual bands as narrow as 0.0019 μm. The CASI 1500 records a 1,500 pixel swath width over the 0.38 μm to 1.05 μm spectrum with individual bands 0.0023 μm wide. The ITRES SASI-640 offers 160 band coverage of the short-wave infrared. GER, an important American manufacturer, produces a number of hyperspectral systems that are widely used. Spectral Imaging Inc. of Finland produces the AISA+ and AISA Eagle along-track hyperspectral scanners, which offers swath widths of 512 and 1024 pixels respectively. They record 244 bands 0.0023 μm wide.

The HYMAP Hyperspectral Scanner (HYMAP) manufactured by Integrated Spectronics Pty. Ltd. in Australia is an across-track scanner like AVIRIS. It collects imagery in 126 bands in the visible and reflective infrared spectrum from 0.45 μm to 2.5 μm with bandwidths of 0.01 to 0.02 μm. Figure 6.17 of a 14 km by 21 km area of the Fish Lake Valley, Nevada, is a color-composite mosaic produced from ten image strips collected at a spatial resolution of 3 m GSD. The imagery was used to map alteration minerals at the earth's surface as part of a program to develop geothermal resources in the state.

Hyperion is an experimental satellite-based hyperspectral scanner carried on the NASA Earth Observing-1 satellite launched in November 2000. Hyperion provides hyperspectral imagery in 220 spectral bands (from 0.4 to 2.5 μm) and can collect images of a 7.5 km by 100 km land area per scene with a 30-meter ground sample distance *(figure 6.18)*.

The European Space Agency in October 2001 launched a hyperspectral scanner, the Compact High Resolution Imaging Spectrometer (CHRIS), aboard the experimental Proba-1 (Project for On-Board Autonomy) satellite. The sensor is capable of imaging up to 200 spectral bands in the visible and near-infrared spectrum (0.415–1.050 μm). Operating at its maximum spatial resolution of 20 m GSD, the sensor can simultaneously collect imagery in 19 spectral bands with a swath width of 15 km *(figure 6.19)*.

Specialized atmospheric corrections are applied to the hyperspectral data collected by Hyperion using the data from other sensors on the satellite.

Figure 6.19 Satellite image of the Mauna Kea volcano, Hawaii, taken by the CHRIS hyperspectral sensor. This color composite was generated from narrow spectral bands centered at 0.466 μm, 0.543 μm, and 0.641 μm. Distance across the image is 15 km with a 20 m spatial resolution.

Source: Image courtesy of the European Space Agency.

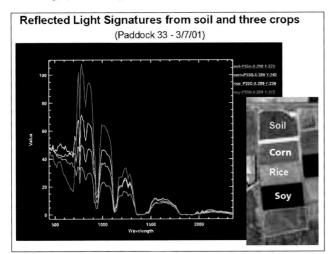

Figure 6.18 Hyperion scanner. Spectral signatures for four land-cover types at a test area were generated from data collected by the Hyperion 220 band hyperspectral scanner.

Source: Imagery courtesy of NASA.

Hyperspectral applications

A number of promising applications of hyperspectral remote sensing have been developed in agriculture, geology, water resource assessment, and military applications, including the following:

- crop condition assessment for precision farming
- military applications, including analysis of terrain access, detection of camouflaged and otherwise concealed military assets, and land-mine detection
- surface material identification, especially for geology and soils in arid areas
- vegetation species identification
- water resource monitoring, including analysis of water quality, chlorophyll content, temperature, sediment load, effluent plumes, and bathymetry

Although technical challenges in handling the high data volumes and the need for sophisticated algorithms remain, applications of hyperspectral imagery are gaining acceptance, and hyperspectral image analysis tools are becoming more widely available in commercial image-processing software.

Stereo scanner imagery

Stereo imaging enables ground elevations and distances to be accurately measured from imagery. To produce stereo images, the sensor must capture a scene from two different sensor positions. As discussed in chapter 5, when the two images, called a stereo pair, are viewed together using a stereoscope, the scene appears three dimensional, called the *stereo effect*.

Scanners are typically oriented to view the terrain directly below, called *nadir viewing*. But the overlap region of imagery collected along adjacent flight lines typically produces a limited overlap and therefore only a weak stereo effect. Instead, stereo scanner imagery is generated by orienting the scanner field of view to the side, front, or rear of the sensor. Off-nadir views increase relief displacements in the image caused by differences in terrain elevation or the height of features. Accurate measurement of the stereomodel is used to produce elevations and generate digital elevation data. Off-nadir imagery is better suited to these photogrammetric operations, while the nadir orientation minimizes these scale distortions, which provides imagery better suited for direct analysis or as a source for producing geometrically corrected imagery. Often one nadir image and one off-nadir image are used to form the stereomodel.

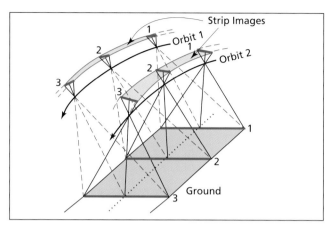

a

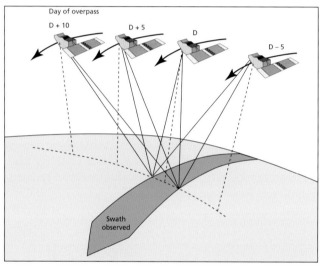

b

Figure 6.20 Across-track stereo imagery. Across-track stereo imagery is produced with two images of the same area collected from different orbit paths (a). For example, a satellite may be able to image an area with a vertical (nadir) view on orbital track D and image the same area five days previous and five days and ten days after from adjacent orbit paths (b).

Source: Image (a) courtesy of G. Petrie and M. Shand. Image (b) © CNES 2004 courtesy of SPOT Image Corporation.

Across-track stereo imagery

Across-track stereo imagery is produced by pointing the sensor at different angles perpendicular to the flight line *(figure 6.20)*. For example, the SPOT satellites have pointable sensors that can record images of areas adjacent to the ground track as well as the terrain directly below. The ability to record images of areas to the side of the orbit path increases the revisit frequency, since the same area can be viewed from several different orbits. However, the time delay in acquiring the second image can be problematic for some applications.

Along-track stereo imagery

Along-track stereo imagery is generated from images scanned at different view angles along the same flight line. Systems that can view fore and aft can produce stereo images acquired at virtually the same time, which is particularly advantageous for automated stereo matching and digital photogrammetric applications. The two images can be obtained by capturing an area with a forward-looking view and then changing the viewing direction during the overpass to image the same area a second time. The IKONOS satellite produces along-track stereo imagery by shifting its viewing direction during the overpass in this way. Dedicated stereo sensors, such as the HRS on SPOT 5, use two telescopes with the different viewing directions at a fixed angle *(figure 6.21)*. See also figure A.28 in appendix A.

High-resolution scanners specifically designed for photogrammetric applications simultaneously collect imagery from multiple along-track viewing directions. Using high-precision satellite positioning and sensor positioning data, these systems can generate digital elevation models, orthorectified imagery, perspective views, and other image products using fewer ground control points than conventional photogrammetric camera systems.

The High Resolution Stereo Camera–Airborne Extended (HRSC-AX) scanner produced by DLR (the German Aerospace Center) can simultaneously capture nine swaths of imagery including three lines of panchromatic data with nadir, forward, and backward-pointing views and four multispectral bands (visible blue, green, red, and near-infrared). From a flying altitude of 2,500 m, the scanner can provide 12,000-pixel-wide, 25 cm (GSD) imagery with 15 to 20 cm photogrammetric accuracy in the horizontal and vertical dimensions. The scanner is capable of producing imagery with spatial resolutions as high as 5 cm GSD.

The Leica ADS40 *(figure 6.22)*, designed by Leica Geosystems GIS and Mapping, LLC, in partnership with DLR, can simultaneously collect 12,000-pixel-wide imagery at three different viewing angles (forward, nadir, and backward) in three panchromatic (one at each view angle) and four multispectral bands (visible, blue, green, red, and near-infrared) *(figure 6.23)*. Forward- and backward-looking

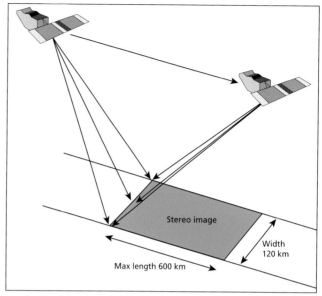

Figure 6.21   The SPOT 5 HRS sensor is able to collect a 120-km-wide image strip in the forward-looking mode and then 90 seconds later collect the corresponding backward-looking view to generate stereo coverage for as much as a 600 km swath. It uses two fixed orientation telescopes looking forward and backward along the flight path.

Source: © CNES 2004/Courtesy of SPOT Image Corporation.

imagery improves the accuracy of photogrammetric measurements and digital elevation data.

The scanner is capable of spatial resolutions as fine as 5 cm GSD and can deliver simultaneously acquired color and color infrared imagery and complete stereo coverage in a single pass. Image data is either compressed online or collected as raw data and stored at a rate of up to 50 MB per second. Onboard data storage uses a mass memory capable of storing 100 GB of data per hour of flight. With a capacity of 540 GB, it provides five hours of acquisition time before changing storage media, which is sufficient for most missions. At a typical flying altitude of 1,930 m, the system can generate imagery with a spatial resolution of 20 cm GSD and horizontal accuracy better than 10 cm that can be used to generate elevation measurements with a vertical accuracy better than 15 cm *(figure 6.24)*.

Thermal infrared scanners

Thermal remote sensing has been successfully used for such diverse applications as military reconnaissance, sea surface

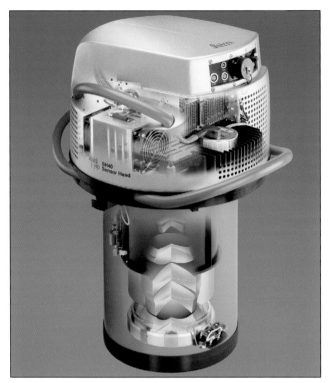

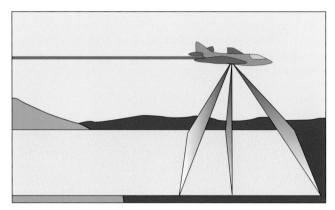

Figure 6.23 Leica ADS40 multiple-view angles. The Leica ADS40 can simultaneously acquire panchromatic imagery at three viewing angles (forward, backward, and nadir), any two of which provide stereo imaging. Forward- and backward-looking imagery improves the accuracy of photogrammetric measurements. The nadir view is preferred as the image source for producing orthorectified imagery using DEM produced from the stereomodel.

Source: Imagery provided by Leica Geosystems GIS and Mapping, LLC.

Figure 6.22 Leica ADS40 large-format airborne metric scanner, partial cut-away view. The ADS40 digital scanner, designed for photogrammetric and remote sensing applications, can simultaneously capture 10 bands of data as a continuous image strip 12,000 pixels wide. The floor-mounted sensor unit shown is 74 cm high, 59 cm in diameter, and weighs 66 kg.

Source: Imagery provided by Leica Geosystems GIS and Mapping, LLC.

Figure 6.24 Large-format scanner image. The ADS40 large-format metric scanner produces a continuous strip (swath) of imagery with a 12,000 pixel swath width. This image strip, covering a 2.4 km by 9.9 km ground area, has been processed using sensor attitude data collected during the flight to remove aircraft pitch and roll effects that occur during normal flight. The wavy top and bottom edge of the image are due to the undulating nature of the terrain. The insert image shows an enlarged portion of the image strip with a ground distance of approximately 400 m by 380 m across with a GSD of 20 cm per pixel.

Source: Imagery provided by Leica Geosystems GIS and Mapping, LLC.

Figure 6.25 Natural color and thermal image forest fire imagery. Imagery flown by the Daedalus Thematic Mapper Simulator Scanner replicates the band characteristics of the seven-band scanner on Landsat 4 and 5. The area is obscured by smoke in the natural color composite on the top but clearly visible in the thermal image on the bottom. Hotter areas are represented by warmer colors, the active fire areas appearing white, yellow, and orange in this image.

Source: Image courtesy of NASA and the Ames Research Center.

temperature mapping, evaluation of building heat loss, geologic mapping and fault detection, soil moisture and type mapping, and delineating the boundaries of active forest fires *(figure 6.25)*.

Naturally occurring earth features like soil, water, and vegetation radiate much less energy than the sun, and consequently their peak emissions are at longer wavelengths. For earth features, the peak energy emission is at a wavelength of about 10 μm, which is in the infrared region. It is for this reason that wavelengths in the region 8 to 14 μm, the *thermal infrared band*, can be used to detect temperature differences

of earth objects. Thermal imaging sensors can resolve radiant temperature differences as small as 0.1° C.

The 8 to 14 μm portion of the spectrum, where the blocking effect of the atmosphere is minimal, is most commonly used for thermal infrared remote sensing. Since the sun emits substantial energy in the thermal infrared region, the energy detected by thermal infrared sensors during the day is a combination of the energy emitted by earth objects and solar energy reflected from them.

Daytime imagery contains thermal shadows—cool areas shaded from direct sunlight by trees, buildings, and other features. In addition, apparent temperatures are affected by the heating and cooling cycle over the course of the day, the orientation of features relative to heat sources such as the sun, and surface characteristics. These effects create thermal patterns unrelated to the nature of the material, making interpretation more difficult. For this reason, thermal infrared imagery is usually collected at night so that only the emitted energy, which is related to the temperature of earth features, is detected. Predawn imagery tends to minimize daytime heating and cooling effects resulting from differential sun exposure.

Thermal emission is affected by both the emissivity of an object (a measure, ranging from zero to one, of an object's efficiency to radiate energy relative to a perfect emitter) as well as its temperature. Accurate temperatures can be determined only if the emissivity is known. Since the emissivity of earth features varies widely, it is impractical to obtain accurate temperature measurements for diverse landscape features. However, the emissivity of water is within a narrow range, with values of 0.98 to 0.99 in the 8 to 14 μm band, and varies little with differences in water quality. For this reason, thermal scanners can produce accurate maps of water surface temperatures. Sea surface temperature is monitored by several meteorological and earth resources satellites for use in weather prediction, habitat analysis, and other applications *(figure 6.26)*. Nevertheless, many useful applications of thermal imagery do not require absolute temperature values. Instead they make use of the relative thermal differences (differences in the apparent temperature of features within a scene).

Thermal infrared scanners generally produce imagery with coarser spatial resolution than sensors operating in the visible and near-infrared bands. In part, this is to compensate for the much lower energy levels available for detection at thermal infrared wavelengths in comparison with the reflected solar energy detected in the visible and near-infrared bands.

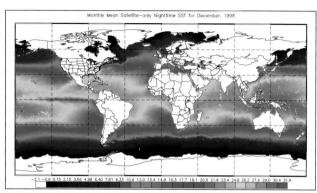

Figure 6.26 Global monthly mean sea surface temperature calculated from AVHRR thermal infrared data at 36 km resolution is based on only nighttime satellite observations.

Source: Courtesy NASA/JPL-Caltech.

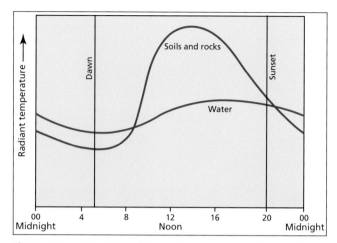

Figure 6.27 Comparison of diurnal temperature variation in soils and rock versus water bodies.

Source: *Remote sensing and image interpretation*. 2nd ed by T. M. Lillesand and R. W. Kiefer. © 1987 John Wiley and Sons, Inc. Reprinted with permission of John Wiley and Sons, Inc.

a

b

Figure 6.28 Thermal infrared imagery of an industrial site at midday (above) and at night (below). Warmer features are lighter in tone. Rectangular ponds of standing water, like those on the left side of the image, are relatively cool in the daytime image and relatively warm at night.

Some multispectral scanners collect data in the visible and near-infrared portions of the spectrum as well as the thermal infrared. For example the Landsat TM sensor includes a thermal band, designated band 6, which detects the 10.4 to 12.5 μm band at a lower spatial resolution than the other Landsat TM bands (60 m instead of 30 m).

Thermal scanner imagery is typically collected in a single band. However, experimental systems, such as the NASA Thermal Infrared Multispectral Scanner (TIMS) system discussed below, detect multiple narrow bands within the thermal infrared region. Color-composite images produced from multiple thermal bands are useful for geological mapping.

Single-band thermal scanner images are typically displayed as black-and-white images, with higher radiant temperatures displayed as lighter tones or as color images by representing the range of values as a progression of hues typically ranging from purple and blue for low values through green and orange to red and white for the highest values (warmest features).

Water tends to gain and lose heat more slowly than soils and rock. As illustrated in figure 6.27, water bodies generally appear cooler than land features on daytime imagery and warmer on nighttime images. Figure 6.28 shows thermal infrared images of an industrial site acquired at midday (when heating is at a maximum) and just before dawn (when the land surface reaches its coolest temperature). Pipes carrying steam and hot liquids are clearly visible on the nighttime image at (Z) and (M) but cannot be reliably distinguished on the daytime imagery. Circular oil storage tanks are clearly visible, as well as the main processing plant at (F). The location where heated effluent is being pumped into a holding pond can be seen at (H).

As noted earlier, interpretation of daytime imagery is more difficult because shadows, orientation of features relative to the sun, and other factors unrelated to the characteristics of a feature can affect its apparent temperature. At night, features cool, which reduces these differential effects.

Thermal imagery has been widely used to study heat loss from buildings. Figure 6.29 shows homes along a city street recorded during a winter night at street level. The colors white, red, yellow, green, blue, purple, and black form a temperature scale. Windows and doors show the highest heat loss, appearing white, red, and yellow. Well-insulated areas appear purple, indicating they are about the same temperature as the surroundings.

Figure 6.30 shows imagery obtained with TIMS. This scanner collects data in six narrow bands between 8.2 µm and 12.2 µm. In this color-composite image, channel 1 (8.2–8.6 µm) is displayed as blue, channel 3 (9.0–9.4 µm) as green, and channel 5 (10.2–11.2 µm) is displayed as red. The geology map of the same area illustrates some of the features recognizable within the image.

Real-time and near-real-time image capture and remote delivery of imagery have been made possible by the availability of small high-speed computers and improved rectification and image-compression algorithms. For example, a system developed by Argon ST to support emergency fire response operations captures a 6 km swath of terrain at a 3.8 m GSD pixel resolution from an aircraft traveling 280 km/hr. The infrared data is rectified and georeferenced in real-time and displayed as a black-and-white image on a topographic map background with 10 m RMSE map accuracy *(figure 6.31)*. A second image layer contains fire detection data generated by processing the imagery from two infrared bands to calculate

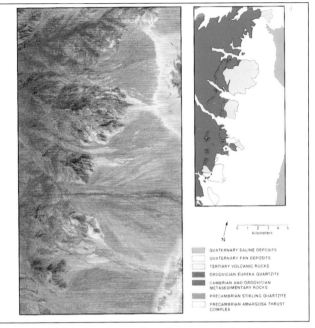

Figure 6.30 Thermal infrared multispectral scanner (TIMS) image (left) and generalized geology map of the Panamint Mountains area in Death Valley, California.

Source: Courtesy NASA/JPL-Caltech.

estimated surface temperature. The temperature data is processed in real time to provide automated detection of fires as small as 0.5 % the area of a single pixel. The imagery can be transmitted to remote receiving sites within 60 km and displayed within two minutes of original image acquisition.

There has been substantial research and development of new sensors for the shortwave infrared (SWIR) 1 to 2.5 µm wavelengths, medium wave infrared (MWIR) at 3 to 5 µm, as well as the long wave (LWIR) or thermal infrared (TIR) at 8 to 14 µm. Much of the funding has been from military sources and these sensors are used extensively for military reconnaissance, targeting, as well as policing activities. They have also found wide use for environmental applications. As previously noted, TIR sensors are used to assess building heat loss, map water temperature, and study the discharge of heated effluents. An important application of airborne thermal infrared imagery is to census animal populations, such as seals. When the animals haul out on shore, they are well camouflaged and difficult to distinguish at visible wavelengths but easily identified and counted on infrared imagery. MWIR as well as TIR sensors have also been found useful for detecting pipeline leakage and the extent of flooding

Figure 6.29 Color-enhanced thermal infrared image of residential buildings during a winter night.

Source: Image courtesy of Argon ST Imaging Group.

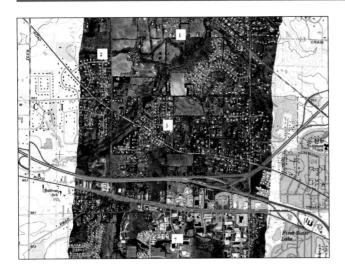

Figure 6.31 Real-time infrared image rectification and fire detection system. This display is generated from infrared scanner data that is rectified and displayed as a black-and-white image overlay on an existing topographic map image as the aircraft acquires the imagery. Fire locations are detected in real time and are displayed as numbered squares.

Source: Image courtesy of Argon ST Imaging Group.

and water damage. With the development of infrared imaging technology and the ongoing concern for environmental issues as well as policing, this remote sensing technology should find increasing use in civilian applications.

Conclusion

Scanners are the predominant sensor used for multispectral remote sensing image acquisition. Advances in detector design have increased their sensitivity and radiometric resolution, which offers greater detail in shadow areas and better performance at low light levels. Refinements in the

manufacture of CCD arrays have improved the geometric accuracy and number of elements in the array, which enables scanners to produce images with swath widths of 12,000 pixels or more and geometric properties sufficient to meet the stringent requirements of photogrammetric applications. Improvements in aircraft attitude and position measurement during image acquisition have reduced the amount of ground control and shortened the processing time needed to produce map and image products.

Large format metric digital sensors are finding growing acceptance for use in large-area high-resolution orthophoto projects. For example, U.S. government image acquisition programs that formerly used 230 mm x 230 mm (9 in x 9 in) format aerial photography exclusively, such as the U.S. Department of Agriculture's (USDA) National Agriculture Imagery Program (NAIP), are making increasing use of digitally acquired imagery from instruments such as the Leica ADS 40. The 1-m resolution digital orthoimagery product generated from this imagery is used for mandated annual compliance review and becomes the updated image base layer for the USDA geographic information system.

As with metric digital cameras, the use of scanner imagery for photogrammetric applications involves very large data volumes and the availability of substantial storage and computing resources to manage and analyze the data. However, digital imagery (from a multispectral scanner or digital camera) offers more accurate radiometric measurements and more consistent imagery than film. Multispectral scanner designs can accommodate simultaneous acquisition of a large number of bands, hundreds in the case of hyperspectral scanners, more easily than digital cameras. As the cost of data storage continues to fall and processing power increases, the trend will continue toward all-digital acquisition and processing of remotely sensed imagery.

References and further reading

Campbell, J. B. 1996. *Introduction to remote sensing.* 2nd ed. New York: The Guilford Press.

Jensen, J. R. 2000. *Remote sensing of the environment: An earth resource perspective.* Upper Saddle River, N.J.: Prentice Hall.

Lillesand, T. M., R. W. Kiefer, and J. W. Chipman. 2003. *Remote sensing and image interpretation.* 5th ed. New York: John Wiley and Sons, Inc.

Mather, P. M. 1999. *Computer processing of remotely-sensed images.* 2nd ed. New York: John Wiley and Sons, Inc.

Petrie, G. 2005. Airborne pushbroom line scanners: An alternative to digital frame scanners. GeoInformatics 8 (1): 50-57.

———. 2003. Eyes in the sky: Imagery from space platforms. *GI News* (January-February): 43–49.

———. 2002a. Optical imagery from airborne and spaceborne platforms: Comparisons of resolution, coverage, and geometry for a given pixel size. *GeoInformatics* 5(1): 28–35.

———. 2002b. Eyes in the sky: Guide to the current state-of-the-art in terrain imagers and imagery. Part I-Airborne imagery. *GI News* 2(7): 36–47.

Petrie, G., and G. Buyuksalih. 2001. Recent developments in airborne infra-red imagers: With military and environmental applications to the fore! *GeoInformatics* 4 (1): 16–21.

Rencz, A. N., and R. A. Ryerson, eds. 1999. *Manual of remote sensing: Remote sensing for the earth sciences.* 3rd ed. New York: John Wiley and Sons, Inc.

Web sites

Canada Centre for Remote Sensing. Natural Resources Canada. Hyperspectral remote sensing. *www.ccrs.nrcan.gc.ca/ccrs/misc/issues/hyperview_e.html*

Center for Space Research. University of Texas at Austin. Hyperspectral remote sensing. *www.csr.utexas.edu/projects/rs/hrs/hyper.html*

Goetz, A. Hyperspectral imaging. University of Colorado at Boulder. *cires.colorado.edu/steffen/classes/geog6181/Goetz/Definition.html*

Integrated Spectronics Pty. Ltd. *www.intspec.com*

ITRES Research Limited. *www.itres.com*

Spectral Imaging Inc. *www.specim.fi*

7 Satellite-based sensors operating in the visible and infrared wavelengths

Stan Aronoff

Availability of continuously collected, complete remote sensing image coverage of the globe dramatically changed earth resource inventory and management. Activities ranging from weather forecasting to forestry and from crop monitoring to mineral exploration depend on this flow of data.

Airborne remote sensing systems offer the flexibility to tailor image acquisition to specific project needs and to fly when conditions are optimal. They do not face the power restrictions and data transmission limitations of a satellite system and so are less restricted in the rate of image acquisition and overall quantity of data they can collect during a mission. The choice of wavelength bands, flight path, and other parameters can be easily modified as needed. The many airborne multispectral scanner systems on the market are frequently upgraded by their manufacturers.

However, satellite-based systems for earth resource monitoring are designed and operated to provide long-term, consistent image acquisition for virtually the entire globe. (The orbit paths used generally cannot provide coverage of small areas near the poles.) Replacement satellites are launched periodically to replace aging systems and ensure continuous coverage. When sensors are upgraded or additional sensors added, provisions are made to ensure compatibility with earlier versions.

Satellite systems have several advantages over airborne remote sensing systems. Satellites have unrestricted access to the entire globe and can collect data continuously from a stable platform. Satellite imagery is necessarily more standardized, making it easier to compare images collected at different times. Also, the need to manage large quantities of image data from a satellite program has led to the development of long-term, continuous image archives, which are themselves a valuable historical record for longitudinal studies that will address issues yet to appear.

Remote sensing satellites are designed to address specific types of applications. However, most applications can make use of more than one type of imagery. There is no one best image type for any one application. A range of options with different trade-offs now exists. These trade-offs are made in satellite sensor characteristics, such as resolution, cost, frequency of coverage, and in speed of image delivery. For example, to acquire more frequent imaging of a large coverage area, a lower spatial resolution may have to be accepted.

The availability of multiple satellite sources with overlapping characteristics has introduced a degree of competition. Consumers now have the flexibility to choose imagery that is less expensive or more quickly available from a different sensor or reseller.

This chapter provides a brief introduction to the satellite sensors supplying image data most widely used for earth resource inventory, management, and visualization. Appendix B provides tables listing important characteristics of each sensor system. Information derived from analysis of satellite imagery is an important data source for geographic information systems. Increasingly, the image data itself is being used within a GIS environment. Thus, it is valuable for practitioners to appreciate the different characteristics of these image sources.

Many factors affect the selection of image products. The price of imagery changes rapidly, and differences in processing techniques and cost enable products with varying degrees of spatial accuracy to be produced from the same imagery. Also, the specific expertise of the analyst and the equipment and tools available can dramatically affect the quality of the information produced. Some sensors, particularly those with high spatial resolution, have pointable sensors that allow areas to be imaged to the side of the ground track (termed across-track viewing) or fore and aft (termed along-track viewing). In this way, an area of interest can be imaged more frequently than the satellite's *return period* (the number of days it takes for the satellite to retrace its orbit path). Off-nadir (i.e., not vertical) viewing is also useful in producing stereo imagery from which digital elevation data can be derived. Along-track stereo images, using images with different viewing angles collected during the same satellite overpass, are generally preferred because conditions in the two images will be virtually the same, improving stereo analysis. Across-track stereo images may be acquired several days apart or more, depending on the satellite orbit and cloud cover when the scene is next within viewing range. During this interval, changes in the landscape can occur (e.g., fields are plowed, leaves change color and fall during the autumn), making some features appear different in the two images of the stereo pair. These differences can make stereo analysis more difficult, particularly for automated stereo correlation programs.

The major satellite systems used for earth resource applications can be broadly divided into three groups based on their spatial resolution: low, medium, and high spatial resolution. In general, high-spatial-resolution sensors produce imagery with higher positional accuracy.

Low-spatial-resolution sensors

The low-spatial-resolution group includes satellites in geostationary orbits that image a portion of the earth repeatedly and those in sun-synchronous orbits that progressively build near-complete global coverage over a period of one to two days. The geostationary group includes the many geostationary meteorological satellites operated by several nations, including the GOES satellites operated by the United States. The low-spatial-resolution, sun-synchronous satellites include the AVHRR, MODIS, and SeaWiFS sensors. There are many other specialized low-spatial-resolution satellites that monitor important global conditions ranging from the thickness of the protective ozone layer in the upper atmosphere to tropical rainfall and air pollutants. Information on these satellites is readily available online from the organizations that operate them.

In many cases, lower resolution imagery has proven to be an advantage over higher resolution data for applications such as regional land-use and land-cover classifications and crop condition assessment for large areas. Much less image data is needed to provide coverage of the area, reducing the cost of processing and handling the data. Perhaps more important, the large area covered by a single image is more consistent than the multiple images collected on different days by high-resolution sensors. A classification analysis of multidate imagery can add substantially to the time and complexity of the analysis. Another advantage of low-spatial-resolution imagery is that small features irrelevant to the analysis that would introduce inaccuracies in a classification are not resolved, so they do not introduce error. For example, drainage ditches, small roads, and trails can reduce the accuracy of regional land-use classification or crop condition assessments.

Low-resolution sensors generate a veritable flood of environmental data. The challenge is to develop reliable, easily accessed, and timely distribution systems for this data to be incorporated into operational applications, many of which are GIS-based.

GOES and other geostationary meteorological satellites

Figure 7.1   GOES satellite platform.

Source: Image courtesy of NOAA.

The Geostationary Operational Environmental Satellite (GOES) *(figure 7.1)* program is operated cooperatively by the National Aeronautics and Space Administration (NASA) and the National Oceanic and Atmospheric Administration (NOAA); NASA is responsible for launch and NOAA for operation. The satellites orbit the earth at an altitude of 36,000 km in the same direction and rate as the earth's rotation, thus maintaining a stationary position relative to the earth, called a *geostationary orbit*.

The GOES series constantly views an entire hemispherical disk. Full earth disk and selected smaller areas are imaged at spatial resolutions of 1, 4, or 8 km GSD, depending on the band *(figure 7.2)*. Repeat coverage is generated every 30 minutes with smaller area or lower resolution repeat coverage available in as little as four minutes.

The GOES series provides continuous monitoring of temperature, humidity, cloud cover, and other data for weather forecasting. One of the instruments is a five-channel multispectral scanner providing measurements and image products, including image maps and animations shown on televised weather forecasts *(table B.1 in appendix B)*. Information derived from the image data includes storm tracking, cloud cover, cloud height, and cloud temperature profiles,

sea surface temperature, water vapor, and hot spot detection (such as volcanoes and forest fires). Designed for meteorological applications, GOES products have not been widely used in the GIS market. However, with the growing interest in image data, efforts have been initiated to develop products more easily imported into a GIS environment.

The GOES satellites are part of a global network of meteorological satellites spaced at intervals around the world. The two GOES satellites that provide coverage of North America are operated by the United States. GOES-East is positioned at longitude 75° west and GOES-West at longitude 135° west. Similar satellites operated by China (Fengyun), the European Space Agency (Meteosat), India (Insat), Japan (GMS), and Russia (GOMS) provide overlapping coverage of the entire globe *(figure 7.3)*. International agreements provide for sharing of the imagery, improving weather forecasting worldwide.

Sun-synchronous satellites

Defense Meteorological Satellite Program (DMSP)
The Defense Meteorological Satellite Program (DMSP), operated by the U.S. Air Force, is a series of satellites used to monitor the meteorological and oceanographic environments and to research solar–terrestrial physics. The Air Force Weather Agency (AFWA) sends data from these satellites to

Figure 7.2   Image of Hurricane Andrew as it makes landfall on the Louisiana coast taken on August 25, 1992, by the NOAA GOES-7 weather satellite.

Source: Image courtesy of NOAA and NASA.

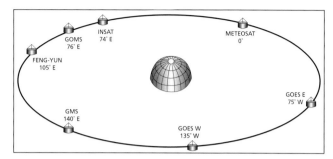

Figure 7.3 Geostationary meteorological satellite families.

Source: *Observation of the earth and its environment.* 2nd ed. by H. J. Kramer. 1994 © Springer-Verlag.

Figure 7.4 This nighttime image of the United States shows the lights from cities and towns, and gas flares off the coasts of Southern California and Louisiana. The composite image was generated from visible band imagery collected by Defense Meteorological Satellite Program (DMSP) satellites between October 1994 and March 1995. Thermal-infrared band imagery was used to identify and remove areas containing clouds from the time series.

Source: Image courtesy of NOAA National Geophysical Data Center.

the National Geophysical Data Center's Solar Terrestrial Physics Division (NGDC/STP) where it is archived. Since 1973, some of the data has been made available to civilian users.

Designed primarily to assess weather and navigation conditions, the DMSP satellites carry a range of meteorological sensors and have a 101-minute, sun-synchronous, near-polar orbit at an altitude of 830 km. Scanners produce imagery in the 0.40–1.10 µm (visible and near-infrared) and 8–13 µm (thermal infrared) bands with a swath width of 3,000 km. These scanners also have a spatial resolution of 0.55 km GSD in high-resolution mode and 2.7 km GSD in low-resolution mode (*table B.1 in appendix B*). Multiple satellites in day-and-night or dawn-and-dusk orbits together provide global monitoring every six hours. Typically, data from four satellites (three day-and-night, one dawn-and-dusk) are added to the archive each day.

The scanning systems onboard the DMSP satellites are uniquely capable of capturing low-illumination images of the earth at night. In order to collect images of clouds at night illuminated only by moonlight, the system is equipped with a photo multiplier for the visible and near-infrared bands to intensify the low illumination. This capability also enables the sensor to collect dramatic images of urban lights (*figure 7.4*). Nighttime imagery has also been used to monitor ephemeral features such as forest fires and volcanoes.

Beginning in 1994, the NGDC developed algorithms to improve the analysis of nighttime urban lights and generate cloud-free composite images with ephemeral light sources and unwanted artifacts removed. Research has shown this data to be a useful predictor of several measures related to the global human population. For example, the pattern of nighttime urban lights closely correlates with human population density, and the illumination intensity correlates with electrical power consumption. Multiyear studies have used

the DMSP nighttime imagery to monitor changes in these and other factors over time (*figure 7.5*).

The NOAA AVHRR sensor

Since 1970, the U.S. National Oceanic and Atmospheric Administration (NOAA) has operated a series of sun-synchronous, near-polar orbiting satellites designated the NOAA satellite series. The orbits of near-polar orbiting satellites are approximately from pole to pole. In the case of the NOAA series, the orbit has an inclination of 98° (i.e., the satellite passes within 8° of each pole). In addition, the orbit is *sun-synchronous* (i.e., the satellite crosses the equator at the same local time relative to the sun on each pass). Satellites collecting imagery in the visible and infrared bands are usually placed in sun-synchronous polar orbits so that the imagery has sun illumination conditions that are as consistent as possible.

The NOAA satellites (*figure 7.6*) were designed primarily as meteorological satellites operating in the visual and thermal infrared band to collect measurements of the earth's surface, atmosphere, and cloud cover. They also carry sensors to receive emergency identification and location signals from ships and aircraft and a tracking sensor to monitor the position of ships, ocean buoys, weather balloons, and even migrating animals.

NOAA satellites began carrying a sensor called the Advanced Very High Resolution Radiometer (AVHRR) when NOAA-6 was launched in 1979. The AVHRR sensors produce image data with a spatial resolution of 1.1 km GSD in five or six

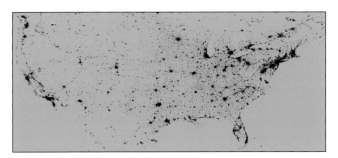

Figure 7.5 This image shows the comparison of nighttime urban lights imagery collected by Defense Meteorological Satellite Program (DMSP) satellites in the 1992–1993 and 2000 periods. Areas with brighter lights (red) or newly lit areas tend to highly correlate with increased human settlement.

Classification key:
  Black = Lights saturated in both time periods.
  Red = Lights brighter in 2000.
  Yellow = Lights only present in 2000.
  Blue = Lights only present in 1992–1993.
  Gray = Dim lighting detected in both time periods, but little change in brightness.

Source: Image courtesy of NOAA National Geophysical Data Center.

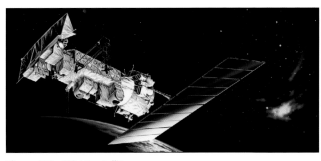

Figure 7.6 NOAA satellite.

Source: Image courtesy of NOAA and NASA.

wavelength bands, depending on the satellite *(table B.1 in appendix B)*. The imagery is received as a continuous image covering a 2,400 km wide area (the swath width). It is generally distributed as individual images with a coverage of 2,400 km by about 1,500 km *(figure 7.7)*. The data is made available in two formats: a local area coverage (LAC) at the full 1.1 km spatial resolution and a global area coverage (GAC) at a 4 km spatial resolution.

The NOAA program maintains two operating satellites, each producing complete global AVHRR coverage every 12 hours. Only the thermal infrared sensors are operated on the nighttime pass, providing daily global AVHRR digital image coverage in the visible and near-infrared bands, and twice daily coverage in the thermal infrared. The orbits of

the two satellites are offset by six hours so that, between the two, every location is imaged every six hours: once in the morning, once in the afternoon, and twice at night.

The data, in both digital and photographic images, is used in a variety of time-critical and large-area applications. Examples of these applications are weather analysis and forecasting; regularly updated maps of global sea surface temperature; search and rescue; global climate research; ice-condition monitoring of navigable waters, assessment of snow coverage, depth, and melting conditions; flood monitoring; forest fire detection and mapping; monitoring of crop conditions, dust, and sandstorms; identification of geologic events such as volcanic eruptions; and mapping of regional drainage networks, physiography, and geology.

The daily coverage of the AVHRR sensor has proven particularly valuable for regional crop-condition monitoring and rangeland assessment. For example, AVHRR data is used as an input to crop models for the measurement of crop area and estimation of yields for important cash and food crops, as discussed in the *Agriculture* section of chapter 12.

Less time-critical regional analyses include drought and desertification monitoring and water-current mapping. The water temperature patterns shown on the thermal infrared imagery and the patterns of water turbidity shown in the visual red channel have been used to detect algal blooms and analyze water currents and water mixing in lakes and coastal areas *(figure 7.8)*.

Global sea surface temperature monitoring data has also been continuously collected by the ATSR-1 and 2 sensors on the ERS-1 and -2 satellites and by the enhanced version of this sensor, the AATSR, on the European Space Agency's Envisat-1 *(table B.2, appendix B)*.

The MODIS sensor on the NASA Terra and Aqua satellites (launched in 1999 and 2002, respectively) has been designed to provide data consistent with that collected by the AVHRR sensor to facilitate historical comparison while offering enhanced capabilities. The MODIS sensor offers higher spatial resolution (as high as 250 m versus 1.1 km for AVHRR), additional spectral bands (36 bands versus 6), greater radiometric sensitivity (12 bits versus 10 bits), and improved geometric correction. AVHRR data is freely available for direct reception or can be purchased at a relatively low cost of less than $100 per scene. MODIS data is free for Internet download.

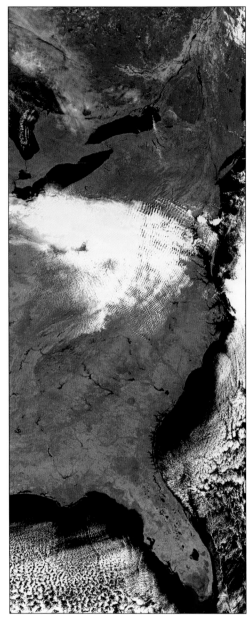

Figure 7.7 This portion of a NOAA 17 AVHRR image of eastern North America represents about 1,000 km by 2,700 km with a spatial resolution of 1.1 km GSD. The color composite was generated by assigning blue, green, and red to channels 1, 2, and 3, respectively. Acquired November 7, 2002, the image shows snow cover in some areas of Canada and the Northern United States (upper portion of the image). In channel 3 (1.58–1.64 µm), snow has a much lower reflectance than clouds and in this color composite appears cyan while the clouds appear white.

Source: © Natural Resources Canada.

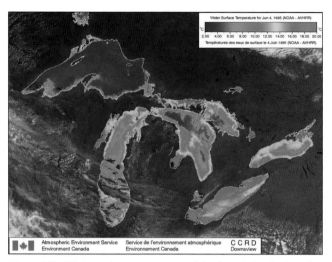

Figure 7.8 Water surface temperature of the Great Lakes in Canada and the United States on June 4, 1995, derived from AVHRR data.

Source: Image courtesy of the Meteorological Services of Canada.

Ocean color sensors

Ocean color sensors are designed to collect imagery in the visible and infrared bands. The sensors detect multiple narrow bands optimized for the measurement of physical and biological oceanographic conditions. The first ocean color satellite sensor was the Coastal Zone Color Scanner (CZCS), which operated from the Nimbus-7 satellite from 1978 to 1986. Ocean color (ocean reflectance in the visible wavelength band) is used to estimate the concentration of chlorophyll and other plant pigments present in the water from which estimates of the different types and quantities of marine phytoplankton can be derived. Marine phytoplankton play a crucial role in the marine ecosystem and in the exchange of gases and elements between the atmosphere and oceans. Though specialized for chlorophyll measurement, the image data recorded by the CZCS also provided information on water depth, ocean bottom characteristics, and other oceanographic information. After the CZCS failed in 1986, there were no other ocean color sensors in orbit until India launched the IRS-P3 experimental satellite in 1996. Several ocean color capable satellites followed: the Ocean Color and Temperature Scanner (OCTS) on ADEOS-I (Japan, 1996), SeaWiFS on OrbView-2 (United States, 1997), the OCM sensor on IRS-P4 (India, 1999), MERIS on Envisat (European Space Agency, 2002), and the MODIS sensor carried on the American NASA Terra (1999) and Aqua (2002) satellites. Though optimized for oceanographic applications, data from the ocean color bands have proven

Figure 7.9   OrbView-2 spacecraft.

Source: Image courtesy of NASA.

valuable in studying land and atmospheric processes as well.

SeaWiFS

Launched in August 1997, SeaWiFS (Sea-Viewing Wide Field-of-View Sensor), the only sensor carried on the OrbView-2 spacecraft (previously designated SeaStar) *(figure 7.9)*, was developed to provide quantitative global ocean monitoring data. The data has also proven valuable for integrated analysis of ocean, land, and atmospheric processes *(figure 7.10)*.

The project was a joint development of NASA and Orbital Sciences Corporation, now ORBIMAGE Inc. ORBIMAGE retains the commercial rights to the imagery and is responsible for the operation of the satellite and distribution of imagery to commercial users. NASA, the largest customer for SeaWiFS data, has a long-term image purchasing agreement with the company to support its requirements for ocean monitoring data and distribution of imagery for scientific research applications. In its first year of operation, more than 800 scientists representing 35 countries accessed SeaWiFS data received by some 50 ground stations throughout the world.

Operating from an altitude of 707 km, SeaWiFS is an across-track, multispectral scanner that produces 1.1 km GSD (nadir) imagery with a 2,800 km swath width. Designed to detect subtle changes in ocean color, the sensor collects imagery in eight bands (six visible and two near-infrared wavelength bands) with a ten-bit dynamic range *(table B.1 in appendix B)*. Individual frames have a 1,500 km by 2,800 km coverage area with a position accuracy of 1 to 2 km at nadir.

SeaWiFS produces two types of imagery: LAC data with a 1.1 km spatial resolution and GAC data that is subsampled onboard to produce 4.4 km spatial resolution imagery. The system produces full global area coverage data every two days with most areas imaged daily. Local-area coverage data is collected selectively, based on research priority. A wide range of information products are available. Products for commercial applications are distributed by ORBIMAGE, and data for scientific research and education is available from NASA.

The SeaWiFS image in figure 7.11 shows algal blooms as light blue. The land areas are shown as true-color imagery, and the water areas have been computer classified to show

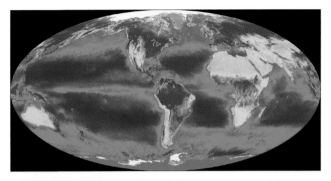

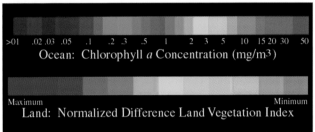

Figure 7.10   This composite biosphere image shows average chlorophyll concentration in ocean areas and average summer vegetation biomass as represented by the Normalized Difference Vegetation Index for land areas. The cloud-free image was generated from SeaWiFS imagery collected between September 1997 and August 1998.

Source: Image provided by ORBIMAGE, Inc. and processed by NASA Goddard Space Flight Center.

Figure 7.11   SeaWiFS image taken May 18, 1998, showing algal blooms in the Celtic Sea. Algal blooms appear yellow, light green, and light blue in this enhanced image.

Source: Image provided by ORBIMAGE, Inc. and processed by NASA Goddard Space Flight Center.

phytoplankton concentration as a color range from dark blue (low) through light blue to green and yellow (high).

Major applications of ocean color imagery include management of coastal zones and fisheries by mapping ocean characteristics such as chlorophyll concentration and water clarity, studies of the ocean thermal regime to better understand climate processes such as El Niño and La Niña, research on the ocean carbon cycle, environmental monitoring and pollution assessment of oceans and lakes, the study of algal blooms, land-use and land-cover mapping, and the detection and monitoring of natural disasters such as forest fires, dust storms, floods, and hurricanes.

Commercial applications have included identifying promising fishing areas, commercial and military maritime operations, coastal monitoring, and agricultural yield estimation. For example, the SeaStar Fisheries Information Service, operated by ORBIMAGE, offers map products to assist in finding high-value fish stocks in open ocean areas. Maps useful in selecting fishing areas are developed from analysis of phytoplankton concentration data derived from SeaWiFS imagery together with information on ocean temperature, currents, sea surface height, and weather information from other sources. Subscribers download map displays and interpretations via satellite to a personal computer daily to identify productive fishing areas for high-value catches such as skipjack and tuna.

ATSR and AATSR sensors

The ERS-1, ERS-2, and Envisat-1 satellites launched by the European Space Agency, in addition to their radar sensors (discussed in chapter 8), carry low-spatial-resolution optical sensors. The Along-Track Scanning Radiometer (ATSR)-1, carried on ERS-1, was a multispectral scanner operating in four infrared bands with a spatial resolution of 1 km GSD. The instrument was a conical scanning device that imaged an elliptical path. One limb of the ellipse provided a nadir (vertical) view, and the other limb scanned through an inclined path across and 900 km ahead of the satellite's track. Only the middle portions of the scans centered on the satellite track were used, providing a 500 km swath width *(figure 7.12)*. The two views of the same scene taken through different atmospheric path lengths make it possible to calculate a more precise atmospheric correction than is obtainable with a single measurement. As a result, sea surface temperature can be calculated with accuracies of 0.3° Kelvin or better.

An enhanced version, the ATSR-2, covers seven visible and infrared bands at 1 km GSD. In order to maintain data

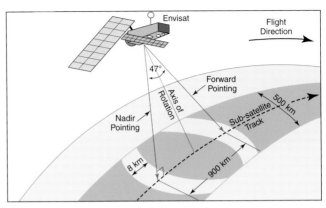

Figure 7.12 ATSR viewing geometry. The conical scanning action of the ATSR and AATSR imagers give two scans of each location—one from a nadir position, the other from an off-nadir, along-track position.

Source: Imagery courtesy of G. Petrie and M. Shand.

continuity, the Advanced Along-Track Scanning Radiometer (AATSR) instrument was included on Envisat-1, providing the same spectral bands and spatial resolution as the ATSR-2 instrument but with full digitization of all the channels *(table B.2 in appendix B)*. The imagery from these sensors is used to collect meteorological and oceanographic data for operational users and for climate research and environmental monitoring. Together these sensors provide an almost continuous 10-year record of high-accuracy sea surface temperature data that is valuable for climatic and oceanographic research.

MERIS (The MEdium Resolution Imaging Spectrometer Instrument)

Launched in March 2002 aboard the European Space Agency's Envisat-1 satellite (discussed in chapter 8), the MERIS sensor collects imagery in 15 spectral bands in the visible and near-infrared at a spatial resolution of 300 m GSD and provides global coverage of the earth every three days *(table B.2 in appendix B)*. Its primary mission is the measurement of ocean color to estimate chlorophyll and suspended sediment concentrations and aerosol levels. The sensor can also provide data on cloud-top altitude and total column water vapor.

IRS-P4 (Oceansat)

Oceansat *(figure 7.13)* was developed by the Indian Space Research Organization primarily for the measurement of physical and biological oceanographic characteristics, including the detection of phytoplankton blooms, chlorophyll and suspended sediment concentration, and aerosol. Launched in May 1999,

Figure 7.13   The IRS-P4
(Oceansat) satellite.

Source: Image courtesy of Antrix.

the satellite carries the Ocean Color Monitor (OCM) and the Multifrequency Scanning Microwave Radiometer (MSMR). The OCM is a solid-state camera operating in eight narrow spectral bands with a spatial resolution of 360 m GSD *(table B.2 in appendix B)*.

The MSMR operates in four microwave frequencies (6.6, 10.65, 18, and 21 GHz), both in the vertical and horizontal polarization. The MSMR provides data on total precipitable water, cloud liquid water, sea surface temperature, and sea surface winds, and offers global coverage once every two days. Other valuable research products generated from MSMR data include soil moisture, Antarctic snow cover, and oceanic rainfall.

MODIS sensor on the NASA Terra and Aqua satellites

The Terra *(figure 7.14)* and Aqua satellites are part of the NASA Earth Observing System, which is designed to collect a comprehensive set of global observations of the earth's land, ocean, and atmosphere in the visible and infrared regions of the spectrum. The program has been designed to support operational programs that require global scale environmental data as well as scientific research programs.

The MODIS (Moderate Resolution Imaging Spectro-Radiometer) instrument, an across-track scanner, is the primary sensor collecting data for global-change monitoring on both the Terra and Aqua satellites. This sensor was designed to measure physical properties of the atmosphere and biological and physical properties of the oceans and land. The data processing and distribution system for this data has been designed to build on the continuous remote sensing record of environmental data collected by earlier sensors, particularly the AVHRR sensor flown on the NOAA series of satellites since 1979. However, the MODIS sensor offers several important enhancements over the older AVHRR technology, including finer spatial resolution, higher radiometric sensitivity, improved geometric rectification, and more accurate radiometric calibration. MODIS imagery is available free of charge and can be received directly from the satellite or downloaded from the Internet.

With its 2,330 km wide swath, the MODIS sensor provides one-to-two-day coverage of the earth in 36 spectral

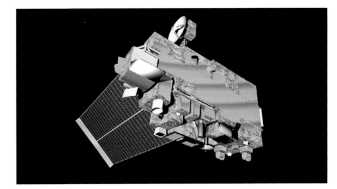

Figure 7.14   Terra satellite.

Source: Image courtesy of NASA.

bands from 0.4 to 14 μm at spatial resolutions of 250 m (2 bands), 500 m (5 bands), or 1 km (29 bands) *(table B.3 in appendix B)*. Overlap between adjacent image swaths increases from the equator toward the poles, and the corresponding repeat period increases from two days to multiple observations per day. The Terra satellite acquires data twice daily, crossing the equator at 10:30 AM and 10:30 PM Equatorial crossing for the Aqua satellite is at 2:30 AM and 2:30 PM Each satellite provides almost complete daily global coverage in the MODIS bands observing reflected solar radiation and almost complete twice-daily coverage in the bands observing emitted thermal radiation.

Normalized Difference Vegetation Index (NDVI) images can be produced from the visible and near-infrared MODIS bands 1 and 2 (with a 250 m spatial resolution) that are consistent with those produced from AVHRR imagery (with a 1.1 km spatial resolution). In addition, an Enhanced Vegetation Index (EVI) MODIS product was specially developed to provide improved sensitivity in high-biomass regions and a reduction in atmospheric and canopy background influences. Whereas the NDVI uses data from the visible red and near-infrared bands, the EVI also uses data from the visible blue band. Vegetation index products have proven valuable in a wide range of regional management applications such as crop condition assessment, crop-yield estimates, and tracking land-use and land-cover change over time *(figure 7.15)*.

A number of initiatives have been implemented to generate regularly updated global datasets for land cover, sea surface temperature, chlorophyll concentration, snow and sea-ice cover, cloud properties, and other environmental parameters *(figure 7.16)*. Over 40 standardized land and water data products have been developed and are distributed over the Internet at no charge *(table 7.1)*. One objective of providing

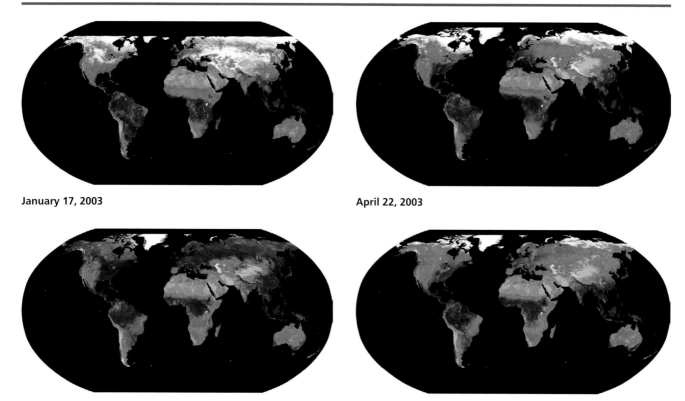

January 17, 2003

April 22, 2003

July 11, 2003

September 24, 2003

Figure 7.15 These enhanced vegetation index images derived from MODIS satellite imagery show the progressive green-up of vegetation in the spring. Comparison with averages from previous years enable the condition of crops and other vegetation to be monitored and stress caused by drought or disease to be identified early.

Source: © The University of Arizona 2004. Courtesy of Terrestrial Biophysics and Remote Sensing Lab, Dr. K. Didan.

standardized products is to facilitate environmental inventory and monitoring, change detection, and condition assessment over time by ensuring consistent processing. Many of these products are 8- or 16-day composite images in which the optimum cloud-free pixels are selected to minimize cloud-covered areas. Several products are made available within hours of reception.

The daily MODIS observations discussed here can also be used to support time-sensitive global monitoring. For example, the U.S. Forest Service in conjunction with NASA and the University of Maryland has implemented a rapid fire detection and response system that includes a dedicated direct reception and processing system to detect forest fires throughout the United States and to provide ongoing image data in support of fire response activities *(figure 7.17)*. MODIS data is used to monitor burn scars, vegetation type and condition, smoke aerosols, water vapor and clouds for overall monitoring of the fire process and its effects on ecosystems, the atmosphere, and climate.

MODIS data is supplied in HDF-EOS format. A utility is available online from NASA to convert from the MODIS data format HDF-EOS to the GeoTIFF format commonly used in GIS (see MODIS in appendix B).

Medium-spatial-resolution sensors

Beginning in 1972 with Landsat 1 (80 m GSD), the deployment of satellite-based, medium-spatial-resolution (5 m to 250 m GSD), multispectral scanners initiated the rapid development of digital remote sensing of earth resources. Since then, myriad earth-resource applications in fields such as agriculture, forestry, geology, archaeology, urban

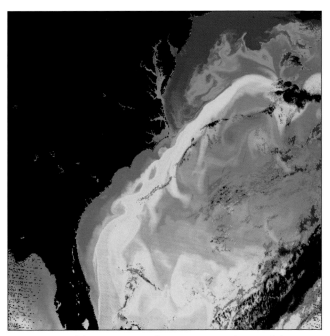

Figure 7.16   This MODIS sea surface temperature image of the east coast of the United States shows the warmer water (shown orange and yellow) of the offshore Gulf Stream current moving north and meeting the colder (blue and purple) northern waters.

Source: Image courtesy of MODIS Ocean Group, NASA/GSFC. SST product by R. Evans et al., University of Miami.

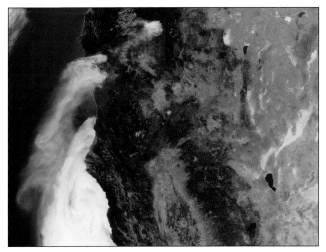

Figure 7.17   MODIS image of forest fires in the Klamath Mountains in Oregon burned over the state line into California in August 14th, 2002.

Source: Image courtesy of NASA.

Table 7.1   Summary of MODIS data products
Land
• Surface reflectance
• Land-surface temperature and emissivity
• Vegetation and land-surface cover, conditions, and productivity
• Land-cover change
• Vegetation indexes
• Thermal anomalies/fire
• Vegetation parameters (e.g., leaf area index)
• Bidirectional reflectance and albedo
• Fire occurrence, temperature, and burn scars
Ocean
• Sea-surface temperature (daytime and nighttime)
• Concentrations of organic matter, suspended solids, chlorophyll
• Ocean aerosol, color, and primary production measures
Atmosphere
• Aerosol, water vapor, ozone, and temperature measurements
• Cloud parameters
• Optical thickness
• Global distribution of total precipitable water
MODIS system
• Calibration datasets

and regional land-use planning, and watershed management have been developed for this data.

Application development, in turn, guided the development of new sensor systems. The placement and width of wavelength bands were optimized for the most widely used applications. The introduction of pointable sensors on the SPOT satellite provided high-quality stereo imagery suitable for measuring terrain elevation and generating digital terrain models. Advances in data storage and transmission systems enabled higher-spatial-resolution sensors to be operated from orbital altitudes.

The spatial resolution offered by the new sensor systems also increased. Ten-meter resolution data first became available with the launch of SPOT-1 in 1986. The IRS-1C satellite launched in 1995 now acquires imagery with a resolution of 5.8 m (which is resampled to provide 5 m resolution image data). Landsat 7 carries the eight-band Thematic Mapper multispectral scanner that includes a 15 m panchromatic band. SPOT 5 acquires 5 m panchromatic imagery with a pointable dual-scanner system that can provide along-track stereo coverage. Until the 1999 launch of IKONOS, producing 1 m GSD imagery, these medium-resolution satellites provided the most detailed publicly available imagery.

Landsat

The Earth Resources Technology Satellite (ERTS-1), later renamed Landsat 1, was the first satellite designed to provide systematic global coverage of earth resources. Launched by the United States on July 23, 1972, it was primarily designed as an experimental system to test the feasibility of collecting earth resources data from satellites. Originally, there were to be six satellites in the ERTS series. Landsats 1 through 3 operated from 1972 to 1983.

In 1975, the ERTS program and satellites were renamed Landsat, highlighting its focus on land-related applications. The Landsat program was government operated until 1985 when operation was transferred to EOSAT Corporation (later renamed Space Imaging EOSAT), a private sector company. In 1999 with the launch of Landsat 7 *(figure 7.18)*, operation of the Landsat program was transferred back to government control and is now managed cooperatively by NASA, NOAA, and the USGS. NASA is responsible for construction of the spacecraft and instruments, NOAA for satellite operation, and the USGS carries out data processing, archiving, and distribution at the EROS Data Center in Sioux Falls, South Dakota. Landsat imagery is available directly from the EROS Data Center and from third-party vendors.

In addition to its scientific aspects, the Landsat program was also an expression of an ideal. The data was made publicly available worldwide. This nondiscriminatory access to data was called the *open skies policy*. The data collected from space, in particular the Landsat data, was to be available to all and not limited by national interests. The resulting worldwide experimentation with Landsat data produced overwhelmingly favorable results. In fact, the experimental design of the satellites led to demands for levels of service that were better than NASA, and later NOAA, had planned to provide—evolving into an operational program.

The first three Landsat satellites carried a multispectral scanner (MSS) and a return-beam vidicon (RBV) camera. The RBV was similar to a television camera but was calibrated to higher standards of geometric accuracy. It produced digital imagery with a pixel resolution of 80 m in Landsats 1 and 2 and 30 m in Landsat 3. However, due to technical malfunctions, the RBV systems quickly became secondary sensors on the Landsat satellites.

The MSS sensor on Landsat 1, 2, and 3 imaged a 185 km wide swath in four bands: two bands in the visible and two in the near-infrared *(table B.4 in appendix B)*. These bands were designated 4, 5, 6, and 7 (the first three bands were assigned to the RBVs). The eighth band flown on Landsat 3 failed shortly after launch and is not indicated in the figure. Each pixel acquired by the MSS sensor represents a ground area of approximately 79 m by 56 m. Imagery is distributed as 185 km by 185 km scenes with 80 m GSD square pixels.

More importantly, the MSS system was the first to provide digital images suitable for computer analysis for most of the earth's surface. Availability of this new high-resolution data source spurred a flurry of application development and the demand for additional Landsat missions with higher spatial resolutions and additional narrower spectral bands positioned to optimize classification results.

The satellite platform used for Landsat 1, 2, and 3 *(figure 7.19)* was about 3 m high, 1.5 m in diameter, and weighed about 800 kg. The satellites were launched into circular, near-polar, sun-synchronous orbits with altitudes of about 900 km. Circling the earth every 103 minutes, the satellites completed 14 orbits per day, crossing the equator at the same local sun time (the time relative to the sun) on each pass over the illuminated side of the earth. The local sun time was about 9:30 AM to 10:00 PM, depending on the location within a time zone. Adjacent swaths were imaged on successive days *(figure 7.20)*, providing complete earth coverage every 18 days (except for the polar regions between latitudes 82° to 92°). The importance of a sun-synchronous orbit is that it ensures repeatable sun illumination conditions for an area at the same time of year, making it easier to mosaic imagery from adjacent tracks and to compare changes

Figure 7.18 Landsat 7 satellite.

Source: Image courtesy of NASA.

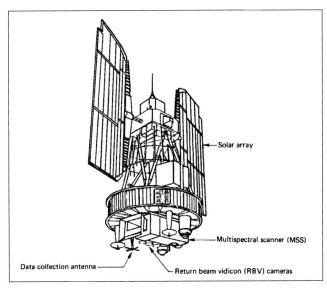

Figure 7.19 The satellite platform used for Landsats 1, 2, and 3.

Source: Image in *Remote sensing and image interpretation*. 2nd ed. by T. M. Lillesand and R. W. Kiefer. © 1987 John Wiley and Sons, Inc. Reprinted with permission of John Wiley and Sons, Inc.

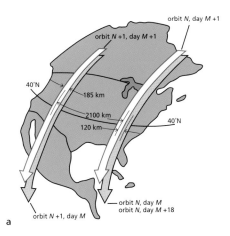

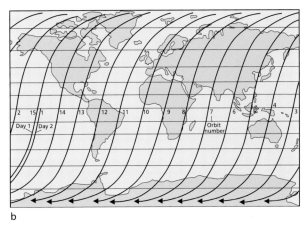

Figure 7.20 Landsat 1, 2, and 3 orbit paths. (a) Adjacent swaths were imaged on successive days with an overlap that increased with latitude. Complete image coverage was achieved every 18 days. (b) To be sun-synchronous, orbit paths were inclined relative to the equator. The descending (north to south) orbit paths for one day are shown; the ascending orbits pass on the night side of the earth.

Source: Image in *Images of the earth: A guide to remote sensing*. 2nd ed. by S. A. Drury. 1998. Reprinted with permission of Oxford University Press.

in land cover. Since sun elevation and intensity changes with the seasons, sun illumination will vary over the year.

As with all polar orbiting remote sensing systems, the overlap between adjacent swaths increased with latitude, ranging from about 14% at the equator to about 85% at 81° north and south latitudes.

Landsat 4 and 5 were launched into lower sun-synchronous polar orbits of about 700 km and have a return period of 16 days. Launched in 1982, Landsat 4 failed shortly after launch due to electrical problems. Landsat 5 was launched in 1984. Both satellites carried the MSS and TM sensors. The MSS sensor was included to provide continuity with the previous Landsat data. MSS bands 1 to 4 correspond to bands 4 to 7 of Landsat 1 through 3. The swath width of Landsat 4 and 5 is 185 km, like previous Landsat satellites, and the data is processed to be compatible with the earlier MSS data. The other sensor, the Thematic Mapper (TM), is a more advanced multispectral scanner than the MSS system. It provides seven bands ranging from the visible blue to the thermal infrared (band 7 is out of sequence because it was added late in the design stage). The TM sensor also provides a 30 m spatial resolution for all bands except the thermal infrared channel (band 6), which has a 120 m resolution. Figure 7.21 is a Landsat 5 TM image of the National Capital Region, Ottawa, Canada.

Landsat 6 failed on launch in 1993. Landsat 7, launched April 15, 1999, carried the Enhanced Thematic Mapper Plus (ETM+) sensor. In addition to the seven 30-meter-resolution bands carried on Landsats 4 and 5, the ETM+ has a 15-meter-resolution panchromatic band. The thermal infrared channel (band 6) has an increased spatial resolution of 60 m, and the instrument's radiometric calibration was improved. Landsat 7 is operated by NOAA with the primary receiving station and data processing facility at the EROS Data Center. In May 2003, the ETM+ scanner on Landsat 7 began to malfunction and production of imagery from this sensor was

Figure 7.21 Landsat 5 Thematic Mapper image of the National Capital Region, Ottawa, Canada. The color composite image was produced from bands 3, 4 and 5.

Source: Radatsat-1 data © Canadian Space Agency 1996. Received by the Canada Centre for Remote Sensing. Processed and distributed by RADARSAT International.

Figure 7.22 SPOT 5 satellite in orbit.
© 2004 CNES.

halted. The Landsat 5 TM sensor continues to operate normally after 20 years in service. As of this writing, a satellite to continue Landsat data collection (designated the Landsat Data Continuity Mission) has not been approved.

SPOT

France created the SPOT (Système Pour l'Observation de la Terre) program in 1978. It was designed from the start to be a long-term, operational system ensuring continuity of data collection while improving the technical capabilities and performance of the sensors. At the time of its inception, SPOT offered the only competition to the U.S. Landsat program for publicly available earth resource remote sensing imagery. It offered fewer spectral channels but higher spatial resolution imagery designed for earth resource applications. Sensor enhancements introduced in SPOT 4 and SPOT 5 *(figure 7.22)* improved the spatial resolution from the original 10 m to 2.5 m and added additional sensors, increasing the range of spectral bands, spatial resolutions, and stereo imaging capabilities. As a result, applications of SPOT imagery have also broadened to include urban and regional mapping, disaster management, flight simulations, and oceanographic

analyses. The high spatial resolution also made SPOT imagery suitable for business applications such as site selection, market analysis, planning, and change detection.

SPOT 1 was launched in early 1986 and has been followed by SPOT 2, 3, 4, and 5. (SPOT 3 failed in November 1996.) SPOT 1 through 3 carried identical high-resolution-visible (HRV), along-track or *pushbroom* scanners using linear array charge coupled device (CCD) detectors. The HRV instruments operating in a panchromatic mode produces 10 m resolution imagery for a single visible band (0.51–0.73 μm). In the multispectral mode, the sensors produce 20 m resolution imagery in three bands: 0.50–0.59 μm (green), 0.61–0.68 μm (red), and 0.79–0.89 μm (near-infrared). The SPOT satellites have a sun-synchronous orbit at an altitude of about 830 km and a return period of 26 days. The viewing direction of each HRV sensor can be varied by a rotating mirror by as much as 27° to either side by ground command. This allows each sensor to image any area within a 475 km strip on either side of the satellite ground track. The ability to image areas from adjacent swaths not only allows areas of special interest to be imaged more frequently, but it also enables full-scene, across-track stereo images to be produced. Also, the overlap between adjacent orbit tracks increases with latitude, so areas at higher latitudes are imaged more frequently.

When both of the SPOT sensors are pointed vertically, they each image a 60 km swath with a 3 km overlap, together producing imagery with a 117 km swath width *(figure 7.23)*. Swath width increases with viewing angle to a maximum of

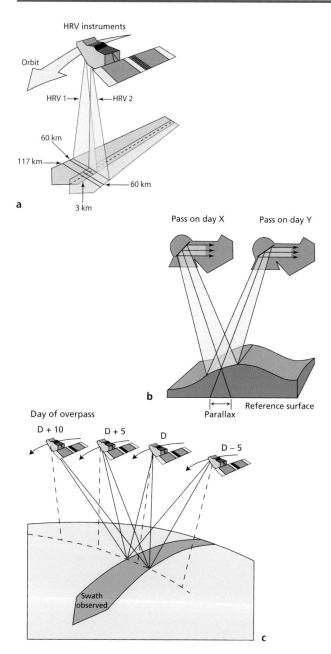

a

b

Pass on day X   Pass on day Y

Reference surface

Parallax

Day of overpass

D + 10   D + 5   D   D – 5

Swath observed

c

Figure 7.23 Image acquisition by the SPOT HRV sensors. The two HRV sensors can be pointed independently. (a) A 117-km swath is imaged when both are in the vertical position. (b) Stereoscopic coverage can be obtained by using the pointable mirrors to image the same area from different orbits. (c) The mirror settings allow the same area to be imaged as many as 11 times during a 26-day cycle, depending on the latitude.

Source: Image in *Images of the earth: A guide to remote sensing.* 2nd ed. by S. A. Drury. 1998. Reprinted with permission of Oxford University Press.

80 km at the maximum oblique[1] viewing angle of 27°. The increase in swath width occurs because, when the viewing direction is pointed away from the vertical, the distance from the sensor to the area on the ground being imaged increases, and the resulting image scale is smaller, so a larger area is captured within the scene. By viewing areas to the side of the ground track[2], areas at latitudes greater than 45° can be imaged as many as 11 times during a 26-day orbit cycle, seven times at the equator.

SPOT 4, launched in 1998, incorporated a number of improvements over previous SPOT missions. The two HRV instruments were replaced with two high-resolution visible and infrared (HRVIR), four-band, multispectral, pushbroom scanners. The additional 20 m resolution channel in the 1.58 to 0.75 μm mid-infrared band (designated the shortwave infrared or SWIR channel) is valuable for vegetation monitoring, mineral discrimination, and soil moisture assessment. The other multispectral channels, pointing capabilities, and swath widths remained the same as those on previous missions.

The 10 m resolution panchromatic band of earlier missions was dropped. Instead, channel 2, the visible red band (0.61–0.68 μm), operated at either 20 m or 10 m resolution, thereby maintaining a 10 m imaging capability similar to the former panchromatic channel. However, dissatisfaction with this narrower band led to reinstatement of the panchromatic band on SPOT 5.

The other major addition to SPOT 4 was the vegetation instrument. Designed primarily for vegetation monitoring, this four-band, pushbroom, multispectral scanner provides a wide 2,000 km image swath with a spatial resolution of about 1 km at nadir. It uses the same visible red, near-infrared, and mid-infrared bands as the HRVIR instrument and a visible blue band (0.43–0.47 μm) useful in oceanographic applications and for atmospheric correction of the other image channels. Global coverage is provided almost daily. (A few zones near the equator are imaged every other day.)

The vegetation instrument uses the same geometric reference system and three of the same spectral bands as the higher resolution HRVIR sensor, which facilitates multiscale

---

1. An image taken with a camera or sensor is said to be an *oblique image* if the axis of the viewing direction is intentionally directed between the vertical and horizontal planes. A high oblique image includes the horizon in the field of view, while a low oblique shows only the ground.

2. A satellite's *ground track* is the vertical projection of the satellite's path onto the ground below.

analyses. SPOT 4 could use these two instruments simultaneously, acquiring coarse-resolution, large-area imagery with the vegetation instrument, while recording fine-resolution imagery of a subarea with the HRVIR. For example, the accuracy of a regional crop analysis based on the lower resolution vegetation instrument imagery could be improved by including a subsample of sites analyzed using the higher resolution HRVIR imagery. Monitoring programs could use a sample of high-resolution images to calibrate the lower resolution, more frequently collected coarse-resolution imagery. It is an approach well suited to applications such as regional assessments of crop condition, crop-yield forecasting, forest cover monitoring, and studies of long-term environmental changes.

Full scene, across-track stereo imagery can be produced from SPOT imagery using several different combinations of viewing angles. For SPOT satellites 1 through 4, across-track stereo imagery can be obtained on *successive days* twice at the equator and more frequently at higher latitudes (e.g., six times at a latitude of 45°) during the satellites' 26-day orbital period. If successive day imagery is not suitable (e.g., due to cloud cover), stereo coverage can be obtained more frequently but with a longer gap between acquisition dates and greater viewing angle. SPOT 5 (see below) can provide along-track stereo coverage, preferred for stereo analysis because the second image of the stereo pair is acquired immediately after the first, so features appear the same in both images.

SPOT 5, launched in May 2002, carries a complement of three sensors. The high-resolution geometric (HRG) sensor provides improved 5-meter resolution panchromatic imagery over wide swaths (up to 60 x 120 km) with the same spectral range (0.51–0.73 µm) as SPOT 1 to 3. The HRV sensor also produces multispectral imagery at 10 m spatial resolution in the green, red, and near-infrared bands, and 20 m in the mid-infrared band. Figure 7.24 is an example of 10 m multispectral data used to assess flood damage along the Ebro River in Spain. The HRG sensor can be operated in a higher resolution mode, which generates 2.5-meter imagery from two 5-meter images obtained simultaneously with a half-pixel offset *(figure 7.25).*

The high-resolution stereo (HRS) instrument incorporates a forward and backward viewing scanner that generates near-simultaneous, along-track, stereo multispectral imagery with 10 m spatial resolution in the same four spectral bands as the HRG instrument. This imagery is designed for production of digital elevation models (DEMs) used for image rectification and diverse GIS applications.

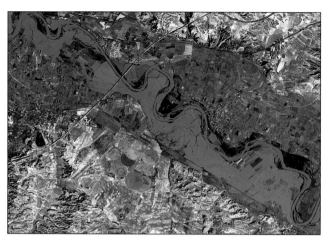

Figure 7.24 SPOT 5 image of flooding in Spain. This SPOT 5 10-meter-resolution image was acquired on February 10, 2003, three days after the Ebro River flooded. The flooding was the worst seen in Spain for 20 years and caused extensive damage to crops in the river valley.

Source: © CNES 2004. Courtesy of SPOT Image Corporation.

Figure 7.25 SPOT 5 panchromatic image of Oakland, California, port facilities. This 2.5-meter image was generated from two 5-meter, panchromatic images offset by half a pixel.

SOuce: © CNES 2004. Courtesy of SPOT Image Corporation.

DEMs with a 30-meter posting (i.e., elevation values are calculated for a grid of points 30 meters apart) can be generated from the high-resolution stereo (HRS) instrument on board SPOT 5 or from a stereo pair of 20 m, 10 m, or 5 m resolution images. These DEMs are available for full scenes or smaller areas. A full-scene product covers an area of 60 km x 60 km minimum (depending on viewing angle).

The vegetation 2 sensor on SPOT 5 offers the same four spectral bands and 1 km resolution as the vegetation sensor on SPOT 4, providing continuity of daily, global, environmental monitoring image data. Figure 7.26 is an example of a standard vegetation index product generated from 10 days of SPOT 4 data collected from May 11 to May 20, 1998.

Figure 7.26   SPOT 4 vegetation index composite image. This cloud-free composite image product was generated from imagery collected over a 10-day period from May 11 to May 20, 1998. The NDVI values were color-coded to show values ranging from green (high index values) through yellow to deep red (low values).

Source: © CNES 2004. Courtesy of SPOT Image Corporation.

Indian Remote Sensing satellites (IRS)

Figure 7.27   IRS satellite.

Source: © 2003 by Euromap GmbH, Neustelitz.

The Indian IRS-1C satellite was launched in 1995, followed by the identical IRS-1D in 1997. The IRS satellites *(figure 7.27)* were designed to provide systematic repetitive coverage of the earth's surface under consistent illumination conditions while repeating its orbital track every 24 days. At the time of their launch, the IRS satellites provided a higher spatial resolution alternative to SPOT and Landsat imagery. Until the launch of the IKONOS satellite in 1999, the IRS satellites offered the highest resolution global satellite image coverage commercially available.

Both satellites carry three sensors: a single-band panchromatic sensor, the Linear Imaging Self-Scanning (LISS)-III scanner, and a wide field sensor (WiFS). The single-band panchromatic sensor provides 5.8 m resolution imagery that is processed to 5 m pixels *(figure 7.28; table B.6 in appendix B)*. The sensor is pointable, providing ±26° off-nadir viewing. This also enables off-nadir imagery to be collected to produce across-track stereo imagery or to increase revisit frequency to as little as five days. The LISS-III is a four-band multispectral scanner operating in two visible bands (green and red) and two infrared bands (near- and shortwave infrared) with a spatial resolution of 23 m. The multispectral

Figure 7.28   IRS-1C multispectral image of the San Francisco Bay area. Bands 1, 2, 3 are displayed as green, blue, and red respectively.

Source: Courtesy of Antrix Corporation Ltd., Bangalore, and Space Imaging Inc.

data collected from this scanner is commonly sharpened using the higher resolution panchromatic data to produce pan-sharpened multispectral products with 5 m pixels. The third sensor, the wide field sensor (WiFS), produces imagery with a spatial resolution of 188 m in two bands, a red and a near-infrared band designed for regional analyses of vegetation vigor, land cover, and land use.

The two IRS satellites are in sun-synchronous polar orbits with a 24-day orbit cycle for the panchromatic and LISS-III sensors and a 5-day revisit cycle for the WiFS sensor. They follow the same groundtracks but are offset by 12 days so that together the return period is halved.

A range of GIS-ready image products are available from third-party distributors around the world.

Figure 7.29 ResourceSat-1 satellite.

Source: Image courtesy of Antrix Corporation Ltd., Bangalore, and Space Imaging Inc.

IRS-P6 (ResourceSat-1)

Launched in October 2003 by the Indian National Remote Sensing Agency IRS-P6 (ResourceSat-1) was planned as a continuation of IRS-1C/1D, offering enhanced capabilities for integrated land- and water-resources management. ResourceSat-1 *(figure 7.29)* provides both multispectral and panchromatic imagery of the earth's surface with additional spectral bands, greater radiometric resolution, broader spatial coverage, and stereo imaging not available on the earlier sensors.

Designed primarily for agricultural applications, the satellite carries three pushbroom CCD scanners. The high-resolution Linear Imaging Self-Scanner (LISS-4) operates in three spectral bands in the visible and near-infrared region (VNIR) with 5.8 m GSD spatial resolution. The sensor is pointable up to ±26° across track to obtain stereoscopic imagery and achieve a five-day revisit capability. The medium resolution LISS-3 operates in three spectral bands in the VNIR and one in the shortwave infrared (SWIR) band with a spatial resolution of 23.5 meter GSD. The Advanced Wide Field Sensor (AWiFS) operates in three spectral bands in VNIR and one

band in SWIR with a spatial resolution of 56 meters GSD. The satellite also carries a solid-state recorder with a capacity of 120 gigabits to store imagery for later transmission when a ground receiving station is within range.

Japanese remote sensing satellites

Japan has built and operated geostationary meteorological satellites since 1977 and has developed several earth resource imaging satellites under the auspices of the National Space Development Agency of Japan (NASDA), now renamed the Japan Aerospace Exploration Agency (JAXA). Two oceanographic satellites, MOS-1a and 1b, were operated from 1987 until the mid 1990s. The JERS-1 satellite, which operated from 1992 until 1998, carried an L-band (23 cm) synthetic aperture radar and a seven-band optical multispectral scanner that operated in the visible, near-, and mid-infrared bands. Along-track stereo coverage was provided by a second forward-looking near-infrared band. The optical sensor, in addition to its relatively high spatial resolution of 18 m GSD for both the optical and radar sensor, was the only civilian satellite producing along-track stereo imagery at that time. With spectral bands similar to those of the Landsat Thematic Mapper, the imagery could be used for a wide range of earth resource applications. Though the satellite is no longer operational, the existing data remains a valuable archive of historical data.

The Advanced Earth Observation Satellite, ADEOS-I *(figure 7.30)*, operated from 1996 to 1997 and carried two optical multispectral scanners covering the visible, near-, mid-, and thermal-infrared bands *(figure 7.31)*. ADEOS II, which, operated from 2002 to 2003, carried a 36-channel optical

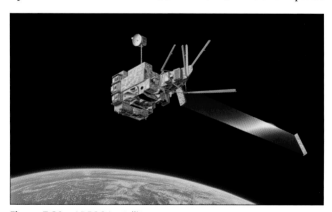

Figure 7.30 ADEOS-I satellite.

Source: © JAXA.

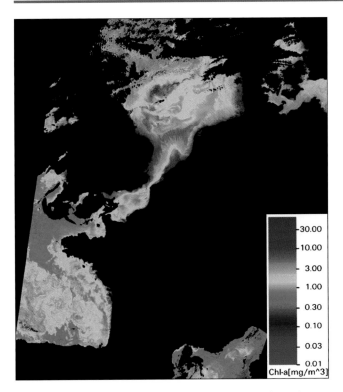

Figure 7.31   Chlorophyl-a concentration in the English Channel generated from ADEOS-I image data.

Source: © JAXA.

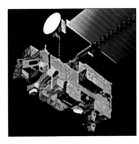

Figure 7.32   Terra spacecraft with the ASTER instrument shown in yellow.

Source: Courtesy of R. Stockli, NASA Goddard Space Flight Center.

sensor similar to the MODIS sensor and a synthetic aperture radar. These missions were designed to collect precise ocean color, sea-surface temperature, and other environmental data for climate change and other research objectives but ended prematurely due to system failures.

The Advanced Land Observation Satellite (ALOS), scheduled for launch in 2005, carries three imaging sensors, a four channel 10 m GSD multispectral scanner operating in the visible and infrared bands, a three-channel 2.5 m scanner with forward-, vertical-, and backward-viewing to produce continuous along-track stereo coverage, and an L-band synthetic aperture radar capable of producing imagery with spatial resolutions ranging from 10 m to 100 m GSD.

ASTER

The Advanced Spaceborne Thermal Emission and Reflection Radiometer (ASTER) is an imaging instrument aboard the Terra spacecraft. ASTER is a cooperative effort between NASA and Japan's Ministry of Economy Trade and Industry (METI) *(figure 7.32).*

The ASTER instrument is designed to measure snow and ice distribution, vegetation types, rock and soil properties, surface temperature, and cloud properties. It consists of three separate instrument subsystems: the visible and near-infrared (VNIR), the shortwave infrared (SWIR), and the thermal infrared (TIR). Each subsystem operates in a different spectral region, with a different pixel resolution, and separate telescope. The four-band VNIR subsystem produces imagery with a 15 m GSD, the six-band SWIR subsystem produces 30 m GSD, and the five-band TIR subsystem produces 90 m GSD *(table B.6 in appendix B).*

Particularly noteworthy for GIS applications is the VNIR subsystem, which consists of two independent telescope assemblies. The first, a nadir-viewing telescope, captures imagery in the visible green, red, and near-infrared bands. The second telescope captures imagery in the same near-infrared band but at a 27.7° view angle, which provides continuous along-track stereo coverage that is well suited to photogrammetric applications such as the generation of digital elevation models. With accurate ground control points, elevation accuracies better than 25 m (CE90) can be achieved—17 m (CE90) under optimal conditions (Toutin and Cheng 2001). The two telescopes can also be rotated ±24° across track, providing a five-day revisit capability.

Figures 7.33, 7.34, and 7.35 illustrate three different image products derived from ASTER image data: a false-color infrared 15 m GSD image, a color-coded image of suspended sediments derived from the brightness values of band 1, and a color-coded temperature image from a thermal infrared band.

ASTER is an on-demand instrument, acquiring imagery only for locations that have been requested. Data already acquired can be ordered from the Earth Observing System Data Gateway (EDG) operated by the USGS.

High-spatial-resolution satellites

Before 1999, satellite images with ground resolutions of 1 m or better were acquired only by military intelligence satellites. The demand for high-resolution imagery coupled with dramatic improvements in computer storage capacity

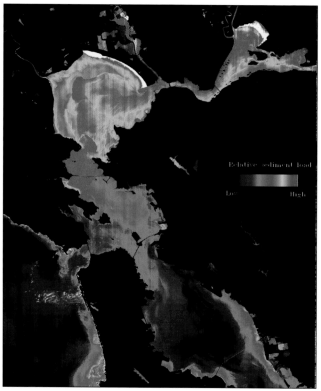

Figure 7.34 This color-coded, suspended sediment ASTER image was created from the visible green band by blacking out the land. Colors were assigned to the water areas according to their relative brightness values ranging from white for high brightness values (indicating high suspended sediment levels) through red, yellow, green, and blue.

Source: Image courtesy of NASA/GSFC/METI/ERSDAC/JAROS.

Figure 7.33 San Francisco Bay color infrared image. This image covers an area 60 kilometers wide and 75 kilometers long in the visible green, red, and near-infrared bands. This combination of bands portrays vegetation in red and urban areas in gray. Sediment in the water shows up as lighter shades of blue. Along the west coast of the San Francisco Peninsula, strong surf can be seen as a white fringe along the shoreline. A powerful rip tide is visible extending westward from Daly City into the Pacific Ocean. In the lower-right corner, the wetlands of the South San Francisco Bay National Wildlife Refuge appear as large dark blue and brown polygons. The high spatial resolution of ASTER allows fine detail to be observed in the scene. With enlargement the entire road network can be easily mapped; individual buildings are visible, including the shadows of the high-rises in downtown San Francisco.

Source: Image courtesy of NASA/GSFC/METI/ERSDAC/JAROS.

With the launch of the IKONOS satellite in 1999, space imagery with 1 m GSD spatial resolution became commercially available. QuickBird, launched in 2001, offers image products with an even higher resolution of 0.70 m GSD[3]. These satellites, together with EROS-A1 and OrbView-3, offer a range of products and coverage and a certain degree of competition. (Specifications for these satellites are provided in *table B.7 in appendix B*.)

and processing speed, wireless data transmission rates, and the GPS satellite positioning system has made the acquisition and rectification of high-resolution imagery for large areas possible.

3. The spatial resolution at which imagery from these satellites is acquired varies with the view angle of the pointable sensor. The maximum spatial resolution is achieved with a nadir (vertical) view. At nadir, the IKONOS panchromatic sensor generates 0.82 GSD imagery and QuickBird produces 0.61 GSD imagery. However, the georectified image products have been resampled to 1 m and 0.7 m GSD respectively.

Figure 7.35 The color-coded temperature ASTER image was created from a thermal infrared band by blacking out the land and assigning colors to the relative temperature values. White is warmest, followed by yellow, orange, red, and blue.

Source: Image courtesy of NASA/GSFC/METI/ERSDAC/JAROS.

These satellites are operated as commercial ventures. Image acquisition programs are tailored to produce commercial products of selected locations to meet the needs of government and private-sector clients. The U.S. government through its ClearView program makes large purchases of IKONOS, QuickBird, and OrbView-3 imagery and receives high priority order fulfillment. While providing important base-level funding for the satellite operators, government priority compromises timely order fulfillment for other customers. Vendors seek to provide customization of scene area and other parameters to suit client needs. As well, for an additional fee, clients can schedule the acquisition of specific areas. In the case of EROS-A1, the client can purchase exclusive and confidential image acquisition time during which the acquired imagery can be directly received at the client's facilities for the area lying within the footprint of the ground station.

The price of the resulting imagery varies with the spatial resolution, number of bands, and level of rectification. The lowest cost products with minimal processing and no correction for relief are typically priced at about $25 per square kilometer for four-band multispectral imagery that has been sharpened using the panchromatic imagery obtained at the maximum spatial resolution. For a 10 km by 10 km area, that would amount to $2,500. Precision-corrected products can cost 5 to 10 times this amount, depending on the level of correction chosen.

High-resolution satellite sensors can be positioned to point along-track and across-track and can be scheduled to acquire stereo imagery or repeat coverage of selected areas. Priority is given to client requests with the unscheduled time being used to build image archives to suit particular commercial objectives. Unlike the low- and medium-resolution satellite programs discussed previously, these high-resolution commercial satellites do not collect complete global image coverage on a regular schedule.

The spatial resolution of the imagery is fine enough to capture much of the infrastructure of urban areas. Buildings, streets, vehicles, and individual trees can be seen. Some features smaller than the resolution of the sensors, such as the white stripes in parking lots and crosswalks, can be distinguished due to their high contrast. Airborne imagery, whether photographic or digital, can be flown at low altitudes to obtain as high a spatial resolution as is required. However, where the accuracy and resolution of satellite image products is suitable or where alternative imagery is not available, imagery from these high-resolution satellites can be a cost-effective solution.

Most of the high-resolution satellites carry both a panchromatic and multispectral sensor. The multispectral imagery typically has a GSD four times that of the panchromatic (e.g., 4 m GSD for IKONOS multispectral image products and 1 m GSD for the panchromatic imagery). To generate color imagery at the finest resolution, the multispectral and panchromatic imagery are combined to generate a pan-sharpened color image. Though this image theoretically does not have the same resolution as a multispectral image generated at 1 m GSD, in practice, for visual image interpretation, there is little perceived difference.

Another characteristic of imagery from the high-resolution satellite sensors to be considered is viewing angle. Aerial photography for mapping is acquired with a nadir (vertical) view direction. Much of the imagery collected by high-resolution satellite sensors are oblique (nonvertical) views

Figure 7.36 Oblique image of Denver, Colorado, July 17, 2002. This satellite image with a spatial resolution of 61 cm GSD acquired by QuickBird illustrates how features such as tall buildings can obstruct the visibility of portions of a scene.

Source: Image courtesy of DigitalGlobe Inc. and OrbImage Inc.

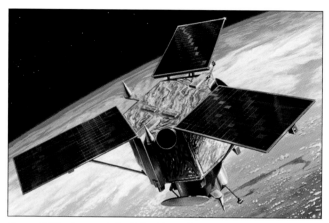

Figure 7.37 Image of the IKONOS satellite

Source: Image courtesy of Space Imaging Corporation.

(figure 7.36) taken with a large across-track pointing and tilt angle[4]. On oblique imagery, significant portions of a scene can be blocked by tall structures such as office towers in urban areas. As a result, for scenes with tall structures, oblique imagery would be less suitable for base mapping, though it may still be quite useful for image interpretation.

A wide range of image products in formats compatible with GIS and image analysis software are available from the system operators as well as from resellers. Customized products are also readily available from these sources and from service companies specialized in remote sensing. GIS applications typically use orthorectified image products as an image underlay over which other geographic information is displayed. This level of rectification significantly improves the registration of image features with the GIS data by correcting for the position

errors caused by variations in terrain relief (relief displacement). GIS data layers can be updated or new data layers created by digitizing features interpreted on the image.

IKONOS

The IKONOS satellite *(figure 7.37)*, developed by Space Imaging Inc. (formerly EOSAT), was launched September 24, 1999. Its along-track scanner captures 4 m GSD multispectral imagery in the visible blue, green, red, and near-infrared bands and 1 m GSD panchromatic imagery. In addition to separate panchromatic and multispectral image products, pan-sharpened color imagery is also available. Combining multispectral and panchromatic imagery, the pan-sharpened product has an effective 1 m GSD. All imagery is collected with 11-bit dynamic range (2048 gray levels). The system is pointable, allowing viewing at angles of up to ±45° in the across-track or along-track directions. Not only does this provide more frequent imaging of a given area, but it also enables collection of across-track (side-by-side) and along-track (fore-and-aft) stereo images. The ground track of the satellite repeats every 11 days, but an area can be imaged at 1 m resolution every three days and at 2 m about every 1.5 days. The system has a swath width of 11 km at nadir, and standard images are 11 km by 11 km in size. However, user-specified image strips and mosaics can be produced to order. IKONOS data is available from Space Imaging Inc. and resellers.

Figures 7.38a and b illustrate the detail that can be seen in an IKONOS pan-sharpened 1 m image.

---

4. High-resolution satellite sensors depend on nonvertical viewing to achieve an acceptable revisit frequency to image a selected area. This is required because these satellites image a narrow swath, the largest being QuickBird with a 16.5 km swath width. If they acquired imagery only using vertical viewing, these satellites would generate complete global coverage only about twice a year. Medium-resolution satellite sensors such as the Landsat Thematic Mapper generate a global coverage every 16 days, and coarse-resolution satellite sensors such as SeaWiFS and MODIS produce almost daily global coverage.

Figure 7.38a   Satellite imagery for tornado damage assessment. The IKONOS satellite image shows a portion of the 19-mile path (orange outline) of a tornado that struck Moore, Oklahoma, on May 8, 2003. The arrow indicates the location and viewing direction from which the ground photo in figure 7.38b was taken.

Source: Image courtesy of Space Imaging Corporation

Figure 7.38b   Motel along Interstate 35 in Moore, Oklahoma, taken May 9, 2003. This photo shows the damage caused by a tornado the previous day within the damage area shown on the IKONOS satellite image.

Source: © 1993 Associate Press. Photo by S. Ogrocki.

EROS-A1

Launched in December 2000, the EROS-A1 is owned and operated by ImageSat International (formerly called West Indian Space). The EIOp pushbroom scanner and the EROS satellite were designed and built in Israel. The original program called for eight EROS satellites. Delays and cost overruns have delayed the launch of additional satellites. The satellite offers worldwide coverage, producing panchromatic imagery with a 1.8 m GSD and a 13.5 km swath width. Oversampling techniques (the process of combining two images of the same scene acquired with a subpixel offset) can produce imagery with a 0.9 m GSD. Selected sites can be viewed as often as two to three times a week. Clients are able to task the satellite at predetermined times and directly receive the acquired imagery, greater control than is offered by other commercial operators. Governments and their intelligence agencies can thus obtain direct, confidential, and secure access to high-resolution satellite imagery without having to own and operate a satellite.

QuickBird

The QuickBird satellite *(figure 7.39)* operated by Digital-Globe Inc. was launched October 18, 2001. QuickBird provides the highest spatial resolution commercial imagery currently available. Originally designed to acquire 1-meter imagery, the planned satellite orbit for QuickBird was modified to produce 0.61-meter imagery after DigitalGlobe Inc. was licensed to receive higher resolutions.

Vertical panchromatic imagery is acquired with spatial resolutions as high as 61 cm GSD and 2.44 m GSD for multispectral imagery (visible blue, green, red, and near-infrared bands). The multispectral imagery is commonly used in the form of pan-sharpened products that incorporate the greater detail

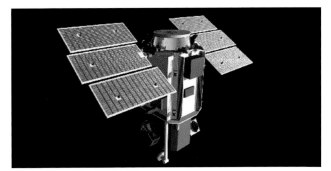

Figure 7.39   QuickBird satellite.

Source: Image courtesy of DigitalGlobe.

of the panchromatic imagery to the lower-spatial-resolution multispectral data. Figure 7.40 is a pan-sharpened 70 cm GSD image of the Sydney Opera complex and surrounding area.

The pointable sensor aboard QuickBird enables along-track and across-track stereo imagery capture, as well as a revisit frequency of 1 to 3.5 days. The sensor's swath width is 16.5 km to 19 km (depending on the viewing angle), the largest offered by the commercial high-resolution satellites. The data is distributed by DigitalGlobe Inc. and a network of resellers.

OrbView-3

Figure 7.41 Image of the OrbView-3 satellite.

Source: Image provided by ORBIMAGE Inc. and processed by NASA Goddard Space Flight Center.

OrbView-3 *(figure 7.41)* was launched in June 2003 and is operated by ORB-IMAGE Inc. A pushbroom scanner system produces 1 m GSD panchromatic imagery and four-band multispectral data at 4 m GSD. The sensor offers the same spectral bands as IKONOS and QuickBird but with a narrower swath width of 8 km. The sensor is pointable ±45° from nadir and can reimage an area in less than three days. Image data can be transmitted in real time to customer ground stations

Suitability for use

In judging the suitability of high-resolution satellite imagery for a specific task, it is important to consider a number of factors in addition to the spatial resolution of the imagery. The interpretability of an image also depends on image quality, which can be affected by atmospheric conditions such as clouds or haze, brightness (low light levels reduce contrast, making similar features more difficult to distinguish), and angle of illumination (low illumination angles produce large shadow areas that may obscure features but can also highlight low-relief features).

Factors not related to imaging conditions and spatial resolution include the availability and suitability of alternate image sources, such as existing aerial photography or airborne digital imagery. The availability and cost of equipment and personnel or contractors with the skills to analyze the imagery and produce the required products is another important consideration. The time frame within which information products must be delivered may also be a deciding factor.

Figure 7.40 QuickBird image of the Sydney Opera Complex, Australia. This 0.70 m pixel pan-sharpened image shows the Sydney harbor area and opera complex.

Source: Image courtesy of DigitalGlobe.

The level of geometric accuracy required depends on the application. It is important to distinguish *mapping applications* required to meet specific standards of geometric accuracy and feature identification from those primarily concerned with identifying the characteristics of selected feature types, often referred to as *remote sensing applications*. Mapping applications commonly demand considerably higher levels of image rectification. National standards, e.g., U.S. National Map Accuracy Standards for topographic mapping, specify the features that must be identified and the positional accuracy with which they must be mapped in order to be considered suitable for use at a specified scale.

By contrast, *remote sensing applications* focused on characterizing specific features, such as estimating the area of forest harvesting, the location and extent of flooding, or the level of plant vigor of an agricultural crop, can accept lower levels of geometric accuracy. Thus, imagery suitable for mapping at scales of 1:10,000 or smaller may be quite suitable for remote sensing applications at larger scales.

The imagery from high-resolution satellites is striking in the detail captured from a sensor orbiting hundreds of kilometers above the earth. The ability to display digital images at any scale can suggest that the imagery is suited to applications beyond what their geometric accuracy can support. Considerable research has been published on the photogrammetric quality of IKONOS and QuickBird imagery. There are differences of opinion within the photogrammetry and remote sensing community on the largest scale of mapping for which high-resolution satellite imagery is suited.

Based on the focal length of the sensors and the flying height of the satellites, the calculated scale of QuickBird and IKONOS panchromatic imagery is about 1:51,000 and 1:68,000 respectively[5]. These are the scales at which the image is projected onto the focal plane of the sensor. In the case of a photographic system, it is the scale at which the image would be recorded on film.

Practitioners have found that image products orthorectified using suitable ground control points and digital elevation data can be used in applications for which airborne imagery at scales of about 1:25,000 and 1:40,000, respectively, are used. Orthorectified IKONOS imagery is generally accepted by practitioners to have a geometric accuracy that meets or exceeds that of the 1:24,000 scale USGS digital orthophoto quadrangles (DOQ). Orthorectified QuickBird imagery is generally considered to meet or exceed the geometric accuracy of the 1:12,000 scale USGS digital orthophoto quarter quadrangles (DOQQ).

In general, the greater the level of geometric accuracy required, the larger the scale of imagery needed. Passini and Jacobsen (2004) suggest a rule of thumb for topographic mapping is that the pixel size at map scale should be 0.05 mm to 0.1 mm. Based on resolution alone, IKONOS or OrbView-3 imagery with a 1 m GSD (i.e., 1 m ground pixel size) would be expected to support mapping at scales of 1:20,000 (0.05 mm/1m) to 1:10,000 (0.1mm/ 1m). QuickBird imagery with a 0.70 m GSD would be expected to support mapping at about 1:14,000 to 1:7,000. However, factors other than resolution affect the geometric accuracy of imagery. Orthorectified images are commonly produced using 8 pixels per mm, which would give a scale of 1:8,000 for 1m GSD imagery and about 1:6,000 for 0.70 m GSD imagery.

These rules of thumb were developed from experience using aerial photography. Satellite imagery offers some advantages that enables its use for mapping scales with a larger magnification than is typically used for airborne photography. Even on a calm day, an aircraft is buffeted by variable air currents that cause tipping and tilting. A satellite orbiting in space has a very smooth trajectory, improving the relative accuracy of the imagery compared with the same scale of airborne imagery. Satellite sensors are calibrated very accurately, and the interior orientation (i.e., the image forming parameters of the instrument) tends to be more stable than that of airborne vehicles subjected to repeated jarring during flight and landing. The high-resolution satellites have a very narrow field of view (e.g., about 1° of arc) relative to airborne imagery [e.g., about 90° of arc for 230 cm by 230 cm (9 in. by 9 in.) photography with a 152 mm (6 inch) lens]. As a result, relief displacement toward the edges of vertical aerial photography is much more significant than for vertical satellite imagery. *Vertical images* acquired by these high-resolution satellites have been found to require little or no correction for relief displacement. Together these factors enable higher accuracy levels to be achieved for satellite imagery than would be obtained for aerial photography of the same scale.

For example, IKONOS imagery is used to generate products that meet NMAS standards for 1:4,800 scale mapping using standard digital elevation data products from the USGS and a few ground control points (GCPs). In areas where no GCPs can be obtained, accuracies that satisfy the National Map Accuracy Standard (NMAS) for 1:50,000 scale mapping can be achieved. By using multiple images collected with different viewing angles, accuracy can be improved to meet NMAS 1:20,000 scale mapping standards.

The scale used to map urban areas and to create vector line data for GIS and LIS applications is typically 1:1,200 in Canada and the United States and 1:1,250 in the United Kingdom. This scale of mapping is generally produced from aerial photography at scales of about 1:4,000 to 1:6,000. For rural areas, mapping at scales of about 1:2,500 is generated from aerial photography at a scale of about 1:10,000. Smaller 1:4,800 scale mapping is often used for statewide mapping in the United States for which imagery at scales of 1:15,000 to 1:20,000 is typical. Areas with little infrastructure development are mapped at smaller scales.

Toutin and Cheng (2002) and others (see references) have shown that with high-quality ground control points and digital elevation data, QuickBird imagery can be rectified using a rigorous 3D parametric model to meet U.S. National Map

---

5. Image scale is equal to the focal length of the sensor lens divided by the flying height. IKONOS uses a 10 m focal length lens and orbits at an altitude of 680 km. The QuickBird sensor has an 8.8 m focal length lens and orbits at an altitude of 450 km.

Accuracy Standards at the 1:2,400 to 1:4,800 scale. Rectifying satellite imagery to meet these standards requires sophisticated processing. Image processing software manufacturers have developed programs that make it relatively straightforward for users to perform their own orthorectification of high-resolution satellite imagery using digital elevation data (such as that available from the USGS for the United States) and a limited number of ground control points (preferably five or six). However, large-scale airborne imagery is more commonly used for mapping at these scales. To achieve the 1:1,200 scale accuracy typically required to generate vector line mapping of urban areas, either large-scale aerial photography or digital imagery with a GSD of 30 cm or better is required. Such imagery is commercially available only using airborne acquisition, though some military satellites can purportedly achieve this resolution level (Petrie 2004).

Existing aerial photography is commonly the most cost-effective image source. For example, in the United States, the National Aerial Photography Program (NAPP) of the USGS provides a standardized set of cloud-free aerial photographs covering the conterminous United States over five-to-seven year cycles *(figure 5.26 in chapter 5)*. Begun in 1987, the program provides a consistent source of attractively priced high-quality 1:40,000 scale aerial photography in black-and-white (B/W) or color infrared (CIR), depending on location and date.

Principal markets for high-resolution satellite imagery

The principal markets for high-resolution satellite imagery have been government acquisitions for classified intelligence gathering and mapping activities, resource exploration and development planning, emergency response, and the mapping and monitoring of natural resources such as forest harvesting and urban infrastructure.

Intelligence gathering is one of the major applications of imagery from the four high-resolution satellites currently in orbit. Governments or their contractors purchase large quantities of imagery over current and potential conflict areas and other regions of interest to which access by airborne sensors is denied. In the United States, government agencies have been instructed to rely to the maximum practical extent on U.S. commercial space remote sensing capabilities to meet their imagery and geospatial data requirements for areas abroad. This strategy also serves to reduce the demands on classified intelligence satellites, limiting their use to acquiring imagery not available from other sources. For most nations, commercial high-resolution satellite imagery is the best imagery available to meet their intelligence requirements.

In most areas of the world, topographic mapping is outdated and inaccurate or nonexistent. For these areas, satellite imagery can offer a means to produce maps at scales as large as 1:25,000 with minimal or no ground control.

In many countries, medium- and large-scale topographic maps exist but are classified and not readily available. High-resolution space imagery is commonly used to generate maps for these areas to support resource development activities such as oil and gas and mining exploration and development activities and for infrastructure development projects such as the siting of microwave and cellular telephone towers.

The high-resolution and frequent coverage have made high-resolution satellites particularly valuable in monitoring events that occur unexpectedly in areas of the globe that are difficult to access. Some examples are assessing damage and planning emergency response to natural and human-caused disasters such as earthquakes, severe storms, floods, forest fires, oil and chemical spills, illegal forest harvesting, and areas of military conflict.

Future high-resolution satellites

The technology exists to build still higher-resolution satellite sensors. Current U.S. military intelligence satellites purportedly achieve 10 cm spatial resolutions with optical sensors and 1 m with radar imagery (Petrie 2004). With funding provided by the U.S. NextView program, DigitalGlobe is developing a satellite (WorldView) to be launched in 2006 that will provide 0.5 m GSD optical imagery. Other high-resolution optical imaging satellites under development include the French Pleiades (0.7 m resolution for launch in 2008), the United Kingdom's TopSat (2.5 m GSD launched in 2004), ImageSat's EROS-B (0.9 m GSD launched in 2004), India's Cartosat-1 (2.5 m GSD launched in 2004), and Japan's ALOS satellite with its PRISM sensor generating continuous 2.5 m GSD stereo imagery launched in 2004.

Increased competition from the larger number of image providers should reduce data costs and improve product quality, making high-resolution satellite imagery more attractive for applications currently served by airborne systems. With government requirements for high-resolution imagery firmly established, it is likely that funding programs and guaranteed image purchases will continue to be provided to ensure continued development of high-resolution satellite systems around the world. Imagery is available from distributors as well as directly from ORBIMAGE.

Conclusion

The sensor systems discussed in this chapter represent the remote sensing image sources most widely used in geographic information systems. They are a sample of a large and growing array of space-based earth observation instruments that have a wide range of imaging and coverage characteristics, spatial resolution being just one. The choice of imagery available to a user depends on the specifics of the application. Cost, timeliness of delivery, availability of wavelength bands, spatial and radiometric resolution, and extent of coverage are some of the more important image parameters considered.

Image acquisition also needs to be considered within the larger context of an organization's longer-term objectives and broader mandate. Where more than one image source can fulfill the technical requirements of an application, other factors may determine imagery selection. Factors to consider include which image source would best fit the organization's overall image acquisition program and the type of imagery with which the analysts have greater experience. An increasingly attractive option is to forego image acquisition in favor of subscription services offering access to imagery on demand, which greatly reduces the need for the user to bear the overhead costs of storing and managing image data. In many cases, an organization may choose to outsource the entire remote sensing analysis process. Receiving only the derived information products, the user may have little direct involvement in the choice of sensor or type of imagery, confining their scrutiny only to the quality of the information products they receive.

References   Ager, T. P. 2003. Evaluation of the geometric accuracy of IKONOS imagery. SPIE 2003 AeroSense Conference, April 21–25. Orlando, Fla.

Campbell, J. B. 1996. *Introduction to remote sensing*. 2nd ed. New York: Guilford Press.

Dare, P. M., N. Pendlebury, and C. S. Fraser. 2002. Digital orthomosaics as a source of control or geometrically correcting high-resolution satellite imagery. Proceeding of the 23rd Asian Conference on Remote Sensing, November 25-29. Kathmandu, Nepal. www.gisdevelopment. net/aars/acrs/2002/vhr/108.pdf

Davis, C. H., and X. Wang. 2003. Planimetric accuracy of IKONOS 1-m panchromatic orthoimage products and their utility for local government GIS basemap applications. *International Journal of Remote Sensing* 24 (22): 4267–88.

———. 2001. High-resolution DEMs for urban applications from NAPP photography. *Photogrammetric Engineering and Remote Sensing* 67 (5): 585–92.

Dial, G., and J. Grodecki. 2002. IKONOS accuracy without ground control. Proceedings of ISPRS Comission I Mid-Term Symposium, November 10–15. Denver, Colo. www.spaceimaging.com/ techpapers/default.htm

Drury, E. S. 1998. *Images of the earth: A guide to remote sensing*. 2nd ed. New York: Oxford University Press.

Elvidge, C. D., K. E. Baugh, E. A. Kihn, H. W. Kroehl, and E. R. Davis. 1997. Mapping of city lights using DMSP Operational Linescan System data. *Photogrammetric Engineering and Remote Sensing* 63: 727–34.

Emap International. 2002. QuickBird aerial product comparison: Prepared by Emap International for DigitalGlobe. Reddick, Fla.: Emap International Inc.

Federal Geographic Data Committee. 1998. Geospatial positioning accuracy standards. Part 3: Standards for spatial data accuracy. *FGDC-STD-007*.3-1998. Federal Geographic Data Committee. Washington, D.C.

Fowler, R. A. 1997. Map accuracy specifications. Part 2: Where did they come from, what are they, and what do they really mean? *Earth Observation Magazine* (November). www.eomonline.com/ Common/Archives/1997nov/97nov_fowler.html

Fraser, C. S. 2000. High-resolution satellite imagery: A review of metric aspects. *International Archives of Photogrammetry and Remote Sensing* 33 (B7/1): 452–59.

Fraser, C. S., and H. B. Hanley. 2003. Bias compensation in rational functions for IKONOS satellite imagery. *Photogrammetric Engineering and Remote Sensing* 69 (1): 53–7.

Fraser, C. S., E. Baltsavias, and A. Gruen. 2002. Processing of IKONOS imagery for submetric 3D positioning and building extraction. *ISPRS Journal of Photogrammetry and Remote Sensing*. 56 (3): 177–94.

Fraser, C. S., H. B. Hanley, and T. Yamakawa. 2002. Three-dimensional geopositioning accuracy of IKONOS imagery. *Photogrammetric Record* 17 (99): 465–79.

Greenwalt, C. R., and M. E. Schultz. 1968. Principles and error theory and cartographic applications. ACIC Technical Report No. 96, Aeronautical Chart and Information Center, U.S. Air Force, St. Louis, Mo. www.fgdc.gov/standards/status/tr96.pdf

Grodecki, J. 2001. IKONOS stereo feature extraction-RPC approach. Proceedings of ASPR 2001 Conference, April 23–27. St. Louis, Mo. www.spaceimaging.com/techpapers/default.htm

Grodecki, J., and G. Dial. 2003. Block adjustment of high-resolution satellite images described by rational polynomials. *Photogrammetric Engineering and Remote Sensing* 69 (1): 59–68. www.spaceimaging.com/techpapers/default.htm

———. 2002. IKONOS Geometric Accuracy Validation. Proceedings of ISPRS Comission I Midterm Symposium, November 10–15. Denver, Colo. www.spaceimaging.com/techpapers/default.htm

———. 2001. IKONOS geometric accuracy. Proceedings of Joint Workshop of ISPRS Working Groups I/2, I/5 and IV/7 on High Resolution Mapping from Space 2001, Sept 19–21. University of Hannover, Hannover, Germany. www.spaceimaging.com/techpapers/default.htm

Grodecki, J., G. Dial, and J. Lutes. 2003. Error propagation in block adjustment of high-resolution satellite images. Proceedings of ASPRS 2003 Conference, May 5–9. Anchorage, Alaska. www.spaceimaging.com/techpapers/default.htm

ITC. ITC online database of satellites and sensors. International Institute for Geo-Information Science and Earth Observation, Enschede, The Netherlands. www.itc.nl/research/products/sensordb/searchsat.aspx

Jacobsen, K. 2003. Geometric potential of IKONOS and QuickBird images. In *Photogrammetric Weeks '03*, ed. D. Fritsch, 101-110. Heidelberg: Wichmann Verlag.

Jensen, J. R. 2000. *Remote sensing of the environment: An Earth resources perspective.* Upper Saddle River, N.J.: Prentice Hall.

Kennie, T. J. M., and G. Petrie. 1993. Analogue, analytical, and digital photogrammetric systems applied to aerial mapping. In *Engineering surveying technology.* Glasgow: Blackie Academic.

Kramer, H. 2002. *Earth observation remote sensing: Survey of missions and sensors.* 4th ed. Berlin: Springer Verlag.

Lillesand T. M., R. W. Kiefer, and J. Chipman. 2003. *Remote sensing and image interpretation.* 5th ed. New York: John Wiley and Sons, Inc.

Narayanan, M. S., and A. Sarkar. 2001. Observations of marine atmosphere from Indian Oceansat-1. Proceedings of the MEGHA-TROPIQUES 2nd Scientific Workshop, July 2-6, Paris, France. meghatropiques.ipsl.polytechnique.fr/documents/workshop2/proc_s5p01.pdf

NASA Ocean Color Sensors Web site. simbios.gsfc.nasa.gov/oceancolor.html

Passini R., and K. Jacobsen. 2004. Accuracy of digital orthophotography from high-resolution space images. URISA, Charlotte. www.ipi.uni-hannover.de/index1.htm

Petrie, G. 2004. High-resolution imaging from space: A worldwide survey. *GeoInformatics* Part I—North America. 7 (1): 22–27. Part II—Asia 7 (2): 22–27. Part III—Europe 7 (3): 38–43. web.geog.gla.ac.uk/~gpetrie/pubs.htm.

———. 2002. Optical imagery from airborne and spaceborne platforms: Comparisons of resolution, coverage, and geometry for a given pixel size. *GeoInformatics* 5 (1): 28–35. web.geog.gla.ac.uk/~gpetrie/pubs.htm

Rencz, A. N., and R. A. Ryerson, eds. 1999. *Remote sensing for the earth sciences.* 3 vols. New York: John Wiley and Sons, Inc.

Tao, C. V., Y. Hu, and W. Jiang. 2004. Photogrammetric exploitation of IKONOS imagery for mapping applications. *International Journal of Remote Sensing* 25.

Toutin, T. 2003. DEM extraction from high-resolution imagery. *Geospatial Today* 2 (3): 31–36. ess. nrcan.gc.ca/esic/ccrspub-cctpub/index_e.php

Toutin, T., and P. Cheng. 2002. QuickBird: A milestone for high-resolution mapping. *Earth Observation Magazine* 11(4):14–18. ess.nrcan.gc.ca/esic/ccrspub-cctpub/index_e.php

———. 2001. DEM generation with ASTER stereo data. *Earth Observation Magazine* 10 (June): 10–13. ess.nrcan.gc.ca/esic/ccrspub-cctpub/index_e.php

———. 2000. Demystification of IKONOS! *Earth Observation Magazine* 9 (7): 17–21. ess.nrcan.gc.ca/esic/ccrspub-cctpub/index_e.php

U.S. Geological Survey. National Aerial Photography Program (NAPP) information. EROS Data Center. U.S.Geological Survey. Sioux Falls, S.D. edc.usgs.gov/products/aerial/napp.html

Vozikis G., C. Fraser, and J. Jansa. 2003. Alternative sensor orientation models for high-resolution satellite imagery, band 12. Publikationen der Deutschen Gesellschaft für Photogrammetrie, Fernerkundung und Geoinformation. Bochum.179–186. www.ipf.tuwien.ac.at/publications/gv_bochum_2003.pdf

8 Active sensors: Radar and lidar

Stan Aronoff and Gordon Petrie

Radar, lidar, and sonar remote sensing systems actively illuminate a scene with an energy source and detect the reflected return signals. Radar uses microwave energy operating at wavelengths between 1 and 100 cm, while lidar systems use lasers operating at much shorter wavelengths in the visible or near-infrared portion of the electromagnetic spectrum. Sonar systems make use of sound (acoustic energy), typically propagated through water, to detect subsurface features and measure depth.

Radar and lidar, the systems that use electromagnetic energy, are discussed in this chapter. Sonar is addressed in chapter 9.

Radar

Radar (radio detection and ranging) was developed during World War II to detect the positions and tracks of enemy aircraft from ground stations and as a navigation aid in bad weather when fitted to aircraft. A radar system directs short pulses of microwave energy toward a target and measures the strength of the reflected signal and the time it takes for the signal (called *backscatter*) to return to the sensor. Because the system provides its own source of illumination and the microwave wavelengths used are not blocked by clouds, radar sensors can provide an all-weather day-and-night imaging capability. Invaluable for military reconnaissance and targeting, radar's all-weather capability has also proved valuable for resource inventory and monitoring in cloud-covered remote locations common in tropical, subtropical, and maritime regions and for time-critical applications such as emergency response. One of the most important applications of radar has been in the production of detailed digital elevation models, widely used in GIS applications and to rectify remotely sensed imagery. Commercial airborne systems can provide horizonatal accuracies of ±1.25 m RMSE and vertical accuracies as high as ±0.3 m RMSE. Satellite-based imaging radar can offer horizontal accuracies of ±10 m to ±20 m RMSE and vertical accuracies as high as ±10 m RMSE. Radarsat-2, scheduled for launch in 2006, will offer spatial resolutions as high as 3 m, while the forthcoming fleet of European radar satellites—including the German SAR-Lupe series and TerraSAR and the Italian COSMO-SkyMed series—will generate images with spatial resolutions of 1 m.

Radar principles

Radar was originally developed during WWII for military applications. For security, the different microwave bands used for radar were designated by letters *(table 8.1)*. When the technology was later declassified, the letter designations continued to be used. In general, shorter wavelengths provide higher spatial resolution but offer less penetration of clouds. Atmospheric interference occurs at wavelengths less than 3 cm. This property is used to advantage to detect radar reflection from water droplets to identify areas of heavy precipitation for use in weather forecasting. Imaging radars for earth observation are normally operated at longer wavelengths that penetrate clouds, haze, smoke, light rain, and snow. Since radar is an active system, generating a microwave signal to illuminate the scene, imagery can be acquired day or night and in virtually all weather conditions.

Table 8.1 Radar band designations and usage		
Band designation	Wavelength range	Sensors and usage
Ka	0.75–1.1 cm	Weather radar systems, RAMSES
K	1.1–1.67 cm	Weather radar systems
Ku	1.67–2.4 cm	Weather radar systems, CryoSat, RAMSES
X-band	2.4–3.75 cm	Used widely for military reconnaissance and commercially for terrain surveys, STAR-3i, SIR-C, SRTM, SAR-Lupe, TerraSAR-X, COSMO-SkyMed, GeoSAR
C-band	3.75–7.5 cm	Used by many spaceborne SAR systems: Envisat; ERS-1 and 2; Radarsat-1, -2, and -3; SIR-C; SRTM
S-band	7.5–15 cm	Used by Russian Almaz radar satellite that operated from March 1991 to October 1992.
L-band	15–30 cm	Spaceborne: SeaSat, SIR-A, SIR-B, SIR-C, JERS-1, PALSAR
P-band	30–100 cm	Airborne: NASA-JPL AirSAR, GeoSAR, AES-3, OrbiSAR

Nonimaging radar

Radar systems may be ground based or mounted on aircraft or spacecraft. Many common radar systems are nonimaging. For example, traffic police use handheld Doppler radar systems that determine speed by measuring the frequency shift caused by the vehicle's motion *(figure 8.1)*.

Plan position indicator (PPI) radars use a rotating antenna to detect targets over a circular area. The distance and direction to features reflecting the signal are presented on a circular display, often overlaid on a map with the position of the radar antenna at the center and the position of detected objects shown as bright spots. A radial sweep continuously updates the display. PPI systems are commonly used in weather forecasting, air traffic control, and in aircraft navigation *(figure 8.2 and figure 8.3)*.

Satellite-based radar altimeters can provide centimeter-level profile measurements of the terrain and ocean surface. Radar scatterometers measure microwave scattering effects to estimate near-surface wind speed and direction with spatial resolutions on the order of 25 km. Radar scatterometer data from several earth resource satellites provide global coverage wind speed and direction data products for meteorology and oceanographic applications.

Figure 8.1 Doppler radar systems measure the difference in frequency or Doppler shift between the transmitted and return microwave signal to determine the vehicle speed.

Source: © Natural Resources Canada.

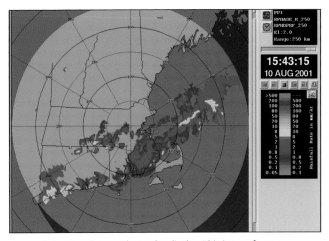

Figure 8.3 Doppler weather radar display. This image from a meteorological radar system operating at 11 cm wavelength shows rainfall rate to a distance of 250 km from the antenna location. The results are displayed over a map base with axes centered on the antenna location and graduated in kilometer distance. Plan position indicator radar (PPI) uses a rotating antenna to emit directional pulses of microwave energy in a circular pattern. By measuring characteristics of the reflected signal, including the time for it to return, the distance and elevation of targets can be calculated. Other information about the target can be derived by analysis of characteristics of the return signal.

Source: Image courtesy of SIGMET, Inc.

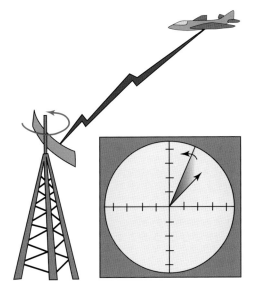

Figure 8.2 Plan position indicator (PPI) radar systems use a rotating antenna to transmit microwave pulses and receive the reflected returns. The distance and direction to features reflecting the signal are presented on a circular display.

Source: © Natural Resources Canada.

Imaging radar

Imaging radar systems used in remote sensing operate at higher resolutions than nonimaging systems to produce imagery with sufficient detail for interpretation and mapping applications.

An imaging radar system consists of a transmitter, a receiver, one or more antennas, and the electronic and computer resources to process and record the data. Imaging radar systems used for remote sensing also collect precise sensor attitude and geographic position data with inertial measurement units and GPS satellite positioning receivers. This data is later used to rectify the geometry of the imagery to match the chosen mapping base.

The transmitter generates short bursts (or pulses) of microwave energy at regular intervals that are focused by the antenna into a beam perpendicular to the direction of travel and downward at an oblique angle. When the burst of microwave energy strikes a target, a portion of the signal is reflected or backscattered from features within the illuminating radar beam and is returned to the sensor. The farther the target, the longer it takes for the reflected signal to be received *(figure 8.4)*. The reflection data is processed to determine, for each return, the signal strength and the time and angle at which it was received to calculate the geographic location of the target. A radar image is generated by showing the strength of the return signals as brightness levels in their correct geographical position.

As the aircraft or spacecraft moves forward, successive swaths of terrain to one or both sides of the platform can be recorded, producing a continuous strip of imagery. Radar systems cannot image the area directly beneath the sensor. This is because when a radar pulse is directed vertically downwards, there are simultaneous returns from those objects that are equidistant on either side of the flight line. Thus, the system cannot distinguish between them, hence

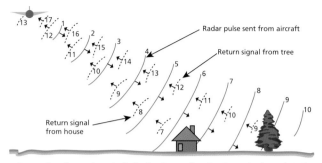

a. Propagation of one radar pulse (indicating the wavefront location at time intervals 1-17)

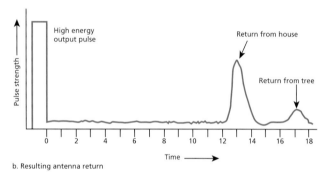

b. Resulting antenna return

Figure 8.4 Operating principle of a side-looking airborne radar.

Source: Image in *Remote sensing and image interpretation*. 2nd ed. by T. M. Lillesand and R. W. Kiefer. © 1987 John Wiley and Sons, Inc. Reprinted with permission of John Wiley and Sons, Inc.

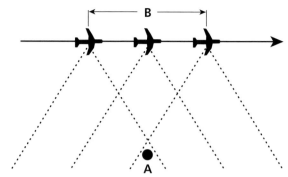

Figure 8.5 Principle of synthetic aperture radar operation. By processing the returns from targets (A) for the entire time they are illuminated by the radar beam, a short antenna can operate as if it was many times longer (B) than its actual length, providing improved spatial resolution.

Source: © Natural Resources Canada.

the gap in coverage. Early systems recorded the image data directly onto film. However, in current systems the digital data is recorded onto magnetic tape or other digital media.

Real and synthetic aperture radars

The first imaging radar systems, designated Side-Looking Airborne Radar or SLAR systems, are usually described as real aperture radars (RAR) to distinguish them from the synthetic aperture radars that were developed later. In a SLAR system, the longer the antenna, the more narrowly focused the microwave beam and the better the spatial resolution. With a 15 m antenna, which is about the longest practical antenna length for an airborne SLAR, imagery flown at an altitude of about 5 km will have a spatial resolution on the order of 15–20 m RMSE. In a SLAR system, the spatial resolution decreases with distance from the sensor. To acquire high-quality imagery from higher altitudes or from space, a different technology had to be developed to overcome the limitations of antenna size.

By contrast, synthetic aperture radar (SAR) uses a short antenna about 1–2 m in length but makes it perform as if it

were hundreds of times longer. The short antenna produces a wide microwave beam that would generate a coarse-resolution image in a SLAR system. A SAR system uses a higher pulse rate and sophisticated signal processing to record all the echoes received from a specific target for the entire time that it is within the radar beam *(figure 8.5)*. The effective antenna length or synthesized antenna becomes the distance the sensor has traveled during the time the feature was in view. Targets farther from the sensor, where the beam is wider, are illuminated for a longer period of time, thereby providing more data, which is used to improve the spatial resolution. When processing the signal, the SAR processor can divide the data into groups of signal samples, each representing an independent representation or *look* of the scene. Images can be formed using different numbers of looks. However, while more looks reduce speckle (discussed below), they also degrade resolution. The trade-off is chosen in accordance with the specific application.

SAR systems have a theoretical spatial resolution limit of approximately one-half the actual antenna length, regardless of range or wavelength; i.e., the shorter the antenna, the finer the resolution. In theory, a 1m long antenna could provide 0.5 m resolution imagery, whether from an aircraft or spacecraft. In practice, other factors, such as the signal-to-noise ratio, degrade resolution. SAR systems, with their capability to provide higher resolution imagery from greater altitudes, have virtually replaced SLAR systems for remote sensing.

Radar imagery

Though superficially similar in appearance, a radar image is very different from a black-and-white aerial photograph (see the sidebar *Radar terminology*). In general, slopes facing the radar antenna, built-up urban areas, bridges, railways, and other metal objects, and features with high moisture levels appear bright on radar imagery.

Whereas a black-and-white aerial photograph shows surface reflectance at visible or near-infrared wavelengths, a radar image shows surface roughness. The orientation of the target surface relative to the radar sensor, its moisture content, and the radar band chosen also have major effects on an object's reflecting characteristics of a microwave signal.

Surface roughness

Apparent surface roughness depends on the average size of the surface irregularities relative to the wavelength of the radar signal. In general, surface variations less than about 1/8th of a wavelength behave as smooth surfaces. Surfaces with height variations greater than half a wavelength generally appear rough. So, for a radar system operating at longer wavelengths of about 15 cm (L-band), a flat surface of 3 cm sized pebbles would appear smooth and produce little backscatter, but to a short wavelength radar operating at about

6 cm (X-band), the surface would be rough and generate significant backscatter.

Relatively smooth, flat features such as lakes and rivers, pavement, dry lakebeds, airport runways, close cut grass, and some agricultural crops appear dark. It is the backscatter from waves that enables radar systems to image ocean currents and wave patterns and to measure wave height and direction. Oil spill detection is made possible because the oil film on the water surface reduces wave heights, thereby reducing surface roughness and reducing radar reflectance. So where there are moderate waves, oil-covered water appears relatively dark *(figure 8.6)*. For this reason, oil spill detection is most effective when surface wind speeds are 3–12 m/sec (6–23 knots).

Shape and alignment

The shape of a feature and its alignment relative to the direction of the radar beam can dramatically affect the strength of the return *(figure 8.7)*. There will be weak backscatter from linear features (such as crop rows, power lines, and fences) if they are oriented parallel to the look direction but strong returns if they are oriented perpendicular to this direction. For example, a plowed field will appear smooth to a radar beam oriented in the same direction as the furrows but will appear rough if the furrows are perpendicular to the beam.

Moisture content and electrical properties

The electrical properties of terrain features have an important influence on the intensity of radar returns. For example, metal (a good electrical conductor) is a much better reflector and will appear brighter on radar imagery than vegetation or soil (which is a poor electrical conductor). Moisture content increases the conductivity of materials. For this reason, wet soil is a stronger reflector and appears brighter on radar

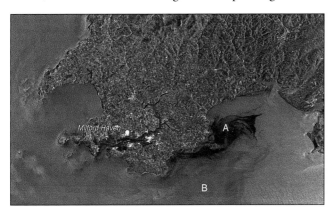

Figure 8.6   Oil spill detection. The *Sea Empress*, a 147,000 ton supertanker ran aground off the coast of Wales on February 15th, 1996, spilling 65,000 ton of crude oil. This Radarsat image, captured seven days later, shows the progress of efforts to control the oil spill. Oil floating on the water surface suppresses the ocean's capillary waves, creating a smoother surface than the surrounding water. Oil covered water appears darker than the surrounding water. Areas where the slick is thinner or has begun to break up appear lighter (B) than the main body of the slick (A). Image area coverage is about 90 x 50 km.

Source: Radarsat-1 data © Canadian Space Agency 1996. Received by the Canada Centre for Remote Sensing. Processed and distributed by RADARSAT International.

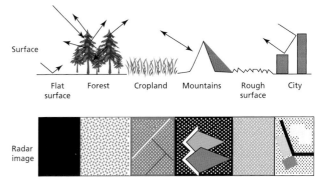

Figure 8.7   Effect of feature characteristics on radar backscatter.

Source: Image courtesy of NASA, JPL.

Figure 8.8 Radarsat image of the Dead Sea. The Dead Sea, straddling the border between Israel and Jordan, appears very dark in tone on this Radarsat image. The side-looking radar illumination highlights topographic features. Steep slopes facing the radar beam produce strong backscatter and appear bright in the image (A), while shadowing can emphasize subtle features such as the ridge (C) marking the old shoreline. The water level of the Dead Sea has been falling as water from the Jordan River (B) has been diverted for agricultural use. Also visible are commercial salt evaporators (D) that extract chemical products from the water and the city suburbs of Jerusalem (E) and Amman (F) that appear very bright due to the strong radar backscatter from urban structures such as roads and buildings. This Radarsat image resampled to 200 m pixels shows an area 160 km across.

Source: © Natural Resources Canada.

imagery than dry soil. Radar imagery has been useful in mapping soil moisture.

Speckle

Radar imagery also has a characteristic grainy appearance called speckle. It is caused by random constructive and destructive interference among return signals, producing random black-and-white pixels in the imagery. Speckle can be reduced by image enhancement techniques but cannot be eliminated.

Side-looking view

Although radar imagery is typically displayed as if viewed from above, the microwave illumination is actually from the side. Viewing angle affects relief displacement and the areas

that the terrain blocks from view, producing data gaps (see the *Effects of radar viewing angle on relief displacement and data gaps* below). Strong backscatter from slopes facing the radar antenna and shadowing produce an edge-enhanced effect similar to viewing a scene illuminated with a low-sun elevation angle *(figure 8.8)*. The surface roughness and slope of landscape features are emphasized, which is particularly valuable for geology. The topographic sharpening also enhances the contrast of smaller features such as surface texture and drainage patterns. Variations in surface roughness appear as tonal patterns on radar imagery. For example, snow and ice are represented as gray tones due to their texture.

Applications

Radar imagery has found important uses in geology (geological structures), hydrology (drainage networks), and oceanography (surface waves and currents), agriculture, land-use planning, coastal management, flood inundation, ship detection, and forestry (particularly to monitor forest clear cutting in tropical rainforest areas). Soon after radar technology was declassified, airborne radar imagery of perennially cloud-covered tropical rainforest areas such as the Amazon Basin revealed major rivers never before mapped. Satellite radar imagery is widely used to monitor sea ice conditions for navigation in polar regions *(figure 8.9)*. Radar imagery

Figure 8.9 Analysis of ice conditions for ship routing in northern Canada. The data ellipses indicate estimated ice thickness and coverage conditions.

Source: © Natural Resources Canada.

Radar terminology

The side-looking geometry of imaging radar systems determines many of the characteristics of the imagery. Figure 8.10 illustrates important features of radar system geometry. The flight direction of the sensor is called the *azimuth direction*. The direction of the side perpendicular to the flight path is called the *range direction*. Microwave energy is transmitted toward the ground in a cone-shaped beam spreading outward and becoming weaker with distance. Thus, the beam width on the ground increases with distance. The portion of the ground illuminated by the beam is called the *swath*. The portion of the swath closest to the sensor is called the *near range* and that farthest from the sensor the *far range*.

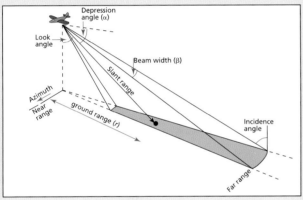

Figure 8.10

Source: *Images of the Earth: A guide to Remote Sensing.* 2nd ed. By S. A. Drury (1998). Reprinted with permission of Oxford University Press.

Radar systems measure the direct line-of-sight distance to a target called the *slant range*. Because this distance is measured at an angle to the ground surface, it must be converted to the corresponding distance along the ground, called the *ground range*. Knowing the angle at which the target is viewed and the slant range, the ground range can be calculated, assuming that the terrain is flat and at a distance below the aircraft equal to the flying height. Features higher than this ground level will be displaced inward (called *relief displacement*) *toward* the sensor in the range direction. In radar imagery, features higher in elevation are displaced *toward* the sensor whereas in optical imagery produced with a lens system, such as a camera or scanner, they would be displaced *away* from the sensor. With the use of digital elevation data, the relief displacement effects in a radar image can be removed to produce an orthorectified image. Most

radar imagery is presented in ground range format, producing a map-like image.

Figure 8.11 illustrates the three angles commonly referenced in radar geometry. The *incident* or *incidence angle* is the angle measured at the earth's surface between the radar beam and a line perpendicular to the earth's surface. The angle of the microwave beam as measured at the sensor is the *depression angle* measured from a horizontal plane passing through the sensor antenna to the beam, and the angle measured from nadir (vertical to the earth's surface) to the beam is called the *look angle* (so named because it is the angle at which the radar looks at or views the terrain). For airborne systems, where earth curvature is negligible, the look angle and incidence angle are equal. For satellite-based radar, the two angles differ slightly.

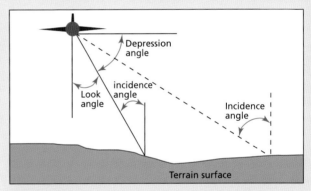

Figure 8.11

Source: *Remote sensing imagery for natural resources* by David Wilkie and John Finn. © 1996 Columbia University Press. Image reproduced with permission from Columbia University Press.

Note that the angle at which the radar beam strikes the target surface, the incidence angle, varies with distance from the flight line. It is largest at the far range and smallest at the near range. Since the incidence angle affects the strength of the reflected signal as well as the relief displacement, having a large range of incidence angles across the swath makes interpretation of the image more difficult because the same type of feature may be represented quite differently depending on its distance from the flight path. The higher the altitude of the radar sensor, the smaller the range of incidence angles needed to image a given swath of terrain *(figure 8.12)*. Satellite-based radar systems can therefore provide a more consistent image by using a narrower range

Radar terminology (cont.)

of incidence angles. Particularly for geologic interpretation, a large range of incidence angles across the swath makes it more difficult to distinguish the differences in backscatter caused by variations in the incidence angle from those related to the structure and composition of the surface materials.

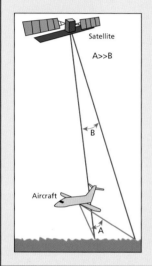

Figure 8.12

Source: © Natural Resources Canada.

Figure 8.13 Dulles International Airport, Washington, D.C. Landsat 7 image, left, has a resolution of 30 meters. The 2.5 meter pixels in the Intermap radar image, center, is much richer in detail-clearly showing the location of buildings, runways and other features. Color and clarity are both preserved in the final merged product, right.

Source: Image courtesy of Intermap Technologies, Inc. Englewood, Colo.

has also proven to be a valuable data source for emergency response such as flood monitoring and oil spill detection where rapid data acquisition, unimpeded by weather conditions or time of day, is important.

The detail provided by high-resolution radar imagery can be used to sharpen other remotely sensed imagery having lower spatial resolution. Figure 8.13 illustrates the use of 2.5 m airborne SAR data to sharpen 30 m Landsat 7 multispectral imagery. Although high-resolution visible and near-infrared satellite imagery from such sources as IKONOS and QuickBird could also be used, archived SAR imagery can be less costly.

Radar penetration capability

The ability of imaging radar systems to penetrate clouds is one of its major advantages over optical remote sensing in the visible and near-infrared bands. Longer wavelength radar systems can also penetrate vegetation to reveal features obscured by a forest canopy and, under certain conditions, by soil as well.

Radar penetration of clouds, vegetation, and soils is increased at longer wavelengths, lower moisture levels, and steeper incidence angles. However, there are trade-offs. Spatial resolution is reduced at longer wavelengths, and steeper incidence angles reduce the shadowing effect. The radar shadows give images a 3D effect that emphasizes subtle terrain relief, useful for such applications as terrain analysis.

X-band radar penetrates a vegetation canopy to some degree, varying from about 10% to as much as 50%, depending on the physical characteristics of the vegetation and degree of canopy closure. P- and L-band radar can achieve substantial penetration of dense vegetation *(figure 8.15)*, P-band systems being capable in some cases of penetrating a tropical rainforest canopy to the ground below. The vegetation penetrating capability of the P-band radar often enables features obscured by vegetation to be imaged such as roads and waterways.

Soil penetration by radar is generally minimal because soil is usually moist. However, in sand-covered desert areas where the soil is very dry, P- and L-band radar have revealed features several meters below the surface. Figure 8.16 shows a Landsat MSS band 7 image of part of NE Sudan and a SIR-A radar image of the same area acquired in 1981. The radar image revealed layers of water-eroded pebbles—the remains of ancient river channels carved into the plain long ago when the region was less arid and now covered by sand.

Effects of radar viewing angle on relief displacement and data gaps

Differences in viewing angle can dramatically affect the radar image of a scene, especially in areas of significant terrain relief. Figure 8.14 illustrates the effect of viewing angle on relief displacement. The same-sized mountain at different distances from the radar sensor is represented on the radar image quite differently, depending on the sensor look angle. At (E), the relative position of the peak and base of the mountain correspond closely to their actual position in the landscape. Here the large incidence angle creates a long shadow area—better described as a gap or void—behind the image of the mountain for which there is no image information. The face of the mountain reflects the microwave energy back to the sensor strongly and is imaged white.

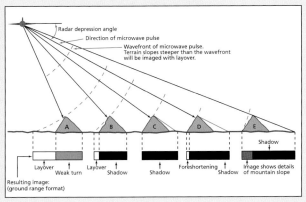

Figure 8.14

Source: Image in *Remote sensing and image interpretation.* 4th ed. by T. M. Lillesand and R. W. Kiefer. © 2000 John Wiley and Sons, Inc. Reprinted with permission of John Wiley and Sons, Inc.

Foreshortening

In foreshortening, the foreslopes of hills and mountains appear compressed and brighter relative to other features in the image. At (D) in figure 8.14, the incidence angle is lower than at (E), which reduces the area of shadow but also causes an increase in the relief displacement, i.e., on the image there is a greater shift in the position of the peak relative to its base. In the image, the position of the peak is displaced to the left toward the sensor. This reduction or shortening of the distance from the base to the peak is called *foreshortening*. An extreme example is illustrated at (C) where the radar wave front strikes the face of the entire mountain at the same time. The entire face is represented as

being at the same distance and appears as a thin white line next to the black shadow area.

Layover

Layover occurs when a feature appears to lean toward the direction of the radar antenna. Illustration (B) in figure 8.14 illustrates this effect. The radar wave front strikes the peak of the mountain before its base, so in the image the position of the peak is in front of the base. In (A), the layover effect is further exaggerated. However, the angle of view is close enough to the vertical (i.e., it has a small enough incidence angle) that a weak return is received from the far side of the mountain.

Shadow

Shadow areas (i.e., gaps or voids) occur in the image in those areas that are blocked from the radar's field of view, such as the far side of high mountains. Although there is little or no detail in these shadow areas, they are valuable cues for depth perception, providing a 3D effect without stereo viewing. Shadow can improve the interpretability of terrain features by accentuating relief. So radar imagery with minimal shadow is not always the best choice.

Thus, it is important to consider incidence angles in relation to the terrain relief and application in selecting radar imagery as they determine the degree of foreshortening, layover, and shadow effects in the imagery. For areas of extreme relief, a larger incidence angle is usually preferred to reduce the effect of relief displacement. For moderate and level terrain, the preferred incidence angle is governed more by the shape and orientation of the features of interest than on the relief displacement effects.

In general, images acquired at small incidence angles (less than about 30°) emphasize variations in surface slope, although the resulting geometric distortions due to layover and foreshortening in mountainous region may be severe. Images with large incidence angles have reduced geometric distortion and emphasize variations in surface roughness, but there are larger areas of radar shadow in which there is no image detail. Thus, the selection of an incidence angle involves trade-offs specific to the terrain and the application.

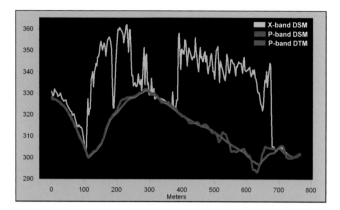

Figure 8.16 Radar penetration of dry sand. Landsat MSS band 7 near-infrared image from NE Sudan (left) and a SIR-A (L-band) radar image (right) of the same area. Distances across the images are 40 km. A large wadi in the mountains emerges onto a sand plain. On the Landsat image the mountains appear dark, and the highly reflective sand is bright. In the radar image, the rough texture of the mountains produces strong radar backscatter so they appear bright, whereas the fine-grained sand behaves as a smooth surface producing little backscatter and appearing dark. However, the radar signal penetrated the sand and was reflected from rough-textured buried features revealing the drainage pattern beneath the sand cover.

Source: Image in *Images of the earth: A guide to remote sensing.* 2nd ed. by S. A. Drury. 1998. Reprinted with permission of Oxford University Press.

Figure 8.15 Comparison of penetration of a forest canopy with X-band and P-band radar interferometry. The color aerial photograph shows the location of the elevation profile in a heavily forested area with varying topography and a 30 to 40 m high forest canopy. A graph of the profile shows digital surface model (DSM) elevations derived from X-band radar to be from near the top of the canopy compared to P-band digital surface model elevation and the derived digital terrain model (DTM) on the ground.

Source: Image courtesy of Intermap Technologies, Inc. Englewood, Colo. From B. Mercer, J. Allan, N. Glass, S. Reutebuch, W. Carson, and H. E. Andersen. 2003. Extraction of ground DEMs beneath forest canopy using P-band polarimeteric InSAR (PolInSAR). ISPRS Workshop on three-dimensional mapping.

However, penetration of vegetation and dry soils is variable, making precise elevation measurements by radar interferometry (see below) difficult. Radar penetration of vegetation and the use of X-band with P- or L-band imagery to measure forest canopy height is a topic of active research.

Radar polarimetry

Synthetic aperture radar polarimetry is used in a wide range of applications including crop identification, soil moisture measurements, biomass estimates, forestry and clear-cut mapping, geologic mapping, snow-cover and wetlands monitoring, terrain classification, ship detection, and oil slick detection.

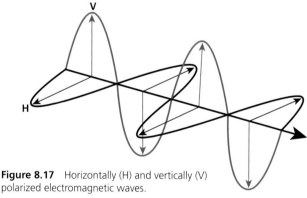

Figure 8.17 Horizontally (H) and vertically (V) polarized electromagnetic waves.

Source: © Natural Resources Canada.

The *polarization* of an electromagnetic wave or *polarimetry* refers to the orientation of the electrical field relative to the wave's direction of travel. When all the radar waves have the same orientation, the radiation is said to be polarized. Most radar systems send and receive microwave energy that is either vertically (V) or horizontally (H) polarized *(figure 8.17).* Some radar systems are capable of receiving and transmitting both. Thus, there can be four combinations of transmission and reception polarization: HH (horizontal transmit and

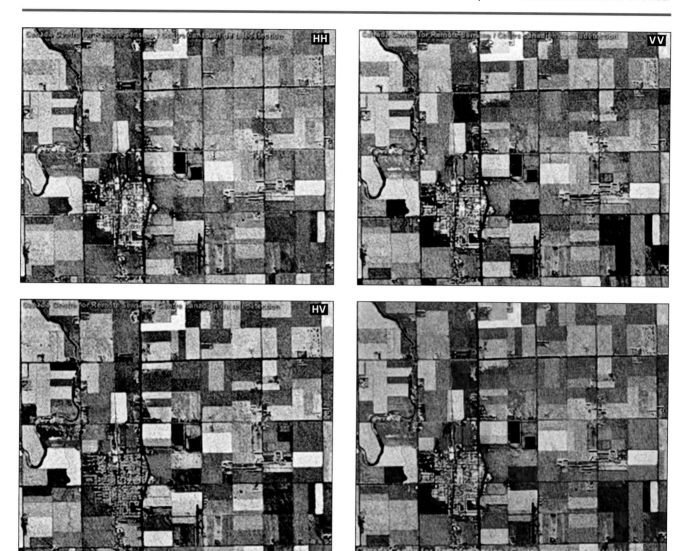

Figure 8.18 C-band radar images of agricultural fields acquired with different combinations of sending and receiving polarization. The color composite image is produced from the three black-and-white images.

Source: © Natural Resources Canada.

receive), VV (vertical transmit and receive), HV (horizontal transmit and vertical receive), and VH (vertical transmit and horizontal receive). HH and VV are said to be like-polarized because the transmit and receive signals have the same polarization while HV and VH are said to be cross-polarized.

Differences in the physical and electrical properties of surface features affect the way in which they depolarize radar signals. Feature discrimination can often be improved by generating color composite images from radar images acquired with different polarizations and wavelengths. Figure 8.18 illustrates

the effect of polarization on radar response for agricultural fields. The top two images are like-polarized, the bottom left image is cross-polarized (HV) and the bottom right image is a color composite produced by displaying the other three images as red, green, and blue.

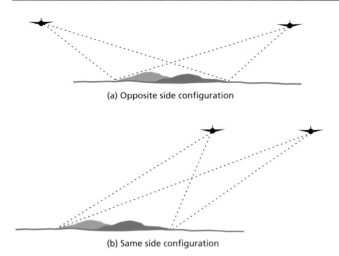

(a) Opposite side configuration

(b) Same side configuration

Figure 8.19   Stereo radar. Stereo radar images can be captured by same-side or opposite-side flight lines. Same-side stereo is often preferred for high relief terrain where the deep shadow effects would make stereo viewing difficult.

Source: Image in *Remote sensing and image interpretation*. 2nd ed. by T. M. Lillesand and R. W. Kiefer. © 1987 John Wiley and Sons, Inc. Reprinted with permission of John Wiley and Sons, Inc.

Digital elevation model and radar interfereometry

The production of digital elevation data has become one of the most important applications of radar imagery for GIS users. Elevation measurements can be generated from radar data in two ways.

Stereo radar is conceptually similar to stereo mapping using aerial photography. Stereo radar image pairs are produced by imaging the same area from well-separated viewing positions. Terrain elevations are derived by measurement of the differential relief displacement (parallax) between the two images. Calculations are somewhat different and more complex than for aerial photography. Whereas in aerial photographs, relief displacement is radially outward from the nadir point (the point in the image directly below the sensor), in radar imagery, displacement is outward and perpendicular to the flight line. Stereo radar images taken from opposite viewing directions can have such high contrast that they are difficult to interpret visually or digitally. In mountainous terrain, the resulting shadowing on opposite sides of a feature can eliminate the stereo effect. For this reason, stereo radar imagery is often acquired from the same side but with viewing angles and altitudes sufficiently different to produce stereo suitable for the application *(figure 8.19)*. The Radarsat satellite, with

Figure 8.20   Radar image map of the Brazeau Area in Alberta, Canada. This is a radar image map produced using the STAR-1 airborne synthetic aperture radar system. The image has a pixel resolution of 6 m and shows a ground area of 10.5 km x 17.8 km. An analytical stereoplotter was used to generate the elevation contours and to produce a digital elevation model from the stereo radar images. The digital elevation data was then used to produce this rectified radar image. The elevation contours with solid lines are drawn at 100 m intervals, which is about three times the RMS elevation accuracy achieved. The intermediate contour lines are dashed to indicate a reduced level of confidence. The dotted contours in areas of shadow have been interpolated. The planimetric accuracy of this image map is ±25 m RMS, and the elevation accuracy is ±30 m RMS. The X-band (3 cm) radar data was flown at an altitude of 9.5 km. Near-range off-nadir look angle was 66° and at far-range was 76°.

Source: Image courtesy of Intermap Technologies, Inc. Englewood, Colo.

its multiple beam modes, is well suited to the acquisition of stereo radar.

Photogrammetric analysis of radar images is sometimes referred to as radargrammetry. Figure 8.20 shows a rectified radar image generated from a SAR stereo pair with elevation contours produced in this way. The other method of producing digital elevation data using SAR imagery is by radar interferometry. Radar interferometry is made possible by measuring multiple characteristics of the backscattered microwave energy, precise knowledge of the sensor position and orientation, and specialized processing of image pairs.

Synthetic aperture radar interferometry

Synthetic aperture radar interferometry or interferometric synthetic aperture radar (abbreviated IfSAR or InSAR) has revolutionized the production of high accuracy digital elevation models (DEMs). First developed to measure subtle deformations of the earth's crust after earthquakes, the technology has been applied to the measurement of other types of deformation including those related to volcanic activity

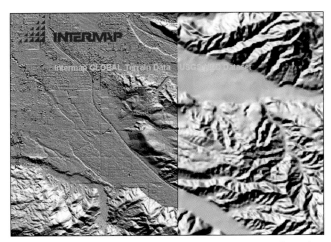

Figure 8.21   High-resolution radar-derived digital elevation data. The shaded relief image (left) of Valle Vista, California, is a radar-derived digital elevation model with postings of 2.5 m and a vertical accuracy of 2 m. The National Elevation Dataset (NED) of the United States Geological Survey for the same area (right) has 30 meter postings with vertical accuracies that vary from 7 to 15 m. Bautista Creek can be clearly seen winding through the street grid of Valle Vista in the left image, but the streets and the creek are not at all apparent in the right image.

Source: Image courtesy of Intermap Technologies, Inc. Englewood, Colo.

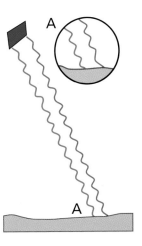

Figure 8.22   Phase shift for terrain elevation measurement. Radar signals generated from two antenna positions a small distance apart will travel slightly different distances and thus a different number of wavelengths to reach a point on the earth's surface. Where the difference is a fraction of a wavelength, as shown at A in the illustration, the returning signals will be out of phase.

Source: Image courtesy of the USGS.

and land subsidence associated with human activities such as oil, gas, and water extraction. The technology has also been applied to velocity measurements of ocean currents and sea ice movement, as well as terrain elevation mapping.

In general, the accuracy of elevations derived from radar interferometry is much better than that derived from stereo pairs of SAR images, and the process is highly automated. Airborne systems typically produce DEMs with accuracies as high as 0.3 m, and satellite-based systems offer vertical accuracies on the order of ±10 to 30 m RMSE. Terrain deformation, the difference in the height of a feature at two points in time, can be measured much more accurately. Vertical displacements can be measured with accuracies on the order of a few centimeters from satellite imagery. Figure 8.21 illustrates the difference in detail between digital elevation data with 2.5 m postings generated by airborne radar interferometry and a standard USGS DEM with 30 m postings.

There are three types of interferometry: differential, across-track, and along-track interferometry which differ in the relative position of the antenna to acquire the two images. *Differential interferometry* uses images acquired on different dates from as close as possible to the same position. *Across-track interferometry*, where the two antenna positions are displaced perpendicular to the flight path, is used to measure terrain elevation. *Along-track interferometry*, in which the two antenna positions are separated parallel to the direction of flight, can measure the velocity of surface features, such as ocean currents.

Principles of radar interferometry

Radar interferometry is based on the measurement of a property of electromagnetic waves called *phase*. Radar systems illuminate a target area with pulses of electromagnetic energy which have the properties of a sinusoidal wave pattern with a regular series of crests and troughs (see *figure 8.17* earlier). The signal repeats itself, varying from an energy level of zero to a maximum, through zero, to a minimum, and back to zero completing a single oscillation. Synthetic aperture radar systems record three items of information for each return: the time for the reflected signal to return, the strength of the return, and the exact point in the oscillation at which the reflection was received, termed the *phase*.

By comparing the phase difference of the reflected microwaves recorded in two radar images acquired from the same position or with a small precisely measured offset, measurements of the deflection of the terrain surface (i.e., its elevation) can be derived with a theoretical precision of half a wavelength *(figure 8.22)*. However, the images must be sufficiently similar that a feature has the same backscatter characteristics, i.e., there must be a high correlation. If the images were acquired days or weeks apart (e.g., on different satellite overpasses), there can be a problem for vegetated areas that may change in appearance in the time between the times the two images were acquired, termed *temporal decorrelation*.

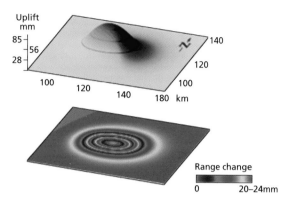

a

b

c

Figure 8.23 Radar interferogram. The radar interferogram (lower image) represents the phase difference between the two images of an interferometric image pair. The color-coding, ranging from blue to red, represents increasing phase difference values. In this illustration, about 10 cm of uplift produces 3 sets of deformation fringes.

Source: Image courtesy of the USGS.

In practice, the most accurate way to measure small phase differences is to precisely register and then, in effect, superimpose and subtract the phase values. The combined image produces an image of the interference pattern termed an *interferogram*. The image is usually color-coded to show phase differences as a series of colored bands which are meant to resemble the interference fringes produced by a thin film of oil or soap on water. Where pixels in the two images have the same phase, they reinforce one another and are color-coded red. Where they are out of phase, they cancel each other out and are color-coded blue. Intermediate phase shifts are coded with intermediate colors. The precise difference in travel distance cannot be determined directly by the phase shift because any distances that differ by a whole wavelength will be in phase. Further processing, termed *unwrapping*, is needed to analyze the repeating pattern of color fringes, in effect counting the number of fringes to calculate the measurement values. (One of the reasons that differential interferometry can offer a much higher vertical accuracy than the across-track interferometry used to generate digital elevation models is that the unwrapping processing can be much more accurate.)

For example, an area that has domed upward between the two image acquisition dates will produce a concentric pattern of color bands, termed *fringes (figure 8.23)*. One complete set of color bands, i.e., one fringe, represents one half a wavelength of surface movement towards or away from the receiver (because it reperesents the round trip increase in the travel path), or about 3 cm for C-band satellite radars such as

Figure 8.24 Radar interferometry of Mt. Vesuvius. Interferometric analysis of an ERS-1 satellite radar image pair (a) was used to generate an interferogram (b) from which terrain elevation data was calculated and used to generate a satellite radar perspective view (c).

Source: © European Space Agency. Courtesy of Eurimage.

ERS-1, ERS-2, and Radarsat. In the illustration, a 10 cm vertical displacement would produce three sets of color bands in a C-band radar image. Figure 8.24 illustrates the use of ERS-1 satellite radar interferometry to generate terrain elevation data and presents a perspective view of Mt. Vesuvius.

Differential interferometry

Differential interferometry compares two images acquired at different times from almost the same position. In principle, if the images were acquired from exactly the same position and with exactly the same viewing parameters, there should be no phase difference between pixels representing the same ground location. But if in the interim the terrain moved slightly towards or away from the sensor, the phases of some of the pixels in the second image will have shifted. This phase shift can be used to measure minute movements of the earth's surface, as caused by crustal deformation.

Satellite imagery is used because a second image can be obtained with almost the same acquisition parameters. In practice, it isn't possible to acquire two images from exactly the same position and with exactly the same viewing geometry. Slight differences in viewing position will result in some of the observed phase shift being due to variations in terrain elevation. If these differences are small, the phase shifts related to elevation can be measured and subtracted out to derive measurements of differential earth movement. However, if there is too great a difference in acquisition location or viewing parameters or there has been too great a change in the appearance of features in the landscape, the two images will not be sufficiently similar for interferometric processing to be successful.

Measurements of terrain deformation on the order of centimeters have been achieved. For example, interferometric analysis of radar images of Mt. Etna acquired over an 18-month period in 1992 and 1993 at the end of its eruptive cycle showed that Mt. Etna subsided an average of two centimeters per month during the last seven months of eruption.

Figure 8.25 shows a satellite radar interferogram of the Los Angeles basin used in a study of land subsidence. The study used radar interferometry and GPS observations to show that the 40 km long Santa Ana Basin rises every winter and falls every summer up to 11 cm as a result of recharge and then widespread pumping of groundwater. Similar analyses have been used to measure subtle topographic deformations such as land subsidence in areas of oil and gas extraction.

Across-track interferometry

Across-track interferometry is also used to generate digital elevation models. Terrain elevation is measured by determining the phase difference between two images acquired a relatively small fixed distance apart in the across-track direction. The required separation distance between the two antenna positions (called the *interferometric baseline*) depends on the radar wavelength.

The phase difference in the radar backscatter received at the two antenna positions from the same target will show a phase shift due to the different distances the reflected signal has traveled. This phase difference is related to both the terrain elevation and the viewing geometry. By removing viewing geometry effects, an interferogram of the remaining phase difference can be generated to measure elevation. Figure 8.26 illustrates different products generated by radar interferometry for terrain mapping.

Shorter wavelength X-band systems can have smaller separation distances on the order of a few meters, short enough for the two antennas to be mounted on an aircraft. For satellite missions using C-band and X-band radar, separation distances of 60 m to 100 m are generally used.

The two images can be acquired by a single antenna radar at different times, called *repeat-pass interferometry,* or with two antennas, the image pair can be acquired simultaneously, called *single-pass interferometry (figure 8.27).* Except for measurements of change over time, single pass interferometry is the preferred approach. Simultaneous acquisition eliminates error due to changes in surface conditions over time, and the viewing geometry of the two images and the antenna positions are precisely known.

Airborne X-band IfSAR systems generally carry two antennas for single pass interferometry. Satellite based radar systems have a single antenna and must use repeat pass imagery.

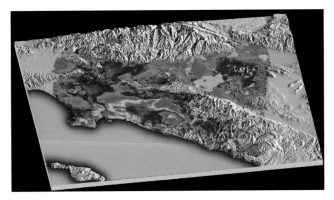

Figure 8.25   Rise and fall of the Santa Ana Basin measured by radar interferometry. A satellite radar interferogram generated from ERS-1 imagery with a 90 m spatial resolution has been draped over a shaded relief Landsat image of the Los Angeles area. The repeating color bands of the large oval showed 5 cm of subsidence in the Santa Ana basin between May and September 1999.

Source: Image courtesy of the USGS.

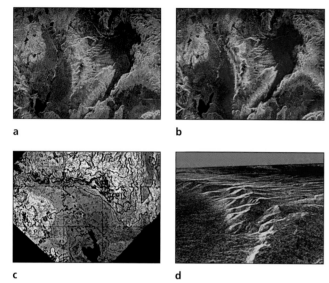

a   b

c   d

Figure 8.26 Radar interferometry. These four images of a 34 by 59 km area of the Long Valley region of California illustrate the steps required to produce three-dimensional data and topographic maps from radar interferometry. Imagery was acquired by a synthetic aperture radar operating in the C and X bands carried aboard the space shuttle during two flights in April and October, 1994.

A. Single radar image. The bright areas are hilly regions that contain exposed bedrock and pine forest. The darker gray areas are the relatively smooth, sparsely vegetated valley floors.

B. The image in the upper right is an interferogram of the same area, made by combining L-band data from the April and October flights. The colors in this image represent the difference in the phase of the radar echoes obtained on the two flights. Variations in the phase difference are caused by elevation differences. Formation of continuous bands of phase differences, known as interferometric fringes, is only possible if the two observations were acquired from nearly the same position in space. For these April and October data takes, the shuttle tracks were less than 100 meters apart.

C. The image in the lower left shows a topographic map derived from the interferometric data. The colors represent increments of elevation, as do the thin black contour lines, which are spaced at 50-meter elevation intervals. Heavy contour lines show 250-meter intervals. Total relief in this area is about 1,320 meters. Brightness variations come from the radar image, which has been geometrically corrected to remove radar distortions and rotated to have north toward the top.

D. The image in the lower right is a three- dimensional perspective view looking toward the northwest. The radar image data is draped over topographic data derived from the interferometry processing. No vertical exaggeration has been applied. Combining topographic and radar image data allows scientists to examine relationships between geologic structures and landforms and other properties of the land cover, such as soil type, vegetation distribution, and hydrologic characteristics.

Source: Image courtesy of NASA/JPL-Caltech.

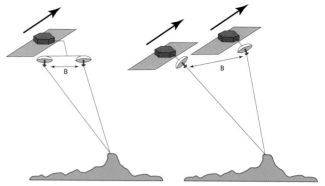

Figure 8.27 Across-track interferometry uses two radar images acquired from different viewing positions oriented perpendicular to the flight line and separated by a precisely known distance termed the interferometric baseline (B). Single-pass systems (left) have two antennas to simultaneously acquire the two images. Repeat-pass systems (right) acquire the images at different times.

Source: Image courtesy of NASA/JPL-Caltech.

However, in February 2000, the 10-day Shuttle Radar Topography Mission (discussed later in this chapter) used a dual antenna SAR for single-pass radar interferometry. The second antenna was mounted on an extended boom providing a 60 m separation.

From July 1998 to March 2000, two similar satellites, ERS-1 and 2 operated by the European Space Agency were flown in tandem with a 24 hour offset and the same imaging parameters to acquire a radar dataset optimized for interferometric applications. The short time interval between the two acquisitions significantly improved image correlation. Radarsat-3 is planned to operate in tandem with Radarsat-2 scheduled for launch in late 2005 (see below *Radarsat-3*).

In addition to the phase differences, the magnitude of the return signal can also be analyzed to distinguish broad terrain classes, such as identifying forested from nonforested areas. Figure 8.28 compares a 1:24,000 scale USGS topographic map with a map generated from airborne radar interferometry data used to extract both elevation and terrain class information. In this case, almost all of the detail in the USGS map could be derived from the radar imagery. Figure 8.29 illustrates an effective visualization technique termed an intensity-hue-saturation image in which the brightness values in the image represent the magnitude of the radar return (called an *amplitude image*), and the color represents the terrain elevation ranging from red (highest elevations) to blue (lowest). Saturation is set to a value that produces a pleasing image.

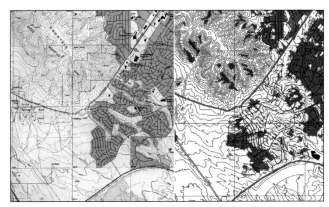

Figure 8.28   Comparison of portions of a USGS 1:24,000 scale topographic map and a similar map produced from airborne interferometry data.

Source: Image courtesy of Vexcel Corporation.

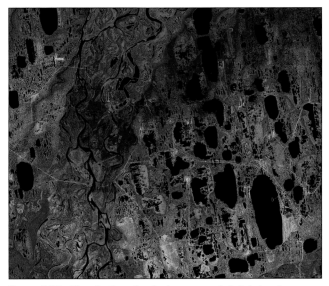

Figure 8.29   Visualization of radar imagery and digital elevation model of the lower Kuparuk River, Alaska. This intensity-hue-saturation image represents the radar amplitude as intensity and radar-derived elevation as color ranging from red (24 m) to blue (sea level). Saturation is set to a value that produces a pleasing image. The amplitude image and DEM have 2.5 m and 5 m postings, respectively. The distance across the image is 17 km.

Source: Image courtesy of M. Nolan of the Institute of Northern Engineering, University of Alaska, Fairbanks, and Intermap Technologies, Inc. Englewood, Colo.

As previously noted, radar penetration of vegetation varies—about 10% to 50% for X-band while P- and L-band radar achieve much greater penetration of closed vegetation canopies, in some cases reaching the ground surface, to image features obscured by foliage. Research continues with the goal of developing algorithms for dual X-band and P-band IfSAR systems to measure vegetation height by simultaneously generating elevation models of both the forest canopy and forest floor.

Along-track interferometry

Along-track interferometry is used to measure the velocity of targets moving towards or away from the radar during a single pass. With the data from a second pass, direction can be calculated as well. Two antennas separated in the direction of the flight path are used *(figure 8.30)*.

The SAR image on the left in figure 8.31 shows the Minas Basin in the Bay of Fundy, Canada, a waterway with very high tides. The area was imaged at peak flood tide, when the water velocity is highest. The long bright area in the lower part of the image is a spit of land. Along-track radar interferometry was used to combine velocity measurements from two passes 15 minutes apart to create an image of the

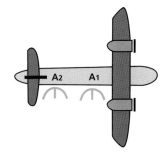

Figure 8.30   Along-track interferometry. Along-track interferometry uses two antennas mounted so the separation distance is parallel to the flight line.

Source: © Natural Resources Canada.

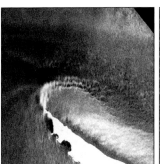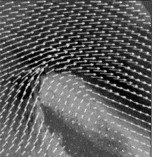

Figure 8.31   Radar inferometry of water currents in the Bay of Fundy, Nova Scotia.

Source: © Natural Resources Canada.

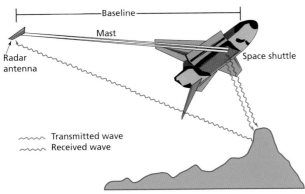

Radar signals being transmitted and received in the SRTM mission
(Image not to scale)

Figure 8.32 The Shuttle Radar Topography Mission carried a dual antenna interferometric SAR. One antenna was mounted in the cargo bay and the other at the end of a 60 m mast extended from the cargo bay when in space. The precise distance between the two antennae was continuously measured.

Source: Image courtesy of NASA/JPL-Caltech.

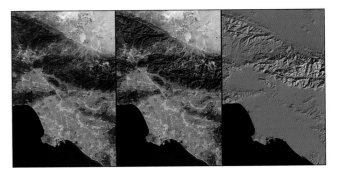

Figure 8.33 Los Angeles, California. Landsat and SRTM digital elevation data. The color 30 m spatial resolution Landsat TM image on the left, and 30 m spatial resolution digital elevation data from the Shuttle Radar Topography Mission (SRTM) shown as a shaded relief image on the right, were merged to generate the topographically enhanced image in the center.

Source: Image courtesy of NASA/JPL-Caltech.

Figure 8.34 San Gabriel Mountains and Los Angeles area, California. Landsat perspective view using SRTM digital elevation data. The 3D perspective view was generated from a Landsat 7 Thematic Mapper image, digital elevation data from the Shuttle Radar Topography Mission (SRTM), and a false sky.

Source: Image courtesy of NASA/JPL-Caltech.

currents shown on the right. The arrows show the current direction and the velocity is indicated by the color, ranging from green for stationary features to purple for the maximum detected current (of about 5 m/sec).

Shuttle Radar Topography Mission (SRTM)

The Shuttle Radar Topography Mission (SRTM) of February 2000 used a dual antenna single pass interferometric SAR operating in both the C-band and X-band. During its 10-day mission onboard the space shuttle *Endeavour*, enough data was obtained to produce a high-resolution digital topographic database of most of the earth (between 60° N and 56°S latitude, covering about 80% of the earth's land surface). The SRTM payload used two antennas, one located in the shuttle's payload bay and the other on the end of a 60 m mast extended from the payload bay after the shuttle was in space *(figure 8.32)*. The high resolution digital elevation data has a 30 m spatial resolution with 10 m relative and 16 m absolute vertical height accuracy. Distribution of the high resolution dataset is restricted. However, a lower resolution dataset with 90 m spatial resolution is publicly available.

Figure 8.33 shows the Los Angeles, California, area with a Landsat image on the left, a shaded relief rendering of the SRTM digital elevation data on the right, and a merge of the two datasets in the center image enhancing the topography visible in the scene. Figure 8.34 shows the Los Angeles area with the San Gabriel Mountains in the background. The 3D

perspective view was generated using SRTM digital elevation data, an enhanced Landsat 7 Thematic Mapper image, and a false sky. In addition to their aesthetic appeal, 3D perspective images such as these can be valuable in communicating geographic information to diverse audiences.

Characteristics of digital elevation models generated by radar interferometry

A *digital elevation model* (DEM) is a grid of data points representing elevation values at regularly spaced intervals over the terrain. The smaller the distance between the data points, termed postings, the greater the detail that can be represented. Airborne digital terrain models generated by radar interferometry typically have 5 m postings and vertical accuracies of ±0.5 m, ±1 m, or ±3 m RMSE. The greater the vertical accuracy, the higher the cost. This data is suited to topographic mapping at scales of approximately 1:4,000; 1:8,000; and 1:24,000 respectively.

DEMs generated from satellite radar such as Radarsat can have 10 to 100 m postings and vertical accuracies on the order of ±10 to 100 m RMSE, depending on the source imagery selected, terrain characteristics and availability of ground control. This data is suited to topographic mapping scales ranging from 1:50,000 to 1:250,000.

There are two types of elevation products. *Digital surface models* (DSM) represent the elevation values derived from the radar return signals. It is a representation of the first surface reflections from any features large enough to be resolved, be they buildings, vegetation, or the terrain surface. It provides a geometrically correct reference surface over which other data layers can be draped, such as maps, aerial photography, or multispectral imagery. It can be used to generate 3D perspective views, fly-through simulations, and viewshed analyses.

A *digital terrain model* (DTM) is a topographic model of the terrain surface removing objects that are on but not part of the surface (often termed the *bare ground* or *bare earth* surface), including vegetation, buildings, and other cultural features. Sophisticated processing algorithms identify measurements of the terrain surface contained in the original radar data (the digital surface model) and infer values for areas obscured by vegetation and other features. An interactive editing process is usually needed to identify and remove blunders and other inconsistencies to ensure the final product satisfies the required specifications. DEMs are used for topographic mapping, and the analysis of terrain characteristics such as slope, aspect, elevation, and cross sections which are then used in a wide range of remote sensing and GIS applications.

Orthorectified imagery is produced in the process of generating elevation datasets and is usually provided at little or no additional charge. The horizontal accuracy is typically on the order of ±2 m RMSE for airborne imagery and as fine as ±10 to 20 m RMSE for satellite imagery in high resolution mode.

Elevation model accuracy

Two commonly used accuracy measures are the Root Mean Squared Error (RMSE) and the 90% or 95% circular error confidence level, commonly abbreviated as CE90 or CE95. The RMSE is calculated as RMSE = SQRT $(\Sigma\delta_i^2)/(N-1)$ where $\delta_i = (h_i - h_{true})_i$ where h_{true} is the height taken to be true, h_i is the height of test point i , δ_i is the calculated

error for test point i , and N is the number of points. It is commonly assumed that the mean error (m) = $\Sigma\delta_i/N = 0$ or is very small and that the errors are normally distributed. When these assumptions are true, the RMSE is approximately equal to the standard deviation and the confidence levels associated with a standard deviation, can reasonably be assigned to the RMSE value.

It is a property of a normal distribution that about 68% of all measurements would be expected to fall within ±1 standard deviation of the mean and about 95% within ±2 standard deviations, often termed the 95% confidence interval. The measurement is a way of expressing the quality of the data in terms of the statistical risk of encountering an error greater than some agreed upon value.

However, these assumptions are often violated. The errors may not be normally distributed nor have a mean of zero. There may be *systematic errors* that cause values to be in error by a constant amount or that vary slowly over the area being sampled. This offset or bias in the values might be caused by GPS errors, for example. The offset can usually be reduced or eliminated using ground control points, depending on the intensity of the sampling and the number of ground control points used.

Although it is common in the mapping industry to attribute confidence levels based on the calculated RMSE, it is not technically valid unless the conditions of zero mean error and normal distribution are met. Intermap Technologies Ltd., the major commercial supplier of digital elevation data from radar interferometry, has found that, for their elevation data products, 95% of the measured errors in reference points generally had errors less than about 2 times the calculated RMSE. This empirical evidence suggests that the CE95 error level is empirically as well as theoretically equivalent to about 2 times the RMSE.

However, this accuracy measure does not take into account the qualities of the sample points. For example, test points falling on a raised structure finer than the post spacing will not be correctly represented. Areas obscured by buildings or vegetation over a large enough area compromise the algorithms designed to interpolate the bald earth terrain surface and may cause errors beyond those expected. Since every remote sensing technology has its limitations, judgment and experience are needed to assess the suitability of a product for the intended use. For this reason, accuracy measures should be viewed as approximate indications of quality, recognizing that the nature as well as the magnitude of the errors may vary significantly with the technologies used.

The accuracy of elevation models and orthorectified imagery produced by radar and other sensors is dependent on the skill of the personnel, the quality of their equipment, the effectiveness of quality assurance procedures, and the limitations of the technology. Remote sensing technologies differ not only in the levels of accuracy achievable but also in the characteristics of those errors that do inevitably occur. Error rates can always be reduced by expending additional time and resources, but at some point the cost of reducing the risk of error exceeds the benefit.

Airborne and spaceborne radar systems

There are significant differences in the characteristics of airborne and spaceborne SAR imagery that affect the way they are used operationally for earth resources applications. Synthetic aperture radar has the advantage of providing high resolution imagery independent of altitude. (However, a more powerful pulse is needed to reach greater distances.) Although spatial resolution may be independent of altitude, the viewing geometry and swath coverage are greatly affected. At aircraft operating altitudes, an airborne radar must image over a much wider range of incidence angles (see *Radar terminology* earlier in this chapter), as much as 60 to 70 degrees, in order to achieve a swath width of about 50 to 70 km whereas a spaceborne radar can image the same swath with a much narrower range of incidence angles, typically a 5 to 15 degree range. As noted previously, the incidence angle significantly affects the amount of backscatter received from a target and the degree of relief displacement (foreshortening and layover) that will occur. The greater the range of incidence angles, the greater the inconsistency in the way features are represented in the image.

Although airborne radar systems must handle more complex imaging geometry, they are much more flexible. They can collect data anywhere an aircraft can reach and at anytime (so long as flying conditions permit). They can acquire imagery from different look angles and look directions, optimizing the geometry for the terrain.

However, aircraft are exposed to variable flying conditions, causing changes in velocity, tipping and tilting of the aircraft. By recording the random motion of the aircraft with an inertial measurement system, sophisticated image processing can use the data to rectify the imagery. Global Positioning Systems (GPS) are also commonly used to continuously record the geographic position and flying height of the aircraft. Satellite platforms are very stable and their orbits are precisely known, but geometric correction of spaceborne imagery must take into account other factors such as the rotation and curvature of the earth.

Research radar systems

Much of the leading edge research on SAR capabilities and system design is conducted by military research organizations and is classified. However, substantial nonmilitary research programs have supported the development of a wide range of earth resource applications.

Both Canada and the United States have operated extensive research radar programs for earth resource applications. The National Aeronautics and Space Administration in the United States has been one of the organizations at the forefront of radar research for earth resource applications. For many years, the Jet Propulsion Laboratory (JPL) in California has built and operated various advanced SAR systems on contract to NASA. The NASA/JPL radar program began in 1978 with the SeaSat satellite (discussed below). The other spaceborne SAR systems developed under this NASA program have been short-duration missions operated from the space shuttle. The Shuttle Imaging Radar-A (SIR-A) was an L-band fixed look angle SAR flown on the space shuttle in 1981. In 1984, a similar radar, SIR-B, with multilook angle capability was flown on the shuttle. SIR-C, developed jointly by NASA, the German Space Agency (DLR), and the Italian Space Agency (ASI), operated in three bands (L, C, and X) with variable look angles and polarizations. The X-band SAR was, in fact, provided by DLR. The system flew on the shuttle in April and October of 1994, providing radar data for two seasons with spatial resolutions ranging from 10 to 25 m. The Shuttle Imaging Radar (SIR) imagery is available from the EROS Data Center (see appendix B). NASA/JPL also operate an airborne system known as AIRSAR/TOPSAR, which is mounted on a DC-8 aircraft. Operating in the L, C, and P bands, the system can simultaneously collect all four polarizations (HH, VV, HV, and VH) in the L- and P-bands while operating as an interferometer at the C-band wavelength to simultaneously generate topographic height data.

Another American government agency that has designed, built, delivered, and operated numerous SAR systems is the Sandia National Laboratory, which is funded principally by the U.S. Department of Energy. Notable among its nonmilitary examples are the SAROS (SAR for Open Skies) operating in X-band and AMPS (Airborne Multi-sensor Pod System) that operates in the K_u-band.

The Canada Centre for Remote Sensing (CCRS) operated the SAR-580, comprising a Convair®-580 aircraft equipped with a C/X-band SAR, in Canada and around the world, developing applications in preparation for Radarsat, the Canadian radar satellite launched in 1995. Operating at both C- band and X-band in both like-polarized and cross-polarized modes and with three different imaging geometries, the system could acquire a wide variety of imagery on a single mission. Outfitted with a second antenna, the system was also used for radar interferometry, oil spill research, and other environmental applications.

Airborne SAR systems have also been developed and operated extensively in various countries in Western Europe. In particular, the German Space Agency (DLR) has been very active in this area. Its E-SAR multifrequency SAR system is mounted on a twin-engined Dornier DO228 STOL (short take-off and landing) aircraft. The SAR sensor operates in five different frequencies—in the X-, C-, S-, L- and P-bands. Single-pass interferometry is carried out at X-band frequencies. The system is calibrated polarimetrically in L-band with a multiple polarization capability. Since the aircraft's operational ceiling is limited to 3,600 m, swath widths are limited to 3 km (narrow swath) and 5 km (wide swath). Besides being used as a research and technology development system, the E-SAR has also been used extensively in a number of application campaigns. These include the ProSMART series carried out in collaboration with numerous partners in Germany. It has also been deployed in other countries such as the SHAC campaign carried out in the United Kingdom.

Several other agencies in West European countries have built and operated airborne SAR systems. They include TNO-FEL in the Netherlands, whose PHARUS (Phased Array Universal SAR) system is a fully polarimetric phased-array SAR operated in the C-band and mounted on a Cessna® Citation® business jet aircraft. TNO-FEL has also developed a miniature lightweight Mini-SAR which is mounted on a motor glider. Another prominent system is EMISAR, which has been developed by the Danish Center for Remote Sensing. EMISAR is a dual-frequency (C- and L-band), fully polarimetric SAR that is also capable of carrying out single-pass interferometric measurements in C-band. It is mounted on a Gulfstream® business jet aircraft. Both the PHARUS and EMISAR systems have been used in a wide variety of applications—in the case of EMISAR, in all the Scandinavian countries.

Yet another European airborne SAR system is the French RAMSES, which is a multifrequency, fully polarimetric, high-resolution SAR that has been developed by the ONERA organization. The RAMSES system is mounted on a Transall C160 transport aircraft and can be operated over a huge wavelength range, including the high-frequency W-, K_a- and K_u-bands as well as the X-, C-, S-, L- and P-bands found on other airborne research radar systems. Just like them, it has been used also to acquire SAR data for a number of application-oriented campaigns.

Finally, in Japan, yet another airborne research SAR system is the Pi-SAR that has been developed by the Communications Research Laboratory (CRL) of the Japanese Ministry of Posts and Communications in collaboration with the national space agency (JAXA, formerly NASDA). Pi-SAR is a dual-frequency SAR operating in the X- and L-bands with polarimetric functionality and with an interferometric capability in the X-band. It is mounted in a Gulfstream jet aircraft. Once again, numerous flights have been made for a wide range of applications.

Commercial airborne radar systems

Commercial airborne radar systems are used operationally in a wide range of disciplines. Topographic mapping, sea ice monitoring, geology, forestry, and the enhancement of lower resolution imagery have been some of the principal applications.

Only in the 1970's did the U.S. military declassify certain imaging radar systems, allowing their commercial use. Since then the capabilities of commercial systems have steadily improved, and current systems are regularly upgraded to provide higher spatial resolution, additional bands, and more sophisticated signal processing to improve the quality of data products.

Among the more advanced commercial systems have been the STAR (Sea ice and Terrain Assessment Radar) series of synthetic aperture radars developed and operated by Intermap Technologies Ltd., based in Calgary, Alberta. Designed primarily for terrain elevation mapping and sea ice monitoring, the system has seen numerous upgrades and enhancements. The current version, the STAR-3i® system, is an X-band radar designed for interferometry with two antennas collecting data simultaneously *(figure 8.35)*. It is mounted onboard a Learjet® 36 business jet aircraft. Standard products include high accuracy digital elevation models and orthorectified radar imagery. An onboard global positioning system and laser-based inertial measurement unit provide high-accuracy attitude and position data sufficient to produce these products without in-scene ground control points. (However, higher accuracies are achieved when ground control points

Figure 8.35 INTERMAP STAR-3i interferometric SAR mounted on a Learjet 36 aircraft. The two SAR antennas separated by the baseline distance are housed within the radar pod beneath the aircraft fuselage.

Source: Image courtesy of Intermap Technologies, Inc. Englewood, Colo.

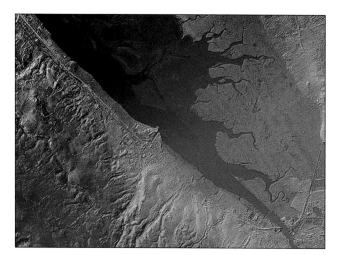

Figure 8.36 STAR-3i digital elevation data, Dee Estuary, northwest England. This shaded relief image, color-coded by elevation, was generated from digital elevation data generated from SAR interferometry as part of Intermap Technologies' NEXTMap Britain project.

Source: Image courtesy of Intermap Technologies, Inc. Englewood, Colo.

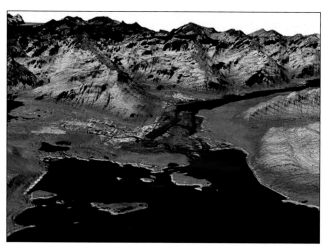

Figure 8.37 Perspective view of Juneau, Alaska. This perspective view, color-coded by elevation, was generated from digital elevation data generated by SAR interferometry collected by the STAR-3i system.

Source: Image courtesy of Intermap Technologies, Inc. Englewood, Colo.

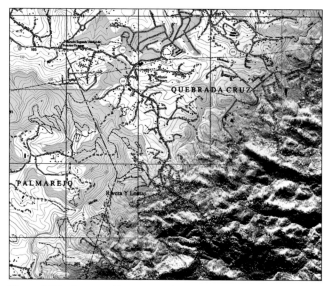

Figure 8.38 Topographic mapping from radar data. The topographic line map (left) of Corozal, Puerto Rico, with an original scale of 1:20,000, was produced by stereo analysis of X-band radar imagery (right) on a digital photogrammetric workstation.

Source: Image courtesy of Intermap Technologies, Inc. Englewood, Colo.

are available.) The system is typically flown at altitudes from about 3,300 m (3 km swath width), to 10,000 m (10 km swath width) producing imagery with spatial resolutions as fine as 1.25 m and vertical accuracies as high as ±0.3 m RMSE *(figures 8.36 and 8.37)*.

During the late 1990s, Intermap carried out a number of mapping projects covering large areas of terrain on behalf of U.S. government agencies using its STAR-3i system. Several of these projects were carried out over tropical areas of Central America (Panama, Honduras) and the Caribbean (Puerto Rico, see *figure 8.38*) where it is difficult to acquire optical imagery due to cloud cover. In 2002, Intermap then embarked on its NEXTMap Britain project. This resulted in the systematic coverage of most of Great Britain with its

IfSAR image data, from which a country wide DEM has been produced. One of the main anchor customers for the data has been the insurance industry, which has utilized the DEM data for flood risk analysis. For the lower and flatter area of southern England, where the flood risk is greater, the

Learjet aircraft was flown at an altitude of 6 km to acquire elevation data with a vertical accuracy of ±0.5m RMSE. The rest of the country was covered from a flying height of 9 km with the elevation data having a reported accuracy of ±1m RMSE.

After the success of NEXTMap Britain, in 2002, Intermap embarked on its NEXTMap Indonesia project—again a tropical country with near-continuous cloud cover. Initially it has concentrated its image acquisition and DEM generation efforts on the eastern parts of the country. Intermap has also announced its NEXTMap USA project which aims to cover the whole of the United States (amounting to 7.8 million sq. km) with IfSAR imagery, from which it will generate a nationwide DEM. This project began in late 2003. To increase its capacity to undertake all these large projects, Intermap has acquired Aero-Sensing Radarsysteme based near Munich, Germany. This was formed originally as a spin-off company from the German Space Agency (DLR). Currently Intermap is also developing a more advanced STAR-4 system.

Several companies operate dual X-band and P-band radar systems, including EarthData International (GeoSAR), Intermap (AES-3) and Orbisat (OrbiSAR-1). The GeoSAR system, for example, is a dual-band, four antenna system built by NASA's JPL lab, which also developed the data processing software. The system is mounted on a Gulfstream jet aircraft and is capable of simultaneously collecting interferometric SAR data with a 10 km swath at both X-band and P-band wavelengths on both sides of the flight line *(figure 8.39)*. The vegetation penetrating capability of the P-band radar improves the accuracy of bald earth digital elevation models

from the data. The system is flown typically at altitudes of 5 to 10 km, generating imagery with spatial resolutions as high as 2.5 m and vertical accuracies of ±2.5 m for X-band and about ±5 m for P-band data.

Military airborne SAR systems

Radar imagery is used extensively for military applications. Although military airborne systems are generally classified and not available for civilian applications, there is some publicly available information about these systems, providing an indication of capabilities likely to appear in future civilian systems.

Military SAR systems operate over the full range of the microwave spectrum and typically generate products in real time or near-real time. Systems operating at the shorter wavelength K_a, K_u, and X-bands offer higher spatial resolution while systems operating at the longer L-band and P-band wavelengths provide superior vegetation penetration. Research is being conducted on the use of even longer UHF and VHF wavelength radar systems to achieve even greater penetration of vegetation and soil.

Small SAR systems have been developed for use in unmanned aerial vehicles (UAVs). For example, the Lynx system (developed by Sandia National Laboratories) used in the Predator UAV *(figure 8.40)* is a K_u band SAR system weighing less than 55 kg. It can generate imagery with spatial resolutions as high as 0.1 m at a slant range distance of 30 km in rain falling at a rate as high as 4 mm/hr *(figure 8.41)*. Northrup Grumman produces the Tactical Endurance Synthetic Aperture Radar (TESAR) also used on the Predator UAV. Its larger Advanced Synthetic Aperture Radar 2 (ASARS-2), designed for the high altitude U.S. Air Force U-2, delivers real-time high-resolution imagery and long

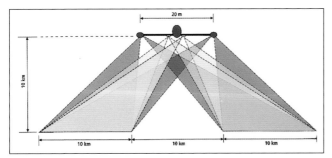

Figure 8.39 Multiband interferometric radar. The GeoSAR airborne interferometric radar operates in both X-band and P-band wavelengths. P-band antennas are located on the aircraft wingtips, and X-band antennas are mounted on the underside of each wing near the fuselage. Each antenna collects data to both sides of the aircraft, allowing interferometric data in two bands to be simultaneously collected for two 10 km swaths.

Source: Image courtesy of Earthdata Holdings, Inc.

Figure 8.40 Predator unmanned aerial vehicle (UAV). The Predator UAV carries two color video cameras, a forward-looking infrared (FLIR) imaging system, and a Lynx synthetic aperture radar (SAR). Eight meters long with a wingspan of 15 m, it can remain airborne for 40 hours delivering real-time imagery day or night in all weather conditions via satellite worldwide.

Source: Image courtesy of General Atomics Aeronautical Systems, Inc.

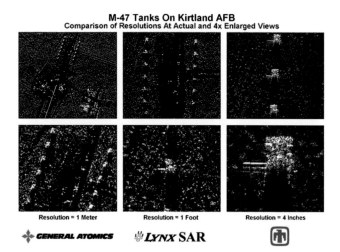

M-47 Tanks On Kirtland AFB
Comparison of Resolutions At Actual and 4x Enlarged Views

Resolution = 1 Meter   Resolution = 1 Foot   Resolution = 4 Inches

GENERAL ATOMICS   LYNX SAR

Figure 8.41 Military radar imagery. These K_u-band SAR images acquired by an airborne Lynx system show M-47 battle tanks at spatial resolutions of 1 m, 0.3 m, and 0.1 m.

Source: Image courtesy of General Atomics Aeronautical Systems, Inc. and Sandia National Laboratory.

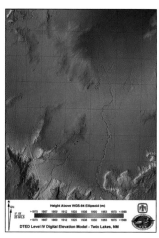

Figure 8.42 Real-time DEM production by radar interferometry. The U.S. Army's Rapid Terrain Visualization (RTV) interferometric radar is capable of producing image and DEM products in real time such as this SAR image (a) and color-coded DEM (b) of the Twin Peaks area of New Mexico.

Source: Images courtesy of Sandia National Laboratory.

Table 8.2	Characteristics of selected radar satellite sensors				
Characteristic	SeaSat-1	ERS-1 and 2	Envisat-1	Radarsat-1	Radarsat-2
Mission dates	27 June 1978–9 Oct 1978 (operational failure) 1–3 year design life	ERS-1 17 July 1991–1998 ERS-2 21 April 1995–current 3 year design life	1 March 2002–current	4 Nov 1995–current 5 year design life	2006 7 year design life
Waveband	L-band (23.5 cm)	C-band (5.7 cm)	C-band (5.6 cm)	C-band (5.6 cm) pointable	C-band (5.6 cm) pointable
Look angle	20°–26°	20°–26°	14°–45°	20°–59° depending on mode selected	10°–60° depending on mode selected
Polarization	HH	VV	HH,HV,VH, HH	HH	HH, VV, HV
Pixel resolution	25 m	30 m	30 m, 150 m, 1 km	10–100 m	3–100 m
Swath width	100 km	100 km	58–405 km	50–500 km	50–500 km
Repeat cycle	17 days	ERS-1: 3–168 days ERS-2: 35 days	35 days	24 days	24 days
Other sensors	radar altimeter radar scatterometer microwave radiometer visible and infrared radiometer	radar altimeter infrared radiometer microwave sounder	radiopositioning,ozone monitoring, multispectral scanners (300 m, 1km), microwave radiometer, radar altimeter	none	none

range mapping capabilities. The system is designed to detect, accurately determine the positions of fixed and moving targets, and transmit high resolution imagery. In addition to its military role, ASARS systems have seen use for civilian emergency response applications such as mapping and assessment of flooding along the Mississippi and Missouri Rivers in 1993 and in Northern California in 1995, and the Northridge California earthquake of 1994.

To satisfy the need for rapid generation of high quality digital elevation data, airborne systems capable of real-time DEM production have been developed. For example, the Rapid Terrain Visualization (RTV) interferometric SAR developed for the U.S. Army can generate DEMs in real time with 1 m absolute vertical accuracy and 2 m horizontal accuracy *(figure 8.42)*.

Figure 8.43   SeaSat.

Source: image courtesy of NASA/JPL.

Spaceborne radar systems

Satellite radar sensors support many operational applications including sea ice monitoring of navigable coastal waters and emergency response to floods, oil spills and other disasters. Unlike the short-duration SAR missions that have been operated from the space shuttle, the following section covers radar systems that have been operated from satellites over a much longer period of time. Table 8.2 summarizes important characteristics of some of these sensors.

American radar satellites

SeaSat

Launched in June of 1978 by the United States, SeaSat-1 featured an L-band (23.5 cm) HH polarization radar. It produced radar imagery with a 100 km swath width and a 25 m spatial resolution. Though it failed only 99 days after launch, it convincingly demonstrated the value of a satellite-based radar system. The imagery proved valuable in showing ocean wave patterns and for mapping ocean currents and arctic ice conditions. Other SeaSat sensors measured wave heights and direction, sea surface temperature, rain rate, and the water vapor content of the atmosphere.

Over land areas, the data demonstrated the value of satellite radar imagery for mapping geology and water resources. Figure 8.44 is a SeaSat image of the Rocky Mountain region of British Columbia, Canada. The image is striking for the texture and details shown in the steeply dipping rock strata of this rugged terrain. It also illustrates extreme layover. The

Figure 8.44   SeaSat image of the Rocky Mountains in British Columbia. Contrasting textures and the bright edges, caused by radar layover, highlight the geology of this mountainous area.

Source: © Natural Resources Canada.

mountain slopes facing the sensor are very bright, and the peaks are displaced so far toward the sensor that they are imaged in front of the base of the slope.

SeaSat's nonimaging radar altimeter measured the sea surface height profile and wave heights with a relative accuracy of 10 cm. Sea surface height is affected by the gravitational force of the earth. Since rock is denser than water and exerts a stronger force of attraction, the ocean level tends to be higher in shallower areas and lower over deep water. Scientists at the Lamont-Doherty Earth Observatory used this relationship to predict ocean depth from the radar altimeter data and generate a map of seafloor topography (see *Mapping seafloor topography with satellite radar altimetry* later in this chapter).

Figure 8.45 Radarsat-1 image of the national capital region, Ottawa, Canada. The image, about 76 km across, includes the Ottawa River with the city of Gatineau, Quebec, to the north (top) and the larger city of Ottawa, Ontario, to the south. The image was acquired on March 29, 1996, at an original resolution of 8 m in SAR Fine 5 beam mode.

Source: Radarsat-1 data © Canadian Space Agency 1996. Received by the Canada Centre for Remote Sensing. Processed and distributing by RADARSAT International.

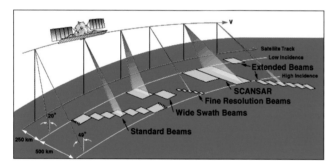

Figure 8.46 Radarsat imaging modes. The Radarsat sensor is pointable over almost a 1000 km range, enabling areas of interest to be revisited more often than the satellite orbit cycle. The range of acquisition modes offers imagery with different coverage, spatial resolution, and viewing geometries.

Source: Radarsat-1 data © Canadian Space Agency. Received by the Canada Centre for Remote Sensing. Processed and distributed by RADARSAT International.

Despite its short mission, SeaSat radar imagery convincingly demonstrated the potential of satellite radar for monitoring a wide spectrum of oceanographic and terrestrial phenomena.

Canadian radar satellites

Radarsat-1

Launched in November 1995, Radarsat-1 is operated by the Canadian Space Agency with reception, recording, and data archiving carried out by the Canada Centre for Remote Sensing. RADARSAT International, a subsidiary of Mac-Donald Dettweiler Associates Inc., has an exclusive world-wide licence to process, market, and distribute Radarsat data. Radarsat carries a C-band, HH-polarized SAR with a radar beam pointable across almost a 1,000 km range *(figure 8.45)*. Imaging swaths can be varied from 35 to 500 km in width, with resolutions from 10 to 100 meters *(figure 8.46)*. Viewing geometry varies from 20 degrees to more than 50 degrees depending on the swath selected. Although the satellite's orbit repeat cycle is 24 days, the pointable radar beam enables selected areas to be imaged more frequently as required. The Radarsat orbit is optimized for frequent coverage of mid-latitude to polar regions and is able to provide daily images of the entire Arctic region as well as view any part of Canada within three days. Even at equatorial latitudes, complete coverage can be obtained within six days using the widest swath of 500 km.

Radarsat-2

Radarsat-2 is a C-band imaging radar satellite with a point-able antenna scheduled to be launched in 2006. Designed as a follow-on to Radarsat-1, it is being built and will be operated by Alenia Spazio for MacDonald Dettwiler Associates Inc. (MDA) as a public/private partnership between MDA and the Canadian government.

Radarsat-2 is significantly more capable than Radarsat-1. Figure 8.47 shows a Radarsat-1 image at the best resolution of 7.5 m and a simulated Radarsat-2 image with 3 m ground resolution. In addition to Radarsat-1's seven beam modes, this new system will offer five new sensor imaging configurations, higher spatial resolution (3 m), more frequent revisits, and increased downlink margin (allowing data reception from lower cost reception antenna systems). It will be able to acquire data at HH (horizontal transmission and reception), VV (vertical transmission and reception), and HV (cross-polarized—horizontal transmission and vertical reception) polarizations. With its dual transmit and dual receive antennas, Radarsat-2 will be able to simultaneously receive two different polarization combinations.

Radarsat-3

The proposed Radarsat-3 spacecraft is a copy of Radarsat-2 and would be equipped with the same type of SAR imager. Its launch is intended to take place two years after that of Radarsat-2. The two satellites would then be operated in a tandem mission that would involve them being flown in

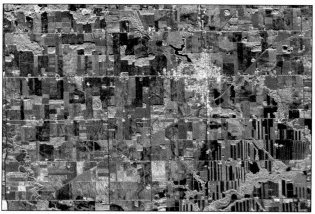

Figure 8.47   Comparison of Radarsat-1 and simulated Radarsat-2 data. Above is a typical Radarsat-1 image of an agricultural area at the best resolution (7.5 m). Below is a simulated Radarsat-2 image of a portion of the area with a resolution of 3 m.

Source: Radarsat-1 data © Canadian Space Agency. Received by the Canada Centre for Remote Sensing. Processed and distributed by RADARSAT International.

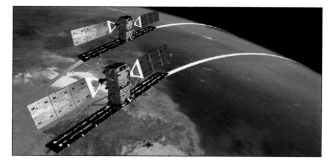

Figure 8.48   Radarsat 2/3 tandem mission. The Radarsat 2 and 3 missions are being optimized to provide repeat coverage for radar interferometry.

Source: Radarsat-1 data © Canadian Space Agency. Received by the Canada Centre for Remote Sensing. Processed and distributed by RADARSAT International

Figure 8.49   ERS-1 satellite.

Source: © ESA-DENMAN Production.

formation with a controlled and measured cross-track separation and baseline of a few km. This would allow the collection of accurate IfSAR data in various different modes. The overall aim of the tandem mission would be to produce a global DEM from space *(figure 8.48)*.

European radar satellites
ERS-1 and 2

Launched in 1991 by the European Space Agency, ERS-1 *(figure 8.49)* carries a radar altimeter, an infrared radiometer and microwave sounder, and a C-band radar which can be operated as a synthetic aperture radar or as a scatterometer to measure reflectivity of the ocean surface, as well as ocean surface wind speed and direction. Eventually, after seven years successful operation, ERS-1 encountered system problems and was switched off in late 1998.

In synthetic aperture radar mode, imagery is collected with a 100 km swath width and 30 m spatial resolution over an incidence angle ranging from 20 to 26 degrees. Polarization is vertical transmit and vertical receive (VV) which, combined with the fairly steep viewing angles, make ERS-1 particularly sensitive to surface roughness. The revisit period (or repeat cycle) of ERS-1 could be varied by adjusting the orbit and has ranged from 3 to 168 days, depending on the mode of operation. Generally, the repeat cycle is about 35 days. A second satellite, ERS-2, launched in April 1995, has the same radar system as ERS-1 but is operated on a 35-day repeat cycle. ERS-2 is currently operational

Designed primarily for ocean monitoring applications and research, ERS-1 and 2 have proved valuable for terrestrial applications as well. Like SeaSat, the steep viewing angles cause severe relief displacement in imagery of high relief regions.

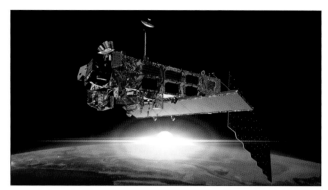

Figure 8.50   Envisat-1 satellite.

Figure 8.51   Envisat radar image of Larsen ice shelf, Antarctica, March 18, 2002. Over the last decade, ice conditions off the coast of Antarctica have been monitored as an indicator of global climate change. Analysis of radar imagery from the earlier ERS-1 and -2 satellites together with this radar image from Envisat-1 document the 100 km retreat of the Larsen B ice shelf. The break-up has now reached the rock faces of the peninsula itself.

Envisat-1

Launched in March 2002, the European Space Agency's Environmental Satellite, Envisat-1 *(figure 8.50)*, was designed as a follow-on to ERS-1 and ERS-2. The satellite's complement of ten instruments chosen to support research and monitoring of the global environment includes a C-band SAR—the Advanced Synthetic Aperture Radar (ASAR) *(figure 8.51)*. The system can operate in high and low resolution modes, offering swath widths of 100 km or 400 km with spatial resolutions of 30 m, 100 m, or 1 km, HH or VV polarization, and look angles ranging from 14° to 45°.

CryoSat

The latest in the series of radar satellites from the European Space Agency (ESA) is CryoSat *(figure 8.52)*. It is designed specifically to carry out observations of the earth's polar areas to determine variations in the thickness of the earth's continental ice sheets—e.g., over Antarctica and Greenland—and of the earth's marine ice cover, especially in the Arctic Ocean. These measurements are needed to help predict the probable effects of global climate change. The satellite is being built by Astrium in Germany and will be placed in non-sun-synchronous orbit at an inclination of 92 degrees (in order to cover the polar regions) and an altitude of 717 km. It will be launched by a Russian Rockot launcher from the Pletesk site in northern Russia in 2005.

CryoSat will use novel measuring techniques based on the use of the SIRAL (SAR Interferometer and Radar Altimeter) instrument built by the French Alcatel company. This comprises twin radar altimeters operating in the K_u-band that are mounted one meter apart on the satellite platform. The two receiving antennas form an interferometer in the across-track

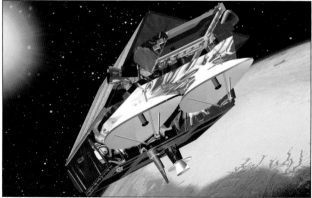

Figure 8.52   CryoSat satellite.

direction, while the return signals in the along-track direction are processed to form a synthetic aperture for enhanced ground resolution. The different operating modes include (i) conventional pulse-limited operation for the measurements over the ice sheet interiors (and parts of the ocean, if required), so continuing the ERS and Envisat measurement series; (ii) synthetic aperture operation over areas of sea ice, including the detection of narrow leads of open water; and (iii) dual-channel SAR/interferometric operation over ice sheet margins. The very accurate determination of the satellite's orbital position and altitude that is required is provided by an on board DORIS receiver, together with triple star trackers to provide precise measurements of the satellite's attitude and the orientation of the interferometric baseline. Further very accurate supplementary measurements of position and altitude will be carried out from ground stations utilizing satellite laser ranging (SLR) techniques. These laser ranging devices measure distances very accurately to retro-reflectors mounted on the satellite.

German radar satellites

Following on from the country's strong interest and successful record in SAR imaging, both from the air and from space, German agencies have now embarked on a very extensive national program concerned with the construction and operation of several spaceborne SAR imagers.

SAR-Lupe

The SAR-Lupe program is now well under way. It involves the construction and launch of five relatively small (770 kg) satellites, each equipped with an X-band SAR. The primary mission of the SAR-Lupe satellites *(figure 8.53)* is to provide imagery on an all-weather basis for national security

purposes. However, a surprising amount of information has been made available regarding the characteristics of the satellites and their SAR systems.

The five satellites will be placed in three different near-polar orbital planes, spaced out to provide world-wide coverage. Orbital plane 1 will contain two satellites; orbital plane 2 will have a single satellite; while the remaining two satellites will occupy orbital plane 3. The satellites are being built by the OHB-System company based in Bremen, Germany, together with a number of other German companies and the supply of components from a number of European companies—Alcatel, Tamex (France); Saab Ericsson (Sweden); and Carlo Gavalli (Italy)—and American electronics companies—e.g., Integral Systems and EMS Tech. By using proven components developed for other space missions—e.g., the antennas supplied by Saab Ericsson are derived from ESA's Rosetta interplanetary mission, and the SAR electronics are based on those developed by Alcatel for the Jason and Poseidon radar altimeters—development has been rapid, while the cost has been kept low. The ground resolution of the SAR images will be better than 1m over an area of 5 x 5 km when the X-band SAR is used in its highest resolution mode. The launch of the first SAR-Lupe satellite is scheduled for 2005, using a Russian Rockot launcher. It is hoped that the launches of all five satellites will be completed by 2007.

TerraSAR

The other major German spaceborne SAR project is the TerraSAR program. This involves a so-called public/private partnership—with DLR contributing $90 million and the German arm of Astrium (formerly Dornier) contributing a further $25 million towards the costs of the program—in a rather similar manner to the arrangements for the Canadian

Figure 8.53   SAR-Lupe satellites.

Source: SAR-Lupe satellite image courtesy of OHB-System AG, Germany.

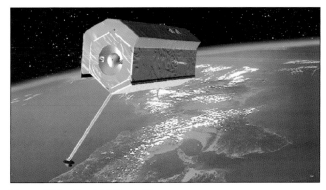

Figure 8.54   TerraSAR-X satellite.

Source: © EADS Space.

Radarsat-2 mission. The first satellite will be TerraSAR-X *(figure 8.54)*, which, as the name suggests, will operate an X-band SAR that will generate images with 1 m spatial resolution. A second satellite, TerraSAR-L—equipped with an L-band SAR—is also proposed. However, as yet, it has not been approved and funded. The TerraSAR-X satellite is scheduled for launch into a near-polar orbit in 2006 using a Russian Dnepr launcher.

Italian radar satellites
COSMO-SkyMed

The Italian government is also funding a national spaceborne SAR program, called COSMO-SkyMed that is for dual use, i.e., for civilian and defense applications. The COSMO part of the title is an acronym derived from the full title of the original project—Constellation of Small Satellites for Mediterranean Basin Observations. Thus originally the satellites were intended to generate imagery for scientific studies of the Mediterranean basin, including disaster monitoring. However, later, it was decided that the images could also be used for national security purposes. Thus, although the Italian Space Agency (ASI) is responsible for the execution of the program, the funding is being shared between the Italian ministries of research and defense. The program comprises four quite large satellites (each weighing 1.7 ton) that are being built by Alenia Spazio with Galileo and Laben as subcontractors. Each of the four satellites will be placed in a sunsynchronous orbit, 90 degrees apart in longitude to provide worldwide coverage. Each satellite will be equipped with an X-band SAR that will produce images with a ground resolution of 1 m over a 10 km wide swath when operating in its high-resolution (spotlight) mode. The first of the COSMO-SkyMed satellites is planned to be launched in 2005.

In 2001, a cooperative intergovernmental agreement was entered into between France and Italy, by which the SAR image data produced by the COSMO-SkyMed program will be exchanged with the high-resolution optical imagery that will be produced by the two French Pleiades satellites that will follow on from the present program of SPOT satellites. Together, the two programs will form the so-called Orfeo system. This will operate on the basis of dual (civilian/military) use—though the defense authorities of the two countries will have priority in the mission planning and operation of the satellites.

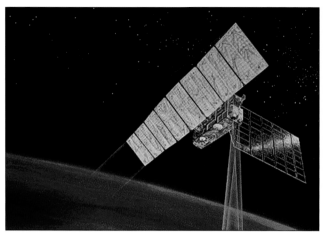

Figure 8.55 JERS-1 satellite.

Source: © JAXA.

Japanese radar satellites

The Japan Aerospace Exploration Agency (JAXA), formerly the National Space Development Agency of Japan (NASDA), has launched and operated a number of meteorological and earth resource satellites. One of the latter group—JERS-1 (Japanese Earth Resource Satellite-1)—was equipped with two quite high-performance imaging sensors—one, the OPS optical scanner; the other, a synthetic aperture radar (SAR).

JERS-1 SAR

JERS-1 *(figure 8.55)* operated at an altitude of 568 km in a sun-synchronous orbit with an orbital inclination of 97.7 degrees. The satellite was launched in February 1992 and operated very successfully from April 1992 till October 1998, when it ceased to function.

The SAR sensor on JERS-1 operated in the L-band—which made the resulting images quite different and complementary to those produced by its contemporaries, ERS-1 and -2 and Radarsat, all of which carried SAR devices operating at C-band wavelengths. The JERS-1 SAR images have a ground resolution of 18m and a swath width of 75 km. During its six years of operation, most of the Earth's land area was covered. The JERS-1 SAR images have been used extensively in studies concerned with the forest mapping of very large areas—especially the Global Rain Forest and Boreal Forest Mapping (GRFM/GBFM) program and the JERS-1 Amazon Multi-season Mapping Study (JAMMS). The L-band images have also been used to provide complete cover in the Siberia Boreal Forest Mapping project in conjunction with the C-band images acquired from ERS-1 and -2. The use

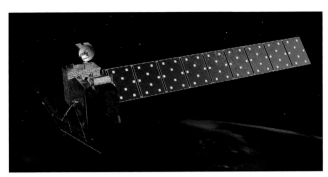

Figure 8.56   ALOS satellite.

Source: © JAXA.

of the dual-frequency images from the two datasets proved to be particularly effective in this particular study in distinguishing different forest types.

ALOS (Advanced Land Observing Satellite)

The Japanese ALOS (Advanced Land Observing Satellite) *(figure 8.56)* is currently being made ready for launch in 2005, when it is planned to be placed into a sun-synchronous orbit at an altitude of 692 km. This large and very heavy satellite (weighing 4 tons) will carry three imaging sensors—(i) PRISM (Panchromatic Remote-sensing Instrument for Stereo Mapping) for DEM generation; (ii) AVNIR-2 (Advanced Visible & Near Infrared Radiometer-2) for land-cover mapping; and (iii) PALSAR (Phased Array L-band Synthetic Aperture Radar) for all-weather land observation. Like the JERS-1 SAR, PALSAR will operate in the L-band part of the spectrum. However, it will be fully polarimetric (with HH, VV, HV and VH polarization modes) instead of the single HH mode that was possible with the JERS-1 SAR. The ground resolution of the PALSAR images will range from 7m (in fine mode) to 100 m (in scanSAR mode) with respective swaths of 40 km to 350 km. With its orbital inclination of 98.2 degrees, it cannot observe the polar areas beyond 81 degrees north and south latitude.

Mapping seafloor topography with satellite radar altimetry

Masked by 3–5 km of seawater, the topography of the deep oceans was, until recently, largely uncharted. Ship-based sonar can produce detailed bathymetric charts, but ships travel slowly; only 3% of the ocean has been mapped with shipborne sensors so far. Using the latest ship-board swath-mapping sonar technology, it would take 125 years to chart the ocean basins at a cost of a few billion U.S. dollars. The first

comprehensive charts of seafloor topography were produced not from data collected at sea but from satellite radar altimeter data collected hundreds of kilometers above the earth's surface. A satellite altimeter mission can cost as little as $60 million (the cost of the Geosat mission) and can generate a global dataset over a period of a few years.

A radar altimeter is a nonimaging radar system that calculates altitude by measuring the time delay between transmission of a microwave pulse towards the earth's surface and reception of its echo. Over the ocean, the altimeter does not see through the water to the ocean floor but instead measures the height profile of the ocean surface. Gravitational attraction causes the ocean surface to bulge over massive sea mounts and dip over trenches. In this way, satellite altimetry allows one to recover gravity anomalies and exploit their correlation with ocean depth variations to make reconnaissance charts of the ocean floor.

The idea that ocean depth could be estimated from measurements of gravity anomalies was first proposed in the 19th century. The feasibility of microwave radar altimetry of the ocean surface from space was first demonstrated aboard *Skylab* in 1973, and the first successful free-flying altimeter spacecraft, Geos-3, flew in 1975. However, it was the SeaSat radar satellite in 1978 which first led to new discoveries of ocean floor features. SeaSat's altimeter could achieve 10 cm accuracy, enough to reveal the continuity of the Louisville seamount chain in the South Pacific. A marine gravity anomaly map produced from SeaSat data in 1987 by William F. Haxby, a geophysicist at the Lamont-Doherty Laboratory of Columbia University, is shown as figure 2.41 in chapter 2. Figure 8.57 is a more recent map of global marine gravity anomalies developed by the NOAA National Geophysical Data Center using the higher quality data sources and more sophisticated processing algorithms developed since that time.

The absolute height of the ocean surface is the sum of a complex mixture of dynamic oceanographic and meteorological effects, tides, and the integrated effect of gravity anomalies. Sea level departs from a reference ellipsoidal shape for the earth by as much as 100 meters. However, the gravity anomalies correlated with depth variations appear as sea height differences of 1 cm to a few meters in height over horizontal distances from a few kilometers to a few hundred kilometers. Each satellite radar altimeter measurement is collected over an area about 2 to 12 km in diameter, large enough to average out local surface irregularities such as waves. A continuous profile of sea surface height is collected beneath the satellite track. Since SeaSat, higher resolution altimeter data has

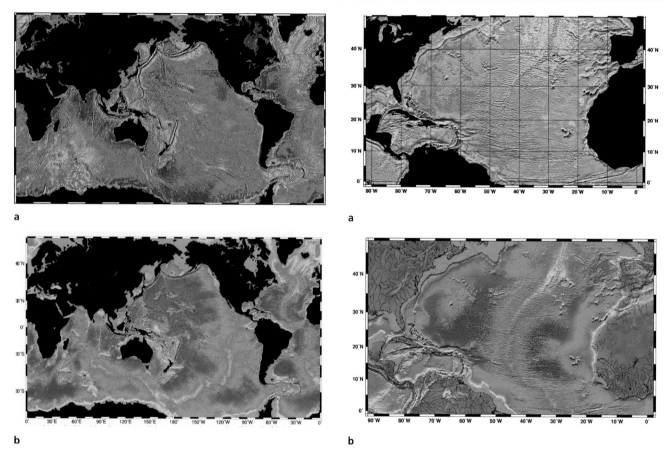

a

b

a

b

Figure 8.57 Global seafloor topography from satellite data and shipboard depth soundings. Gravity field data estimates (a) derived from ERS-1 and Geosat satellite radar altimeter data corrected for tides and atmospheric effects were combined with shipboard depth surveys to produce this 1997 map of seafloor topography of the world's ice free oceans (b). Gravity anomaly data on its own *(figure 8.58b)* can be a misleading indicator of bathymetry. Note that in the gravity anomaly map, many of the shallow coastal areas and inland seas such as Hudson's Bay are coded purple, the same as many deep water ocean areas whereas using shipboard data for calibration, these areas are represented as shallow (orange) in the combined dataset.

Source: Images courtesy of W. H. F. Smith of NOAA and D. Sandwell of Scripps Institute of Oceanography.

Figure 8.58 Map of gravity anomalies and seafloor topography predicted from gravity data without shipboard depth soundings, , North Atlantic Ocean. Gravity anomaly data over ocean areas can be mapped (a) and is useful in predicting large topographic features of the seafloor such as seamounts, fracture zones, and ridges (b). Predicted ocean depths can have very large errors, and other data sources, such as ship-based bathymetric survey data is needed to calibrate their accuracy for a specific area. The data is primarily used by geologists to visualize the seafloor structure. This image shows the Mid-Atlantic Ridge as a sinuous feature trending in the north-south direction in the center of the image with its crosscutting fracture zones.

Source: Images courtesy of W. H. F. Smith of NOAA and D.Sandwell of Scripps Institute of Oceanography.

been available from the ERS-1 (European Space Agency) and Topex/Poseidon (NASA and CNES) satellites as well as declassified data from the U.S. Navy's Geosat satellite providing vertical accuracies as high as 3 cm. A new technology called delay-Doppler altimetry could achieve 1 cm.

The U.S. Navy uses the Geosat radar altimetry data to correct navigational errors in inertial navigation and guidance systems. These instruments track the position of a moving vehicle by precise measurements of acceleration. The minute variations in the gravitational field that produce bumps and dips in sea surface height also affect inertial measurement instruments causing navigational errors that can be significant over long distances such as a flight path across the Pacific Ocean or an extended submarine mission.

The correlation between gravity and ocean depth is imperfect, varying with local geological conditions. Over large horizontal distances on the order of tens or hundreds of km, the scale of seamounts, fracture zones, and ridges on the seafloor, the variations in gravity can be correlated with variations in depth. Satellite altimeter data is now processed using more sophisticated algorithms and more precise measurement of satellite orbits. The use of statistical techniques to combine the profiles from several satellites collected over many years together with shipboard gravity measurements and sonar bathymetry have greatly improved the accuracy and detail of charts of seafloor topography predicted from gravity anomalies *(figures 8.57 and 8.58)*.

Over areas of thin sediments, such as the Mid-Atlantic Ridge region shown in figure 8.58, accuracies may be as high as about 65 m. But in sedimented areas, there is little or no correlation. For this reason, the charts are not suitable for navigation. Their primary application has been for geologists to visualize the seafloor structure. They have also been useful in identifying seafloor structures such as constrictions that influence the major ocean currents, locating promising fishing areas, e.g., by finding shallow sea mounts where fish and lobster are abundant, planning undersea cable and pipeline routes, and in petroleum exploration to locate offshore sedimentary basins in remote areas. As an indicator of previously uncharted seafloor structures, satellite altimetry data is used in planning shipboard bathymetric surveys.

Satellite altimetry data products are available free by download from Web sites or on CD for a nominal fee. However, the data generally requires some additional processing before it can be input to a GIS. GIS-ready datasets are under development and will be available from NOAA's Laboratory for Satellite Altimetry and National Geophysical Data Center.

Lidar

Lidar is the acronym for Light Detection and Ranging. Like radar, it is an active remote sensing system using laser[1] light instead of microwaves to illuminate the target area. Lidar systems produce accurate measurements of surface elevation to generate digital elevation models with vertical accuracies as

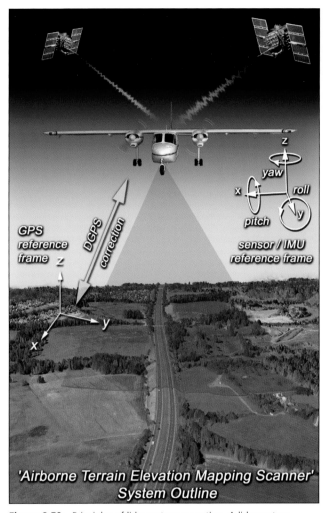

Figure 8.59 Principles of lidar system operation. A lidar system includes a scanning instrument that generates laser pulses and records the reflected returns, an on-board GPS to record sensor position, and an inertial measurement unit (IMU) to measure sensor orientation. A second GPS located at a known ground position collects data at the same time as the overflight for later differential correction of the on-board GPS data (DGPS).

Source: © Precision Terrain Surveys, Ltd.

high as about 15 cm. Specialized systems have also been developed for the measurement of water depth (see below).

Lidar systems incorporate three technologies: (i) laser ranging for accurate distance measurement, (ii) satellite positioning using the Global Positioning System (GPS) to determine the geographic position and the height of the sensor, and (iii) aircraft attitude measurement using an inertial measurement unit (IMU) to record the precise orientation of

---

1. A laser is a device that generates a stream of high-energy particles, termed photons, in a narrow wavelength band. Low power lasers are used in a variety of consumer products such as CD players. High power lasers are used in industrial cutting tools and weaponry. Lidar systems use medium power lasers.

Figure 8.60 Pattern of laser reflections. The irregular pattern of laser reflections in lidar data is a result of aircraft motion, variation in the reflectivity and orientation of ground targets, and the mosaicking process.

Source: Image courtesy of Earthdata Holdings, Inc.

a

b

Figure 8.61 Elevation and intensity lidar images. The two principal datasets generated from a lidar survey are elevation data, shown here as a color coded image (a) and intensity data shown as a monochrome image (b).

Source: Images courtesy of Optech Incorporated and the U.S. Army JPSD-PO.

the sensor. (The IMU is commonly a component incorporated into an inertial navigation system, INS, which typically includes a GPS as well.) In addition, data acquisition and processing depend on the availability of high performance computers and high capacity data storage *(figure 8.59)*.

Lidar operation and measurement

Lidar systems emit pulses of laser light and precisely measure the elapsed time for a reflection to return from the ground below. Knowing the speed of light, which is a constant, and the elapsed time, the distance to the target can be calculated. This is the *slant range distance*, the distance to the target at the angle it was viewed by the scanner. With the additional data of the scanner viewing angle at the time each pulse reflection was received and measurements of the sensor position, height, and orientation from the GPS and IMU, the three-dimensional position of the target can be calculated. Since not all pulses will generate detectable reflections, an irregular pattern of points may be acquired *(figure 8.60)*.

A scanning mechanism directs the pulses of light and collects reflection data in successive lines perpendicular to the flight path of the aircraft. Terrain mapping systems have pulse rates as high as 75,000 pulses per second, and the typical ground spacing of the raw data points ranges from 0.5 m to 4 m. For engineering applications such as highway mapping, up

to 10 points per square meter are collected generally using helicopter-mounted lidar systems.

Accurate sensor position and height measurements are obtained with differential GPS and an inertial measurement unit that records the precise orientation of the aircraft. (With differential GPS, a GPS ground station at a known position and elevation records data at the same time as the acquisition flight. This data is later used to further correct the GPS data collected onboard the aircraft.)

In-flight measurements are made simultaneously and recorded for the scanner's viewing angle (generated by the scanner system), aircraft position (differential GPS), and aircraft attitude (inertial measurement unit). After the flight, this data is processed to generate elevation datasets and related products *(figure 8.61a)*.

Postprocessing the very large datasets produced from one day's flying typically requires several days or more. However, small areas can be processed in a matter of hours if required. Though highly automated and capable of generating digital elevation data more quickly than by photogrammetric methods, lidar processing requires qualified personnel to inspect and correct data products and implement quality assurance and control procedures similar to those of traditional surveying.

The accuracy of lidar distance measurements depends on the flying height which affects the strength of the reflected return, the precision of the lidar instrument and processing, accuracy of the sensor position, height and attitude data, and characteristics of the target area (e.g., more accurate measurements can be obtained from a solid level surface than from an area of shrubs on undulating terrain). The vertical accuracy of commercial lidar systems is usually ±15–50 cm depending on terrain characteristics, instrument specifications, and the quality of processing. Some helicopter-mounted systems flown at altitudes as low as 100 m can provide accuracies as fine as ±5 cm RMSE for hard surfaces such as road pavement.

Lidar reflections and returns

As well as the digital elevation datasets, the intensity of the lidar returns can be used to generate an image useful for orienting and analyzing the dataset *(figure 8.61b)*. Since the intensity image represents an interpolation of the laser returns from irregularly distributed points, it does not have the detail or geometric precision of an aerial photograph or scanner image. However, both the lidar intensity and shaded relief imagery are produced from x,y,z survey points and are therefore orthoimages, corrected for terrain variations.

Unlike radar, lidar systems cannot penetrate clouds and fog. The target must be visible. Furthermore, laser signals cannot penetrate vegetation. However, if the vegetation canopy is somewhat open, there will usually be sufficient openings to obtain measurements to the ground, effectively providing a means to measure the elevation of the ground surface beneath a forest canopy. Lidar systems cannot produce

ground measurements through fully closed canopies such as tropical rainforests.

Features with an open structure, such as a forest canopy, can produce multiple reflections per pulse from structures at different heights above the terrain *(figure 8.62)*. Some lidar systems are capable of simultaneously recording the first return signal, the last return signal, or multiple reflections per pulse depending on the application. Not every pulse will generate multiple returns, and the number of returns will

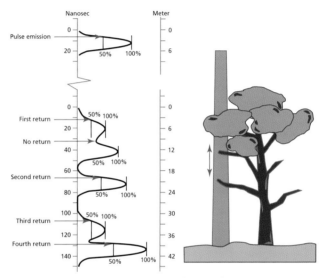

Figure 8.62 Multiple lidar returns. Sophisticated processing algorithms define multiple reflective surfaces by analyzing intensity variations within each pulse reflection.

Source: Image courtesy of Earthdata Holdings, Inc.

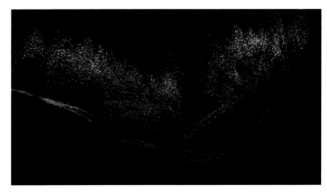

Figure 8.63 Multiple lidar returns over a forest area. Multiple lidar returns collected over a forest area are displayed as a perspective view. A maximum of five returns, coded in the image by color, were collected for each point.

Source: Image courtesy of S. B. Gross, Inc.

a

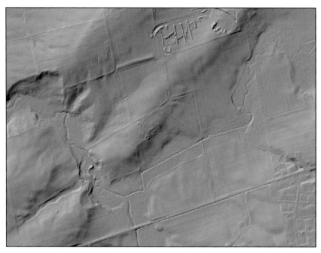

b

Figure 8.64 Surface feature removal by filtering lidar returns. These shaded relief presentations of a last-pulse lidar dataset show (a) all elevations returned from the last pulse or last return and (b) the bare earth or ground points derived by filtering the returns from this dataset.

Source: Images provided by Mosaic Mapping Systems Inc./TerraPoint USA, Inc.

generated. A surface representing the ground level with surface features removed is commonly termed a *bald earth, bare earth* or *ground level* surface. Figure 8.64 illustrates shaded relief or hill-shade presentations of a lidar dataset processed to show (a) all elevations returned from the last pulse or last return and (b) the bare earth or ground points derived by filtering the returns from this dataset.

Characteristics of certain features on the surface can then be calculated, such as the sag in a power line or the height of a tree canopy above ground. Similarly, estimates of vegetation biomass can be derived by subtracting the surface representing the ground level from the surface representing the forest canopy. Biomass estimates have been valuable for forest management and to the oil and gas industry to estimate stumpage fees payable for land clearing. A ground level surface is also useful in visualizing the appearance of land clearing activities such as highway construction, subdivision development, and forest removal.

Multiple returns can also be used to distinguish different land-cover classes. For example, multiple returns are only found in vegetated areas. The difference between the highest and lowest elevations in a small area can distinguish treed from shrub areas. The vertical distribution of reflections can be used to classify the type of forest or its maturity *(figure 8.65)*.

Figure 8.65 Lidar products. The data stack shows, from bottom to top, a digital orthophoto rectified using a DEM derived from lidar data, lidar derived bare-earth surface color coded by elevation, lidar derived canopy elevation, and lidar derived land-cover classification.

Source: Images courtesy of Earthdata Holdings, Inc.

vary so the resulting data is in effect, a three-dimensional cloud of points *(figure 8.63)*.

Sophisticated algorithms are used to filter the data and generate surfaces representing the lowest elevation (the ground), the highest elevation (e.g., a tree canopy or the top of a building) and in some cases, intermediate levels as well (e.g., a shrub layer within or beneath the forest canopy) can be

Airborne topographic lidar systems

Lidar systems being used for topographic applications use lasers operating at wavelengths of about 1 µm in the near-infrared band with high pulse rates. A major commercial manufacturer of such systems is the Optech Inc., based in Toronto, Canada, which has built a wide range of models. As the technology has developed, successive models have been produced with ever higher pulse rates and the capability of being operated from ever greater flying heights. Another substantial supplier is Leica Geosystems, which, in 2001, purchased the Azimuth Corporation based in Massachusetts. Currently the company offers its ALS40 and ALS50 lidar systems. In Europe, airborne lidar systems are manufactured by Saab (TopEye) in Sweden and Reigl in Austria. However, other companies have built their own airborne lidar systems, which they also operate and offer as a commercial service. American examples are TerraPoint, Spectrum Mapping (RAMS and DATIS) and Fugro-Chance (FLI-MAP). European examples are TopoSys (Germany) and Precision Terrain Mapping (U.K.).

The use of airborne lidar systems is now widespread, especially for engineering applications. Numerous aerial survey and mapping companies operate these systems, acting as service providers on a competitive commercial basis, both in North America and in Europe. An interesting trend in recent years is to combine the operation of an airborne lidar system generating elevation data with that of a medium-format (4,000 x 4,000 pixel) digital frame camera to generate corresponding images of the terrain. On the basis of the lidar produced DEM, the digital frame images can then be orthorectified.

Airborne hydrographic lidar systems

Lidar systems have been developed for the measurement of water depth as well as the measurement of topography. Specialized dual frequency lidar methods, termed airborne lidar bathymetry (ALB) or airborne lidar hydrography (ALH), have been developed to measure the depth in shallow waters. Hydrographic lidar systems use pulse rates as low as 400 Hz and require higher powered lasers operating in the visible wavelength band to penetrate the water column and produce enough of a reflection from the seabed to be detected.

Although the method was first conceived in the 1960s, it was not until the 1980s that the first prototype systems were actually built and put into operation. These included the Airborne Oceanographic Lidar (U.S.), Larsen 500 (Canada), Whelads (Australia) and Flash (Sweden) systems. During the 1990s, fully operational systems such as SHOALS (U.S.), LADS (Australia) and Hawkeye (Sweden) came into use.

One of the most widely used hydrographic lidars is SHOALS (Scanning Hydrographic Operational Airborne Lidar Survey) manufactured by Optech Inc. and operated by the Fugro-Chance company on behalf of the U.S. Army Corps of Engineers (USACE). It has been used not only by the USACE but also by NOAA for coastal mapping. Another SHOALS system is operated by Japan's Coastguard for the hydrographic mapping of coastal waters.

SHOALS has the unique capability to map shallow waters, shoreline, and topography (ground elevations) simultaneously, thereby integrating land and water measurements in the same dataset. It fires co-aligned infrared and blue-green laser pulses at the water. The infrared wavelength pulse is reflected by the water surface while the blue-green wavelength pulse penetrates the water surface and is reflected from the seabed (*figure 8.66*). The water depth is derived from the time difference between reception of the two reflection signals after accounting for system and environmental factors.

Depth penetration and accuracy depend on water quality. In clear water, depths of 40 m to 50 m can be determined with accuracies as high as 15 cm (*figure 8.67*). The system is

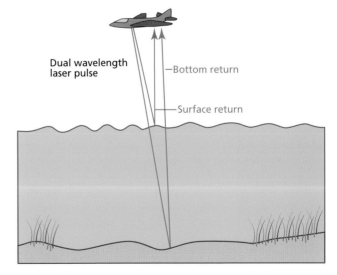

Figure 8.66 Hydrographic lidar. The SHOALS hydrographic lidar system use two lasers, operating in the visible blue-green (0.532 µm) and infrared (1.064 µm) wavelengths. The elevation of the water surface is calculated from the infrared reflection and the seabed elevation from the blue-green.

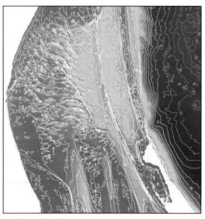
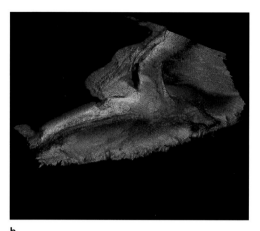
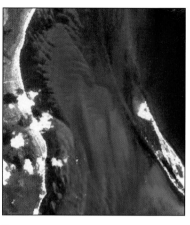

a   b   c

Figure 8.67 Nautical charting. A 900 km² area of the Yucatan Peninsula, Mexico, was surveyed by the U.S. Army Engineers over a 2 1/2 month period using a SHOALS hydrographic lidar. The project developed new charts for an area of coral reefs and other shallow water obstructions. Lidar data color-coded to depth with contour lines (a) and a perspective view (b) are shown for the area in the color airphoto (c).

Source: Images courtesy of NAVO (U.S. Naval Oceanographic Office).

also used to locate submerged features such as ship wrecks and other objects on the ocean floor as small as about a 2 m cube at 50 m depths. Smaller features can be detected at shallower depths.

Hydrographic lidar is used principally for mapping depths in shallow water and the water's edge in coastal areas and to identify obstructions. At a flying height of 400 m, the SHOALS system collects data over a 230 m wide swath. Sonar systems can measure much greater depths but the shallower the depth, the narrower the swath width. In shallow waters, sonar bathymetry requires so many passes to cover an area that hydrographic lidar becomes a more cost-effective solution *(figure 8.68)*.

Another well-known airborne hydrographic lidar system is the Tenix LADS (Laser Airborne Depth Sounder) built in Australia. One is operated by the Royal Australian Navy; a second example is operated commercially by the Tenix Corporation on a worldwide basis. Similarly, the Swedish Hawkeye system is operated by the Swedish Maritime Authority, while a second system is being used in Indonesia, operated by the Norwegian Blom mapping company.

U.S. research lidar systems

A number of lidar systems have been built and operated by NASA, principally for research purposes. These have done much to develop the technology. They have also been used in extensive application programs carried out in collaboration with other U.S. government agencies.

NASA airborne lidar systems

One of the best known of these systems is the Airborne Topographic Mapper (ATM) which was developed from the original Airborne Oceanographic Lidar (AOL) project from the late 1970s. The current ATM-1 and -2 instruments use a powerful laser transmitter operating in the blue-green spectral region in conjunction with a reflector telescope, a scan mirror, and a photo multiplier tube. A survey grade GPS and

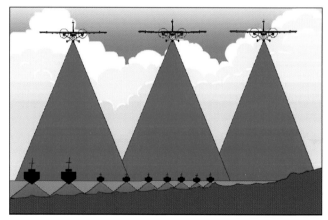

Figure 8.68 Multibeam sonar (also known as *multibeam echo sounding*) technology can measure deeper water depths than hydrographic lidar. In addition, with multibeam sonar, the width of the area measured in a single pass (i.e., the *swath width*) is dependent on the angular spread of the sonar beam, and so the deeper the seabed, the wider the swath width. The limitation is the maximum depth from which a particular instrument can obtain a reliable return signal.

Source: Image courtesy of Optech Incorporated.

a ring laser gyro supply the required position, altitude and attitude data to the system. These instruments have been used extensively on the airborne laser beach mapping and coastal erosion programs of NOAA and the USGS carried out along substantial stretches of the Atlantic and Pacific coasts of the United States They have also been deployed on measurements of parts of the Greenland ice sheet to determine changes in the elevation of the ice sheet. NASA has also built and operated the RASCAL (Raster Scanning Airborne Lidar) system which has been mounted on a T-39 jet aircraft. The system has been used on a variety of research projects, including the mapping of an earthquake-prone area near Los Angeles and the repeated mapping of the Long Valley caldera, also located in California.

Another research area involving airborne lidar systems that has been pursued by NASA has been the mapping of vegetation height and the underlying topography below the canopy. This has also included the determination of the tree canopy structure, enabling overall biomass to be estimated.

An earlier instrument constructed specifically for this purpose was SLICER (Scanning Lidar Imager of Canopies by Echo Recovery). This instrument used a laser operating in the near-infrared part of the spectrum to measure and analyze the time-varying distribution of the return pulses from the multiple targets in the canopy that have been illuminated by the scanning laser. A later, improved instrument that carries out the same type of measurements is the LVIS (Laser Vegetation Imaging System) which has been used to carry out research programs in California and Costa Rica.

NASA spaceborne lidar systems

NASA has also operated lidar instruments on short duration flights of the space shuttle. The experimental Shuttle Laser Altimeters (SLA-01 and -02) were operated from the spacecraft in January 1996 and August 1997 respectively. The SLA lidars were constructed mainly from spare parts of components used in the Mars Observer Laser Altimeter (MOLA). They used infrared laser pulses to measure the elevations of land surfaces and vegetation canopies. SLA-02 gave a measuring resolution of 1.5 m in elevation within its 100 m diameter footprint on the earth's surface. These measurements were used to sample and characterize the vertical roughness of different land-cover classes and landscapes on a global basis—though obviously coverage was quite limited both by the short duration of the flights and by the orbital inclination of the shuttle spacecraft which restricted the latitudinal extent to 56 and 28.5 degrees respectively.

Based on the technology developed for the SLICER and LVIS airborne lidars, NASA also developed its Vegetation Canopy Lidar (VCL) mission with the aim of providing the first global inventory of the vertical structure of forests across the earth. The VCL mission was built around a Multi-Beam Laser Altimeter (MBLA) that featured three beams arranged in a circular configuration 8 km across with each beam tracing out a separate ground track 4 km apart. However, doubts about the instrument package resulted in the cancellation of the mission in 2001. The satellite bus was then used for another completely different and unrelated mission.

NASA has been more successful with its ICESat satellite which was launched in January 2003 *(figure 8.69)*. This uses the GLAS (Geoscience Laser Altimeter System), which features twin lasers. One of these operates in the infrared part of the spectrum to measure the surface altimetry of the earth and the heights of dense clouds; the other operates in the green part of the spectrum to measure the distribution of clouds and aerosols. In fact, the satellite is equipped with three of these laser units to provide redundancy. The value of this was shown when the first laser system stopped working after 36 days; operations continued using the other two systems. As the satellite's name suggests, the primary mission of ICESat is to measure the surface elevations of the large ice sheets covering Antarctica and Greenland, the dataset being built up from the topographic profiles being generated by the GLAS lidar instrument.

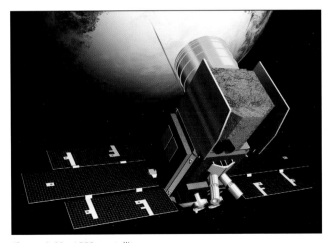

Figure 8.69 ICESat satellite.

Source: Image courtesy of NASA.

Lidar applications

Numerous firms collect airborne lidar data and many more provide processing and other consulting services. Lidar derived digital elevation data is widely used in geographic information systems. Integrated with other datasets, a wide range of derived products can be generated, from forest canopy closure and stand volume estimates to bathymetric maps, flood hazard maps, and 3D visualizations.

The principal applications of lidar data can be broadly categorized as follows:

- engineering and construction (e.g., detailed elevation drawings for highway construction, cut-and-fill estimates, environmental suitability, assessment of power line sag and obstructions, optimizing placement of telecommunication towers)
- topographic mapping
- urban infrastructure mapping (e.g., flood risk mapping, insurance assessment, airport exclusion zones)
- wildland resource management (e.g. forest stand data, wetlands mapping, watershed analysis)
- hydrographic applications (nautical charting, shoreline mapping, and identification of natural hazards)
- emergency response (e.g., rapid data collection has made lidar valuable for emergency response and recovery operations.)
- image rectification (e.g., lidar data is one of several sources of digital elevation data that can be used to ortho-rectify remote sensing imagery.)
- development of 3D visualization products in combination with photographic and digital imagery to create perspective views and fly through animations

Following is a sample of the scope of applications to which lidar data has been applied.

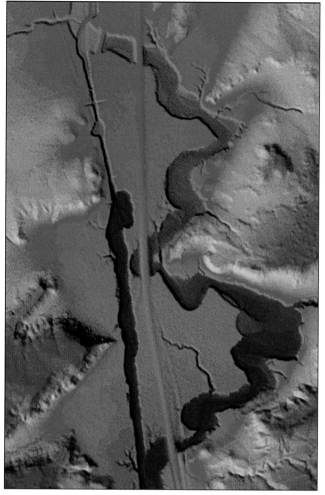

Figure 8.70 Digital terrain model generated from lidar data. Elevation ranges are color-coded from blue (low), through green and yellow, to red (high).

Source: Image courtesy of Earthdata Holdings, Inc.

Engineering, construction, and seismic survey support

Engineering surveys for roadway design and other construction projects have made use of lidar data in the assessment of alternative route locations and cost estimates such as cut-and-fill calculations. Data collected during and after construction can track progress and provide as-built surveys to document the project as delivered.

Figure 8.70 is a portion of the DEM generated for a 670 m ft wide by 200 km long highway corridor acquired for a road widening engineering project for New Mexico Highway 44. Lidar data for the entire corridor was collected in a single day producing elevation points spaced 3 to 5 m apart with an elevation accuracy of 15 cm. Figure 8.71 is a perspective view

of a portion of the highway created by draping an orthorectified color aerial photograph over the lidar derived DEM.

Lidar derived bald earth surface models are used in producing perspective views to assess proposed highway construction, subdivision development, and other urban infrastructure projects. Urban infrastructure modeling with lidar data is used in such activities as optimizing telecommunications tower locations, property assessment, and real estate development.

Lidar can be flown over areas where ground access is denied or is undesirable. In addition, the remote sensing approach

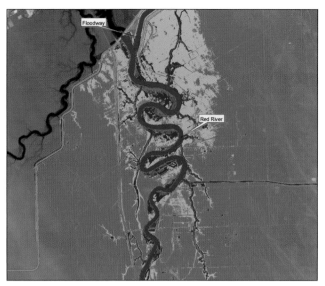

Figure 8.71   Color-infrared photography draped over a digital elevation model derived from lidar data.

Source: Image courtesy of Earthdata Holdings, Inc.

Figure 8.72   Flood risk mapping. Digital elevation models generated from lidar data of the Winnipeg Floodway and Red River were used in hydrologic models to identify areas that would be inundated at different flood levels. Key: 1 m flood level—light green, 2 m—dark green, 3 m—brown, 4 m—purple, 5 m—beige. The accuracy of the dataset is 15–30 cm.

Source: Image provided by Mosaic Mapping Systems, Inc./TerraPoint USA, Inc.

reduces land disturbance and human presence on the ground. The limited ground verification required can generally be done outside the sensitive area. In the oil and gas industry, ground survey activities for seismic data collection and other exploration activities can often be reduced or replaced by the use of lidar surveys, thereby reducing tree cutting and ground disturbance as well.

Flood risk mapping

Hydrologists use digital elevation models to predict the extent of flooding and to plan mitigation and remediation strategies. The reliability of flood-risk modeling is dependent on the accuracy of the digital elevation data. In broad flat river basins, such as the Red River in Manitoba and North Dakota, small differences in topography dramatically affect the predicted area of flooding. These subtle variations in topography are difficult and costly to map by photogrammetric methods. Lidar surveys are often the most cost-effective and quickest means of generating digital elevation data with the high density of elevation points and high accuracy required.

Figure 8.72 illustrates a flood risk map for the floodway control system in the city of Winnipeg, Canada. Highly susceptible to flooding by the Red River, a floodway was constructed to channel floodwaters past the city. The illustration shows areas that would be inundated by different flood levels.

Figure 8.73   Flood simulation. Lidar data was used to generate a bald earth surface, the 100- and 500-year flood inundation limits, and 3D model of buildings in an urban area for this flood simulation.

Source: Image courtesy of Earthdata Holdings, Inc.

Figure 8.73 illustrates the use of lidar data within a GIS to model an urban flood event. Models such as these are used to calculate the risk and to develop predictive damage assessment and response plans for a range of flood events. Planners can then assess the costs of alternative mitigation measures against the risk and potential cost of different flood events. The insurance industry uses these methods to set flood insurance rates.

Corridor mapping and seismic survey support

Narrow pipelines and power lines can be efficiently surveyed using helicopter-mounted lidar systems. Sensitive enough to detect and measure the height of power lines above the ground and simultaneously detect vegetation canopies, automated analyses of the lidar data can identify trees and other potential hazards too close to the power line that will require their removal *(figure 8.74)*. Measurements of power line sag and tower locations are used in optimization programs to minimize power loss.

Airborne lidar corridor surveys are often flown with simultaneous recording of the data from digital cameras, video cameras, or other sensors for line inspection and to document terrain conditions.

Natural resource surveys

Lidar can simultaneously provide elevation information for the forest canopy and the ground below, valuable in characterizing forest stands as well as in planning harvesting activities. Stand statistics, including percentage canopy closure, vegetation height, and percentage vegetation cover at different heights above ground, can be calculated from lidar data. When used with field data, stand basal area and volume estimates can also be developed *(figure 8.75)*.

Airborne lidar surveys have proven valuable in supporting the mapping of environmentally sensitive areas, such as wetlands and protected wildlife areas. Commonly having limited or no road access and dense vegetation, these areas can be difficult and more costly to map by ground survey or photogrammetry.

Coastal mapping

Some coastal zones are highly dynamic. Dunes, shorelines, and shallow water shoals can be significantly reshaped by storm events, creating shallow water obstructions that are navigational hazards. The limited contrast and relatively featureless terrain of broad, level, coastal areas can make traditional photogrammetric mapping difficult and expensive. Lidar surveys can provide a cost-effective means for mapping and monitoring coastal areas. The use of hydrographic lidars such as the SHOALS system can simultaneously record topographic elevation and near-shore bathymetry to locate ship wrecks and other obstructions at depths of up to 50 m. Hydrographic lidar data has also proven valuable in optimizing the placement of undersea cables.

Emergency response

Lidar data is used for hazard management planning and disaster recovery. Disaster events can so radically alter an area that it is difficult for rescue and recovery personnel to orient themselves. This was the case during the September 11, 2001, disaster in which the twin towers of the World Trade Center in New York were destroyed. Rescuers depended on rapid production of detailed image maps to plan operations and direct personnel. A wide range of image and map products was generated, tailored to the specific needs of each operations group *(figure 8.76)*. Among the applications of lidar data was the estimation of debris volume to plan the logistics for transporting and placement of the material.

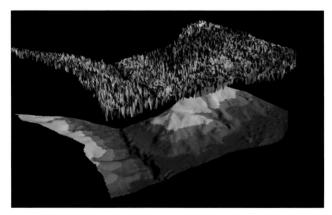

Figure 8.75 Forest biomass estimation. The forest biomass volume can be calculated from multiple return lidar data using first returns to identify the canopy elevation and last returns to define the bald earth surface.

Source: Image courtesy of Earthdata Holdings, Inc.

Figure 8.74 Profile view of a section of powerline corridor. A specialized classification algorithm isolates the area of the powerline within the 3D digital model of the landscape derived from lidar data. The powerlines and parts of the towers are visible, as well as the encroaching vegetation growing immediately beneath the lines.

Source: Image courtesy of Optech Incorporated.

Figure 8.76 Emergency response. Coordination of emergency response and recovery operations at the World Trade Center in New York required rapid production and updating of accurate highly detailed topographic image maps. Airborne lidar, panchromatic digital camera, and thermal infrared imagery were combined in this perspective view to show high temperature areas within the context of the site.

Source: Image courtesy of Earthdata Holdings, Inc.

Comparison of remote sensing-based elevation measurement

The different remote sensing technologies used to generate digital elevation data are photogrammetric analysis of stereo aerial photographs or digital images, stereo radar, radar interferometry (IfSAR), and lidar. The technology that will be most cost-effective depends on the accuracy requirements, time-frame, nature of the terrain, and weather conditions as well as other factors specific to an application.

The most accurate elevation measurements are generally produced by photogrammetric analysis of aerial photographs or visible and near-infrared airborne digital imagery. Photogrammetric analysis tends to be more labor intensive, slower, and more costly than IfSAR or lidar. The level of detail needed for planimetric mapping generally requires photogrammetric analysis of aerial photography or comparable digital imagery. The photogrammetric process can also generate a digital elevation model of high accuracy (0.1 m) but requires additional time and cost. Ortho rectification of imagery does not require an elevation model of such high accuracy, and if the elevation data is required only to generate image mosaics, it may be less costly and faster to use existing elevation data or acquire lidar or IfSAR data for this purpose.

IfSAR offers all-weather capability, more rapid collection, and lower cost per unit area. Acquisition rates for airborne systems with a swath width of about 8 km are typically about 1,000 km^2/hr and as high as 4,000 km^2/hr at a cost on the order of $60/km^2. Existing data, where available, can be pur-

chased for a fraction of the price. Spatial resolutions as high as 1.25 m are commercially available for X-band radar with elevation accuracies as high as about ±1 m RMSE reliably achieved over nonvegetated terrain. X-band radar penetrates vegetation to a variable degree. As a result, there is a significant accuracy loss in elevation measurements over vegetated areas. Depending on the vegetation structure and terrain conditions, elevation accuracy may be reduced to ±10 m RMSE or more.

P-band radar is capable of penetrating the forest canopy to detect features hidden from view. However, methods to reliably measure the ground surface level beneath a closed forest canopy with P-band radar interferometry are still under development. IfSAR from space can provide elevation data of any location on the globe with accuracies as high as ±10 m–20 m RMSE.

Lidar surveys require clear flying weather and are flown at lower altitudes than IfSAR and with narrower swath widths of about 0.2 km to 1 km. Lidar can provide higher elevation accuracies of about ±10–15 cm RMSE and in some cases ±5 cm RMSE with customized systems flown at low altitude. While not able to penetrate vegetation, if the canopy is somewhat open, analysis of ground and canopy reflections enable bald earth, canopy top, and sometimes intermediate strata in the vegetation to be accurately measured. Lidar is capable of measuring fine details such as power line heights and water depth that radar cannot provide. Lidar surveys are generally more expensive than IfSAR at about $150/km^2. However, for smaller areas or narrow corridors, the cost differential may be less than the cost per area suggests. Radar systems must operate at high coverage rates whereas a lidar system may be able to fly a circuitous corridor or a small area in a single pass at a comparable cost to radar acquisition.

The pros and cons of different remote sensing-based technologies for generating elevation data is controversial, in part due to commercial competition. Continuing improvements in the technologies and project specific nature of the results make comparisons difficult and quickly dated. Purchasing archived data, if available, is significantly less expensive than contracting for customized data acquisition. Competitive bids, stringent quality assurance, and knowledge of the capabilities of the different technologies are important in ensuring a cost-effective solution.

Good communication between client and contractor is perhaps as important in reducing cost as understanding the precise accuracy requirements that a project justifies. Consider a project to map a roadway with 10 cm elevation accuracy through shrub-covered terrain. Lidar can readily achieve 10 cm accuracy for a hard level surface, but for a vegetated area, it is much more difficult and substantially more costly. By accepting a lower accuracy for the vegetated areas, the project could be completed at substantially less cost while meeting the objectives. But to develop that project specification would require effective communication. The client must know exactly what they need, i.e., 10 cm accuracy only for the road areas, and the contractor must be willing to take the time to understand the client's application in detail. In this respect, it is little different from other remote sensing applications.

References and further reading

Radar

Bawden, G. W., W. Thatcher, R. S. Stein, K. W. Hudnut, and G. Peltzer. 2001. Tectonic contraction across Los Angeles after removal of groundwater pumping effects. *Nature* (412): 812–15.

Coleman, D. 2001. Radar revolution: Revealing the bald earth. *Earth Observation Magazine* (November): 30–33.

Drury, S. A. 1998. *Images of the earth: A guide to remote sensing.* Oxford: Oxford University Press.

Dutra, L., M. T. Elmiro, C. C. Freitas, J. R. Santos, J. C. Mura, and B. S. Soares. 2002. The use of multifrequency interferometric products to improve SAR imagery interpretability and classification by image fusion. www.npdi.dcc.ufmg.br/workshop/wti2002/pdf/dutra.pdf

Henderson, F. M., and A. J. Lewis, eds. 1998. *Principles and applications of imaging radar, Manual of remote sensing*, Vol. 2. 3rd ed. New York: John Wiley and Sons, Inc.

Intermap Technologies Inc. 2003. *Product handbook and quick start guide.* Calgary, Alberta: Intermap Technologies Inc. istore.intermaptechnologies.com/handbook.cfm

Li, X., and B. Baker. 2003. Characteristics of airborne IfSAR elevation data. In *Proceedings of the ASPRS annual conference, Anchorage, Alaska, May 5–9.* Bethesda, Md.: American Society of Photogrammetry and Remote Sensing.

Maune, D. F. 2001. *Digital elevation model technologies and applications: The DEM users manual.* Bethesda, Md.: American Society of Photogrammetry and Remote Sensing.

Mercer, B., J. Allan, N. Glass, S. Reutebuch, W. Carson, and H. Andersen. 2003. Extraction of ground DEMs beneath forest canopy using P-band polarimetric InSAR (PolInSAR). In Proceedings of ISPRS WGI/3 and WGII/2 joint workshop on three-dimensional mapping from InSAR and lidar. June 17–19. Portland, Ore.

Moreira, J., M. Schwäbisch, C. Wimmer, M. Rombach, and J. Mura. 2001. Surface and ground topography determination in tropical rainforest areas using airborne interferometric SAR. In *Photogrammetric Week 2001*, ed. D. Fritsch and R. Spiller, 341. Heidelberg: Wichmann Verlag. www.ifp.uni-stuttgart.de/publications/phowo01/Moreira.pdf

Rosen, P. A., S. Hensley, I .R. Joughin, F. K. Li, S. M. Madsen, E. Rodriguez, and R. M. Goldstein. 2000. Synthetic aperture radar interferometry. *Proceedings of the IEEE* 88(3): 333–82.

Sabins, F. F. 1997. *Remote sensing: Principles and interpretations.* 3rd ed. New York: W. H. Freeman.

Tapley, I. J. 2002. Radar imaging. In *Geophysical and remote sensing methods for regolith exploration.* CRCLEME open file report 144, 22–32. leme.anu.edu.au/Pubs/OFR144/05Radar.pdf

Tsunoda, S. I., F. Pace, J. Stence, and M. Woodring. 1999. Lynx: A high-resolution synthetic aperture radar. SPIE Aerosense Conference. www.sandia.gov/RADAR/files/spie_lynx.pdf

Radar tutorials and educational Web sites

Canada Centre for Remote Sensing Tutorial on Radar and Stereoscopy. *www.ccrs.nrcan.gc.ca/ccrs/learn/tutorials/stereosc/stereo_e.html*

Center for InSAR Monitoring of Deforming Aquifer Systems (CIMDAS). U.S. Geological Survey. *ca.water.usgs.gov/insar*

Jet Propulsion Laboratory information on imaging radar principles and radar missions. *southport.jpl.nasa.gov/index.html.*

USGS IfSAR information: *quake.usgs.gov/research/deformation/modeling/InSAR/whatisInSAR.html* *volcanoes.usgs.gov/About/What/Monitor/Deformation/InSAR.html*

Companies operating commercial SAR systems

EarthData International, Inc. *www.earthdata.com*

Intermap Technologies. *www.intermaptechnologies.com*

Orbisat da Amazonia S.A., Campinas, Brasil. *www.orbisat.com.br*

Military SAR system manufacturers

Northrup Grumman, Inc. *www.capitol.northgrum.com/programs/tesar.html*

Raytheon, Inc. *www.raytheon.com/products/asars_2*

Sandia National Laboratories. *www.sandia.gov/RADAR/sar.html*

Satellite radar altimetry for mapping seafloor topography

Gierloff-Emden, H. G. 1997. Measurements of ocean surface topography using the ERS-1 radar altimeter. *ESA Earth Observation Quarterly* 56-57: 36–39. esapub.esrin.esa.it/eoq/eoq56/eoq56pp36-39.pdf

Haxby, W. F. 1987. Gravity field of the world's oceans 1:40,000,000. A portrayal of gridded geophysical data derived from Seasat radar altimeter measurements of the shape of the ocean surface. Office of Naval Research and NOAA. NGDC, report MGG-3; col. poster, smooth gravity gradients, 2 inset maps.

Sandwell, D. T. 1990. Geophysical applications of satellite altimetry. *Reviews of Geophysics Supplement* 1990, 132–37.

Sandwell, D.T., and W. H. F. Smith. 2000. Bathymetric estimation. In *Satellite altimetry and earth sciences: A handbook of techniques and applications*, ed. L. Fu and A. Cazenave. San Diego: Academic Press.

Smith, W. H. F., and D. T. Sandwell. 1994. Bathymetric prediction from dense satellite altimetry and sparse shipboard bathymetry. *Journal of Geophysical Research-Atmospheres* 99 (B11): 21803–24.

————. 1997. Global seafloor topography from satellite altimetry and ship depth soundings. *Science* 277: 1956–62.

Smith, W. H. F. 1998. Seafloor tectonic fabric from satellite altimetry. *Annual Review of Earth and Planetary Sciences* 26 (May): 697–747.

Satellite radar altimetry Web sites

Exploring the ocean basins with satellite altimeter data. D. T. Sandwell and W. H. F. Smith. NOAA National Geophysical Data Center.
www.ngdc.noaa.gov/mgg/bathymetry/predicted/explore.HTML

Global seafloor topography from satellite radar altimetry. NOAA National Geophysical Data Center.
www.ngdc.noaa.gov/mgg/image/seafloor.html

Laboratory for Satellite Altimetry. NOAA Oceanic Research and Applications Division.
ibis.grdl.noaa.gov/SAT

NOAA National Geophysical Data Center.
www.ngdc.noaa.gov

Ocean gravity anomaly maps. NOAA National Geophysical Data Center.
topex.ucsd.edu/marine_grav/gif_images/grav8.gif

Predicted and measured seafloor topography. W. H. F. Smith, D. T. Sandwell, and S. M. Smith. Geodynamics Laboratory (NOAA), Institute for Geophysics and Planetary Physics (SIO), Geological Data Center (SIO).
www.ngdc.noaa.gov/mgg/bathymetry/predicted/predict_images.html

Satellite Geodesy. Scripps Institutution of Oceanography. University of California San Diego.
topex.ucsd.edu/WWW_html/mar_grav.html

Lidar

Davis, P. A., S. N. Mietz, K. A. Kohl, M. R. Rosiek, F. M. Gonzalez, M. F. Manone, J. E. Hazel, and M. A. Kaplinski. 2002. Evaluation of lidar and photogrammetry for monitoring volume changes in riparian resources within the Grand Canyon, Arizona. Proceedings of the Pecora 15/ Land Satellite Information IV/ISPRS Commission I/FIEOS 2002 Conference.

Irish, J. L., J. K. McClung, and W. J. Lillycrop. Airborne lidar bathymetry. Joint Airborne Lidar Bathymetry Technical Center of Expertise. U.S. Army Corps of Engineers. Mobile, Alabama. shoals.sam.usace.army.mil

Mercer, B. 2001. Comparing lidar and IfSAR: What can you expect? In *Photogrammetric Week 2001*, ed. D. Fritsch and R. Spiller. Heidelberg: Wichmann Verlag.

Mosaic Mapping Inc. 2001. Lidar mapping. Ottawa, Ontario: Mosaic Mapping Inc. www.mosaic-mapping.com/library/LidarWhitePaper.pdf

Schmidt, L., J. R. Jensen, M. E. Hodgson, D. Cowen, and S. R. Schill. 2001. Evaluation of the utility and accuracy of lidar and IfSAR derived digital elevation models for flood plain mapping, final report. Commercial Remote Sensing Program Office. NASA Stennis Space Center.

Lidar Web sites

Lidar news and information.
www.airbornelasermapping.com

Lidar shoreline topographic change mapping.
www.csc.noaa.gov/crs/tcm/index.html

Optech Inc. lidar systems, including SHOALS.
www.optech.on.ca

The SHOALS hydrographic lidar system, managed by the Joint Airborne Lidar Bathymetry Technical Center of Expertise (JALBTCX), U.S. Army Engineers.
shoals.sam.usace.army.mil

The U.S. National Digital Elevation Program (NDEP).
www.ndep.gov

9 Sonar

Stan Aronoff

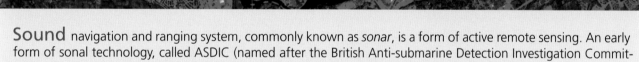

Sound navigation and ranging system, commonly known as *sonar*, is a form of active remote sensing. An early form of sonal technology, called ASDIC (named after the British Anti-submarine Detection Investigation Committee), was developed during World War II to detect enemy vessels such as submarines and surface ships. Sound pulses emitted by the system hitting a metal hull were reflected back, giving a characteristic pinging sound. By measuring the time for a pulse to return and the direction from which it was received, the location of the submarine or ship could be determined.

Since World War II, technology has been developed to image the ocean floor and to measure water depth and subsurface features. GIS-compatible bathymetric[1] data is available for many coastal regions, and sonar imaging is used for geologic mapping of the seafloor, search and recovery of ships and other vessels, and for environmental monitoring.

Sonar, together with satellite radar altimetry (discussed in chapter 8), has enabled scientists to map seafloor bathymetry

1. *Bathymetry* is the measurement of water depth in relation to sea level whereas *hypsometry* is the measurement of land elevation in relation to sea level. *Relief* is a general term often used to include descriptions of both land and water elevations. The term *topography* refers to the configuration of a surface and the relations among its man-made and natural features. It is often used to mean the surface relief of land areas. Though frowned upon by the hydrographic community, the term *seafloor topography* is used by some disciplines when referring to the relief of the seafloor and, by extension, the water depths.

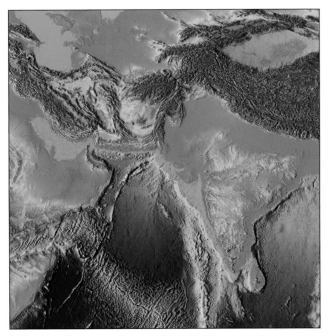

Figure 9.1 Integrated representation of bathymetric and topographic digital data. Datasets such as ETOPO2 developed by NOAA combine bathymetric measurements from satellite radar altimetry and sonar with digital elevation data to provide a digital model of the planet's relief. This shaded relief image centered on the Indian subcontinent, color-coded by elevation, was generated from ETOPO2 two-minute data.

Source: Image courtesy of NOAA, National Geophysical Data Center.

for most of the planet. Combined with global digital terrain data, the earth's relief, both topography and bathymetry, are available as integrated gridded digital datasets at low cost. An example of one of these datasets is the ETOPO2 dataset available from NOAA's National Geophysical Data Center, which provides global coverage with a spatial resolution of 2 minutes of arc (about 3.6 km) *(figure 9.1)*.

Principles of sonar

Because sound waves are mechanical vibrations, they can only travel through an elastic medium such as air, water, or the earth. Unlike electromagnetic energy, sound cannot be transmitted through a vacuum. Sound waves are measured by their frequency in cycles per second or *hertz* (Hz) in the same way as electromagnetic radiation. The human ear can detect frequencies ranging from about 20 Hz to 20,000 Hz.

Sound waves travel much slower than electromagnetic energy. In water, sound waves travel at about 1,500 m/sec versus 3×10^8 m/sec (the speed of light). Since earth and water absorb acoustic energy far less than electromagnetic energy, sonar can be used to collect information from deep below the surface.

Seismic surveys use small explosions at the surface of the land or water to generate a short, intense sound wave that travels through the earth or water. The reflected sound is then recorded by a series of detectors on the surface. Information about subsurface features, such as the depth to different rock formations, can be obtained by means of sophisticated signal processing. The data can then be presented in the form of a cross section shaded to distinguish materials with different acoustic reflection properties. Seismic surveys are used extensively in petroleum exploration *(figure 9.2)*.

A form of sonar used in medical imaging is ultrasound, which uses sound waves at frequencies of 1 to 15 MHz. A transducer (a device that converts electrical energy to sound and sound to electrical energy) is moved across the skin surface, intermittently emitting ultrasound and receiving the signal reflected by body tissues. The pattern of the reflected signal strength can then be displayed as an image. For example, ultrasound is routinely used to examine an unborn child while in its mother's womb *(figure 9.3)*.

The three major types of sonar instruments are side-scan sonar, single-beam sonar, and multibeam sonar. Side-scan sonar systems typically use frequencies in the 100 kHz to 400 kHz range. Specialty side-scan systems such as SeaMarc® I

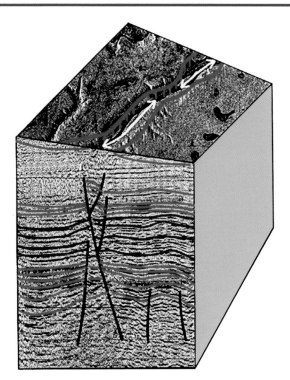

Figure 9.2 Seismic surveys record the time and strength of sound waves reflected from different rock strata beneath the earth's surface to a set of receivers placed in the ground. The illustration shows a typical presentation of this data as a profile. The alternating light and dark gray bands represent reflecting features, such as boundaries between different types of rock. The annotations highlight geologic structures suggested by this image.

Source: Courtesy of Image Interpretation Technologies, Calgary, Alberta. © IITECH.

and II go as low as 30 and 10 kHz, respectively. Single-beam and multibeam sonar systems operate at lower frequencies—from about 12 kHz to 200 kHz. The lower frequencies achieve greater depth but at the expense of spatial resolution. Depending on the frequency and power, multibeam sonar can reach depths of more than 6,000 m. More recently, a fourth type of sonar has been introduced. Portable high-resolution sonar imaging systems using acoustic lenses have been developed to provide real-time video imaging with a range of 1 m to more than 30 m.

Side-scan sonar

Side-scan sonar systems generate images of the ocean floor that can be used to detect objects and map ocean-bottom characteristics. The sonar imagery shows the shape, size, and

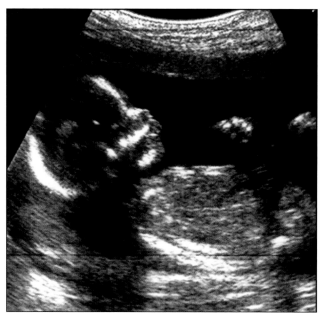

Figure 9.3 Ultrasound imaging is a standard diagnostic tool for examining the unborn fetus.

texture of features. The length, width, and height of target features can also be determined; some side-scan sonar systems also provide depth information.

Side-scan sonar systems are carried aboard survey ships. The sound generation and receiving equipment is usually mounted on a sled or contained within a torpedo-shaped device or *towfish* that is towed behind the survey ship and close to the ocean floor *(figure 9.4)*. The towfish has two banks of sonar transducers[2] that point in opposite directions. Sound pulses sweep the ocean floor perpendicular to the direction of travel.

Side-scan sonar towfish are generally positioned close to the sea bottom, about 10% of the maximum acoustic range away. The sonar beam is directed at a low grazing angle to emphasize surface relief the way a low sun angle emphasizes topography on land.

As the towfish moves forward, successive swaths are recorded, building up a continuous image with spatial resolutions as high as 3 cm. The sonar system generates image swaths on each side of the towfish from the received echoes. These swaths range from 100 m to 500 m wide. To generate high-quality mosaics, side-scan surveys commonly acquire

---

2. A sonar transducer is a device that converts sound into an electronic signal.

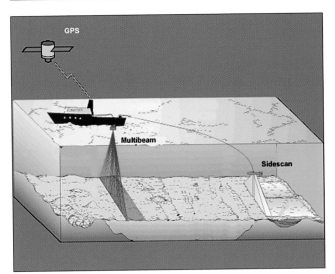

Figure 9.4 Operation of side-scan and multibeam sonar systems. The sensing device in a side-scan sonar system is normally towed behind the survey vessel close to the seafloor. Multibeam systems are usually hull-mounted. Depending on the survey requirements, both systems may be operated simultaneously. Geographic positioning is provided by GPS or differential GPS.

Source: Image courtesy of D. K. Able of the Institute of Marine and Coastal Sciences, Rutgers University. Image modification courtesy of the U.S. Geological Survey and S. Aronoff.

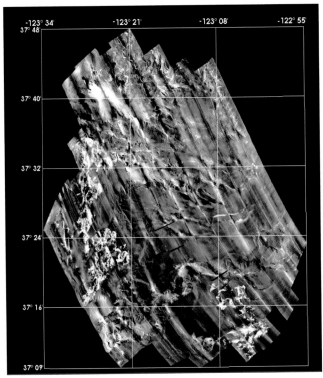

Figure 9.5 Side-scan sonar bathymetric mosaic of the Gulf of the Farallones.

Source: Image courtesy of the U.S. Geological Survey.

200% coverage (i.e., the overlap between swaths is great enough that double coverage of the entire area is acquired). The extra coverage is necessary since side-scan sonar cannot image the area directly beneath the sensor. Also, the quality of imagery acquired at the near and far portions of the sonar systems range is degraded. Individual image swaths can be mosaicked to produce a composite image. Rectified to a chosen map projection, the mosaic can be presented as an image map as shown in figure 9.5.

This side-scan sonar image mosaic, produced from over a dozen individual image strips, shows a 50 km by 75 km area of the ocean floor in the Gulf of the Farallones region west of San Francisco. The image map was produced for a study of seafloor geology to assess sites for deep-ocean disposal of dredge material.

In sonar imagery, features producing stronger echoes appear brighter. The strength of sonar returns is influenced by several characteristics, including slope and surface roughness. Slopes facing the imaging system will produce strong returns and appear bright in the image while those facing away produce weak returns and will be dark. Relatively smooth surfaces, such as clay and silt, will have a lower backscatter return and appear darker while rough surfaces, such

as gravel, will have a higher backscatter return and appear brighter. Objects that protrude above the seabed, such as shipwrecks, generate strong echoes (appear bright) and also stop the signal from spreading any further, which creates an acoustic shadow (a zone of no return) extending away from the object. This shadow appears black in the image. Figure 9.6 is a side-scan sonar image of the SS *Portland*, a paddle-wheel steamer about 100 m long that sank in a vicious storm on November 27, 1898. The long, black feature in the center of the image is the shadow of the vessel's superstructure.

Major applications of side-scan sonar include the following:

- **Engineering:** Side-scan sonar imagery is used in planning undersea pipelines and telecommunications cables, surveying pipeline maintenance, finding stable seabed areas for placement of offshore drilling rigs, and monitoring impacts of construction on the seabed environment.
- **Dredging:** Analysis of side-scan sonar imagery helps maintain a clear underwater channel for marine vessels

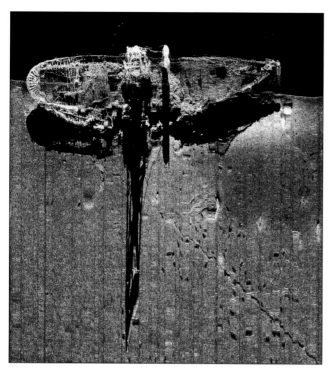

Figure 9.6   Side-scan sonar image of the SS *Portland* shipwreck. Located in the Stellwagen Bank National Marine Sanctuary, the SS *Portland* was a 291-foot paddle wheel steamer with a beam of 42 feet. The ship sank in a vicious storm on November 27, 1898. The bow is to the right, and the ship's superstructure casts long shadows. The small, bright, elongate objects to the right of the wreck are fish swimming near the wreck. Most are imaged separately and cast shadows on the bottom.

Source: G. Kozak, Klein Associates Inc.

by identifying areas of underwater sediment build-up or rock outcrops for removal by dredging or blasting.

- **Environmental applications:** Using side-scan sonar imagery, effluent plumes and dumpsites can be inspected to monitor pollution and other impacts on marine or freshwater environments.
- **Fisheries:** Side-scan sonar is used by the fishing industry to locate fish schools, thereby increasing catches. Individual fish as small as 10 cm long have been imaged using this technology.
- **Mapping seabed characteristics:** Interpretation of side-scan sonar images can distinguish seabed materials such as sand, fine sediments, and rock formations.
- **Mine detection:** Side-scan sonar is one of the primary tools for detecting explosive mines and finding areas of safe passage through mined areas, termed Q-routes.

- **Target location:** The capability to image large areas of the seabed at resolutions as high as 10 cm has made side-scan sonar ideal for locating objects on the seafloor and assessing their characteristics and conditions. Shipwrecks, aircraft, and other man-made and natural features are studied using this technology.

Acoustic lens sonar

In the mid-1990s a group of high-resolution imaging sonars were developed to provide sharp images at close range in turbid water. Operating at frequencies between 1 MHz and 2 MHz, the technology uses plastic lenses instead of electronic means to focus multiple sonar beams onto a linear array of minute acoustic transducers from which a digital image is generated. The sonars have ranges of 1 meter to 30 meters and produce video images with frame rates of 6 to 20 frames per second. The technology was developed for military applications in poor visibility such as imaging in turbid water or at night.

Principal military applications of acoustic lens sonar are underwater surveillance, detection of limpet mines on ship hulls, and forward-looking video for remotely operated vehicles (ROVs). The technology has also been adopted for a number of nonmilitary applications including underwater salvage and construction and monitoring fish behavior and population density *(figure 9.7)*.

Because the beam-forming acoustic lens component consumes no power, the sonar units can operate on about 30

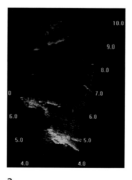
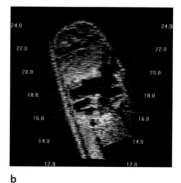

a   b

Figure 9.7   Acoustic lens sonar imagery. These images of (a) divers and (b) a shipwreck were captured with acoustic lens sonar systems. The annotations are distances in yards.

Source: Images courtesy of Applied Physics Laboratory, University of Washington.

Figure 9.8 Diver-held acoustic lens sonar. The portable sonar system enables a diver to view video sonar images at frame rates of six to nine frames per second at ranges of 1 m to 60 m.

Source: Images courtesy of Applied Physics Laboratory, University of Washington.

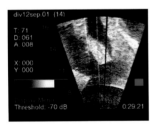 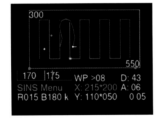

a   b

Figure 9.9 These diver mask displays show (a) a sonar image of an upside down rowboat about 14 m from the diver and (b) a search grid with the diver's current location and search track to be followed. When the diver marks a target with the sonar, the location is marked with a "+" on the display.

Source: Images courtesy of Applied Physics Laboratory, University of Washington.

watt. These systems are designed to be mounted on riverbanks, under hulls of ships, on ROVs, or carried by divers. The portable diver-operated units *(figure 9.8)* include a tracking module and a display in the diver's mask. The

diver can view the sonar image as well as data on his position *(figure 9.9)*. A search grid can also be shown with the diver's location and a trace of his movements. The diver can then identify and mark targets during a dive, saving the geographic position coordinates of identified features for later retrieval.

Single-beam sonar

Used primarily for mapping channels and bathymetry for hydrographic and engineering applications, single-beam sonar was the first sonar technology developed for ocean bathymetry and is the source of most publicly archived ship-track bathymetry data. A single-beam sonar fires acoustic pulses from a single source and measures the time it takes for the single hull-mounted transducer to receive a reflection signal from the ocean bottom. Depth is calculated by taking the time to receive the reflection and correcting it for the water characteristics at the time of the survey. Environmental effects and instrument noise affect the accuracy of sonar bathymetric data. The water environment is by nature dynamic and variable. Sound velocity will be affected by temperature, salinity, and pressure in many locations, varying with the season and time of the day. Data processing procedures are designed to identify and correct errors, optimize depth measurements, and validate the integrity of the final data product.

Single-beam sonar produces sparse bathymetry coverage, so more passes are required to map an area than for multibeam systems *(figure 9.10 and see below)*. Also, the spatial resolution is generally not sufficient to distinguish textural differences on the seabed. However, the data is easily interpreted and requires less storage capacity than multibeam data.

Multibeam sonar

Developments in sonar technology and efforts to increase and refine bathymetric coverage led to the development of the multibeam sonar. This sensor uses an array of sound sources and listening devices, usually mounted on the hull of the survey ship.

A burst of acoustic energy, termed a *ping*, is emitted every few seconds in deep water and as frequently as 10 pings per second in shallow water. (The acoustic energy pulse travels at the speed of sound, about 1500 m/sec.) The energy is

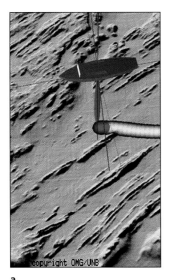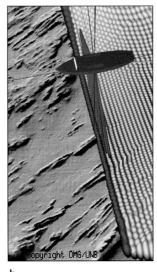

a   b

Figure 9.10   Single-beam (a) and multibeam (b) sonar echo sounding of the seafloor. Single-beam systems measure depth along a single line of the seafloor whereas multibeam systems can measure an entire area.

Source: Images courtesy of J. H. Clarke. © OMG, UNB.

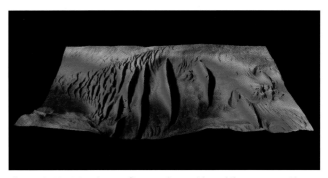

Figure 9.11   Mapping seafloor geology with multibeam sonar. The image shows sand and gravel waves about 25 m high and 200 m to 500 m between peaks. These waves occur at a depth of 90 m off the coast of Victoria, British Columbia. Created by large tidal currents passing through a channel near a group of islands, the waves are present year round and do not migrate significantly. Color coding indicates depth.

Source: Image courtesy of the Canadian Hydrographic Service.

focused to reach a narrow strip of seafloor perpendicular to the ship's direction of travel *(figures 9.4 and 9.10b)*. The array of detectors simultaneously records the sound reflected from the ocean bottom. From the multiple echoes, a series of depth measurements can be generated at regularly spaced intervals along a profile perpendicular to the ship track. As the ship moves forward, bathymetric measurements for a swath of seafloor are successively collected.

Swath width depends on the distance from the sonar transducer to the ocean floor. Because multibeam sonar systems are usually attached to the survey vessel (rather than being attached to a sled towed at depth), the swath width is dependent on the depth of the water and is usually two to four times the water depth. Multibeam sonar has dramatically improved the detail of bathymetric charts *(figure 9.11)*.

Older multibeam systems only recorded the travel time of the strongest reflections. New multibeam systems also record the backscatter intensity, thereby producing both bathymetry from the calculated water depth and imagery (often termed pseudo-side-scan imagery) from the measured signal intensity.

Multibeam systems use differential GPS and precise measurements of the ship's attitude to achieve horizontal accuracies of 5 m to less than a meter, depending on the differential GPS technology used. Accuracy of depth measurements

range from about 0.1% to 0.7% of the water depth with accuracy declining the greater the distance from the ship's track (i.e., greater scan beam angles).

Because of the large lateral extent of the multibeam swaths, multibeam sonar surveys can provide 100% coverage of the seafloor at a nominal resolution. Hydrographic surveys using multibeam sonars exploit the lateral extent to overlap neighboring survey lines by 20% to 100% to ensure complete coverage, provide redundant measurements to verify accuracy, and eliminate imaging problems at high beam angles.

The denser array of measurements reveals more subtle textural characteristics than single-beam sonar. Image enhancement and visualization techniques can be applied to generate perspective views and shaded relief models. Features such as sand ripples, iceberg scours, erosional features, rock outcrops, isolated boulders, and shipwrecks can be distinguished.

Figure 9.12 is a perspective view generated from multibeam sonar data of the Heceta Bank, a large, rocky shoal of the outer continental shelf off the central Oregon coast. The image shows an area of seafloor about 48 km north to south and about 16 km east to west with colors representing elevation. In the top image, the southern part of the bank has a clearly defined western edge, marking an ancient shoreline. A distinctive sediment slide scar can be seen on the southern slope of the bank. The bottom image is a larger-scale perspective view of the middle northern portion of the shoal, displayed as if side-lit to emphasize the geology. The fracture patterns are caused by internal stresses in the earth. The curvilinear

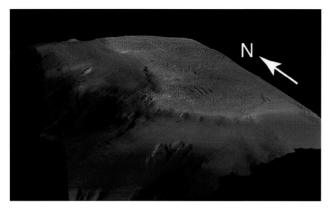

a

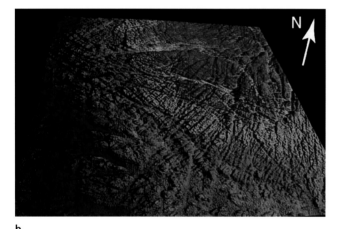

b

Figure 9.12 Perspective view of Heceta Bank from multibeam sonar data. Overview (a) and detailed view (b).

Source: Image courtesy of NOAA. Oceanexplorer.noaa.gov/explorations/lewis_clark01/background/seafloormapping/media/heceta_orientations.html

Figure 9.13 Perspective view of the Los Angeles basin. Digital elevation data for the land surface has been merged with bathymetry data from multibeam sonar to generate this view of the seafloor and surface topography.

Source: Image courtesy of USGS, Western Coastal and Marine Geology Team.

troughs in the northeast section of the view are caused by variable rates of erosion of outcropping rocks.

Figure 9.13 is a shaded relief perspective view of the Los Angeles basin. Bathymetry data collected by multibeam sonar was combined with digital elevation data for the surface topography to create this continuous surface model extending from the ocean floor to the hilltops surrounding the bay.

The large data volumes generated by multibeam sonar have prompted the development of new software and processing methods that offer powerful semi-automated filters, batch processing, and visualization techniques. New sonar technologies are being developed to increase the spatial resolution of the data. For example, autonomous underwater vehicles and synthetic aperture sonar systems (similar in principle to synthetic aperture radar) are being developed for applications such as mine detection. Improvements continue to be made in georeferencing sonar data, allowing more accurate integration with other spatial datasets such as nautical charts, satellite and airborne digital imagery, and aerial photography.

Acoustic seabed classification

Recent technological developments now permit single-beam and multibeam sonars to measure acoustic diversity within the seafloor. This diversity is often directly related to geology, which, in turn, provides habitat to marine life such as bottom fish, crustaceans, and shellfish. Sonar echoes from the seafloor are modified by seafloor composition in a unique manner. Sophisticated signal processing algorithms can analyze this uniqueness, which provides an estimate of the bottom type and, more importantly, can map the breadth and range of each habitat. GIS systems map these seabed types combining layers of geology, marine habitat, and even sites impacted by human industrial intervention *(figures 9.14 and 9.15)*.

Global seafloor topography from satellite radar altimetry

Satellite-based radar altimeters have provided accurate measurements of sea-surface height along tracks only 3 km to 4 km apart. From this data, accurate maps of the marine gravity field can be derived with spatial resolutions as high as 15 km. (The gravitational pull of the earth varies over the

Figure 9.14 Multibeam sonar image of the Race Rocks Marine Protected Area near Victoria, British Columbia. This oblique angle view looking southwest shows submerged rocks and sediments. The image was created with a GIS from multibeam sonar data.

Source: Image courtesy of the Canadian Hydrographic Service.

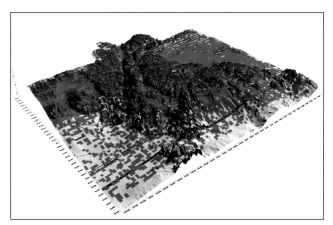

Figure 9.15 Perspective view of the Race Rocks Marine Protected Area near Victoria, British Columbia. A seabed classification of multibeam sonar data is displayed draped over bathymetry. The colors represent different sediments and rock structures. The classification was produced using specialized software for processing multibeam sonar backscatter data. The image was then generated on a GIS.

Source: Image courtesy of the Canadian Hydrographic Service.

earth's surface, tending to lower the sea surface where the gravitational pull is greater.) Where sediment cover on the ocean floor is thin, the gravity field mimics large seafloor structures and can be used to predict ocean depth and map the topography of the seafloor. However, the inconsistent correlation of gravity and bathymetry made the accuracy of depth predictions highly variable, from accuracies of plus or minus 100 m to no correlation.

Walter Smith and David T. Sandwell (1997) developed sophisticated algorithms that take into account physical ocean dynamics such as tides and currents and also integrate shipboard sonar bathymetry data to produce a global gridded dataset of predicted seafloor topography of the world's ocean areas not permanently covered by ice. This data is valuable to geologists and useful in applications ranging from petroleum exploration to the routing of undersea cables and pipelines; however, it does not have the spatial resolution or accuracy for navigational use. The data is a valuable means of extending the limited sonar bathymetry coverage in deep ocean areas and regions beyond intensively charted global shipping routes to visualize seafloor structures. Satellite radar altimetry is discussed in greater detail in chapter 8.

Sonar data acquisition programs

Sonar data acquisition programs in the United States and Canada have produced detailed coverage of coastal areas and coarser resolution global coverage of the world's oceans. Beginning with the United States in 1981, the world's maritime nations declared the waters and seafloor within 200 nautical miles of their shores to be *exclusive economic zones*. In 1986, the United Nations Convention on the Law of the Sea (UNCLOS) produced a definitive document that aimed to define all maritime boundaries. Article 76 of UNCLOS provides the rules allowing a maritime nation to extend its jurisdiction beyond the 200 nautical miles for the purpose of exploiting the resources on and below the seabed of the land's natural extension. To begin assessing these underwater regions, maritime nations have implemented a variety of offshore mapping programs to study these underwater territories. In order to ratify UNCLOS, a nation must map its 2,500 m isobath[3], locate the foot of the slope of its continental margin, and provide evidence of the legitimacy of its claim. Multibeam sonar surveys have provided some of the necessary bathymetric coverage.

In the United States, the National Oceanic and Atmospheric Administration and the U.S. Geological Survey have mapped hundreds of thousands of square kilometers of the seafloor off the coast of the Atlantic and Pacific oceans and the Gulf of Mexico. Data from the Geological Long-Range

3. An isobath is a contour line connecting all points of equal depth below the water surface.

Inclined Asdic (GLORIA) mapping program of the USGS and subsequent surveys of the U.S. continental shelf can be obtained from the USGS directly or through the EROS Data Center. The ETOPO5 dataset from the NOAA National Geophysical Data Center was generated from a database of land and seafloor elevations on a 5-minute latitude/longitude grid. Other global datasets, such as ETOPO2 (2-minute arc latitude/longitude grid), are available. These were assembled from several data sources, including bathymetry derived from satellite altimeter data and from ship surveys from various research institutes (using mostly single-beam sonar systems) provided by a number of countries. The data offers almost complete global coverage at coarse spatial resolution.

The coastal relief model, also being developed by the National Geophysical Data Center, is a gridded elevation database of the coastal regions of the coterminous United States that will eventually include Alaska, Hawaii, and Puerto Rico. It extends from the land area of the coastal states across the shorelines and out toward the edge of the continental shelf, which provides a continuous representation of the relief of the land, shore, and seafloor with a spatial resolution (postings) of 3 arc seconds (about 90 m). The dataset merges topographic data from the USGS and the National Geospatial-Intelligence Agency (NGA, formerly the National Image Mapping Agency, NIMA) with hydrographic data, mostly from sonar, collected primarily by the National Ocean Service (NOS) and from various academic institutions (*figure 9.16*).

The International Bathymetric Chart of the Arctic Ocean (IBCAO) is an initiative begun in 1997 to develop a digital database of all available bathymetric data for the region north of 64° north. Designed for researchers and practitioners requiring detailed and accurate information on the depth and the shape of the Arctic seabed, the project has drawn support from eleven government and university institutions in eight countries. The first version of the IBCAO gridded dataset is freely available for downloading over the Internet along with bathymetric contours in digital vector form and an updated version of the map suitable for plotting.

The General Bathymetric Chart of the Oceans (GEBCO) is an international project jointly overseen by the Intergovernmental Oceanographic Commission (IOC) and the International Hydrographic Organization (IHO) of UNESCO. Its objective is to provide authoritative, publicly available bathymetry for the world's oceans. The fifth edition world-coverage, hard-copy map set at a scale of 1:10 million was published in 1982. In 1994, a digital version, the GEBCO Digital Atlas

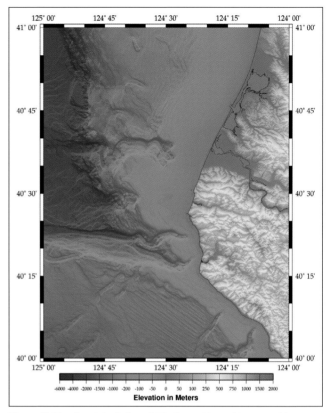

Figure 9.16 Coastal relief model bathymetric and topographic chart from multibeam sonar data. The Coastal Relief Model dataset provides digital bathymetric data from multibeam sonar surveys and U.S. Geological Survey digital elevation models over land areas with a spatial resolution of three arc seconds (about 90 m).

Source: Image courtesy of NOAA, National Geophysical Data Center.

(GDA) on CD-ROM was produced. The digital atlas was generated by digitizing the bathymetric contours, coastline, and ship tracks from the printed sheets of the fifth edition. A second release was published in 1997, and in April 2003, the GEBCO Digital Atlas–Centenary Edition (GDA-CE) was released. This dataset includes a one arc-minute global bathymetric grid, digitized bathymetry contours, the gazetteer of undersea feature names, coastline databases, and a software interface. The one arc-minute grid corresponds to a 1.85 km square area at the equator that becomes progressively narrower with latitude. At 72° latitude the cell dimensions are 0.57 km in the east–west dimension and 1.85 km in the north–south dimension. The standard GEBCO bathymetric contour interval is 500 m with intermediate contours at the 200 m or 100 m level for selected areas. The GEBCO dataset is traditionally used for the study of deep-water areas. Compiled from the best available data sources, the dataset has long been a standard reference for oceanographers and marine scientists.

Conclusion

Sonar technologies are the major remote sensing data sources for imaging and bathymetric measurement of water depths too deep or with turbidity levels too high for sensors dependent on bottom reflection of visible light, such as lidar. Satellite radar altimetry has been a valuable complement to sonar by providing seafloor structure and coarse level bathymetric data where suitable calibration with other data sources is available.

Sonar imaging and classification technologies are being used to map the geology, material composition, and habitat of coastal and deep ocean regions. The increased availability of sonar data and awareness of its capabilities has also led to the development of integrated gridded elevation datasets of topography and bathymetry designed for GIS/LIS applications. These digital coverages provide a unifying framework upon which other geographic information can be overlain and analyzed.

Coastal areas near population centers often experience rapid growth. Demand for attractive living and recreational sites and resource development within coastal regions has led to rapid population growth, conflicting demands, and environmental pressures. Overdevelopment of beaches increases the economic costs of storm damage and flooding, in addition to the environmental costs associated with the loss of wetland areas, contamination of estuaries, dredging, and overfishing. Also, society's demand for energy and other resources has increased natural resource exploration and extraction such as oil and gas exploration and production in coastal areas.

In an effort to better manage coastal resources, planners are using geographic and land information systems (GIS/LIS) to plan future land development, assess coastal and offshore resources, mitigate pollution, prepare for natural or human-caused emergencies, and monitor environmental change within the coastal zone over time.

Sonar references and further reading

Blondel, P., and B. J. Murton. 1997. *Handbook of seafloor sonar imagery*. New York: John Wiley and Sons, Inc.

Fish, J. P. 2002. *Acoustics and sonar primer*. Cataumet, Mass.: American Underwater Search and Survey, Ltd.

Fish, J. P., and H. A. Carr. 1990. *Sound underwater images: A guide to the generation and interpretation of side scan sonar data*. Cataumet, Mass.: American Underwater Search and Survey, Ltd. (Portions of the text are available online at the Institute for Marine Acoustics Web site. www.marine-group.com/SonarPrimer/ SideScanSonar.htm)

Gardner, J. V., P. B. Butman, L. A. Mayer, and J. H. Clark. 1998. Mapping U.S. continental shelves: Enabled by high-resolution multibeam systems, advances in data processing, USGS begins systematic mapping program. *Sea Technology* (June): 10–17.

Hughes Clarke, J. E., J. V. Gardner, M. Torresan, and L. Mayer. 1998. The limits of spatial resolution achievable using a 30 kHz multibeam sonar: Model predictions and field results. *IEEE Oceans 98, Proceedings* 3:1823–27. www.omg.unb.ca/omg/papers/OC98-042.pdf

Tretheway, M., M. Field, and D. Cooper. Making the most of investment in multibeam sonar. Internal technical paper. VT TSS, Ltd. (U.K.). Watford, Hertfordshire, United Kingdom. www.tss-realworld.com/pdf/mtmoiims.pdf

U.S. Geological Survey. 2002. Gulf of the Farallones regional bathymetry and sidescan-sonar digital mosaics. terraweb.wr.usgs.gov/projects/Farallones

U.S. Geological Survey. 1997. Sidescan sonar data collected during U.S. Geological Survey cruises. Open-file report 97-485. USGS cruise 97009. edc.usgs.gov/glis/hyper/guide/sidescan_pending

Sonar image, bathymetry data source, and research organization Web sites

Canadian Hydrographic Service. *www.chs-shc.dfo-mpo.gc.ca/pub*

GLORIA mapping program. *kai.er.usgs.gov/gloria/index.html*

General Bathymetric Charts of the Oceans (GEBCO). *www.ngdc.noaa.gov/mgg/gebco/gebco.html*

Institute for Marine Acoustics. *www.instituteformarineacoustics.org/home.html*

International Bathymetric Chart of the Arctic Ocean (IBCAO). *www.ngdc.noaa.gov/mgg/bathymetry/arctic/arctic.html*

International Bathymetric Chart of the Caribbean Sea and the Gulf of Mexico. *www.ngdc.noaa.gov/mgg/ibcca/ibcca.html*

Lamont-Doherty Earth Observatory. *imager.ldeo.columbia.edu*

National Geophysical Data Center Hydrographic Survey Database System. *www.ngdc.noaa.gov/mgg/bathymetry/hydro.html*

National Geospatial-Intelligence Agency (NGA), formerly the National Image Mapping Agency (NIMA). *www.nima.mil*

National Oceanic and Atmospheric Administration (NOAA). *www.noaa.gov*

NOAA National Geophysical Data Center. *www.ngdc.noaa.gov/mgg/image/images.html*

NOAA National Geophysical Data Center: Coastal Relief Model. *www.ngdc.noaa.gov/mgg/coastal/coastal.html*

NOAA Global ETOPO2 dataset. *www.ngdc.noaa.gov/mgg/fliers/01mgg04.html*

NOAA National Ocean Service. *www.nos.noaa.gov*

NOAA Pacific Marine Environmental Laboratory. *www.pmel.noaa.gov/vents*

NOAA Ocean Explorer Gallery. *oceanexplorer.noaa.gov/gallery/gallery.html*

NOAA Seafloor Mapping. *oceanexplorer.noaa.gov/explorations/lewis_clark01/ background/ seafloormapping/seafloormapping.html*

Scripps Institution of Oceanography. *gdcmp1.ucsd.edu/gdc*

University of Hawaii School of Ocean and Earth Science Technology. *www.soest.hawaii.edu/HMRG*

University of New Brunswick, Canada, Ocean Mapping Group. *flagg.omg.unb.ca/omg*

University of New Hampshire Center for Coastal and Ocean Mapping Joint Hydrographic Center. *www.ccom-jhc.unh.edu*

University of Rhode Island Graduate School of Oceanography. *www.gso.uri.edu*

USGS Coastal and Marine Geology Program. *marine.usgs.gov*

Woods Hole Oceanographic Institution Multibeam Bathymetry Data Archive. *mbdata.whoi.edu/ mbdata.html*

Acoustic lens sonar

Belcher, E. O., and D. C. Lynn. 2000. Acoustic, near video quality images for work in turbid water. In Proceedings of the Underwater Intervention 2000 Conference. January. Houston, Texas.

Sound Metrics Inc. Commercial supplier of acoustic lens sonar, DIDSON (Dual Frequency Identification Sonar). *www.soundmetrics.com*

University of Washington Applied Physics Laboratory Web site for acoustic lens sonar technology. *www.apl.washington.edu/programs/DIDSON/DIDSON.html*

Seabed classification

Anderson, J. T., R. S. Gregory, and W. T. Collins. 1998. *Digital acoustic seabed classification of marine habitats in coastal waters of Newfoundland.* Sidney, B.C.: The International Council for the Exploration of the Sea. Canadian Department of Fisheries and Oceans—Science Branch and Quester Tangent Corporation.

Collins, W. T. and J. L Galloway. 1998. Seabed classification and multibeam bathymetry: Tools for multidisciplinary mapping. *Sea Technology* 39 (September): 45–49.

Galloway, J. L. 2000. Integration of acoustic seabed classification information and multibeam bathymetry. In Proceedings of the European Conference on Underwater Acoustics. July. ECUA2000. Lyon, France.

Galloway, J. L., and W. T. Collins. 1998. Dual frequency acoustic classification of seafloor habitat. In IEEE Proceedings Oceans '98 Conference. September. Nice, France.

Quester Tangent Corporation. Sidney, B.C., Canada. Supplier of seabed classification software and hardware. *marine.questertangent.com/m_overview.html*

Web tutorials on sonar

Acoustics and sonar primer. Institute for Marine Acoustics. Excerpts from *Sound underwater images: A guide to the generation and interpretation of side scan sonar data.* Fish and Carr 1990. Web site. *www.marine-group.com/SonarPrimer/SideScanSonar.htm*

Underwater acoustics. NOAA Pacific Marine Laboratory. *www.pmel.noaa.gov/vents/acoustics/tutorial/tutorial.html*

Sonar system manufacturers

EdgeTech Ltd. *www.edgetech.com*

Klein Associates Inc. *www.l-3klein.com*

Kongsberg Maritime. *www.kongsberg-simrad.com*

Reson A/S. *www.reson.com*

Triton Elics International Inc. *www.tritonelics.com*

Satellite radar altimetry

(Also see references in chapter 8)

Smith, W. H. F., and D. T. Sandwell. 1997. Global sea floor topography from satellite altimetry and ship depth soundings. *Science* 277: 1956–62.

Global seafloor topography from satellite altimetry. NOAA. *www.ngdc.noaa.gov/mgg/announcements/announce_predict.html*

10 Visual interpretation of aerial imagery

James B. Campbell

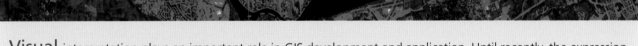

Visual interpretation plays an important role in GIS development and application. Until recently, the expression *visual interpretation* conveyed a very clear meaning—the interpretation of aerial photography, as transparencies or paper prints, by a photointerpreter. For many, visual interpretation was regarded as a somewhat useful but old-fashioned technique likely to be replaced by digital analyses that could examine images and extract required information.

Visual interpretation is now seen in a new context. First, the essential skills that the photointerpreter applies to paper photographs apply equally well to digital images viewed on computer displays. Thus, the photointerpreter's realm has extended into the digital format with only minimal differences. However, because a digital image cannot, strictly speaking, be properly designated as a photograph, the title *photointerpreter*, although still in use informally, has been replaced by titles such as *image analyst* and *image interpreter*.

Secondly, the realm of fine-detail imagery is not limited to aerial photography. It now includes an increasing variety of satellite and aircraft data that introduce new possibilities for GIS applications. Whereas in earlier eras satellite images were characterized by coarse detail, inconvenient formats, or uncertain geometric qualities, the GIS manager can now find convenient access to a variety of satellite image data in forms that lend themselves to the GIS environment. Therefore, although civilian applications of photointerpretation were previously linked almost exclusively to conventional aerial photography, the scope of these applications now extends to a much wider range of image products, including satellite images and digital aircraft data.

Finally, the practice of computerized image analysis has not yet matured to the extent that it can provide the information that clients expect to derive from fine-resolution data. Currently, exploitation of fine-resolution image data for GIS development depends largely upon the human interpreter. Although automated analyses have increased in their sophistication, they serve as companions, not replacements, for the human analyst. Therefore, today's image interpreter must understand and exploit the synergy between the digital domain and the realm of the human interpreter. Image interpreters must use the superior functions digital systems offer for accurate measurement (i.e., multispectral analysis, change detection, and automated searches) in addition to human judgment and analytical insight.

This chapter surveys selected aspects of the field of photointerpretation as it applies to the concerns of the GIS manager, who must assess aerial imagery as a possible source of information for geographic information systems. It describes characteristics of aerial imagery relevant for the GIS manager, sources of aerial imagery, the elements of image interpretation and their application, and image interpretation keys. Other topics of significance include uses of field observations, accuracy assessment, and applications of visual interpretation in forestry, wildlife management, and environmental assessment.

Characteristics of aerial imagery

Visual interpretation, as considered here, is used to extract information from fine-resolution imagery at scales of 1:40,000 or larger. Such images can be examined as conventional, or analog, images (i.e., as paper prints or film transparencies), as numeric images acquired by digital sensors, or scanned from analog images. This definition includes almost all aerial photography, as well as fine-scale imagery from satellite systems such as Space Imaging's IKONOS system. It excludes coarse-resolution imagery produced by systems such as the broad-scale sensors of SPOT, Landsat, or IRS. Although this discussion refers specifically to imagery collected in or near the visible spectrum (the spectrum normally used for black-and-white, color, and color infrared photography), it applies equally well to imagery from the nonvisible spectrum.

Fine-resolution aerial imagery is most useful in GIS analysis when the following conditions are met:

- Geometric errors are removed to provide a planimetrically correct presentation (i.e., features are presented in their correct geometric relationships).
- Detailed spatial information is provided (i.e., small parcels and fine detail are delineated, as is required for urban land use, precision agriculture, or transportation planning).
- Detailed taxonomic detail is provided (i.e., fine levels of detail are classified, such as separation of corn from wheat, oak from maple, or residential property from commercial property).
- Spatial relationships between varied realms of information are examined (i.e., relationships between vegetation distributions and water bodies or between agricultural land and land devoted to residential development are examined).
- Patterns of change from one date to the next are monitored (i.e., changes in land cover or water quality are monitored).
- Equipment and experienced staff are at hand to interpret imagery, or off-site contract services are available to perform the analysis.

- Phenomena that require a historical perspective are examined (i.e., land-use change, growth of communities, or losses of agricultural land).

Fine-resolution aerial imagery is not likely to be useful for the GIS manager when the following conditions exist:

- The data cannot, for either technical or administrative reasons, be brought into a planimetrically correct geographic base.
- The imagery is to be used in a project where the data derived from the imagery is to be highly generalized or filtered to match datasets with coarser levels of detail.
- The imagery is to be used in a project where large regions require characterization at only coarse levels of detail (i.e., *forest*, *farmland*, or *water*).
- The sponsoring organization is not prepared to invest in training or services to assure that necessary imagery, skills, equipment, and field support are available to conduct the interpretation.

Sources of imagery

When deciding to include aerial imagery in a project, the GIS manager will likely choose between two sources of imagery: archival imagery previously acquired for another purpose and held in storage or custom imagery acquired specifically for the project at hand.

Archival imagery

Archival imagery is held by organizations with responsibility for acquiring and holding images for their jurisdictions, such as the U.S. Geological Survey, the U.S. National Archives and Records Administration, or any of the many state agencies with responsibility for acquiring aerial images. In the United States, the responsible agency differs from state to state—in some instances, the state transportation department may have responsibility for acquiring imagery, while in others it may be the department of natural resources or another agency.

The user must purchase copies of the archived imagery for use in a GIS project. Although older images may be available only as paper prints, more recent images may be held in digital format. Images are indexed by date, region, format, and other attributes. Some agencies have computerized databases that permit online searches, while others may require users to submit annotated maps to identify the coverage desired. In some states, the user can view digital copies of photographs online. Johnson and Barber (1996) and Larsgaard and Carver (1996) list principal sources for archived aerial photography, mainly for the United States. Other sources of imagery are listed in appendix C of this text.

For the GIS manager, archival imagery can provide low-cost, accessible, useful imagery. Many states update photography on a regular basis, so it is not difficult to acquire archival imagery that is recent enough for immediate purposes. Archival imagery is especially valuable in applications where it is necessary to examine change, as it serves as a reference for earlier conditions.

Among the disadvantages of using archival imagery are identifying which archives hold the imagery required to complete a project; accepting someone else's choice of scale, format, date, and emulsion; and accepting that the archive may not cover all areas of interest on the desired date since image coverage was determined by decisions made years before. In addition, users requesting large orders from archives may experience delays arising from the difficulty in preparing large numbers of duplicates.

Custom acquisition

Custom acquisition means that imagery is acquired according to a user's specifications. The date, scale, emulsion, coverage, and other characteristics are determined by the user's requirements. Disadvantages include high cost, the requirement for planning well in advance, the effort devoted to preparing and monitoring specifications, and the delays and cancellations that can result from unfavorable weather or equipment malfunctions (including both the aircraft and the imaging instruments).

GIS managers who plan to acquire custom photography should prepare a *contract*, or *statement of work*. The contract specifies in detail the products and services required from the firm or organization with respect to costs, deadlines, and products to be prepared.

The contract is the primary means for controlling the completeness and quality of the imagery. The photointerpreter may participate in making decisions regarding scale, date, time of day, deadlines, choice of film, and coverage. Once such details become part of the contract, the contractor is responsible for acquiring the imagery as specified, but the photointerpreter, and others who prepared the contract, are responsible for delivering imagery that fulfills project needs. Therefore, it is essential that the photointerpreter accurately anticipate the needs of the project and assure that these needs are reflected in the specifications. To do this, the photointerpreter must pose the following questions during mission planning:

- Will the imagery be adequate for the purpose of the project, especially with respect to scale and coverage?
- Is the choice of film emulsion satisfactory for the task at hand?
- Is the season suitable?
- Will the planned time of day provide appropriate shadowing?
- If stereoscopic photography is necessary, is the forward overlap adequate?

Elements of image interpretation

To interpret images, analysts use the same visual clues people use in their everyday experience. However, because the context for image interpretation differs so much from human everyday experience, the practice of image analysis requires the explicit recognition of eight *elements of image interpretation* that form the framework for understanding the meaning of an image.

Shape

Shape refers to the outline of a feature *(figure 10.1)*. Some features have distinctive outlines that indicate their identities or functions. It is important to note that shape, as seen on an aerial image, depends upon perspective. The shapes of objects as viewed from the overhead perspective of an aerial photograph differ greatly from shapes as seen from ground

level. Likewise, the overhead perspective also introduces a scale effect; aerial images allow us to see large features at scales that we would not otherwise be able to see because of the sizes of such features relative to our normal perspective.

Size

Size refers to the dimensions of a feature, either in relative or absolute terms. *Relative size* is determined by comparing the object with familiar features that may be nearby. Usually relative size is sufficient to assign an object to a general class of features (e.g., a vehicle). *Absolute size* refers to the use of the aerial image to derive measurements in standard units of length, usually to assign objects to specific classes (e.g., Ford Aerostar), or for calculation of distances, volumes, or areas.

Tone

Tone refers to the average brightness of an area *(figure 10.2)*, or in the case of color or color infrared (CIR) imagery, to the dominant color of a region. Tone as observed in aerial imagery depends upon the nature of the surface and the angles of observation and illumination.

Surfaces that are smooth relative to the wavelength of the radiation that illuminates them behave like *specular* reflectors—they tend to reflect radiation in a single direction, either towards or away from the lens. In these circumstances, tone depends not only upon the way an object reflects radiation but also upon the way the object is illuminated and observed. Images of these objects appear either very bright

Figure 10.1 Elements of image interpretation: shape. Athletic fields adjacent to a school are readily identified by their distinctive and familiar shapes.

Source: Image courtesy of the U.S. Department of Agriculture.

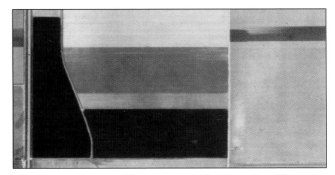

Figure 10.2 Elements of image interpretation: tone. The different crops in this image appear as several distinct gray tones. A cover of alfalfa, which appears dark green to the human eye, is the darkest gray in this image. A field of sorghum or maize, which appears a medium green to the eye, is a slightly lighter shade of gray than the alfalfa. Even lighter than the maize is a field of wheat approaching harvest (light green to the eye). A field of wheat ready for harvest (light tan to the eye) is lightest of all.

Source: Image courtesy of the U.S. Department of Agriculture.

or very dark, depending upon the relationship between the angles of illumination and observation.

Objects that are rough relative to the wavelength of the radiation that illuminates them act as *diffuse* reflectors—they tend to scatter radiation more or less equally in all directions. These objects (such as fields of grain, forests, or lawns) appear in medium gray tones—tones that do not vary much as angles of illumination and observation change.

Texture

Texture refers to the variation in tone over a surface or the apparent roughness of the surface as seen in an aerial image *(figure 10.3)*. Texture is created by microshadows—tiny shadows created by minute irregularities in the surface. Pavement appears as a smooth texture because the irregularities are too small to create noticeable shadows at the scales normally used. A forest, on the other hand, is characterized by a rough

texture created by microshadows of leaves and branches that form a coarse pattern of alternating light and dark patches.

Shadow

Shadow refers to the large, distinctive shadows that reveal the outline of a feature as projected onto a flat surface *(figure 10.4)*. In any instance, the nature of a shadow depends upon the nature of the object, angle of illumination, perspective, and slope of the ground surface. For example, a shadow that falls on a surface that slopes uphill or downhill will appear longer than a comparable shadow that falls on level terrain. Likewise, low sun angles create long shadows and high sun angles create short shadows.

Figure 10.4 Elements of image interpretation: shadow. The significance of shadow is represented well in this image. The shadow of the water tower reveals structure. Shadows of buildings indicate their identity and heights. Shadows in the forests and pastures create the image texture that defines their appearance.

Source: Image courtesy of the Virginia Department of Transportation.

Site

Site refers to a feature's position with respect to topography and drainage. Many features must, by their inherent function, occupy distinctive topographic positions. For example, water towers and telecommunications facilities such as microwave and cell phone antennae are positioned at the highest topographic locations. Sewage treatment facilities are positioned at the lowest feasible topographic position. Coal and nuclear power plants are located in proximity to water bodies in order to acquire sources of water for cooling. The history of photointerpretation is characterized by many instances in which site contributed to successful identification of facilities of unknown function.

Figure 10.3 Elements of image interpretation: texture. Textures in this agricultural scene create distinctive signatures for varied crops. Dark, even textures indicate alfalfa crop ready for harvest. Rougher textures indicate cropping patterns and use of agricultural machinery to cultivate and harvest crops. Mottled patterns indicate variation in drainage and moisture patterns within fields.

Source: Image courtesy of the U.S. Department of Agriculture.

Association

Association refers to distinctive spatial interrelationships between features, usually due to functional connections between the components in question *(figure 10.5)*. For example, often the most distinctive visual clues for identification of primary and secondary schools as recorded on aerial imagery are the athletic fields associated with these facilities. Likewise, the large parking lots associated with shopping malls are often their most distinctive visual indicators. Other examples include the identification of grain elevators positioned near rail sidings and warehouse complexes positioned near transportation facilities.

a

Figure 10.5 Elements of image interpretation: association. A shopping mall can be identified by the association of large parking lots surrounding a cluster of massive buildings.

Source: Image courtesy of the Virginia Department of Transportation.

Pattern

Pattern refers to distinctive arrangements of features *(figure 10.6)*. Obvious examples include orchards, in which trees are aligned in rows, and the systematic positioning of headstones in cemeteries.

Patterns usually have origins in functional constraints that lead to systematic positioning. For example, trees in orchards are positioned in rows to minimize competition between adjacent plants for light and nutrients and to facilitate the movement of agricultural machinery between trees. Patterns of structures in residential neighborhoods derive from constraints to minimize costs of constructing the infrastructure of roads and utilities.

b

Figure 10.6 (a) At the left, the regular pattern of long, rectangular, white shapes is formed by the scattered placement of mobile homes in a sales lot; nearby, to the right, the placement of mobile homes in a trailer park forms a more ordered pattern, and near the periphery of the image, can be seen a few fragments of the more open pattern formed by single-family dwellings, lawns, streets, and driveways. On the right, an abandoned outdoor drive-in movie theater can be identified by the distinctive fan-shaped pattern of stepped level grass-covered benches large enough to accommodate cars. The projection booth is visible near the center of the fan. (b) Another image shows again the distinct pattern of a mobile home park, and, at the right of the image, the contrasting pattern of a single family residential neighborhood.

Source: Images courtesy of the Virginia Department of Transportation.

The eight elements of image interpretation are usually illustrated by examples that attempt to isolate each element in order to clearly emphasize its significance. However, the interpreter must recognize that in the practice of image analysis, the elements work together subtly in ways that are seldom as dramatic as the examples presented in texts.

Image interpretation tasks

Image interpretation tasks apply the elements of image interpretation in a way that helps the interpreter derive useful information from the imagery.

Classification

Classification assigns objects to classes based upon the elements of image interpretation. Often the distinction is made between three levels of confidence and precision. *Detection* is the determination of the presence or absence of a feature. *Recognition* implies that the interpreter can derive a more specific knowledge of the feature in question so that it can be assigned to a particular category. *Identification* means that the interpreter can place the feature in a very narrowly defined class. Interpreters convey their confidence in the interpretation using terms such as *possible* or *probable*.

Enumeration

Enumeration lists or counts discrete items visible on an image. For example, housing units can be classified as detached single family, multifamily complex, mobile home, multistory residential, and then reported as numbers present within a defined area.

Measurement

Measurement, or *mensuration*, is the use of image scale to derive measurements such as lengths, widths, distances, and volumes. Measurements can contribute to the accurate classification of features characterized by distinctive size, or they may form the objective of the interpretation. The practice of making accurate measurements from photographs forms the subject of *photogrammetry*, which applies knowledge of image geometry to the derivation of accurate distances.

Delineation

Delineation is the demarcation of regions as they are recorded on aerial images. The interpreter separates regions characterized by distinctive tones and textures and finds the edges between adjacent regions. Examples include classes of forest and land use, which both occur only as areal entities (rather than as distinct objects). Problems associated with delineation include selecting the appropriate level of generalization when boundaries are intricate or when many small but distinct parcels are present and defining boundaries when a transition (rather than a sharp edge) between two regions exists.

Image interpretation strategies

Knowledge of the basics of image interpretation is not in itself useful without a sense of how to apply them to extract information from an image. It is useful, therefore, to discuss some of the strategies by which image analysts use basic interpretation skills

Field observations

Field observations are required when the image and its relationship to ground conditions are so imperfectly understood that the interpreter is forced to go to the field to make an identification. In effect, the analyst is unable to interpret the image from the knowledge and experience at hand and must gather field observations to ascertain the relationship between the landscape and its appearance on the image. Field observations are, of course, a routine practice in any interpretation as a check on accuracy or as a means of becoming familiar with a specific region.

Direct recognition

Direct recognition means the interpreter can apply his or her experience, skill, and judgment to derive information directly from inspection of the image. In essence, the interpreter conducts a qualitative, subjective analysis of the image using the elements of image interpretation as visual and logical clues. For example, most people can easily recognize dwellings, roads, and forests as represented on aerial photography, even though they may not be able to describe their characteristics in technical terms. Thus, in everyday experience, people can apply direct recognition intuitively. In the context of image interpretation, direct recognition must be a disciplined process based upon systematic examination of the image.

Inference

Inference means the interpreter examines distributions visible in the image to derive information not visible in the image. The visible distribution forms a surrogate, or substitute, for the mapped distribution. Soils, for example, are defined by vertical profiles that cannot be directly observed by aerial imagery. Yet aerial imagery, together with other information, can be applied to the mapping of soils because soil patterns are closely related to distributions of landforms, vegetation, drainage, and other features recorded on images. Therefore, these patterns form surrogates for the soil pattern. The interpreter infers the invisible soil distribution from features that are visible. Although inference can form one of the most powerful and rigorous strategies for deriving information, its success depends upon a comprehensive knowledge of the link between the surrogate and the mapped distribution.

Interpretive overlays

Interpretive overlays are useful when relationships between visible patterns on the image are used to reveal patterns not directly visible. For example, soil patterns may be revealed by distinctive relationships between separate patterns of vegetation, slope, and drainage.

An interpreter using this method examines individual overlays for each image of interest. The first overlay might show the major classes of vegetation, perhaps consisting of dense forest, open forest, grassland, and wetlands. A second overlay might delineate slope classes, including level, gently sloping, and steep slopes. Another overlay might show drainage, and still others might show land use and geology. In all, the interpreter may prepare as many as five or six overlays, each depicting a separate pattern. By superimposing these overlays, the interpreter can derive information from the coincidence of several patterns. For example, the interpreter may know that, in a particular terrain, certain soil conditions can be expected where steep slopes and dense forest are found together and that others are expected where dense forest coincides with gentle slopes.

Photomorphic regions

Photomorphic regions identify regions of uniform appearance on an image. In contrast with interpretation using interpretive overlays, interpretation using photomorphic regions does not attempt to resolve individual landscape components, but instead looks for their combined influence on image pattern. For this reason, application of photomorphic regions may be most effective for small-scale imagery in which the coarse resolution tends to average separate components of the landscape. Photomorphic regions are then simply image regions of relatively uniform tone and texture.

In the first step of applying photomorphic regions, the interpreter delineates regions of uniform image appearance using tone, texture, shadow, and the other elements of image interpretation as a means of separating regions. In the second step, the interpreter must be able to match photomorphic regions to classes of interest useful to the interpreter. For example, the interpreter must determine if specific photomorphic regions match vegetation classes. This step obviously requires field observations, or collateral information, since regions cannot be identified by image information alone. As the interpretation is refined, the analyst may combine some photomorphic regions or subdivide others.

The interpreter must apply this procedure with caution, as the relationships between photomorphic regions and information can vary from one region to another and even from one image to another (depending upon illumination, season, and other factors). Therefore, photomorphic regions do not always fall neatly into categories defined by the interpreter. The appearance of one region may be dominated by factors related to geology and topography whereas that of another region on the same image may be controlled by the vegetation pattern.

Image interpreters, of course, may apply a mixture of several strategies in a given situation. For example, interpretation of soil patterns may require direct recognition to identify specific classes of vegetation, then inference to associate the vegetation patterns with soil classes.

Image interpretation keys

Interpretation keys present reference information designed to enable rapid identification of features recorded on aerial images *(figure 10.7)*. A key usually consists of a collection of annotated images or stereograms and a description, which may include sketches or diagrams. These materials are organized in a systematic manner that enables retrieval of desired images by date, season, region, subject, or other characteristics. Keys are often structured to follow a dichotomous logic that requires the user to make a sequence of simple yes or no decisions that lead to a correct identification *(figure 10.8)*.

Interpretation keys are valuable aids for summarizing complex information portrayed as images and have been widely used for visual interpretation. Such keys typically serve either of two purposes: (a) to train inexperienced personnel in the interpretation of complex or unfamiliar topics, or (b) to form reference aids to help experienced interpreters organize information and examples pertaining to specific topics.

Keys have long been used in the biological sciences, especially botany and zoology, to form a system for complex information. These disciplines rely upon taxonomic systems that are so complex even specialists cannot retain the entire body of knowledge. The key therefore forms a means of organizing the essential characteristics of a topic in an orderly manner. It is important to note that scientific keys of all forms require that the user master the basic foundations of the topic, so a key is not a substitute for knowledge, but a means to systematically order information so that informed users can quickly reach correct identifications.

Often it is useful to classify keys by their formats and organizations. Keys designed solely for use by experts are referred to as *technical keys*. *Nontechnical keys* are those designed for use by those with a lower level of expertise. *Essay keys* consist of extensive written descriptions, usually with annotated images as illustrations. A *file key* is essentially a personal image file with notes; its completeness reflects the interests and knowledge of the compiler. Its content and organization suit the needs of the compiler, so it may not be organized in a manner that seems logical for use by others. A *dichotomous key* organizes relevant information in a way that guides the user through a series of structured, simple yes or no decisions that eventually lead to a correct identification.

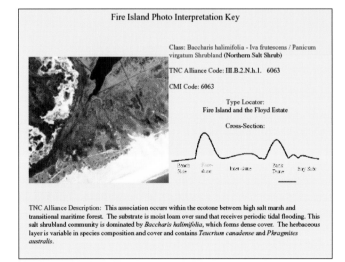

Figure 10.7   An image interpretation key. For clarity, only a section from the complete version of the key is presented. A portion of the aerial photograph has been annotated to indicate the distinctive tone, texture, and shadow characteristic of specific vegetation/terrain units, indicated here by the alphanumeric codes that link to more complete descriptions. Ground views and topographic profiles provide additional information for the user. Fire Island, New York.

Source: Image courtesy of Conservation Management Institute, the National Park Service: Fire Island National Seashore, and the USGS National Park Service Vegetation Mapping Program.

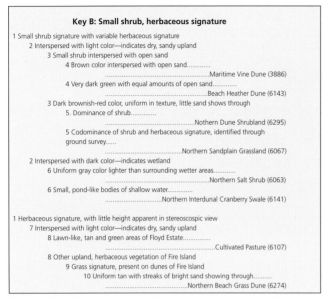

Figure 10.8 Example of a partial photointerpretation key. This text illustrates the logic of a word key to guide photointerpreters in distinguishing different vegetation classes. Fire Island, New York.

Source: Courtesy of Conservation Management Institute, the National Park Service: Fire Island National Seashore, and the USGS National Park Service Vegetation Mapping Program.

Aerial mosaics and image maps

A map is characterized by its planimetrically accurate representation of the earth's surface and its use of symbols to represent selected features. By this standard, an aerial image is not really a map because tilt and relief displacement introduce geometric errors, and its features are represented without selection or symbolization. Therefore, aerial images lack the essential qualities expected in a true map of the earth's surface.

Nonetheless, aerial images form the basis for useful maplike representations of the earth's surface such as mosaics and orthophotos. *Aerial mosaics* are produced by assembling adjacent aerial photographs to form a single image representing a regional view beyond that provided by any single photograph *(figure 10.9)*. Conventional black-and-white panchromatic and natural color films detect only the visible portion of the spectrum. Atmospheric scattering causes high-altitude imagery taken with these films to be of poor quality because haze effects produce low contrast and poor color discrimination. Mosaics were once the only way to provide high-quality airphoto coverage of a large area. Only after World War II, with the availability of infrared films equipped with superior

Figure 10.9 Portion of an uncontrolled aerial mosaic.

Source: Image courtesy of U.S. Environmental Protection Agency.

haze penetration characteristics, could large areas be imaged in a single photograph from high altitudes and from space.

Uncontrolled mosaics are formed by assembling adjacent photographs without strict concern for geometric integrity *(figure 10.9)*. By using the central regions of vertical aerial photographs (which form the most accurate geometric representation of the earth's surface), it is possible to assemble a map-like view of a region. However, an uncontrolled mosaic does not present features in their correct geometric relationships, so it lacks an essential quality of a true map—it cannot be used as the basis for reliable measurements of distance, direction, or area.

A *controlled mosaic* presents the detail of an aerial photograph in its planimetrically correct position. It uses only the most accurate portion of the photograph or has projected the image in a manner that removes positional errors inherent to the vertical aerial photograph. A related form of imagery, an *aerial index (figure 10.10)*, is created by placing adjacent aerial photographs together in their approximate relative positions without representing planimetric relationships. An aerial index is intended to be a guide to identifying those photographs that represent a specific area without the need to search an entire collection in sequence.

Figure 10.10   An aerial index.

Source: Image courtesy USGS.

Orthophotos are aerial photographs prepared using stereoscopic parallax[1] and photogrammetric principles to remove effects of relief displacement and tilt to provide a planimetrically correct aerial image. An *orthophoto quadrangle* is an orthophoto that represents the same region on the equivalent USGS topographic map quadrangle, using scale, symbols, and format consistent with the equivalent topographic quadrangle. Orthophoto quadrangles can be thought of as *image maps*, in the sense that the image detail of the aerial photograph is presented in a geometrically correct format. A

digital orthophoto quadrangle (DOQ) is a digital version of an orthophoto quadrangle that presents photographic detail with planimetric accuracy *(figure 10.11)*. Relatively inexpensive DOQs are available from the USGS and other organizations for large areas of the United States. Because of their planimetric qualities, their availability, low cost, digital format, and fine detail, DOQs are valuable for a wide variety of practical applications, especially as base data for geographic information systems.

Figure 10.11   Digital orthophoto quadrangle (DOQ) of the Platte River Valley near Kearny, Nebraska.

Source: Image courtesy of the U. S. Geological Survey.

Field observations and accuracy assessment

Field observations

Although aerial imagery provides a valuable source of information, it is a mistake to think of it as infallible or as a self-contained source of information. The information that it provides is derived from the interpretation processes described here, which, like any human endeavor, are susceptible to imprecision and error.

We can think of any map or GIS as a statement concerning the conditions that apply to a specific place on the ground. The more precise we can make that statement, by delineating small parcels and providing specific labels, the more useful the GIS is to the user. The larger the parcels and the more general the labels, the less useful the GIS is to the user. For example, a broadly delineated area labeled as *forest* is less useful than a series of smaller, more precisely delineated parcels labeled as *oak*, *pine*, and *fir*. Thus, image interpretation is a process that reduces uncertainty by generating the most precise labels consistent with the client's requirement and the detail available in the imagery.

---

1. *Stereoscopic parallax* is the difference in the way an object appears when viewed, or imaged, from two different perspectives. Stereoscopic parallax permits humans to view our surroundings in three dimensions and permits specialists to derive elevation information from stereoscopic aerial photography and to remove many of the most severe positional errors from aerial photographs.

Analysts further reduce uncertainty by applying field data to confirm and refine interpretations by comparing boundaries and labels derived from the imagery to the actual conditions encountered in the field. *Field data* is first-hand observation collected on the ground. Although each organization will develop its own procedures, it is common for analysts to carry images and maps to the field and annotate them as observations are made to confirm or correct preliminary interpretations. Photocopies of images often serve well for this purpose—they show necessary detail and are inexpensive enough to annotate without damaging original prints.

The timing of the fieldwork must permit responses to questions as the imagery is being interpreted so that analysts can confirm and refine their interpretations in order to guide the development of the project. Fieldwork conducted before the interpretation process begins or after it has concluded will not permit interpreters to refine their work and improve the quality of the product.

For some, the field component may be seen as a distraction to the main agenda for the project. Fieldwork can create delays in schedules. Difficulties offered by logistics, weather, and travel can increase expenses, and fieldwork may identify inconvenient difficulties in the work already completed. However, the field component is essential for establishing the validity and credibility of the product and for confirming analysts' confidence in their own work. From the GIS manager's perspective, the key concern is not whether or not to include a field component, but how to best focus the finite resources available for fieldwork in a manner that yields the most significant benefits for the project.

On-screen digitizing

Using classic photointerpretation procedures, analysts can record their interpretations by drawing on translucent overlays placed over the imagery to show annotations and outlines of features such as buildings or boundaries of vegetation units. Ultimately, such overlays can be digitized and then transferred and registered to a planimetrically correct base or redrafted for publication.

In the context of digital imagery and GIS, the task requires the transfer of information from the raster (grid-like) structure of the pixels that form a digital image into a vector (outline) structure, the format typically used for digital maps (e.g., soils and geology maps). This transfer from one format to another can be greatly simplified if the analyst's interpretation can be recorded directly onto the image as a digital file. This file can then be transferred in its digital form to a GIS,

where it can be registered to other overlays or displayed as a separate map with its own titles, labels, and symbols.

For photointerpreters, this process of *heads-up digitization* (so designated to distinguish it from the *heads-down table digitization* employed to digitize outlines from paper maps or prints) is an important means of recording and translating information from a photographic image into a GIS context. On-screen digitization is an important means for preparing and updating databases, transferring interpreted information from an image to a GIS layer, or for presenting information cartographically.

It is important to emphasize that on-screen digitization is practical only if the image to be digitized has already been georeferenced to remove positional errors inherent in any photographic image and to assure that it matches to a coordinate system. For example, DOQs are one common form of georeferenced image. On-screen digitization of a raw, uncorrected image will provide outlines that represent incorrect areas or positions and will not register to other data, resulting in very limited usefulness.

Image processing and GIS software offer programs that facilitate on-screen digitizing, some with special features to assist in the labeling of features and the delineation of polygons. Once a digital image has been displayed on the computer monitor, photointerpreters use the computer mouse to trace outlines of the desired features on the image as points, lines, or areas *(figure 10.12a)*. Annotations can then be added and the outlined image saved to be converted later to a vector file that can be manipulated as needed.

On-screen digitization is not inherently difficult, but as for any photointerpretation activity, it requires consistency and attention to detail—especially to assure that polygons are completely closed and that outlines are traced accurately and labeled consistently. The task can be more challenging than it initially appears. Edges of features can be difficult to identify in areas of shadow or low image contrast and hard to trace where they assume convoluted forms. Some of these difficulties can be addressed by digital image processing techniques such as digitally enlarging the image to assist in tracing of fine details and complex shapes *(figure 10.12b)* or adjusting brightness and contrast. When outlines of features are obscured in shadow, or when a feature contrasts minimally with an adjacent feature, the interpreter can use image enhancement to adjust image brightness and contrast to improve the visual qualities of the image.

a

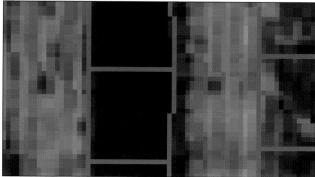

b

Figure 10.12   (a) On-screen digitizing. The outlines of test plots (red) at an agricultural experiment station were annotated on the display monitor. Red borders show outlines of test plots. (b) Enlargement of a subarea showing the additional detail available for on-screen digitizing.

Source: Images courtesy of NASA Stennis (Affiliated Research Center Program), IGF Insurance, Des Moines, Iowa CALMIT, University of Nebraska-Lincoln.

Accuracy assessment

Image analysts have developed procedures for evaluating the accuracy of interpretation products. For interpretations that yield map-like results, in which each point on the map has been assigned to a specific set of categories, it is possible to evaluate the overall accuracy of the map as well as the accuracy of the individual classes.

Ultimately, the reason to be concerned with accuracy is that interpretation errors lead to undesirable consequences—e.g., a suboptimal decision may be made that leads to a cost, monetary or otherwise, for individuals or society as a whole. The higher the cost, the greater the justification to invest in attaining higher accuracy and in improving the accuracy assessment. Interpretation accuracy can be improved by changing methods or materials—for example, using larger-scale imagery, increased numbers of field observations, or improved interpretation strategies might reduce errors in specific classes. These options can be considered only if the error matrix (discussed later in this section) is compiled and interpreted.

Accuracy assessment requires the collection of a set of *reference data* consisting of points of known identity that are dispersed throughout the area to be studied and allocated across the classes used for the map. Ideally, these points should be selected at random; however, a purely random selection will create clusters within the map rather than a desirable even distribution. Further, a purely random selection of points will include only a few samples from classes of small extent on the image. Therefore, to ensure that all classes are represented by adequate numbers of samples, and because samples should be more or less evenly dispersed across the image, reference points are usually collected using a strategy that includes a random component but modified to force reference points to be allocated among classes and spread across the study area to avoid clustering.

Analysts collect reference data in much the same way they collect field observations—they gather first-hand observations in the field in a manner that assures absolute confidence in their accuracy. However, they *do not* use data collected to compile the interpretation as reference data—it is important that the reference data for accuracy assessment be used only for evaluation, not as input for the interpretation.

The assessment proceeds by compiling a table that matches each reference point to its corresponding point on the map. This enables the analyst to determine if the map label for each reference point is correct, and if not, which label was incorrectly assigned. The summary of these tables is known as an *error matrix (figure 10.13 and 10.14)*. The error matrix is the standard form for reporting the accuracy of an interpretation. It takes the form of an *n* by *n* array, where *n* represents the number of categories. It permits identification not

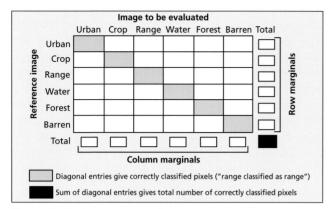

Figure 10.13 Sample error matrix.

Source: Reprinted with permission of The Guilford Press © 2002 by The Guilford Press.

		Urban	Crop	Range	Water	Forest	Barren	Total
	Urban	150	21	0	7	17	30	225
	Crop	0	730	93	14	115	21	973
Reference image	Range	33	121	320	23	54	43	594
	Water	3	18	11	83	8	3	126
	Forest	23	81	12	4	350	13	483
	Barren	39	8	15	3	11	115	191
	Total	248	979	451	134	555	225	1,748

Image to be evaluated

Note: Percentage correct = sum of diagonal entries/total observations = 1,748/2,592 = 67.4%

Figure 10.14 Sample error matrix.

Source: Reprinted with permission of The Guilford Press © 2002 by The Guilford Press.

only of overall errors for each category, but also misclassifications (due to confusion between categories).

Usually, the left-hand side (*y*-axis) is labeled with the categories of the reference (correct) classification. The upper edge (*x*-axis) is labeled with the same *n* categories; however, these labels refer to the categories on the map to be evaluated. Values in the matrix represent numbers of samples for which the analyst has been able to compare the evaluated and reference images.

Inspection of the matrix shows how the classification represents actual areas on the landscape. For example, in figure 10.14 there are 225 observations of urban land (the far right value in the first row). Of the 225 samples of urban land, 150 were classified as urban (row 1, column 1); these, of course, are the samples of urban land correctly classified. The succeeding values along the first row are incorrectly classified urban samples and the categories to which they were assigned: crop, 21; range, 0; water, 7; forest, 17; and barren land, 30. Each row also shows for each class how the interpretation assigned the verified observations. The diagonal from upper left to lower right gives numbers of correctly interpreted observations.

Further inspection of the matrix reveals a summary of other information. Row totals on the far right give the total number of samples in each class, as recorded on the reference image. The column totals at the bottom show the number of samples assigned to each class as depicted on the image to be evaluated.

Errors of omission and errors of commission

Because the error matrix permits assessment of the individual accuracies of specific classes, it is possible to examine errors on a category-by-category basis to reveal errors of omission and errors of commission for each category. An example of an error of omission is the assignment of an area of forest on the ground to the agricultural category on the map (in other words, an area of real forest on the ground has been omitted from the map). Using the same example, an error of commission would be to assign an area of agriculture on the ground to the forest category on the map. The analyst's error in this instance has been to actively commit an error by assigning a region of forest to a wrong category. (This error of commission for one category will, of course, also be tabulated as an error of omission for another category.) The distinction is essential because the interpretation could otherwise achieve 100% accuracy in respect to forest by assigning all pixels to forest. By tabulating errors of commission this bias would be meaningless because the low omission error would be compensated by high commission errors.

The error matrix must be interpreted to understand the varied costs associated with specific kinds of errors. For example, in mapping forest fuels, certain vegetation types are particularly high risk. Not recognizing all occurrences of these types when managing high fire risk conditions could be dangerous. However, misclassifying low-risk fuels may have minimal consequences. Therefore, the interpretation process could be designed to trade omission errors for commission errors—to minimize omission errors for high-risk fuels type at the cost of a higher rate of commission errors, causing over-preparation for some areas thought to be high risk but were actually misclassified. In other words, the interpretation could be structured to overestimate the areas of high-risk fuels because the consequences of missing these areas have such high costs (see, for example, Aronoff 1984).

By examining relationships between the two kinds of errors, the analyst gains insight into the varied reliabilities of classes and learns about the reliability of the interpretation process. Examined from the client's perspective, the matrix reveals *consumer error* (or *user error*), the risk that a category on the map will be incorrect. Examined from the analyst's point of view, the matrix reveals *producer error*, the risk that a reference point was incorrectly interpreted. The difference between the two lies in the base from which the error is assessed. For producer error, the base is the area in each class as represented within the reference data, which is the ultimate authority from the producer's perspective. For consumer error, the base for the percentage is the area for each class as represented on the final classified map, which forms the basic reference for the user of the classification.

Thus, for the example in figure 10.14, the producer error for forest is 350/483, or 73%. For the same class, the consumer error is 350/555, or 63%. Consumer error forms a guide to the reliability of the map as a predictive device—it tells the user of the map that (in this example), of the area labeled as forest, 63% actually corresponds to forest on the ground. Producer error informs the analyst who prepared the classification that, of the actual forested landscape, 73% was correctly interpreted. In both instances, the error matrix enables identification of the classes erroneously labeled as forest and forested areas mislabeled as other classes.

Percentage correct

One of the most widely used measures of accuracy is the *percentage correct*—the overall proportion of correctly classified pixels. The number correct is the sum of the diagonal entries. Dividing this value by the total number of pixels examined gives the proportion that has been correctly classified (*figure 10.14*). This value estimates the unknowable *true* value—the extent to which this value approaches the (unknowable) true value depends, in part, on the sampling strategy. The percentage correct can be reported with a confidence interval (Hord and Brooner 1976).[2]

---

2. A confidence interval estimates a range of values that is likely to include an unknown population parameter (such as a mean). The estimate is calculated from a given set of sample data. The width of the confidence interval indicates the degree of uncertainty associated with the estimated value of the mean. A wide interval indicates the mean is likely to be only approximately correct, whereas a narrow range indicates that the value of the mean is more precise.

Often the percentage correct is used alone, without the error matrix, as a simple measure of accuracy. By itself, the percentage correct may suggest the relative effectiveness of a classification, but in the absence of an opportunity to examine the full error matrix, it cannot produce convincing evidence of the accuracy of the interpretation. A full evaluation must consider the categories used in the classification. For example, it is relatively easy to achieve high percentage correct values by classifying a scene composed chiefly of open water—a class that is easy to interpret.

The kappa statistic

Understanding the meaning of an error matrix can present a challenge because there are so many values subject to arrangement in so many different patterns. People often feel more comfortable interpreting a single measure that summarizes significant aspects of the error matrix. The kappa (κ) statistic offers this opportunity.

The κ statistic measures the difference between the observed agreement between two maps as reported by the diagonal entries in the error matrix and the agreement that might be attained solely by chance alignment of the two maps. That is, not all agreement can be attributed to the success of the classification process since chance matches between maps may exist. The statistic provides a measure of agreement adjusted for chance agreement. κ is estimated by $\hat{\kappa}$ (k hat):

$$\hat{\kappa} = \frac{\text{Observed} - \text{expected}}{1 - \text{expected}}$$

This form of the equation, given by Chrisman (1980), is a simplified version of the expression form given by Bishop et al. (1975). Here *observed* refers to the accuracy reported in the error matrix, and *expected* refers to the correct classification that can be anticipated by chance agreement between the two images. *Observed* is the overall value for percentage correct—the sum of the diagonal entries divided by the total number of samples. *Expected* is an estimate of the contribution of chance agreement to the observed percentage correct. Products of row and column marginals (sums across rows and down columns) are used to estimate the pixels assigned to each cell in the error matrix, given that pixels are assigned by chance to each category.

The κ statistic, in effect, adjusts the *percentage correct* measure by subtracting the estimated contribution of chance agreement. Thus, κ = 0.83 can be interpreted to mean that the classification achieved an accuracy that is 83% better than would be expected from chance assignment of pixels to categories. As the percentage correct approaches 100 and as the contribution of chance agreement approaches 0, the value of κ approaches +1.0, indicating perfect effectiveness of the classification. On the other hand, as the effect of chance agreement increases and the percentage correct decreases, the values of κ become negative.

As an example, for a wildlife application such as that discussed below *(figure 10.20)*, an evaluation of interpretation accuracy was conducted by collecting map points by direct observation of the area interpreted. A total of 47 points were collected to assess the spatial accuracy of the vegetation map, and 495 points were used to assess its thematic accuracy. The initial analysis revealed that the map achieved only a relatively low overall accuracy of 57.6%.

Investigation revealed serious discrepancies between vegetation as classified in the field and its representation on the map. After reviewing the validity of the points, 428 points were available for use in the accuracy assessment. Of these, 329 were located in polygons larger than 0.25 hectares and 99 were found in smaller polygons. The overall accuracy (and kappa statistic) for the map was 66.3% (64%). At more general designations in the classification system, accuracy reached 78.1% (78%) and 87.5% (88%), depending upon the level of specificity. As a generality, it is more difficult for interpreters to attain high accuracy as categories in a classification become more precise. For example, higher accuracies can usually be achieved assigning broad classes such as *forest* or *nonforest* than when assigning classes of different forest types such as *oak*, *pine*, or *maple*.

Interpretation equipment

Image interpreters use a variety of equipment, ranging from the most basic to some of the most advanced and sophisticated. An image interpretation facility is designed for effective storage and handling of photographs as paper prints, film transparencies, and digital images. Paper prints are most frequently nine by nine inch contact prints, often stored in some sort of sequence in file cabinets. Larger prints and indexes must be stored flat in map cabinets. Transparencies, usually in the form of long rolls of film wound on spools, are stored in sealed canisters as protection from dust and moisture. Transparencies in this form are viewed using light tables equipped with holders for the spools. Individual frames or

other segments can be cut from the roll as needed for convenience. If images are removed from a roll, the removal date and name of the person who retains the missing segment should be noted and kept with the roll.

The image interpreter should be able to work at a large, well-lit desk or worktable with convenient access to electrical power and the ability to control lighting with blackout shades or dimmer switches. Maps, reference books, and other supporting materials should also be available as required.

If the interpretation is made from paper prints, special attention must be devoted to prevent folding, tearing, or rough use that will cause wear. If transparencies are used, special care must be given to protect the surface of the transparency by using transparent plastic sleeves or handling them only with clean cotton gloves. Moisture and oils naturally present on unprotected skin may damage the emulsion, and dust and dirt will scratch the surface. Although such damage is invisible to the unaided eye, it can later emerge as a problem under high-power magnification.

Magnification

Magnification equipment can be said to be the photointerpreter's most important tool. This equipment permits detailed examination and measurement of features difficult to discern or that are imperceptible to the unaided eye.

For small or modest levels of magnification, photointerpreters use a simple handheld reading glass. For precise work, *tube magnifiers* provide low-power magnification (2x to about 10x) using lenses mounted in transparent tube-like stands *(figure 10.15)*. For higher levels of magnification, as

Figure 10.15 Image interpretation equipment, including scale and tube magnifier.

Source: J. Campbell.

Figure 10.16   Image interpreter using a binocular microscope.

Source: Image courtesy of the U.S. Air Force.

is needed to examine film transparencies, it is necessary to use the *binocular microscope (figure 10.16)*. Binocular microscopes provide two separate eyepieces, each with individual focus, and separate light paths from the object to the eyepieces. Binocular microscopes reduce eyestrain and provide relatively wide fields-of-view and large depth-of-field. Some models can also be equipped with stereo arms—adjustable prisms used to enable stereo viewing of roll film (which cannot be repositioned in the same manner as can paper prints used with ordinary stereoscopes).

Image interpretation microscopes are similar in principal to conventional microscopes but differ in several important respects. A microscope is an assembly of optical components designed to provide high levels of magnification. Although some microscopes can change the degree of magnification by changing lenses, it is typical of photointerpretation microscopes to use *zoom lenses* that permit the operator to vary the degree of magnification by altering the relationships between optical components within the microscope without changing lenses. Such instruments may have adjustable magnification of 40x, which will approach or exceed the limits of resolution for most photographic images. As magnification increases, the field-of-view decreases; so the interpreter views a smaller region in greater detail.

Another difference between conventional and image interpretation microscopes is the light source. Whereas conventional microscopes require their own light sources to illuminate an object, most photointerpretation microscopes are used to view transparencies backlit by a light table.

Although the ability to magnify is a very important tool for the photointerpreter, magnification in itself is not a solution to the problem at hand. High levels of magnification restrict the field-of-view, so the interpreter loses a sense of the surrounding context. Further, extreme levels of magnification can approach the limits of the resolution for the emulsion—this level of magnification reveals little of interest. The image analyst should use the minimum level of magnification that permits completion of the task at hand.

Flatbed digital scanners

Paper or film photographs can be scanned to create raster images. Among the most accurate instruments for image scanning are *scanning densitometers*, which are designed to accurately and precisely measure image density by systematically scanning across an image, creating an array of digital values to represent the image pattern. Such instruments have been designed to very precisely measure position and image density at resolutions of up to a few microns.

Although instruments designed for such precise tolerances are expensive, many others are less precise but still serviceable. Desktop scanners, although designed for office use, can be used to scan maps and images for visual analysis. If more precise scanning is required, then it is usually advisable to arrange for custom scanning by a professional service that can use high-quality equipment.

When an image is scanned, the analyst must specify a density for the scan. Scan density is measured in dots per inch (dpi)—a designation that pertains to printing but has also traditionally been applied to other portions of the scanning process. In more relevant terminology, dpi could be referred to as pixels per inch. The dpi value instructs a printer how closely to space pixel values as it prints an image to paper. By changing the dpi, the analyst can change the scale and quality of the image. Values of 150 to 300 dpi might be reasonable for printing an image, in the sense that such a value would portray a legible version of the image. (A more detailed discussion of digital images can be found in chapter 4.)

In theory, high scan rates record increasing levels of detail, so it would be desirable to always select a high scan rate. But high scan rates greatly increase the size of the image file, increasing storage requirements and transmission time. In addition, at some level (about 3,500 dpi), the additional detail is composed of variations in the structure of the emulsion rather than genuine features of the image.

Light tables

Light tables are simple but indispensable equipment for photointerpretation. A light table is a translucent surface illuminated from behind to permit viewing of film transparencies. In its simplest form, a light table is a box-like frame with a frosted glass surface. The size of the viewing area can vary from desk size to portable models the size of a briefcase. Larger light tables are usually equipped with brackets to hold film spools for viewing roll film and rollers at the edges to permit the film to move freely.

More elaborate light tables have dimmer switches to control the intensity of the lighting, high-quality lamps to control spectral properties of the illumination, and power drives to wind and unwind long spools of film. Heavyweight frames minimize vibrations, which can inhibit interpretation of imagery viewed under high magnification. These features, and the ability to adjust the height of the table and the tilt of the viewing surface, are important in situations in which interpreters must work continuously for long intervals.

Densitometers

Densitometry is the science of making accurate measurements of photographic density. A densitometer is an instrument that measures image density by directing a light of known brightness through a small portion of the image, then measuring its brightness as altered by the film. Often the objective is to reconstruct estimates of brightness in the original scene or sometimes merely to estimate relative brightness on the film. Typically, the light beam passes through an opening about one millimeter in diameter. Use of smaller openings (sometimes measured in micrometers) is known as *microdensitometry*. An interpreter might use a densitometer to make quantitative measurements of image tone for selected regions within an image. For color or CIR images, filters are used to make three measurements, one for each of the three additive primaries.[3]

Stereo imagery equipment

In everyday experience, we can perceive our surroundings in three dimensions because of our ability to acquire two

Figure 10.17 Image interpreter using a pocket stereoscope for interpretation of aerial photographs.

Source: Image courtesy of the U.S.D.A. Forest Service.

independent views of features—one separate view from each eye. Our brains are experienced in fusing these two independent views and translating the differences (these differences are known as *stereoscopic parallax*) into depth information. This process is so innate that it requires no conscious effort.

We can simulate the effect of this stereovision with a pair of photographs of the same scene as observed from suitably different perspectives. A pair of photographs for stereo viewing is known as a *stereopair*. A stereogram is a stereopair mounted together on a rigid material such as cardboard in the correct position for stereo viewing. An example is shown in figure 5.10 in chapter 5. *Stereoscopes*, instruments for viewing stereopairs, range from a simple stand with adjustable lenses (*figure 10.17*) to more complex models that provide zoom and panning controls to shift the area being viewed without moving the images (see *figure 5.9* in chapter 5).

To understand stereovision in the context of aerial photography, envision an aircraft flying along a straight, level flight path acquiring photographs at an interval such that each photograph shows at least half of the area shown by the previous photograph. This creates the overlapping photographs that provide two independent views of each region in the flight line (see *figure 5.8* in chapter 5).

There are several strategies in addition to the stereoscope that present each eye with a separate image to simulate the stereo effect. The stereoscope is an example of the *optical separation* technique, which presents left and right images side-by-side. An optical device such as the stereoscope separates the analyst's view of the left and right images.

---

3. The additive primaries (red, green, and blue), when combined in equal proportions, form the color white. The additive primaries are significant when discussing the behavior of light or sensitivities of films. In contrast, the subtractive primaries (cyan, yellow, magenta), which combine to form black, are significant in any discussion of dyes or pigments.

Figure 10.18 Polarizing glasses used for stereo viewing of digital aerial photography.

SOurce: Image courtesy of MD Atlantic Technologies, Inc.

Figure 10.19 Portion of a DOQ depicting part of the Eastern Rivers National Wildlife Refuge Complex.

Source: Image courtesy of Conservation Management Institute, the National Park Service: Fire Island National Seashore, and the USGS National Park Service Vegetation Mapping Program.

The *red/blue anaglyph* presents the image intended for each eye in its own color. A common strategy is to represent the image intended for the left eye in blue, the image for the right eye in red, and to use magenta for those portions of the image common to both eyes. When the analyst views the image using special glasses, with a red lens for the left eye and a blue lens for the right eye, the colored lenses cause the image intended for the opposite eye to merge into the background, and the image intended for its own eye to appear normally. The anaglyph has been widely used for novelties but less often as an analytical display.

An alternative method for stereovision employs polarizing lenses to create the stereo effect. The image intended for each eye is projected through separate polarizing filters (e.g., horizontal for the left eye, vertical for the right eye). The image must be viewed using special glasses with left and right lenses that allow only the light with one polarization to pass to the corresponding eye to simulate the stereo effect *(figure 10.18).*

There are other techniques for stereovision that are effective to varying degrees including random dot stereograms, Magic Eye® images, and others. Most are much less effective for scientific and analytical applications than the methods mentioned above.

Application examples

Details of the practice of visual interpretation tend to be very specific to the subject matter at hand. Therefore, the follow-ing brief examples illustrate the kinds of information that can be derived from the visual interpretation of imagery and provide a glimpse into some of the practices and procedures used in different disciplines. The examples illustrate the role of visual interpretation in the fields of wildlife habitat analysis, forestry, and the use of historical aerial photography for environmental quality analysis.

Wildlife habitat analysis

Aerial photography is an important source of wildlife habitat information because it can record vegetation patterns—one of the most important components of habitat. Figure 10.19 shows a color infrared digital orthophoto quadrangle (CIR DOQ) for a portion of the Eastern Rivers National Wildlife Refuge Complex (NWRC), which is located on the Chesapeake coastline of Virginia on the banks of the James and Rappahannock Rivers.

This image was interpreted by the staff of the Conservation Management Institute, Virginia Tech, for the U.S. Fish and Wildlife Service. The Institute produced habitat information in a GIS format using the DOQ image as a base. The first view *(figure 10.19)* shows a portion of the DOQ representing one area of the NWRC. (Remember that a DOQ is an aerial photograph that has been rectified to provide a planimetrically correct base, then formatted to match the USGS

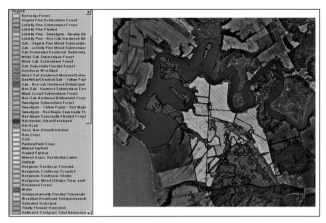

Figure 10.20 Vegetation patterns and habitat interpreted from the image shown in figure 10.19.

Source: Image courtesy of Conservation Management Institute, the National Park Service: Fire Island National Seashore, and the USGS National Park Service Vegetation Mapping Program.

Figure 10.22 Bird sightings plotted on image base shown in figures 10.20 and 10.21.

Source: Image courtesy of Conservation Management Institute, the National Park Service: Fire Island National Seashore, and the USGS National Park Service Vegetation Mapping Program.

Figure 10.21 Sample points positioned using GPS and plotted on a DOQ.

Source: Image courtesy of Conservation Management Institute, the National Park Service: Fire Island National Seashore, and the USGS National Park Service Vegetation Mapping Program.

topographic quadrangle corresponding to this region). This photograph shows the open water of the river, the wetlands and marshes bordering the shoreline, and the agricultural patterns occupying neighboring higher grounds.

A subsequent view *(figure 10.20)* shows the interpretation of vegetation classes using aerial photography. Because vegetation

cover forms one of the most important components of wildlife habitat, it forms one of the primary themes interpreted from the aerial photography. The classification shown here has been developed to apply to several diverse areas, so it has more classes than are present in this specific area. Note also that the classification system uses very detailed classes that require close integration of fieldwork with interpretation of the image, as mentioned previously.

Field points are represented in figure 10.21; these record locations of field observations used to develop and verify the vegetation classes. GPS locations were recorded at each field observation point to assure that data could be accurately positioned on the image.

Finally, the last view *(figure 10.22)* shows locations of points used to observe birds and the numbers of sightings at each point. Sightings were not, of course, made using the aerial photography, but this map clearly illustrates the value of aerial photography as a base to plot other data. Because of the wealth of data conveyed by each image, the image base presents valuable context for interpretation of the meaning of the nonphotographic data. In this example, the numbers of sighting can be easily related to the vegetation classes and to the positions of edges between classes.

Figure 10.23   Aerial photograph showing undelineated forest stands.

Source: Image courtesy of the U.S.D.A. Forest Service.

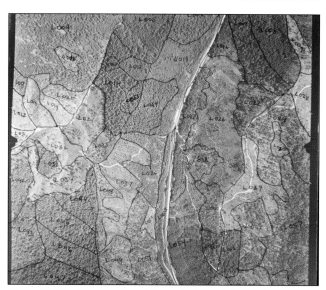

Figure 10.24   Interpretation of figure 10.23 to delineate forest stands based upon tone, texture, and other elements of image interpretation.

Source: Image courtesy of the U.S.D.A. Forest Service.

Forestry

Foresters systematically monitor forests to assess the extent, productivity, health, and marketability of trees. Although ground-level observations form an essential component of forest inventory, aerial photographs form one of the most valuable tools not only to plot information acquired at ground level but also to provide an independent source of information about the extent and status of forest resources.

Foresters use aerial photographs as one of the principal means of delineating *stands*—i.e., areas of trees with relatively uniform composition, age, and ecology. Figures 10.23 and 10.24 illustrate the use of aerial photography to delineate stands based upon the eight elements of image interpretation—shape, size, tone, texture, shadow, site, association, and pattern. Varied tree species and stands are characterized by distinctive textures, tones, and patterns as viewed on the aerial imagery. When interpreters experience difficulties in confirming interpretations, they can use image interpretation keys and field observations to resolve uncertainties. Furthermore, a sequence of aerial photographs collected over time can permit foresters to monitor the development of stands over time and assess their growth and response to environmental events such as drought, fire, or insect infestations.

Figure 10.25 shows the Monte Cristo Peak area on the Wasatch-Cache National Forest (about fifty miles east-northeast of Ogden, Utah), in early October 1991. Elevations here range from about 7,000 to 9,000 feet above sea level. Forest

Figure 10.25   Color aerial photography of the Monte Cristo Peak area on the Wasatch-Cache National Forest (about fifty miles east-northeast of Ogden, Utah) taken in October 1991.

Source: Image courtesy of the U.S.D.A. Forest Service.

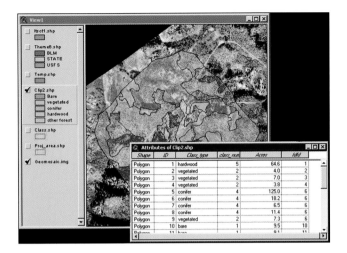

Figure 10.26 Information derived from rectified areal photography can form input for a GIS and contribute to statistical reporting, analysis, and modeling.

Source: Image courtesy of the U.S.D.A. Forest Service.

cover includes aspen and conifers such as Douglas fir, white fir, and subalpine fir. At this date, the aspen are in various stages of fall coloration or leaf-off.

Figure 10.26 illustrates the close integration of GIS technology with photointepretation, as seen in previous examples. When the interpretation has been prepared from rectified imagery, then the GIS format permits flexibility in display and symbolization of the results of the interpretation and convenient tabulation of areas and other attributes for compilation of reports and change summaries.

Environmental assessment of waste disposal sites

Over the decades, in industrial regions of the United States, hazardous industrial wastes have been deposited in varied disposal sites including ponds, lagoons, and landfills not designed to safely store dangerous substances. Although many of these areas were once positioned in rural or sparsely populated regions, urban growth has brought many of these sites close to residential neighborhoods. As a result, many such sites are now near populated regions, and in some instances, these sites have been converted to other uses including residential housing, schoolyards, and recreational areas. Deteriorating containers at these sites have released hazardous materials into nearby soils and groundwater systems where they migrate beyond the original site. Often, the use of the site for waste disposal is not properly recorded, or

the nature of the wastes is never documented, so remediation is limited by ignorance of the specifics of the situation.

Because of the difficulty of establishing the historical pattern of use at such sites, and due to the absence or inaccuracy of official records, aerial photography has formed one of the most valuable resources for investigating such sites. Using imagery stored in archives of aerial photography that extend, in many cases to the 1930s, it may be possible to assemble a sequence of snapshots that record the use of the site over time. When combined with field data, official records, and health surveys, such images can facilitate the development of an understanding of the sequence of events and assessment of the risks to the environment and to nearby populations. In some instances, use of photography also facilitates estimation of the amount of waste present and specification of the class of materials by identifying the kinds of containers used. Aerial photography can bring about the development of an understanding of the relationship between each site and nearby drainage, population, economic activities, and other waste disposal sites.

The sequence of photographs shown in figure 10.27 illustrates the Environmental Protection Agency's (EPA) ability to track the history of waste disposal sites. The most recent photograph, shown first, illustrates how current land use can mask evidence that reveals the history of the site's use for waste disposal.

These photographs show the Elmore Waste Disposal site and the Sunnyside Dump, Spartanburg County, South Carolina. The sites are outlined on each photograph. The Elmore Waste Disposal site is about half an acre in area, situated near a residential neighborhood near Greer. The Elmore site has been designated a Superfund site, so it was monitored by the EPA. (In 1980, the U.S. Congress established the Superfund program to locate, investigate, and clean the very largest and most hazardous of the many industrial waste sites that had been abandoned during previous decades. The EPA administers the Superfund program in cooperation with individual states and tribal governments.)

It is now known that a large number of drums containing liquid wastes were placed at this site between 1975 and 1977. In 1977, the property owner signed a consent order from the state of South Carolina, which demanded cleanup of the site. However, state health and environmental officials judged the cleanup actions taken by the property owner to be inadequate and ordered the owner to stop use of the site. Between 1981 and 1984, state and EPA officials investigated the site and found arsenic, chromium, and other heavy metals, as

a

b

c

d

Figure 10.27   Sequence of aerial photographs illustrating land-cover changes at a waste disposal site in South Carolina. (a) 1981. (b) 1944. (c) 1955. (d) 1965.

Source: Image courtesy of the U.S. Environmental Protection Agency.

well as a number of volatile organic compounds (VOCs) in site soils.

At various times during this interval, there were between 150 and 300 drums present on site, as well as a six-thousand-gallon, partially buried tank containing contaminated waste oil. The owner died in 1983; ownership passed to his heirs, one of whom continued to operate the site and accept waste drums. After many failed efforts to compel this owner to clean the site, in June 1986, state officials completed a state-funded removal of 5,500 tons of contaminated soil and 16,800 pounds of liquid wastes from the site, which were taken to an appropriate hazardous waste facility.

After this removal, groundwater monitoring wells and the EPA's remedial investigation from 1991 to 1992 established that a groundwater plume extending some seven hundred feet north from the site was highly contaminated by two VOCs, trichloroethylene and tetrachloroethylene. The estimated area over this plume (north of the Elmore property) is six to ten acres. Although no private water wells are located near this plume, the groundwater discharges to a creek behind neighborhood homes. Additionally, surface soil in a quarter-acre area at one end of the Elmore property was found to be contaminated by lead and arsenic at levels exceeding health-based residential standards for those metals.

Finding image interpretation services

Usually professionals or organizations that provide image interpretation skills are engaged in related activities that form the principal focus of their enterprise. Therefore, it is rare for an organization to offer image interpretation as its primary service—usually it will form one of the secondary services that supports, for example, GIS development, photogrammetric analysis, or cartography.

In directories, enterprises that provide image interpretation services can be listed under headings such as:

- engineering, architecture, survey
- digital orthophotography
- aerial survey
- aerial mapping
- digital cartography
- aerial photography
- mapping—topographical services
- mapping—geographic information systems

Not all such organizations will offer image interpretation services, so inquiries will be necessary to identify those organizations that offer the required services.

Professional consultants in surveying and photogrammetry may have related expertise and experience in image interpretation. For example, foresters, geologists, or agronomists trained in image interpretation may have expertise in applying image interpretation to related fields. Professional societies such as the American Society for Photogrammetry and Remote Sensing (ASPRS) offer professional directories that list specialists who can apply image interpretation in specific fields. *Photogrammetric Engineering and Remote Sensing*, a journal published by the ASPRS, lists examples of these directories.

Clients should ensure that the organization or individual responsible for interpreting their imagery has expertise in the project field. An image interpreter's individual expertise and experience does not necessarily apply to every project simply because image analysis skills are required. For example, an image interpreter experienced in forestry applications might have developed expertise in related fields such as habitat analysis or land-cover analysis; however, this interpreter would not necessarily be experienced in or qualified for applications related to surface materials or hydrology. Furthermore, because the practice of image interpretation may be closely linked to related aspects of the project, it is important to clearly define the scope of the expertise required. For example, will the project require specific image interpretations that focus specifically and exclusively on interpretation that defines, say, hydrology for siting septic systems? Or will it require more comprehensive image interpretation tasks such as interpreting regional hydrology and drainage; editing; verifying data; and conducting fieldwork, accuracy assessments, and other supporting activities?

Supporting an in-house photointerpretation capability incurs the expense of maintaining staff, equipment, and software for what might be a requirement peripheral to the core mission of the organization. Management must decide if the requirement for visual interpretation satisfies a recurring need that justifies the allocation of resources that could be devoted to other requirements.

If there is a recurring requirement for a visual interpretation capability, these costs may be balanced by the advantages of a stable staff familiar with the requirements and procedures of the organization and its customers. Consistency and familiarity with specific subject matter are other important considerations that should be weighed when making decisions concerning staffing an in-house visual interpretation capability. Organizations that can cross-train staff to perform visual interpretation in coordination with their other duties can achieve efficiency and flexibility, both in completing tasks in-house and in administering outside contracts if needed.

Use of an outside contractor to provide visual interpretation capability can be effective if the project warrants the effort invested in selecting the contractor and launching the project. The most satisfactory results are achieved when both organizations have a long-standing relationship, when there is continuity in staffing, and when the GIS firm has the experience and expertise to administer the contract efficiently. Again, a GIS enterprise seeking a contractor to supply visual interpretation services should ensure that the prospective contractors have experience not only in visual interpretation in general but also in the specific topics required.

Conclusion

Image interpretation can provide information not easily obtained from other sources. When applied to conventional or digital aerial imagery, image interpretation can provide spatially detailed and taxonomically explicit information that complements data acquired from other sources. It can provide an ability to retrospectively examine issues that were not of concern at the time the images were originally acquired. Image interpretation thereby forms a source of information that displays changes from one date to the next.

Although, in its original form, aerial imagery contains positional errors that prevent direct use for applications that require planimetrically correct data, positional errors can be rectified to provide a geometrically correct base compatible with the requirements of GIS applications. In the United States, one of the most widely available planimetric image products is the digital orthophoto quadrangle, offered by the U.S. Geological Survey and other organizations. The DOQ, and related products, provide positionally correct image data in a digital format. Likewise, the increase in the production of fine-resolution satellite imagery in GIS-friendly formats greatly expands the range of imagery available for GIS projects. From such products, visual interpretation can provide access to a broad range of information for GIS development including land use, transportation, agriculture, hydrography, and vegetation.

References

General

Campbell, J. B. 2002. *Introduction to remote sensing.* 3rd ed. New York: Guilford Publications.

Lueder, D. R. 1959. *Aerial photographic interpretation: Principles and applications.* New York: McGraw-Hill.

Philipson, W. R., ed. 1996. *Manual of photographic interpretation.* 2nd ed. Bethesda, Md.: American Society for Photogrammetry and Remote Sensing.

Spurr, S. H. 1960. *Photogrammetry and photo-interpretation.* 2nd ed. New York: The Ronald Press.

Availability and costs of aerial imagery

Falkner, E., and D. Morgan. 2002. *Aerial mapping.* 2nd ed. New York: Lewis Publishers.

Johnson, G. E., and M. W. Barber. 1996. Major imagery collections. In *Manual of photographic interpretation,* ed. W. R. Philipson, 653–57. Bethesda, Md.: American Society for Photogrammetry and Remote Sensing.

Larsgaard, M. L., and L. G. Carver. 1996. Sources of aerial photographs. In *Manual of photographic interpretation,* ed. W. R. Philipson, 659–78. Bethesda, Md.: American Society for Photogrammetry and Remote Sensing.

Lyon, J., E. Falkner, and W. Bergen. 1995. Estimating cost for photogrammetric mapping and aerial photography. *ASCE Journal of Surveying Engineering* 121: 63–86.

Paine, D. P., and J. D. Kiser. 2003. *Aerial photography and image interpretation.* New York: John Wiley and Sons, Inc.

Historical aerial photography for environmental assessment

Erb, T. L., et al. 1981. Analysis of landfills with historic airphotos. *Photogrammetric Engineering and Remote Sensing* 47: 1363–69.

Lyon, J. G., 1987. Use of maps, aerial photographs, and other remote sensor data for practical evaluations of hazardous waste sites. *Photogrammetric Engineering and Remote Sensing.* 53: 515–19.

Mata, L., and Fanelli, D. 1991. Environmental property assessments utilizing aerial photography. Proceedings, Association of Engineering Geologists, 34th Annual Meeting. 301–10.

Accuracy assessment

Aronoff, S. 1984. An approach to optimized labeling of image classes. *Photogrammetric Engineering and Remote Sensing* 50(6): 719–27.

Bishop, Y. M. M., S. E. Fienber, and P. W. Holland. 1975. *Discrete multivariate analysis: Theory and practice.* Cambridge, Mass.: MIT Press.

Chrisman, N. R. 1980. Assessing Landsat accuracy: A geographic application of misclassification analysis. Second colloquium on quantitative and theoretical geography. Trinity Hall, Cambridge, England.

Cohen, J. 1960. A coefficient of agreement for nominal scales. *Educational and Psychological Measurement* 20 (1): 37–40.

Congalton, R. G. 1983. The use of discrete multivariate techniques for assessment of Landsat classification accuracy. Master's thesis, Virginia Polytechnic Institute and State University.

———. 1984. A comparison of five sampling schemes used in assessing the accuracy of land cover/land use maps derived from remotely sensed data. PhD diss., Virginia Polytechnic Institute and State University.

———. 1988. A comparison of sampling schemes used in generating error matrices for assessing the accuracy maps generated from remotely sensed data. *Photogrammetric Engineering and Remote Sensing* 54: 593–600.

———. 1988. Using spatial autocorrelation analysis to explore the errors in maps generated from remotely sensed data. *Photogrammetric Engineering and Remote Sensing* 54: 587–92.

Congalton, R. G., and R. A. Mead. 1983. A quantitative method to test for consistency and correctness in photointerpretation. *Photogrammetric Engineering and Remote Sensing* 49: 69–74.

Congalton, R. G., R. G. Oderwald, and R. A. Mead. 1983. Assessing Landsat classification accuracy using discrete multivariate analysis statistical techniques. *Photogrammetric Engineering and Remote Sensing* 49: 1671–87.

Congalton, R. G., and K. Green. 1999. *Assessing the accuracy of remotely sensed data: Principals and practices.* New York: Lewis Publishers.

Hord, R. M., and W. Brooner. 1976. Land use map accuracy criteria. *Photogrammetric Engineering and Remote Sensing* 46: 671–77.

Webster, R., and P. H. T. Beckett. 1968. Quality and usefulness of soil maps. *Nature* 219: 680–82.

11 Digital image analysis

Joseph M. Piwowar

Digital image analysis is a set of specialized techniques and computer processing tools used to enhance the visual appearance of and extract information from remotely sensed imagery. It is also used to rectify imagery to match a selected map base. Though there is some overlap in capabilities, digital image analysis methods have become an important complement to visual interpretation (discussed in chapter 10). The most cost-effective means of generating quality information from remotely sensed data commonly makes use of both analysis methods *(figure 11.1)*.

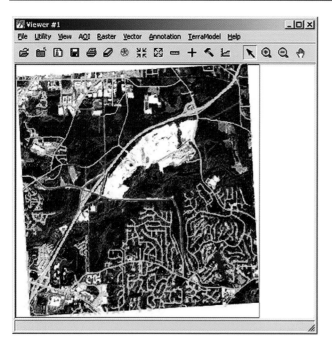

Figure 11.1 Current remote sensing imagery can be used to update road networks in an older dataset. New roads can be found visually by interactively tracing over them on the displayed image, automatically with digital linear feature extraction routines, or a combination of both.

Source: © CNES 2004. Courtesy of SPOT Image Corporation.

Most geographic information systems now provide at least basic image analysis functions to facilitate the use of digital images with other geographic data. Specialized image analysis software offers a wider range of more sophisticated image analysis operations that can be used to improve the visual interpretability of images as well as generate information for input to a GIS. Whereas basic image enhancement tasks can be easily accomplished by the lay person, more sophisticated analyses are generally undertaken by in-house specialists or outside contractors. This chapter provides an overview of the digital image analysis procedures commonly used in the analysis of remotely sensed data for GIS applications.

Image analysis functions can be broadly grouped into five categories: image restoration and rectification, image enhancement, calculating indices, classification, and modeling. *Image restoration and rectification* operations correct degraded image data, remove systematic geometric distortions, and change the image geometry to that of a convenient map projection. *Image enhancement* techniques improve visual interpretation by increasing the visual distinction among features for more effective display. *Indices* calculated from the remote sensing image data are used to generate measures of image characteristics, such as texture, and biophysical measurements, such as sea surface temperature, vegetation condition, and chlorophyll concentration. *Classification* operations automate the identification of features in a scene. The pixels in a digital image are categorized into one of many classes or *themes* representing different types of features useful to an application, such as land-cover types. This data can then be displayed as a thematic map or tabulated to determine the area of each class. *Modeling* procedures may incorporate several geospatial datasets derived from remotely sensed imagery and other sources to identify areas with specific characteristics, map current conditions, or predict future developments.

In addition to these five functions, remote sensing analysis also makes use of other sources of geospatial data, or *ancillary data*, to calibrate processing algorithms, for geometric correction, or to verify classification accuracy. Ancillary data may also be directly integrated into the analysis procedure, in effect modeling the characteristics of terrain features. For example, vegetation types that cannot be distinguished by their spectral response may be separated if the slope, aspect, elevation, or other characteristics of their environment differ. Classification procedures that use information derived from remotely sensed imagery together with other datasets useful in discriminating classes are often termed *hybrid classifications*.

In addition to providing valuable information on the current state of mapped phenomena, one of the most powerful applications that remote sensing offers is the ability to monitor how conditions have changed over time. Since the 1972 launch of Landsat 1, the first civilian remote sensing satellite designed to repetitively capture images of the globe, a vast record of earth imagery has been collected and archived. *Change detection* techniques have been developed to compare current imagery with archival data from the past.

The data used to calibrate, verify, or support remote sensing analyses is frequently stored within a GIS. The ease with which this data can be applied to remote sensing analyses depends on the data integration capabilities of the software, the geometric and classification accuracy of the datasets, and the creativity of the analyst. As with visual interpretation, in digital image analysis the quality and reliability of the results is in large measure dependent on the creative implementation of systematic analysis procedures by experienced and well-qualified personnel.

Image restoration and rectification

Image restoration and rectification operations adjust image values and geometry to correct for errors and limitations in the sensor system, minimize atmospheric effects, and change the image geometry to match a convenient geographic reference system. This section describes a variety of restoration and rectification techniques that may be used, including radiometric correction, geometric rectification, and resampling. The procedures differ for each sensor and the level of correction required. Most remote sensing data providers offer corrected image products as well as the uncorrected or raw data. For GIS applications, the corrected products are generally preferred because they are designed to be easily input and should register with other datasets in the GIS.

Radiometric correction

Ideally, the radiance measured by a remote sensing system would represent only the solar energy reflected by (or in some cases, emanating from) the features of interest on the ground. However, the radiance measured by a sensor is affected by other factors such as variations in scene illumination (e.g., areas in shadow), atmospheric conditions (e.g., atmospheric haze), sensor characteristics, and viewing geometry (e.g., haze effects are reduced when the viewing direction is away from the sun). Many of these factors are systematic and can be corrected using techniques such as atmospheric correction, while others can be compensated for by combining multiple bands of data. For example, the effect of shadow can be reduced by using a ratio of two bands.

Early electronic sensors had more technical problems than current systems. Inconsistencies in sensor operation frequently produced image errors caused by periodic malfunctions in the components of the sensing system or electronic interference as the data is transmitted to earth. Termed *image noise*, these defects may appear as striping along scan lines or as *salt and pepper* or *snow* artifacts in the image. These defects represent pixels with incorrect values. Image correction techniques eliminate the appearance of these errors by estimating a replacement value based on the neighboring pixels *(figure 11.2)*. Fortunately, technological improvements in sensor calibration, reliability, and data transmission have largely eliminated these problems.

Current systems, however, are not without their own technical problems. For example, anyone who has tried to photograph distant landscapes on humid days has inevitably experienced the frustration of trying to get a clear picture

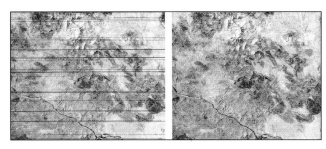

Figure 11.2   Striping shown in the left image can be corrected by replacing the missing scan lines with the average of the two adjacent lines.

Source: Image in *Images of the earth: A guide to remote sensing.* 2nd ed. by S. A. Drury. 1998. Reprinted with permission of Oxford University Press.

with vivid colors and distinct detail. On these hazy days, water droplets and other particulate matter in the atmosphere interfere with the sun's light rays that illuminate the landscape and are reflected toward the camera. This *atmospheric interference* changes the spectral properties of the light energy received by an optical sensor, reducing contrast and obscuring detail, particularly in the blue and green visible bands. A mountain forest that appears verdant one day may take on a dusty khaki hue the next. The same problem exists when trying to capture images of the earth from space. It can be particularly troublesome during digital analysis of remote sensing imagery, where the goal is to identify and discriminate features based on subtle tonal differences.

Atmospheric correction is the name given to a group of image processing methods used to reduce the degradation of image quality caused by atmospheric interference. The correction methods range from physical modeling of energy transmission through the atmosphere to empirical approximations of spectral changes. Cracknell and Hayes (1991) and Jensen (1996) provide good overviews of the different strategies that have been tested. Many atmospheric correction algorithms require costly measurements of atmospheric conditions on the date of image acquisition. For most applications, atmospheric corrections are not done as other image analysis functions account for atmospheric interference indirectly (e.g., contrast enhancements can be used to improve the clarity of hazy images). Where accurate measurements of physical phenomena are required, such as vertical temperature profiles of the atmosphere generated from the digital imagery captured by meteorological satellites, rigorous atmospheric corrections are implemented.

Geometric rectification

The fundamental property of all data included in a GIS is that they can be linked to a specific coordinate or region on the ground; that is, they are *georeferenced*. Before remotely sensed imagery or the information derived from it can be brought into a GIS, the images must be *georectified*, a process which enables the geographic coordinates of each pixel to be determined at a level of accuracy suited to the intended use. In image analysis terminology, this process is called *geometric rectification* or *geometric correction*. A detailed discussion of rectification is provided in appendix A. This discussion provides an overview of the process.

Uncorrected digital images normally contain geometric distortions so significant they cannot be used as maps. These distortions arise from variations in the altitude, attitude, and velocity of the sensor platform; characteristics of the scanning mechanism; relief displacement on the ground; and other factors. Some systematic distortions are well understood and easily corrected by applying the appropriate mathematical formula. For example, skew distortion is caused by the eastward rotation of the earth beneath the satellite, causing each sweep of the scanner to collect images of an area slightly to the west of the previous sweep. This distortion is corrected by slightly shifting each successive scan line to the west, which produces the parallelogram-shaped imagery typical of satellite scanner data. These systematic distortions are routinely removed by the satellite image distributor before releasing the imagery.

The remaining image distortions are generally caused by random effects that cannot be systematically accounted for and require some level of user input to remove. Satellite image distributors generally provide this service (usually for an additional charge) or customers can purchase the uncorrected or raw imagery and georectify them in-house.

Fundamentally, geometric correction is a two-step process. First, a mathematical transformation function is developed to calculate geographic coordinates for each pixel in the image. Then a resampling procedure generates a new image matched to the desired geographic base map or coordinate system. In some cases, the unrectified image may be stored with the georeferencing equations, and a rectified image is only generated as required, such as for a background image over which other geographic layers are displayed.

The transformation function

The driver behind the geometric correction process is the mathematical transformation used to register the original image grid to a specified map projection. The different methods for developing the transformation function can be broadly categorized into three approaches: the use of ground control points (GCPs) alone, the use of sensor parameters to calculate a transformation function (usually using GCPs as well), and the additional use of a digital terrain model with one of these methods to remove relief displacement effects.

GCP approach

A GCP-based transformation is the most straightforward approach. The correction process involves identifying numerous control points and determining their location in the image and their geographic coordinates from field observation or from a reference source such as a map having a suitable level of accuracy. (Where an image is to be registered to another image, GCP selection involves identifying a set of features visible on both images.)

Features easily identified on the imagery whose ground location can be accurately determined, such as highway intersections and distinctive shoreline features, are used as GCPs. Identification of GCPs on the image is usually done interactively on a display *(figure 11.3)*.

The number of points required depends on the image and terrain characteristics and the level of geometric accuracy required. The GCP values are analyzed by least squares regression to generate mathematical functions that calculate a location in the input image corresponding to a location in the reference source. The translations applied are usually first- or second-order polynomial functions that are automatically generated in software from a set of user-specified control points.

Geometric rectification using GCPs alone is best suited to imagery acquired with a nadir viewing perspective (i.e., viewing vertically down to the landscape) over relatively level terrain in which there are well-defined, permanent features visible for GCP location. Landsat ETM data with 30 m spatial resolution over a flat to moderately rolling landscape is often rectified this way.

Rectification from sensor parameters

A second rectification method is to calculate a transformation function based on the position and orientation of the sensor when the image was acquired and the imaging geometry of the sensor system (termed the *sensor model*). Several ground control points are generally used as well to improve accuracy. This method makes use of complex algorithms and requires detailed sensor parameter information, but it can provide

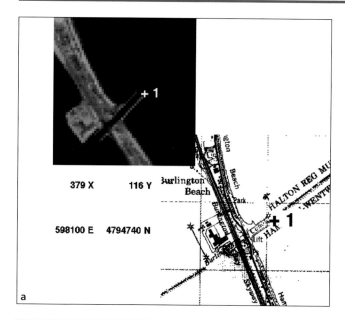

379 X   116 Y

598100 E   4794740 N

a

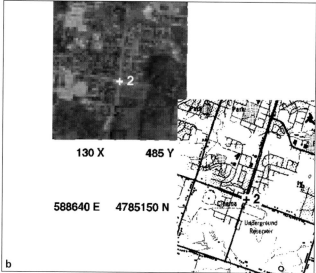

130 X   485 Y

588640 E   4785150 N

b

Figure 11.3   Interactive collection of ground control points. Well-defined features such as the end of a pier (a) or a road intersection (b) are chosen as ground control points because they are easily recognized and have sharply defined edges.

Source: J. Piwowar

Neither of these two methods takes into account relief displacement—scale variations resulting from differences in terrain relief that cause the position of features in the image to be shifted. Corrections for relief displacement can be added to either transformation method using digital elevation data for the scene *(figure 11.4)*. The elevation data may be an existing dataset purchased from a supplier or may be generated using a second image that provides suitable stereo coverage.

The magnitude and direction of relief displacement also depends on the characteristics of the sensor. For example, in a vertical aerial photograph, points closer to the sensor (or with higher elevation) are displaced radially outward from the center of the image; in optical scanners they are displaced outward perpendicular to the flight path. In a radar image, relief displacement is towards the sensor and perpendicular to the line of flight (see chapter 5).

There are no set rules to indicate when relief must be accounted for in the georeferencing process. For areas of low relief or for small-scale imagery with pixel resolutions on the order of 1 km, relief displacement errors are small enough not to require additional terrain information. For mountainous

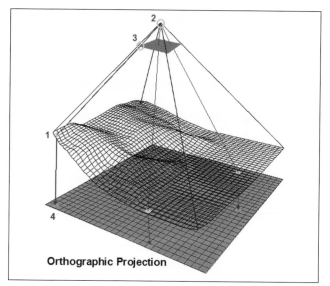

Orthographic Projection

Figure 11.4   A digital elevation model (DEM) is used to remove distortions from the imagery arising from variations in topographic relief. For each pixel location in the source image (3), an elevation value is retrieved from the DEM (1). Using this data and parameters that define the sensor's position and orientation at the time of image capture (2) and the sensor model (the image formation characteristics of the sensor, e.g., view angle), equations are calculated that are used in resampling the source image to construct the orthographic image (4).

Source: Imagery provided by Leica Geosystems GIS and Mapping, LLC.

greater accuracy than by using GCPs alone. This method is preferred, especially when rectifying oblique imagery. Most imagery collected by high-resolution earth resource satellites, such as IKONOS and QuickBird, are acquired obliquely and hence are frequently georectified in this way.

areas or high-resolution images, however, relief displacement can be significant, and elevation data is generally required to achieve suitable registration of the imagery to a map. As previously indicated, medium-resolution imagery (i.e., data with 10–50 m spatial resolution) acquired over a flat to moderately rolling landscape is often accurately rectified without the need to account for relief displacement.

Resampling

Once an acceptable set of transformation equations has been derived, the rectified image is generated by *resampling*, the process of generating a new image with different geometric characteristics.

Conceptually, resampling can be thought of as taking each pixel from the original scene, calculating its projection coordinates using the transformation equations, and then saving the pixel data in that georeferenced location. In practice, however, this procedure would be complex and generate poor results since a pixel may expand or contract to cover more or less than one cell of the output image. Instead, a new empty grid in the desired map projection is generated and then the value for each cell in the new grid is calculated from the one or more pixels in the original scene that should contribute to the new pixel values. Based on the application, one of three resampling algorithms is generally used: nearest neighbor, bilinear interpolation, or cubic convolution.

The *nearest neighbor* method calculates the center of each pixel in the output image (the rectified image), finds the corresponding location in the original image (the input image), and uses the value of the pixel whose center is nearest. In figure 11.5, the pixel with the bold outline in the output image would be assigned the value of the red pixel in the source image.

Bilinear interpolation calculates a distance-weighted mean from the four nearest neighbors (both the red and green pixels in the figure), and *cubic convolution* calculates a distance-weighted mean from a block of 16 pixels in the input image surrounding the output pixel (the red, green, and blue pixels).

Of these methods, nearest neighbor resampling is the only one that does not change the original pixel values. If calculations are to be made from the rectified image, such as a classification or generation of biophysical measurements, nearest neighbor resampling is generally used as it retains the subtle spectral detail present in the original pixel values. However, images generated by nearest neighbor resampling tend to have a blocky appearance because some pixel values may be used more than once and others not used at all, depending

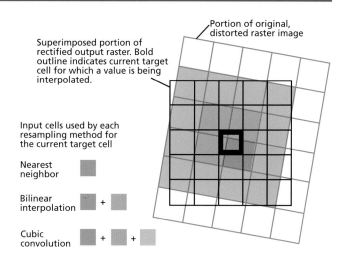

Figure 11.5 Matrix of geometrically corrected output pixels (black grid) superimposed on the pixels of the source image (blue grid).

Source: Image in *Getting started: Rectifying images*. 1999. Lincoln, Neb.: MicroImages, Inc. www.microimages.com

on the alignment and cell spacing of the two grids *(figure 11.6b)*.

Bilinear interpolation produces a smoother image than the nearest neighbor method *(figure 11.6c)*. Cubic convolution frequently yields the most visually appealing results but requires more processing *(figure 11.6d)*. Cubic convolution is generally preferred for images that are to be interpreted visually.

Most image analysis systems and many GIS have the software tools to perform basic geometric corrections. Precision corrected imagery is available from remote sensing image distributors at a higher cost than the uncorrected imagery. The choice to purchase preprocessed imagery or do the corrections in-house may be influenced by several factors including availability of the necessary reference data and sensor parameters, the capabilities of the software and knowledge of the image analyst, the terrain characteristics in the area of interest, the accuracy and precision required, financial resources, and the time available to carry out the work and perform the necessary quality assurance testing.

Image inhancement

Image enhancement refers to a number of image processing procedures that improve the visual interpretability of an image by applying algorithms that change the contrast,

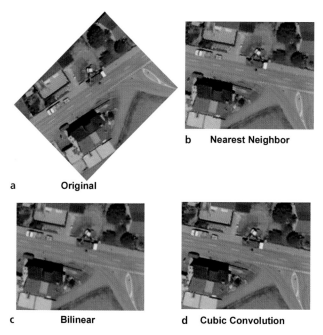

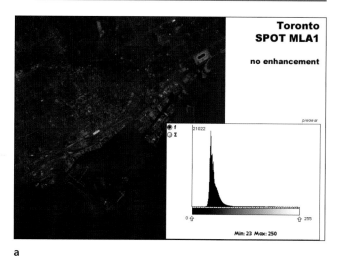

a

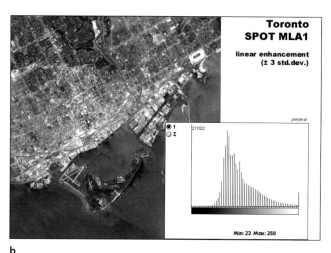

b

Figure 11.6   Comparison of image resampling methods. The original digital image was rotated 45 degrees using nearest neighbor, bilinear interpolation, and cubic convolution resampling.

Source: Imagery provided by Leica Geosystems GIS and Mapping, LLC.

Figure 11.7   Brightness and contrast enhancement of a SPOT HRV Band 1 image of downtown Toronto. The original image (a) appears dark because most of the pixels have low brightness values, as seen in the histogram for the image. By increasing the brightness and contrast using image enhancement techniques that distribute the pixel values over the full range of the output device, an image is produced that is much easier to interpret (b).

Source: © CNES 2004/Courtesy of SPOT Image Corporation.

brightness, sharpness, and color rendition of features in the image. Enhanced images contain no more information than the original data (and in some cases may contain less), but they are more easily interpreted because subtle differences in tone and color have been exaggerated, which makes it easier to for us to visually recognize important features.

Enhancements are applied to remotely sensed images to facilitate *visual* interpretatations. Unenhanced image data is preferred when the values are to be used in *digital* analyses, such as classification, because they are a truer representation of the energy received by the sensor.

Brightness and contrast enhancement

Digital remotely sensed imagery (particularly satellite imagery) produced by a sensor (called *raw image data*) generally appears dark overall with little contrast. This is because the sensor must be calibrated to collect images with the full range of brightness levels that might be encountered, from high reflectance features such as bright snow or sand to low reflectance objects like clear water or terrain shadows. The range of values in a typical image is often less than half the brightness range that can be captured on the display device

such as a computer monitor. Figure 11.7a shows a raw image and the accompanying histogram that graphically represents the distribution of pixel values in the image. Brightness values are plotted on the horizontal axis, and frequency (the number of pixels that have a given digital value) is plotted on the vertical axis. Note that most pixel values fall at the low end of the graph (the left side) which is why the image appears dark.

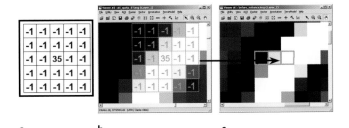

a   **b**   **c**

Figure 11.8 Edges are enhanced in digital imagery by averaging small sets of pixels across the image in a process called convolution filtering. A small matrix of weights (a) is iteratively placed over every pixel in the image (b). The value from each pixel that the matrix falls over is multiplied by its corresponding weight. These multiples are then summed, and the resulting value is used as the pixel value for the central matrix position in the enhanced image (c). In the edge-enhanced image (c), notice how the white linear feature has been made brighter and its neighboring cells darker, thereby giving it a sharper appearance when viewed at normal magnification.

Source: Imagery provided by Leica Geosystems and Mapping, LLC.

The most common contrast enhancement is a contrast stretch, whereby each pixel is multiplied by a function that in effect stretches the narrow range of digital values to extend over the full dynamic range of the display, as shown in figure 11.7b. The resulting image is visually much easier to interpret, and the corresponding histogram shows the pixel values now extend over the entire output range, the full width of the graph. To enhance color imagery, contrast enhancements are applied to each of the three bands making up the red, green, and blue components of the image. Linear stretch, piecewise linear stretch, and histogram equalization are commonly used contrast stretching algorithms.

Edge enhancement

Whereas a contrast enhancement focuses on the spectral properties of the data to improve visual interpretation and analysis, an *edge enhancement* is applied to improve the appearance of *spatial* patterns present in the data. In image terms, an *edge* is any place where there is an abrupt change in pixel values. An edge enhancement works to change the pixel values on either side of the edge; it makes the light tones lighter and the dark tones darker. This emphasizes the tonal change in the image *(figure 11.8)*. By exaggerating localized tonal changes in an image, edge enhancements have the effect of highlighting linear features, making images appear sharper *(figure 11.9)*. For this reason, they are sometimes referred to as *sharpening enhancements*.

Edge enhancements have been useful for a variety of applications, including geologic mapping, urban planning, and cartographic display. Lineaments are linear features that are used by geologists to identify structural properties in bedrock. These topographic or tonal features are sometimes so subtle that they may only be noticeable from the vantage of space. Figure 11.10 shows how these features may be highlighted through edge enhancement.

Figure 11.11 illustrates the application of an edge enhancement to extract the spatial structure in a rapidly developing urban center. Malcolm et al. (2001) show how this type of information can be used in conjunction with GIS by planning departments in developing countries to identify new squatter settlements that are in need of basic urban infrastructure and support services.

Edge enhancements are also applied to enhance the visual appeal of spaceborne imagery. In particular, edge enhancement is frequently one of the steps in *pan-sharpening*, or combining high-spatial resolution data with lower-resolution multispectral imagery to create a pseudo high-resolution color image that preserves the spectral information and facilitates better visualization and interpretation. Pan-sharpened images are quickly becoming the *de facto* standard for remote sensing image presentation to the general public *(figure 11.12)*.

b

Figure 11.9   The effect of applying an edge enhancement to a SPOT scene of Waterloo, Canada. Figure (a) is the original image, and figure (b) is the edge-enhanced image.

Source: © CNES 2004. Courtesy of SPOT Image Corporation.

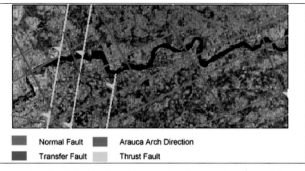

	Normal Fault		Arauca Arch Direction
	Transfer Fault		Thrust Fault

Figure 11.10   A Radarsat SAR image showing a series of parallel lineaments related to faulting in a tropical environment.

Source: Radarsat-1 data © Canadian Space Agency 1996. Received by the Canada Centre for Remote Sensing. Processed and distributed by RADARSAT International.

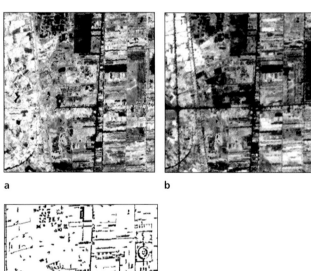

a   b

c

Figure 11.11   New linear elements identified along the fringe of a rapidly growing urban area (Bangkok, Thailand). The edges in (c) were extracted from an analysis of the changes from the 1995 (a) and 1998 (b) data. The images in (a) and (b) are both NDVI composites (see *Biophysical measurements and indices*, below).

Source: Images courtesy of Space Imaging Corporation.

a

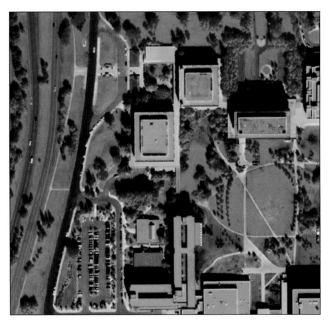

b

Figure 11.12 Pan-sharpening of a QuickBird image of the University of Regina campus. The original image (a) has a spatial resolution of 2.4 m. The image in (b) has been visually enhanced with the inclusion of the spatial detail from the 0.61 m panchromatic band.

Source: Images courtesy of DigitalGlobe.

Band selection for color composite generation

As discussed in chapter 4, a natural color composite image[1] can be captured by recording the brightness values in the visible blue, green, and red spectral bands for each pixel as a set of three images, one for each band. Displaying the images using blue, green, and red light, or using the corresponding color guns of a computer monitor, will then generate an image with a color rendition close to that of the original scene *(figure 11.13a)*.

Unfortunately, atmospheric scattering at the shorter visible wavelengths (blue and, to some extent, green) can severely degrade image quality. Thus, a natural color composite image generated from the blue, green, and red visible bands generally has a bluish cast with low contrast and poor color discrimination. Although brightness and contrast enhancements can substantially improve the visual quality of natural color images, they can never fully repair severely degraded data.

Another approach is to use spectral bands with minimal atmospheric degradation, such as those in the red and infrared (IR) spectral regions. The typical false-color infrared imagery, with its characteristic red vegetation, is but one of several commonly used color composites. Here the near-IR, red, and green bands (e.g., Landsat ETM bands 4, 3, and 2) are displayed as red, green, and blue, respectively *(figure 11.13b)*. Since chlorophyll compounds found in the leaves of green plants strongly reflect near-IR radiation and since radiation in near-IR wavelengths travels through the atmosphere with minimal interference, healthy green vegetation appears bright red in these images. This particular combination of colors and spectral bands is used so commonly in remote sensing that it is known as a *standard false color* or *color infrared* composite.

There are no set rules governing color composite displays, and the image bands can be displayed in any order to enhance different features in the imagery. For example, the particular band combination shown in figure 11.13c is useful for highlighting urban density. Some researchers advocate showing different spectral regions (e.g., visible, near-IR, mid-IR) in separate bands, such as in figure 11.13d. Although natural color and standard false color composites are used most often, the choice of band combination used in image displays is largely left to the user's discretion.

---

1. A natural color image is one that represents features in colors similar to those perceived by the human eye.

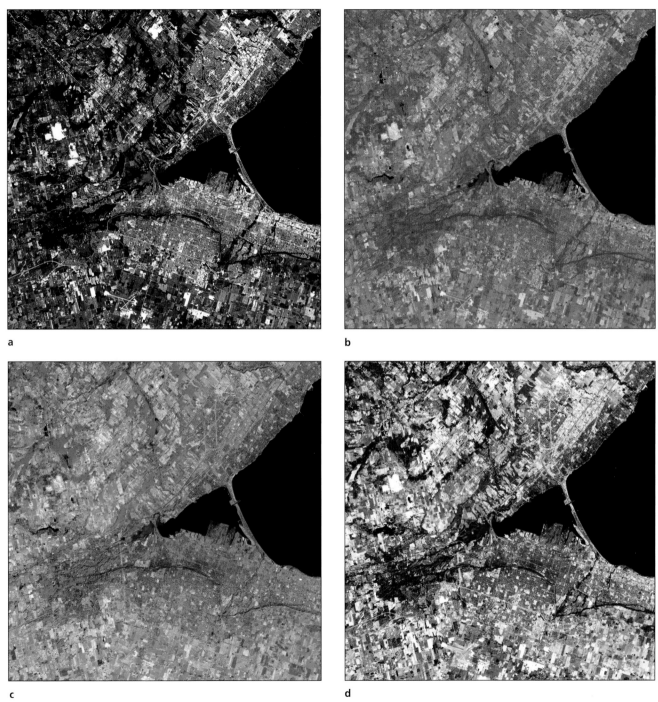

Figure 11.13 Landsat ETM color composite images of Hamilton, Ontario.
(a) Natural color (red=ETM3, green=ETM2, blue=ETM1)
(b) Standard false color (red=ETM4, green=ETM3, blue=ETM2)
(c) False color (red=ETM3, green=ETM4, blue=ETM5)
(d) False color (red=ETM7, green=ETM5, blue=ETM3)

Source: © 1999. Government of Canada with permission from Natural Resources Canada.

a

b

Figure 11.14 Band ratio to reduce terrain shadows. The Landsat ETM Band 3 image in (a) has been divided by ETM Band 4 to produce the shadow-reduced image in (b). The Landsat ETM image was acquired near Banff, Alberta, on October 10, 2000.

Source: © 1999. Government of Canada with permission from Natural Resources Canada.

Indices to enhance visual image quality

Indices calculated from the brightness values of each pixel can also be used to enhance image quality. For example, a ratio of two bands is sometimes used to reduce the effect of shadows in the landscape *(figure 11.14)*. The Normalized Difference Vegetation Index (see *The normalized difference and enhanced vegetation indices*, below) is also used to emphasize the vegetation characteristics in a scene.

Image analysts sometimes feel limited in their ability to effectively display all the data from a scene: a Landsat ETM image, for example, has seven spectral bands, but we can only view three of them at once (since there are only red, green, and blue display colors). A principal components analysis transformation (see *Principal components analysis*, below) is one method of reorganizing the information content between the spectral bands of an image so that more detail can be viewed at once.

Cloud-free image composites

In most cases, analysts would like cloud-free images for earth resource applications. In regions with intermittent cloud cover, cloud-free composite images can be generated by combining several images acquired over a few days and selecting only the clearest pixels from each scene. To produce the composite, a cloud-cover index is calculated for each pixel of each of the images *(figure 11.15)*. The images are then registered, and a composite image is generated by using, for each pixel location, the values in the image with the lowest cloud-cover index for that pixel. In this way, a single image is built up, with cloud pixels replaced by cloud-free data that covers the same ground area but is taken from a slightly later or earlier date.

Weekly or 10-day composites are routinely generated from images acquired by lower spatial resolution instruments such as the AVHRR, MODIS, SeaWiFS, and SPOT Vegetation sensors. They are widely used for climate change monitoring, as well as for specialized analyses such as crop condition assessment as discussed in the *Agriculture* section of chapter 12.

Biophysical measurements and indices

The spectral reflectance values recorded as remote sensing images can be used to measure environmental parameters. In meteorology, complex algorithms are used to calculate vertical temperature profiles of the earth's atmosphere from imagery with multiple narrow bands in the near- and mid-infrared. Estimates of water surface temperature, chlorophyll

a

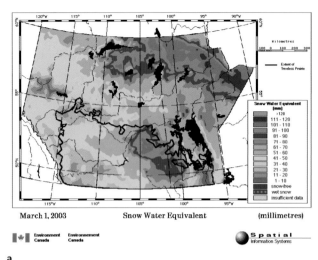

a

March 1, 2003   Snow Water Equivalent   (millimetres)

b

Figure 11.15   Cloud masks. The NOAA-AVHRR image was obtained over Canada on July, 3, 1993. It shows the dark boreal forest (center), lighter tundra to the northeast, and various cloud formations (brightest pixels). The cloud mask in (b) highlights clear-sky pixels in red tones and the cloud-contaminated pixels in shades of blue and gray.

Source: Courtesy of the Canada Centre for Remote Sensing, Natural Resources Canada.

b

Figure 11.16   Cryospheric parameters derived from remote sensing imagery. Snow water equivalents (a) and sea ice concentrations (b) are calculated from measurements of microwave energy sensed by the Special Sensor Microwave Imager (SSM/I) instrument. The SSM/I is a passive microwave radiometer flown aboard Defense Meteorological Satellite Program (DMSP) satellites and made available for civilian use through NOAA.

Source: (a) Image courtesy of Environment Canada. Reprinted with the permission of the Minister of Public Works and Government Services Canada, 2004. (b) Daily ice analysis chart provided by the Canadian Ice Service of Environment Canada.

concentration, suspended sediment *(figures 7.8, 7.10, and 7.34 in chapter 7)*, snow water equivalent *(figure 11.16a)*, and sea ice types *(figure 11.16b)* are other environmental measurements commonly produced.

Vegetation—its presence, type, and condition—is of key importance to human activity. Vegetation indices provide measures of the amount, structure, and condition of vegetation. They are also used to derive biophysical parameters such as leaf area index (LAI), percentage of green cover, biomass, and the fraction of absorbed photosynthetically active radiation (fAPAR). Discussions of the range of vegetation indices developed for remote sensing can be found in Jensen (2000) and Lyon et al. (1998).

Typically calculated from the spectral reflectance values of two or more bands, vegetation indices tend to generalize differences in species composition while enhancing the spectral reflectance characteristics typical of vegetation. In so doing, vegetation indices have proven valuable for monitoring changes in vegetation condition and seasonal development over time and facilitate comparison of vegetation in different regions. The vegetation index values for multiple dates over the growing season have also been used as seasonal profiles in digital image classifications (see below) to distinguish land-cover classes difficult to discriminate by their spectral reflectance on single-date imagery.

Due to their relative ease of application, vegetation indices are used in a wide range of global- and regional-scale biophysical studies, including climate, biogeochemical, and hydrologic modeling; natural resource inventory and monitoring; land-use planning; agricultural crop monitoring and forecasting; deforestation; assessments of rangeland condition and grazing capacity; desertification and drought forecasting; and pollution and public health issues.

For example, a vegetation index has been used to model the predicted spread of West Nile virus in North America. Mosquitoes and birds that carry West Nile virus require very specific environmental conditions to thrive. As temperature and precipitation levels rise and fall throughout the year, these conditions migrate. Satellite instruments measure land surface temperature and vegetation distribution from space, giving scientists a unique perspective from which to monitor seasonal changes in habitat. Combined with observations of infected birds and mosquito population density, and other disease control data, the risk of infection can be estimated and mapped *(figure 11.17)*.

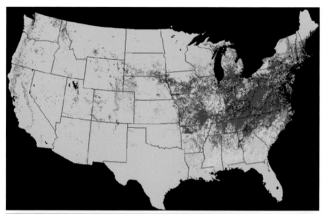

○ uninfected crow   ● infected crows   ☐ no data   no risk ■☐☐■ increased risk

Figure 11.17   A West Nile virus sample risk map for the United States. Data was derived from NASA satellite imagery, and infection rates were derived from the Centers for Disease Control and state health departments.

Source: Image courtesy of NASA.

The normalized difference and enhanced vegetation indices

One of the most widely used vegetation indices is the normalized difference vegetation index (NDVI). The NDVI is calculated from a sensor's visible red (X_{red}) and near-infrared (X_{nir}) bands as follows:

$$\text{NDVI} = \frac{X_{nir} - X_{red}}{X_{nir} + X_{red}}$$

Originally developed for the early Landsat systems, the NDVI calculated from NOAA AVHRR imagery has been used for more than two decades for a broad range of applications by researchers and practitioners worldwide (see the example in the *Agriculture* section in chapter 12). As a result, its use is supported by a large body of research and expertise.

In addition to producing useful measures of vegetation growth and productivity, the NDVI also has practical application as a visual enhancement technique. Piwowar and Millward (2002) show how the NDVI calculated from a Landsat TM/ETM or SPOT HRV scene can be a useful single-band composite of the multispectral data that each sensor produces *(figure 11.18)*. It is also possible to use NDVI imagery as an input to digital classification.

As sensor capabilities improve, researchers have continued to modify the capabilities of the NDVI. The *enhanced vegetation index* (EVI) developed for the MODIS sensor, for example, differs from NDVI by using data from the visible blue band to correct for atmospheric and background effects. The EVI appears to better discriminate subtle differences in areas of high vegetation density, situations in which NDVI tends to saturate *(figure 11.19)*. The NDVI value is predominantly influenced by the chlorophyll content of the vegetation whereas the EVI, in addition to chlorophyll sensitivity, is also more responsive to canopy structural variations,

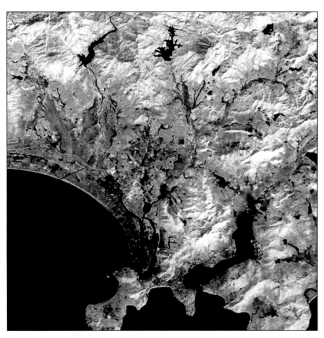

b

a

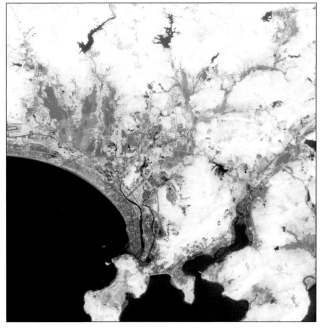

c

Figure 11.18   The application of NDVI as a single-band composite. The NDVI image in (c) was created from the red band (a) and near-infrared band (b) images. SPOT HRV image of Sanya City, China, acquired November 1997.

Source: © CNES 2004. Courtesy of SPOT Image Corporation.

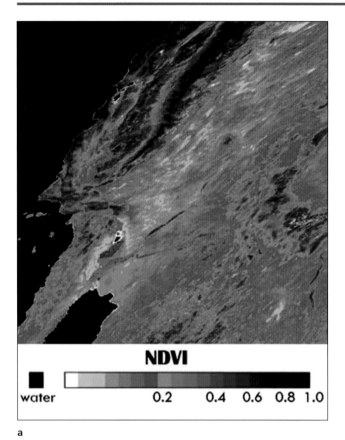

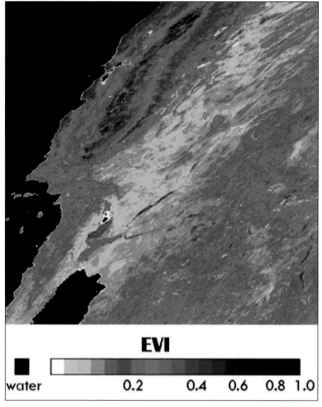

a

b

Figure 11.19 Comparison of vegetation index images. NDVI and EVI images of southwestern North America were generated from the same MODIS sensor data acquired on July 12, 2002. Notice the greater detail provided by the EVI image in areas with high NDVI index values (dark green).

Source: Image courtesy of the U.S. Geological Survey.

including leaf area index (LAI), canopy type, plant physiognomy, and canopy architecture. In some applications, such as global vegetation studies, the two indices complement each other to improve the detection of vegetation changes and the extraction of canopy biophysical parameters.

Because the NDVI has been used for several decades, it continues to be used especially for change detection studies. Figure 7.15 in chapter 7 illustrates the way the EVI vegetation index represents seasonal changes. Both the NDVI and EVI are standard MODIS products produced at 16-day and monthly intervals and available at 250 m, 500 m, 1 km, and 25 km resolutions.

Indices of image characteristics

Image texture measures

Texture, a characteristic easily observed visually and an important element of visual image interpretation, is more difficult to assess by digital image processing. Texture in an image results from the local variability in pixel values. Areas with a wide variation in pixel values appear rough while relatively homogeneous areas appear smooth *(figure 11.20a)*. Whereas brightness and contrast enhancement operate on pixels individually, texture measures are assigned to a pixel based on its value and those of pixels in its immediate neighborhood.

Texture is an inherent property of image features. The use of texture measures has the potential to enhance visual and digital interpretation because they recognize the reality that a pixel does not exist in isolation in the image array. This is particularly true for fine-resolution imagery where the improved scale of measurement may mean that an individual surface feature (e.g., a tree) is spectrally represented in several adjacent cells. Marceau et al. (1989), Coulombe et al. (1991), and Treitz et al. (1993) give different examples of how texture data can be included to enhance traditional multispectral classifications.

Several approaches to quantifying image texture have been developed based on pixel value statistics, the Fourier power spectrum, fractals, and others. Of these, the measures based on pixel value statistics, such as the gray-level co-occurrence matrix (GLCM) method, have proven to be the most robust for remotely sensed imagery *(figure 11.20b)*.

Principal components analysis

Principal components analysis (PCA) is a statistical-mathematical technique used to find trends and patterns in highly correlated datasets. The technique has been valuable for image enhancement, as a means to reduce the number of channels in multispectral imagery (Taylor 1974, Singh and Harrison 1985), and for change detection studies (Fung and LeDrew 1987, Millward and Piwowar 2002). Its use in remote sensing is founded on the observation that there is considerable redundancy in multispectral image data, both between spectral bands and between images of the same area acquired on different dates. A PCA will identify and remove this redundancy and concentrate the scene information in fewer bands.

For example, the bands shown in figure 11.21a all appear very similar, i.e., the information portrayed by the images is highly correlated. A PCA transforms the data so that each image—or *component*—is statistically independent from the others; that is, all of the inter-band correlation has been removed. The first component carries the maximum information content from the original images, typically the overall brightness *(figure 11.21b)*. The second component maximizes the remaining image information, and the third what remains after that, and so on. Typically, over 90% of the variability in pixel values in an image is captured in the first two or three components.

a

b

Figure 11.20   Smooth and rough image texture. Urban areas typically exhibit rough texture, as in the residential subdivision in (a), while the plowed field to the right of the highway has a smooth texture. Note the high variance among the pixel values in the sample grid for the residential area. The pixels in the plowed field, on the other hand, tend to have more similar values. An example of a texture image created from this data is shown in (b). Regions with high texture are shown in brighter tones.

Source: Images courtesy of DigitalGlobe.

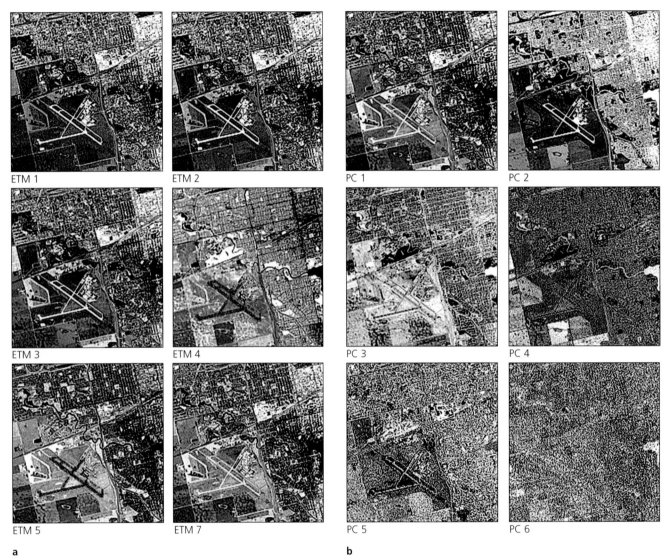

ETM 1   ETM 2   PC 1   PC 2

ETM 3   ETM 4   PC 3   PC 4

ETM 5   ETM 7   PC 5   PC 6

a

b

Figure 11.21 Principal components analysis to remove multispectral correlation. Note how many of the image details remain the same in (a), particularly among the three visible bands (ETM 1-3). This data is highly correlated with each other. The principal components (PCs) calculated from these three bands are shown in (b). Note how the first two components (PC 1 and PC 2) capture the bulk of the information content of the ETM bands. There is increasingly less detail in the higher components until PC 6, which is mostly noise.

PCA can also be applied to multiple images collected on different dates to identify change over time. Just as the previous example showed how there was significant spectral correlation between the bands in a multispectral image, there will also be a considerable amount of temporal correlation between images of the same region acquired on different dates. The challenge in monitoring change over time with multidate images is to identify the relatively small proportion of pixels that actually change. In this situation, the first component image will show features that did not change over time while the other components will highlight features that exhibit different degrees of change. For example, multidate imagery *(figure 11.22a)* was processed by PCA to generate a set of principal component images *(figure 11.22b)*. Much of the urban expansion that occurred between 1991 and 1999 is emphasized in the second component (PC2).

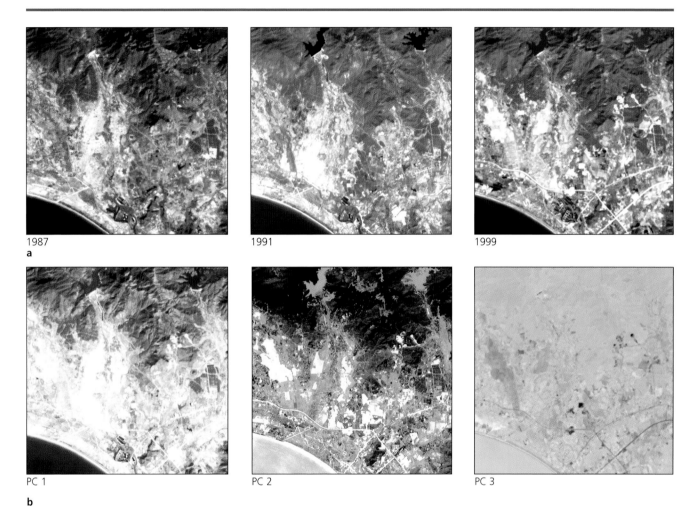

1987

a

PC 1

b

Figure 11.22   Temporal analysis of multidate imagery using principal components analysis. A principal components analysis of three images from 1987 to 1999 was done to study changes in landscape features over time. Note how most image features remain the same each year in (a). The principal components (PCs) calculated from these three years are shown in (b). The first component (PC 1) identifies where

the source images are highly correlated, i.e., it shows what has not changed between the years. The second component (PC 2) highlights the areas of major change, e.g., some major roads were built between 1991 and 1999.

Source: Images courtesy of Space Imaging Corporation.

By using a PCA transformation, most of the information in multiband or multitemporal images is compressed into the first few component images. Not only can this be an effective means to reduce data volumes without losing significant information, but the component images can also be used to generate enhanced color-composite images for visual interpretation or in a classification. Comprehensive discussions of PCA can be found in Jensen (1996) and Richards and Jia (1999).

Classification

We understand our world by classifying the features that we observe. Looking out at a landscape or at an aerial photograph, we recognize roads, buildings, rivers, and other features by their shape, color, texture, size, and other characteristics.

The term *classification* is used both in GIS and remote sensing image analysis to mean the categorization of data into useful groupings. However, the process of how these groupings are formed is significantly different in remote sensing. The data in a GIS is in effect *information classes*, i.e., they are classes that represent categories of information considered useful to the user of the information. The information is commonly stored in the form of objects to which a label representing the information class has already been assigned. Roads, rivers, or land parcels are examples of common information classes stored in a GIS.

In order to be entered into a GIS, both the class and geographic extent of each object must be known. Thus, in a GIS, the process of classification involves assigning objects *whose identity and boundaries are already defined* to groupings based on characteristics stored with the object or readily calculated from the dataset (such as the distance to the nearest school).

In remote sensing, the boundaries and often the information classes themselves are not known; they have to first be generated from the image data. The data to be classified represents a physical phenomenon, most commonly an object's spectral reflectance in one or more bands, presented in the form of an image. Unlike in a GIS, the brightness values in an image do not represent meaningful geographic objects such as a lake or a road; i.e., they do not represent information classes. An analyst may identify as an *image class* areas within an image that exhibit distinctive characteristics, such as a smooth-textured, dark-blue colored feature believed to represent water with low sediment concentration. A computer algorithm can then be used to identify all areas with similar image characteristics. However, these image classes may or may not correspond closely to useful information classes. For example, it may not be possible to accurately separate pine forest from other types of conifers using the particular image and analysis techniques available. The boundaries between different image classes may not be sharp; there may be a gradation from one to the other.

Though a remotely sensed image may be displayed from within a GIS, until useful information classes are derived from it and, in the case of a vector-based GIS, the classification is converted into geographically referenced elements such as points, lines, and polygons and assigned a class label, there is no way for them to be recognized within the GIS domain.

In visual image interpretation, the classification is produced by a more or less systematic process of recognizing distinctive features, such as an aircraft or a river, or delineating areas with similar characteristics such as a smooth-textured, dark-toned region later identified as *mixed conifer forest* and assigned that class label.

Automated classification methods depend primarily on defining rules to assign pixels to a class based on their spectral reflectance values. Some algorithms analyze the values of neighboring pixels as well. Once these image classes have been formed, they can then be combined to generate the groupings (i.e., the information classes) best suited for a specific application.

Some desired information classes may not be distinguishable on the imagery, and other data may be needed to augment the image analysis. Information classes that cannot be distinguished in the imagery may have to be identified from other sources or their inclusion within other classes accepted. For example, to correctly label a body of water found through multispectral classification as a reservoir rather than a natural lake, land zoning data taken from an existing GIS database may be required.

In a sense, the end product of a classification of remotely sensed imagery, a set of labeled objects, is the starting point for GIS classification operations. Thus, it is not surprising that much of the data used in a GIS is generated from remote sensing data.

Computer-based classification algorithms depend primarily on the spectral characteristics of a pixel. Shape, texture, and size, which are easily used in visual interpretation, are much more difficult to incorporate into digital classification algorithms. Computer-based classification offers rapid processing of large volumes of data and can also provide more consistent and repeatable results than visual interpretation, although it is not necessarily more accurate. Also, area calculations that are difficult and time consuming to do manually are easily computed from a computer-classified image.

Visual interpretation is limited to the spectral bands represented in the image, a single band for a black and white representation and normally three bands for a color image. Computer classification can make use of as many bands of image data as are available. What's more, computer-based classifiers can distinguish the smallest brightness difference represented in the image data whereas a human interpreter cannot reliably distinguish small differences in color or tone. However, a skilled visual interpreter is often better able to identify features based on their spatial characteristics and logical association with other nearby features than automated classifiers. For this reason, the two analysis methods are best viewed as complementary.

Image analysis software has made it relatively easy to generate a classification for a digital image. However, considerable expertise, planning, and resources are needed to reliably produce digital image classifications optimally suited to a specific application with a level of accuracy suited to the intended use.

A rigorous classification analysis can be costly, involving not only processing of the image data but also such tasks as the collection of reference data and assessment of accuracy. In many cases, alternatives other than a comprehensive image classification may be more cost effective. For example, in urban GIS applications, imagery is often used as a background over which other information such as lot boundaries and street center lines are overlaid. A great deal of useful information, such as the presence of new structures on a site, can be readily interpreted by simply viewing the image without the need for it to be classified.

Defining the classes

Development of a rigorous and complete set of class definitions is critical to the success of a classification analysis. Adequate planning at this stage requires that the information needs are rigorously defined. In so doing, ambiguities and inconsistencies can be recognized and resolved early, when it is least costly to change the analysis design. Then the information classes required to satisfy those needs can be defined, reviewed, and refined by those who will ultimately use the classification results.

The classification must be sufficiently detailed to distinguish characteristics necessary to the application. Care must be taken, however, not to overspecify the classes because more detailed classes are generally more costly to analyze.

An efficient classification system should be complete and hierarchical. That is, it should be possible to assign a class to every observation, and more detailed classes should be organized as subdivisions of more general classes. A *Forest* class, for example, can be grouped with *Shrub, Grassland* and *Other Vegetation classes* to form a *Vegetation* category at a more general level or at a more detailed level, split into *Coniferous* and *Deciduous*, which may be further divided into classes representing individual tree species *(figure 11.23)*. A hierarchical system permits the specification of detail where it is needed and the generalization of classes that are less important for the intended use.

Class definitions must be rigorous. This can be difficult for classes that grade into each other. For example, a conifer forest may include deciduous trees, but at some point the proportion is high enough for the class assignment to change from *Conifer Forest* to *Mixed Forest*. A consistent method of distinguishing these two classes must be defined in order for them to be accurately classified. Some classes may be defined by regulation or by observation, and a choice must be made. For example, how are residential, commercial, and industrial

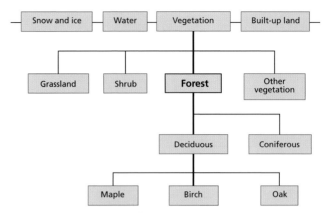

Figure 11.23   An example of how a Forest class may fit into a hierarchical classification scheme.

areas of a city to be distinguished? Will they be delineated according to zoning bylaws or by building characteristics? Class definitions, together with such project constraints as time, budget, and the availability of data, personnel, and other resources, determine the range of feasible analysis methods.

Remote sensing methods need not satisfy all information needs in order to be of value. Some classes may not be uniquely identifiable from remote sensing imagery but may be separable with additional information such as slope, aspect, and elevation obtainable from a digital elevation model. In some cases, they may have to be assessed by direct field observation, but remote sensing may be useful for finding approximate feature locations, thereby reducing the number of expensive site visits that might be required.

Some classes may not be directly observable by remote sensing. However, they may have a strong enough correlation with detectable classes to be confidently classified. For example, soil types are commonly deduced by the type of vegetation they support. In an urban setting, population density can often be inferred with reasonable accuracy from the number of dwelling units observable in a given area.

Land-use versus land-cover mapping

Remote sensing classification represents *land cover*, i.e., the features present on the land surface at the time the imagery was acquired. *Land use* refers to the activities for which a given area of land has been designated. Land use can often be inferred from remote sensing imagery but is not directly captured because the current land cover may not reflect its intended use. For example, in figure 11.24a, a remote sensing

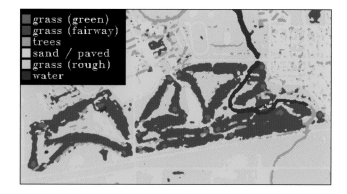

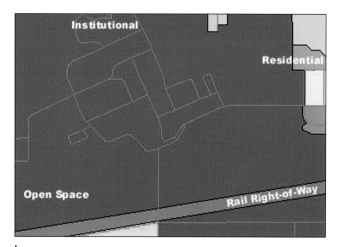

b

Figure 11.24 Land cover versus land use. (a) A golf course that has been digitally classified into grass, trees, water, and sand. (b) A land-use map of the same area.

Table 11.1	USGS Land-use / Land-cover classification framework (Anderson et al., 1976)
Level I	**Level II**
1 Urban or built-up land	11 Residential
	12 Commercial and service
	13 Industrial
	14 Transportation, communications, and utilities
	15 Industrial and commercial complexes
	16 Mixed urban or built-up land
	17 Other urban or built-up land
2 Agricultural land	21 Cropland and pasture
	22 Orchards, groves, vineyards, nurseries, and ornamental horticultural areas
	23 Confined feeding operations
	24 Other agricultural land
3 Rangeland	31 Herbaceous rangeland
	32 Shrub and brush rangeland
	33 Mixed rangeland
4 Forest land	41 Deciduous forest land
	42 Evergreen forest land
	43 Mixed forest land
5 Water	51 Streams and canals
	52 Lakes
	53 Reservoirs
	54 Bays and estuaries
6 Wetland	61 Forested wetland
	62 Nonforested wetland
7 Barren land	72 Dry salt flats
	72 Beaches
	73 Sandy areas other than beaches
	74 Bare exposed rock
	75 Strip mines, quarries, and gravel pits
	76 Transitional area
	77 Mixed barren land
8 Tundra	81 Shrub and brush tundra
	82 Herbaceous tundra
	83 Bare ground tundra
	84 Wet tundra
	85 Mixed tundra
9 Perennial snow or ice	91 Perennial snowfields
	92 Glaciers

Source: Courtesy of USGS. Professional paper 364. Washington, D.C. U.S. Government Printing Office.

image classification would identify the *land-cover* classes of trees, grass, water, and sand, whereas a land-use map may designate these areas as open space *(figure 11.24b).*

Wherever possible, a classification should conform to standard class definitions used by the organization or within the discipline where the data will be used. High-quality data is expensive to create. However, it can more easily be integrated with other datasets and reused in applications that were not foreseen at the time it was created if the class definitions conform to some generally accepted standard.

Anderson et al. (1976), in an effort to provide such a standard and also address the land-use/land-cover dichotomy, proposed a hierarchical classification framework *(table 11.1).* This framework has proven to be remarkably resilient in use for a wide range of land issues and data types. It has become

a *de facto* standard for image classification, especially when the data is to be integrated with other GIS layers. The class structure beyond Level II has not been standardized because definitions of finer classes depend largely on the characteristics of the region and intended use of the classification. When used together with the sensor and resolution information in table 11.2, this classification framework can serve as a useful guide to the types of classes typically distinguished from remotely sensed data.

If the digital classification is to be compared with existing mapping (e.g., for map updating or land-use change), some

Table 11.2 Recommended USGS land-use and land-cover levels for classifying remote sensing data. (after Jensen and Cowan 1999; Lillesand et al. 2004)			
USGS level	**Representative mapping scale**	**Nominal spatial resolution**	**Remote sensing instruments (digital multispectral only)**
I	1:500,000	> 20 m	Landsat ETM, SPOT HRV/HRG, SPOT VMI, NOAA AVHRR, IRS LISS, Terra ASTER, Aqua/Terra MODIS
II	1:50:000	5–20 m	Terra ASTER (visible bands), Landsat ETM*, SPOT HRV/HRG*
III	1:25,000	1–5 m	IKONOS, QuickBird, OrbView 3
IV	1:10,000	< 1 m	Fine-resolution airborne
* Under ideal conditions (clear imagery, strong surface feature contrast), it is possible to obtain a satisfactory Level II classification for most classes from ETM and HRV/HRG data.			

Source: Reproduced with permission of the American Society for Photogrammetry and Remote Sensing. J. R. Jensen and D. C. Cowen. 1999. Remote sensing of urban/suburban infrastructure and socioeconomic attributes. *Photogrammetric Engineering and Remote Sensing* 65: 611–12.

manual processing of the classified imagery may be needed to reconcile differences in class definitions. Where the classification will be a new dataset, there is more latitude in defining classes so long as they suit the intended use. While it is possible to create classes uniquely suited to a particular use, for example, *urban—well treed* or *forest—predominantly red pine and jack pine*, it is generally more prudent to use standardized class designations.

Digital classification: Per-pixel classifiers

Digital classification is the semiautomated process of deriving information classes from pictorial images *(figure 11.25)*. There are two fundamental approaches to this process: *supervised classification*, where the analyst defines the classes to be extracted from the imagery, and *unsupervised classification*, where classes are automatically defined based on statistical clustering of pixel values found in the data. Several procedural variations within each approach have been developed and *hybrid classifiers* that use techniques from both methods are also used.

Supervised, unsupervised, and hybrid classifications can be implemented using algorithms that operate on one pixel at a time, termed *per-pixel classification algorithms*, or by other algorithms that examine larger regions within the image simultaneously. Per-pixel classifiers are currently the most commonly used techniques and provide a context for discussing other classification approaches.

Supervised classification

In a supervised classification, an analyst defines areas in the image (called *training areas*) that are representative of each information class and then, after refining the class specifications, has the computer generate a classification of the image. The accuracy of the classification is then assessed using independent

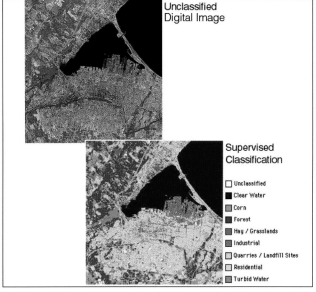

Figure 11.25   Transformation from continuous to categorical pixel data during digital classification. Remotely sensed image data represents the variation in brightness within the scene as a set of pixel values. A classification assigns pixels to categories, such as land-cover classes, based on those values.

Source: © 1999. Government of Canada with permission from Natural Resources Canada.

verification data, i.e., a set of locations that were not used in the classification process for which the class is known.

Where the image classes do not represent the required information classes with sufficient accuracy, additional processing is done to improve the classification. This may include more careful delineation of the training sets, further refinements to the classification, the use of other data such as elevation data, or the use of a different algorithm. This iterative process of improving the classification continues until the required accuracy is achieved or time and budget constraints necessitate settling for less-accurate or less-detailed information classes than desired.

Training area definition

The specification of training areas is the most time-consuming task in a supervised classification. The time required depends on the level of detail of the classification and the source information that can be used to define the training sites. In some cases, the classes to be identified can be reliably interpreted from other remotely sensed data of higher spatial resolution, such as large-scale aerial photography. In that case, imagery of selected areas can be acquired and interpreted to generate the training sites. In other cases, suitable training areas may be obtained from data originally collected for other purposes, such as existing maps or field observations from other studies. Often, training datasets must be verified in the field, which may take days, months, or longer.

Training areas are usually input by interactively delineating their boundaries on the image *(figure 11.26)*. Where training sites are stored in a GIS, it may be possible to input them directly to the classification program. As each set of training data is entered, the analysis system generates summary statistics, known as *spectral signatures*, from the image pixel values at the identified locations. This process is summarized in figure 11.27.

One way to visualize how the spectral signatures are represented in the multispectral data is to plot the spectral correspondence from each training pixel on a scatter plot, as in figure 11.28a. For simplicity, this illustration shows the

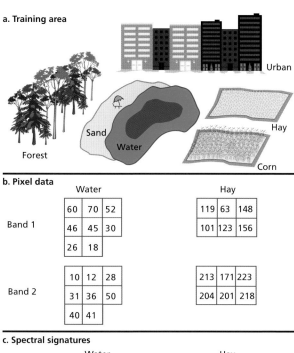

Figure 11.27 Training data and spectral signatures training areas outlining features of interest are drawn on the image (a). The image analysis system extracts the multispectral values for all pixels within the training area boundaries (b) and calculates their spectral signatures (c).

spectral correspondence from just two image bands. In reality, however, the spectral correspondence is compared across all of the image bands sampled during training data collection. Statistical analysis tools and data plots such as these are then used to refine the class specifications.

Unfortunately, the spectral characteristics of a feature as detected by a sensor are only valid for a particular place at a given moment. The same feature sensed in another image (or even a different part of the same image) may exhibit a significantly different spectral pattern due to variations in illumination geometry, atmospheric effects, and site conditions. Consequently, a new set of training areas are required for each image that is to be classified.

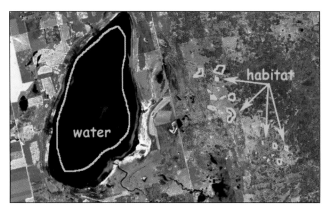

Figure 11.26 Simple and complex training data. Water is easily identified on this Landsat ETM image, but accurately identifying habitat comprised of a specific group of vegetation species is much more difficult.

Source: Image created by J. Piwowar. Data provided by © 1999. Government of Canada with permission from Natural Resources Canada.

Classification

The refined specifications are then used to classify every pixel in the image. An algorithm is chosen to assign each pixel to a class by comparing its spectral data with the signature statistics developed for all classes. The most common classification algorithms for supervised classification are *minimum distance to means, parallelepiped,* and *maximum likelihood.*

Minimum distance to means

The minimum distance to means approach is conceptually the simplest: The spectral values of each image pixel are compared to the arithmetic means for each class. The pixel is assigned to the class to which its values are closest (i.e., the one that minimizes the differences between the average pixel value for the class in each spectral band and the pixel's value in each spectral band). For example, in figure 11.28b, pixel P_1 is closest to the mean values of the Corn training set (distance = $\sqrt{(217 - 190)^2 + (166 - 150)^2}$ = 31.4) and would be assigned to the Corn class. The minimum distance for pixel P_2 is to the Sand class mean. A visual examination of the training area feature clusters would suggest, however, that pixel P_2 is more likely a member of the Urban class. This highlights one of the potential problems with the minimum distance strategy: it doesn't take into account that some features have a broader range of spectral reflectances than others, which can lead to some misclassification. In spite of this drawback, the algorithm can be useful for classifying very large image files because of its simplicity and speed.

Parallelepiped classifier

In a parallelepiped classifier, a pixel is assigned to a class if its spectral values fall within the minimum–maximum range measured in the training data for that class. This is shown, for a two-band case, in figure 11.28c, where P_2 is now appropriately included in the Urban class. In two-dimensions (representing two spectral bands), the region of class inclusion appears as a rectangle; if three bands were used, the region of class inclusion could be drawn as a box. More than three bands are frequently used in multispectral image classifications, however, and although we cannot adequately draw such a region of class inclusion, the shape of an *n*-dimensional rectilinear feature is geometrically known as a *parallelepiped.*

The parallelepiped strategy works best when the minimum–maximum ranges are approximately equal in each band, as they are for the Urban class in figure 11.28c. When the range in one band is very different than that in another, the region of class inclusion expands to include a lot of pixel values

that may not, in fact, be valid representations of that feature. Point P_1 would be misclassified, in this example, as Hay. This can be further complicated if multiple regions of class inclusion overlap, as they do between Hay and Forest. The classification algorithm would be uncertain as to the class membership of pixel P_3. Simple parallelepiped classifiers would arbitrarily assign P_3 to one of the two classes (typically either the first or last valid class it encounters). More advanced systems would use a secondary decision rule (such as minimum distance to means or maximum likelihood) to break such a tie or prompt the analyst for a manual class assignment. Like the minimum distance to means strategy, the parallelepiped method is algorithmically simple and computationally fast, so it can be a practical approach to use with very large images.

Maximum likelihood

The *maximum likelihood* classification strategy uses probabilities to overcome some of the limitations of the minimum distance to means and parallelepiped approaches. Based on a comparison of each pixel's spectral values with the signature statistics from each training set, the maximum likelihood algorithm calculates the conditional probabilities of membership in each class (see *A simplified view of maximum likelihood calculations*). A class assignment is made to the class to which the pixel is most likely to belong, i.e., the one with the highest probability. This is illustrated in figure 11.28d, where *equiprobability contours* are drawn around each training set mean to define class membership likelihood zones. Since the probabilities are derived from the spectral variances within each training set, they are able to account for the range differences in spectral values in the classes, resulting in correct class assignments for pixels P_1, P_2, and P_3. The maximum likelihood strategy typically produces the most accurate classifications, hence it is favored by many image analysts. Although the probability computations can be complex and computationally slower, the increase in processing time is negligible for all but the very largest images.

Unsupervised classification

Different land-cover types tend to produce distinct groupings or clusters of pixel values in remotely sensed imagery. Figure 11.29a is a two-dimensional scatter plot of pixel brightness values obtained from known test sites. The plots illustrate the tendency of the spectral values for a land-cover class to be clustered. In an unsupervised classification, the analyst does not provide examples of the classes to be found. Instead pixels

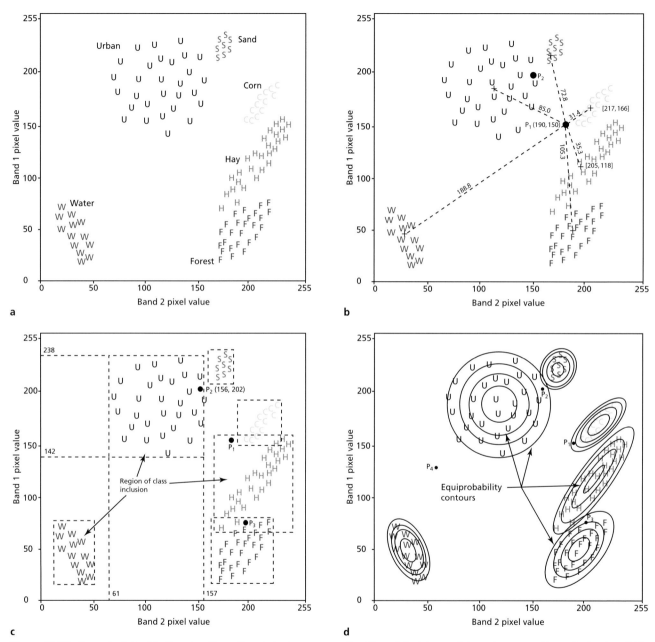

a

b

c

d

Figure 11.28 Supervised classification algorithms. The pixel values from test sites for six classes were plotted for two bands (a). The way these data would be classified are shown for three classification algorithms: (b) minimum distance to means, (c) parallelepiped, and (d) maximum likelihood.

Source: Adapted from *Remote sensing and image interpretation*. 4th ed. by T. M. Lillesand and R. W. Kiefer. © 2000 John Wiley and Sons, Inc. Reprinted with permission of John Wiley and Sons, Inc.

A simplified view of maximum likelihood calculations

The probability of a pixel belonging to a Corn class (P_c) can be determined by the following expression (Jensen 1996, Richards and Jia 1999):

$$P_c = 1n|V_c| - (X - M_c)^T V_c^{-1} (X - M_c)$$

The probability calculation relies heavily on the class covariance matrix calculated from the training data (V_c). The covariance matrix describes the amount of variation in pixel values for all spectral bands sampled. In just a single band, the amount of variation can be described by the standard deviation. In figure 11.27c the standard deviations of Hay training set values were calculated to be 33.7 in Band 1 and 18.6 in Band 2. Observe the shape of the Hay distribution in figure 11.28d. It has a wider spread along the Band 1 (vertical) axis than the Band 2 (horizontal) axis, mirroring the individual standard deviation values.

Maximum likelihood classification calculations are easier to understand if we think in terms of standard deviations instead of covariance matrices. The equiprobability contours of figure 11.28d can be thought of alternatively as *equistandard deviation* contours. That is, each contour represents one standard deviation from its class mean. Recall that pixel P_2 was incorrectly classified as Sand by the minimum distance to means method. If we classify P_2 using the minimum distance to means algorithm again but this time using standard deviations (instead of pixel values) as the metric, we can count the equistandard deviation contours and see that P_2 is 3.5 standard deviations from the Urban mean and about 4.5 from the center of the Sand class. Accordingly, the maximum likelihood method would designate this pixel as Urban. A similar analysis would show how P_1 belongs to Corn and that P_3 is a Forest pixel.

are classified using algorithms that find statistical groupings in the spectral data with the assumption that some or all of those groupings will correspond to useful information classes. The resulting classes are more or less well-defined spectral groupings but may or may not correspond well to the desired information classes. Typically the analyst generates a large number of image classes and then uses different methods to refine and then group the classes to create the required information classes.

Classification

Although there have been several approaches developed for automated image clustering, the most widely implemented method is the *migrating means* algorithm, also known as *k-means*, ISODATA, and *iterative optimization* (Ball and Hall 1965, Jensen 1996, Lillesand et al. 2004, Tso and Mather 2001). The only user input this method requires is a specification of the number of clusters (i.e., classes) to be found. The procedure begins by randomly seeding the desired number of cluster means into the multidimensional spectral space defined by the multiple image bands. A minimum distance to means algorithm (discussed above in the *Supervised classification* section) is then used to assign each image pixel to one of these initial classes. After the first pass through the image, the cluster means are recalculated (hence *migrate*) based on the averages of the pixels that have been assigned to each class. The migrating means procedure thus repeats until either the class means no longer need adjusting at the end of each iteration or a user-specified number of iterations have been completed.

Assign class labels

At the end of the first stage of an unsupervised procedure, a classified image is produced with one class for each spectral cluster found in the image data. Unsupervised algorithms produce classes based solely on the pixel values of the image. However, there are many factors other than the spectral characteristics of the object being sensed that contribute to the pixel values found in an image. Some of the other factors that may contribute to the electromagnetic energy detected by a sensor include the growth stage and health (for vegetation features), sensor viewing angle, degree of solar illumination, background (e.g., soil) characteristics, and atmospheric conditions.

Initially, these classes have no meaning and are assigned arbitrary labels, such as Class 1, Class 2, etc. *(figure 11.29b)*. The process of assigning useful descriptions to these classes is left to the analyst and is a time-consuming process. Some of the features represented by the classes of an unsupervised classification can be quite complex and difficult to label. For example, an unsupervised classifier may represent a stand of trees by multiple classes that differ by the proportion of different tree species and by soil and understory characteristics as well. In agricultural areas, the groupings may not distinguish individual crop types. In addition to separating Corn from Hay, the classifier might also produce a class that represents Corn and Hay Growing on Wet Soils, as in figure 11.29c. By iteratively grouping classes and performing additional classifications to split others, a set of classes may be obtained that is suited to the intended use.

Although the initial clustering stage of an unsupervised classification can be processed quite quickly, the labeling of classes can be difficult and time consuming. However, by

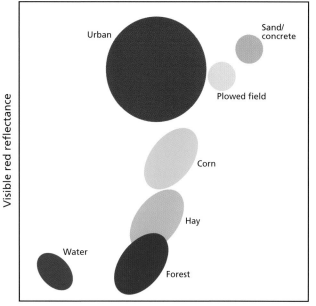

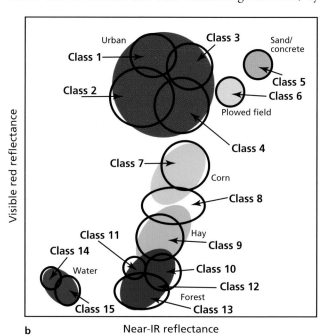

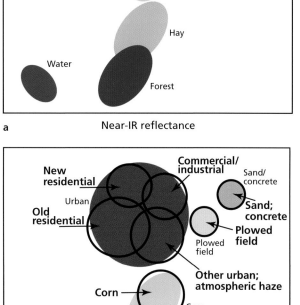

Figure 11.29 Supervised and unsupervised classification. Scatter plots of pixel values in the visible red and infrared bands compare supervised and unsupervised classification: (a) shows the grouping of pixel values for seven land-use classes as might be obtained from training areas for a supervised classification; (b) shows how these clusters compare with the initial 15 classes generated by an unsupervised classification; (c) illustrates how labels were assigned to the unsupervised classes.

not imposing an initial set of classes, an unsupervised classification is more likely to identify unexpected groupings in the data that may suggest unanticipated information in the scene worthy of further investigation.

Accuracy assessment

Accuracy assessment is driven by a need to identify and understand the nature of the errors in a map or image (where they are and why they have occurred) in order to optimize the classification as well as to assess its reliability (Gopal and Woodcock 1994).

Accuracy assessment is an integral part of supervised and unsupervised classification, as it provides a means of assessing the quality of the result, i.e., how well the classified pixels represent the desired information classes and categorize actual features on the ground. For automated classifications to be successful they must be iterative, using quality measures such as accuracy assessment to refine class specifications until a satisfactory result is achieved.

A classification accuracy assessment generally involves the selection of a statistically valid set of sample test sites for which the assigned class is compared with independently collected verification data considered to be of high accuracy. The verification data may be field data collected specifically for the assessment, it may be taken from existing data such as maps or published statistics, or it may be interpreted from other imagery typically of higher spatial resolution if the classes to be assessed can be reliably interpreted.

There is always some degree of uncertainty in how well a class label describes the area represented by a pixel. Figure 11.30 shows the same multispectral image classified by four different methods. Intuitively, there is more confidence in the class assignment of a pixel when different classifiers assign the same class than when they assign different classes. Vieira and Mather (1999) advocate generating multiple classifications using different methods and then assigning classes according to *voting rules*, such as the class assigned by the majority of classifiers.

Imprecise class definitions, poor feature spectral representations, and mixed pixel effects add to the challenge of providing a reliable accuracy assessment (Aronoff 1982a, 1982b). Natural features rarely fall into neatly defined groupings. A coniferous forest will contain some deciduous trees, and there often will be soil-covered areas within a region classified as bare rock. The accuracy of a classification depends to a great extent on how well the class definitions fit the naturally occurring features they are to categorize.

Classification assessments can also be used to intentionally bias a classification so as to reduce the risk of errors that have more serious consequences. For example, in a fire hazard map, it would be more costly to misclassify a high fire risk area as low risk than the reverse because underestimating the hazard could result in an inadequate emergency response and the development of a devastating wildfire. Misclassifying a low-risk area as high risk could result in overresponse to an emergency, a much lower price to pay than the risk of a wildfire. By recognizing that the error present in any classification can be to some extent controlled though not eliminated, the opportunity can be taken to optimize a classification so as to minimize the risk of encountering the most costly errors instead of treating all errors the same (Aronoff 1984).

Classification accuracy assessment data is commonly summarized in the form of an error matrix (also known as a *contingency table* or a *confusion matrix*) that shows the correct and incorrect class assignments for a sample of verification sites (discussed in chapter 10). Although other methods of accuracy assessment have been developed (see Congalton and Green 1998 for a thorough treatment), the error matrix provides a useful summary and is widely accepted.

Accuracy measures that assess each class separately, such as percentage correct, are straightforward to calculate, but in limiting the comparison to whether an individual pixel is correctly or incorrectly classified, this measure does not consider the overall performance of the classification across all classes taken together. Other assessment methods, such as the kappa statistic (see chapter 10), provide a measure of quality for the classification scheme, taking into account the different probabilities of error related to the area of each class present in the scene.

The ability to produce quantified accuracies from image classifications creates a paradox: in remote sensing, an accuracy assessment is essential to judge the suitability of a classification for use in a specific application; however, many GIS users are not aware of the accuracy of the other data they use. They may discount a classification in which the Urban Land class has an accuracy of 80%, a remote sensing accuracy level commonly considered acceptable. Yet the city zoning map they use daily without knowing its accuracy may contain numerous errors such as pockets of nonconforming activities, incorrectly drawn zone boundaries, and land uses that have changed since the map was originally drafted. A greater awareness of the accuracy of GIS data layers would provide a context by which to judge the accuracy of remote-sensing-derived data.

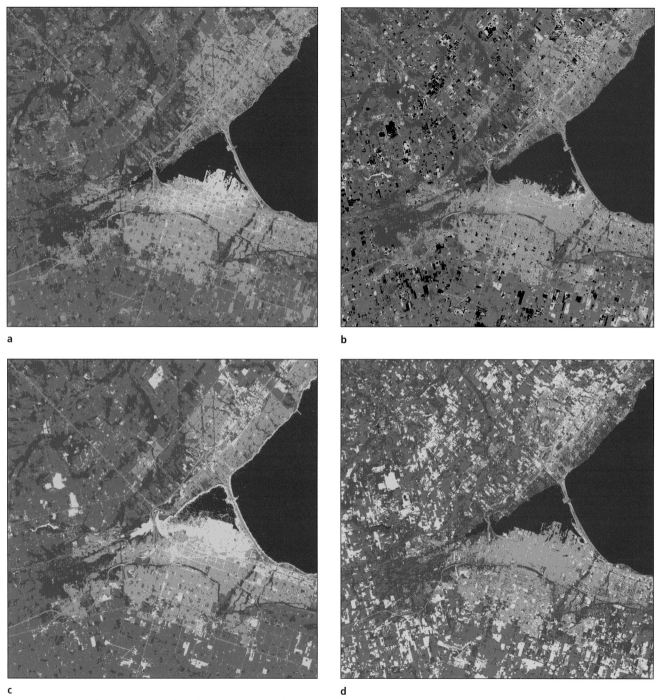

a

b

c

d

Figure 11.30 Landsat ETM image of Hamilton, Ontario, classified by four different methods: (a) minimum distance to means supervised classification, (b) parallelepiped supervised classification, (c) maximum likelihood supervised classification, and (d) k-means unsupervised classification.

Other classification approaches

Although per-pixel supervised and unsupervised classifications are widely used, they have an inherent limitation in being based solely on the spectral characteristics of each individual pixel. Yet information about a pixel's context, such as the class of neighboring pixels or its elevation or climate, could be used to complement the spectral data and improve the classification accuracy. Two vegetation classes that are spectrally similar may be separable using slope, aspect, and elevation data if they are found in topographically distinct conditions. Similarly, a suspect pixel in the middle of a corn field is more likely to be corn than young trees. By using additional data sources and contextual information, classification accuracy can be substantially improved for some applications.

Hybrid classification

There are no hard and fast rules for selecting the classification procedure that will give the most accurate results. Depending on the difficulty of the problem and experience of the image interpreter, some experimentation with different combinations of methods may provide unique and valuable information. For example, in a supervised classification, the analyst first defines training sets and then classifies the image. An unsupervised classification, on the other hand, starts out by classifying the image, after which the user must decide what the classes mean. In both cases, the user-interaction portion—class training or cluster identification—is the most time consuming. *Hybrid classifiers* seek to optimize a classification by using both supervised and unsupervised classification methods.

One such method applies a second unsupervised classification to refine classes created by an initial classification (either supervised or unsupervised). Jensen (1978) and Bauer et al. (1994) provide instructive examples.

Alternatively, an unsupervised classification may be followed by a supervised approach to assist in the class labeling process. The initial unsupervised classification would be intentionally broad, dividing the image into a half-dozen classes. Supervised training and classification is then done only within those classes to be further subdivided. This approach frequently reduces the number of misclassified pixels in an image (Brodley and Friedl 1996, Behera, et al. 2001, Bachmann, et al. 2002).

These hybrid approaches are most useful where the required classes are not of equal importance. For example, in a detailed vegetation analysis, further refinement of aquatic zones or built-up areas beyond generalized *water* and *urban* classes may not be required. By defining some broad classes, and then reclassifying only within the classes of greater interest, the level of effort is optimized.

Contextual classification

All of the classification procedures discussed so far classify an image based on the spectral properties of each pixel without regard for the characteristics of neighboring cells. *Contextual classification* seeks to improve results by examining the spatial context of each pixel, i.e., the values of neighboring pixels. Contextual information can be included at two points in the classification sequence: as an additional input *channel* to a per-pixel classifier or by applying a neighborhood function to improve the output results of a spectral-only classification.

Contextual measures can be particularly beneficial when applied to previously classified imagery that has a significant number of pixels from other categories scattered throughout. In this case, the objective of the contextual classification is to reassign these scattered pixels to the class of their neighboring cells, depending on their context. Figure 11.31 shows several ways in which context can be defined. A simple way to implement an *inclusion* context *(figure 11.31d)* is to examine the class of each solitary pixel in relation to its immediate neighbors and reassign its class to match the most common neighboring category.

Fuzzy classification

Classifications tend to simplify complexity, creating a set of idealized groupings that ignore or downplay the gradations actually present. Inherent in the definition of a class is a *segmentation* of space (i.e., dividing the space into discreet uniform areas considered to belong to one class). In so doing, the extent of each class is defined with a *hard* boundary, e.g., *park* on one side of the boundary, *road* on the other.

This usually fits well with the built environment (e.g., urban planning zones are distinctly defined) but may not be entirely suitable for *soft* boundaries frequently found in the natural environment. For example, natural forest stands don't usually have a sharp edge; they commonly have a gradation such as a zone of shrubs and scattered tree seedlings on their margins. Other natural regions are even less distinct, such as the migratory range of the monarch butterfly. Yet classification typically demands that a line be drawn to separate a *forest* from a *nonforest* area.

Traditional approaches to image classification are designed to produce categorical data with hard boundaries where

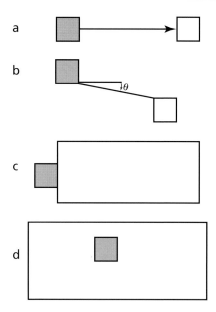

Figure 11.31 Contextual classification from Gurney and Townshend (1983). The shaded regions depict pixels to be classified; the open regions represent other pixels used in the classification decision. (a) Distance is considered. (b) Direction is considered. (c) Contiguity forms a part of the classification process. (d) Inclusion is considered.

Source: Reproduced with permission of the American Society for Photogrammetry and Remote Sensing. C. M. Gurney and J. R. Townshend. The use of contextual information in the classification of remotely sensed data. *Photogrammetric Engineering and Remote Sensing* January 1983: 55–64.

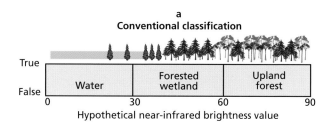

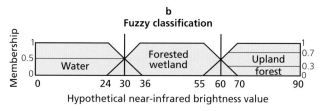

Figure 11.32 Fuzzy classification. Figure (a) shows the hard boundaries established between classes in a conventional classification. The fuzzy classification in (b) shows degrees of class membership depending on the spectral properties of the pixel.

Source: Adapted from Jensen, J. R. *Introductory digital image processing: A remote sensing perspective.* 2nd ed. © 1996. Reprinted by permission of Pearson Education, Inc., Upper Saddle River, N.J.

each pixel can be assigned to one, and only one, class. This approach gives good results for many applications but has its limitations. One is that there is no way to account for multiple feature contributions to a pixel's spectral information. Thus, a pixel that covers an area of 60% *water* and 40% *forest* would be classed as *water*; the *forest* information would be discarded. When this classification is used in subsequent analyses, it would be assumed that this *water* cell represented an area that was completely flooded. *Fuzzy* logic attempts to compensate for this shortcoming by allowing proportional class membership.

A fuzzy classifier creates *membership functions* from the spectral responses gathered during interactive training. Figure 11.32a shows how an image might be segmented into hard classes by a conventional classifier. In figure 11.32b, there are varying degrees of class membership depending on the spectral characteristics of a pixel. Thus, although we might be able to agree that a pixel with a value of 55 belongs to the *Forested Wetland* class, there would be some uncertainty

for pixels with brightness levels of 30 or 60 *(figure 11.32a)*. Pixels of value 30 would be classed by traditional methods into either *Water* or *Forested Wetland* (depending on the particular algorithm used to arbitrate ties). Similarly, cells with values of 60 would be assigned to *Forested Wetland* or *Upland Forest*. In the fuzzy approach, however, the pixels with a value of 30 would be assigned as 50% *Water* and 50% *Forested Wetland*, and the cells with values of 60 would be classed as 70% *Forested Wetland* and 30% *Upland Forest (figure 11.32b)*. Further discussion of fuzzy classification and its applications can be found in Jensen (1996), Tso and Mather (2001), and Wang (1990).

The debate on whether to use fuzzy classification is not exclusive to remote sensing. Any data that varies continuously over space may be best represented as surfaces to show gradations rather than as categories with abrupt boundaries. Although fuzzy logic may seem like an appropriate method of extending image classification, it may not always be suitable, particularly if the remote sensing data is to be used in a GIS. Many GIS analysts and systems assume their data has hard boundaries. Even inherently fuzzy categories (such as population density) are frequently lumped into uniform zones (such as census tracts) to facilitate manipulation. Thus, there is always some degree of uncertainty when using continuous data that has been grouped into hard classes, such as from remote sensing.

Spectral mixture analysis

The basic premise behind remote sensing is that the identity of features within a sensor's field-of-view can be retrieved, based on their unique spectral or spatial properties. Thus, the ability to detect a surface feature is limited by its size relative to the spatial resolution of the sensing instrument and the uniqueness of its spectral attributes in the recorded spectral bands. However, a large proportion of remotely sensed data is spectrally mixed because the scales of spatial variation of natural phenomena are often smaller than the spatial resolutions of sensors (Gong et al. 1992). Thus, a fundamental, yet often overlooked, concept is that each single reflectance value recorded by a sensor during image acquisition actually contains contributions from several individual surface features (Peddle et al.1995a, 1995b).

Spectral mixture analysis (SMA) is a procedure that attempts to extract the fractional reflectance components from the pixels in an image. SMA *unmixes* the reflectance values for a pixel to reveal its composition from individual surface features. For example, a pixel acquired over a known forest region could be unmixed to show that it represents an area that contains spectral contributions from vegetation (85%), shade (10%), and nonvegetated fractions (5%). Inherent in the concept of a fractional component is the idea that there is some set of reflectance values that represent a pure feature. In other words, we must be able to define the spectral response for *pure* (100%) vegetation, *pure* shade, and *pure* nonvegetated surfaces. These pure spectral patterns are known as *endmembers* because they represent the extreme ends of the image data distribution *(figure 11.33)*. Endmember labels are frequently different than those used in conventional classification since they relate to a pure spectral distribution that may not even be present in the data. The object is to find all such endmember classes so that they encapsulate all of the other image data. Any point in the image can then be defined as a linear combination of these extrema, as shown for the category Forest in figure 11.33.

For each pixel in the image, a linear mixing model can be defined as:

$$DN_p = (EM_1 * F_1) + (EM_2 * F_2) + ... + (EM_n * F_n) + \varepsilon$$

Where
  DN_p is the pixel's digital number

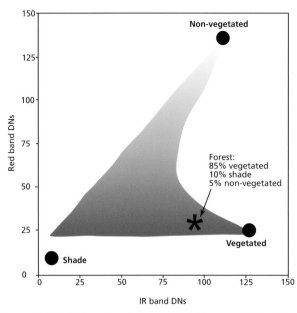

Figure 11.33   Endmembers (shown as bullets) represent the extremes of the spectral data. Image pixels are classified according to the relative contributions of the endmembers to their spectral values (as shown for Forest).

EM_i is the spectrum of the *i*th endmember class (where there are *n* endmember classes in the model)
F_i is the fractional contribution of the *i*th endmember class
ε is any residual contribution not accounted for by the endmember set (Adams et al. 1993)

In practice, we are given DN and we specify EM, so the equation is inverted to solve for F. In other words, given a multispectral image, a set of image or reference endmember classes is defined and then each pixel is decomposed into its fractional spectral components.

SMA produces one *fraction image* for each endmember entered into the mixing model. A fraction image shows the spectral contribution of that endmember to each pixel *(figure 11.34)*. Since pixel reflectances are mixtures of several materials, each pixel can be represented in several endmember fractions. For example, a pixel which is known to be in a wooded area may be represented by a 60% vegetation fraction, a 30% soil fraction, and a 10% shade fraction (assuming that vegetation, soil, and shade endmembers were used in the mixing model).

Although it was originally developed for the analysis of *hyperspectral* images, SMA has been a beneficial tool when used with sensor data having fewer bands. Further, Adams

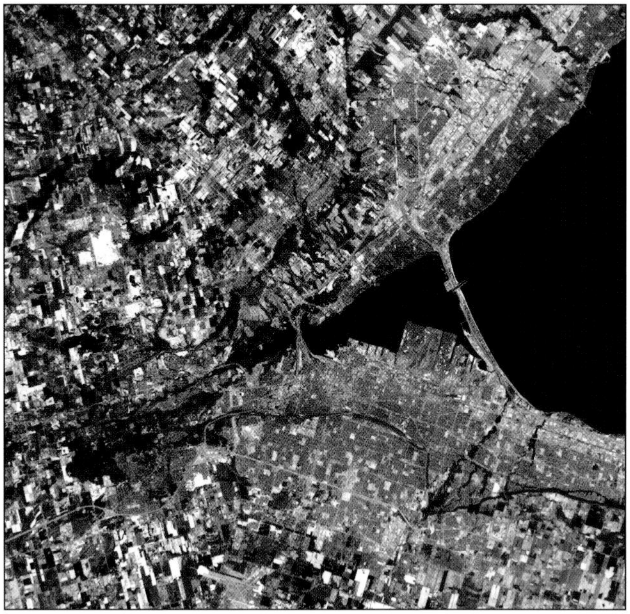

a

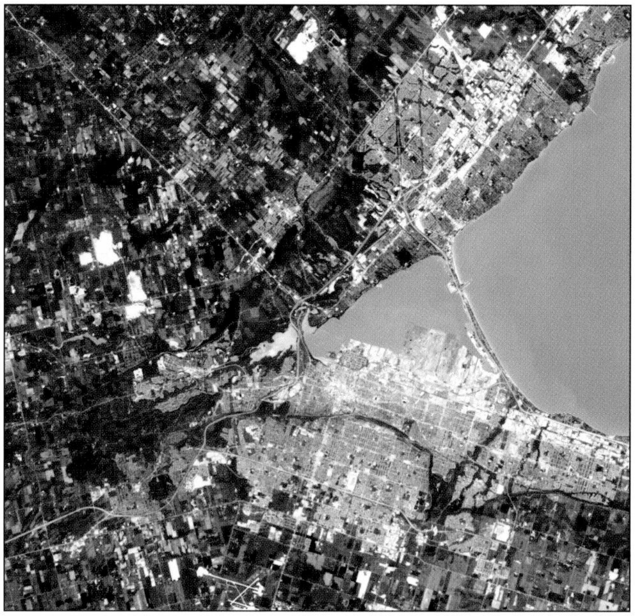

b

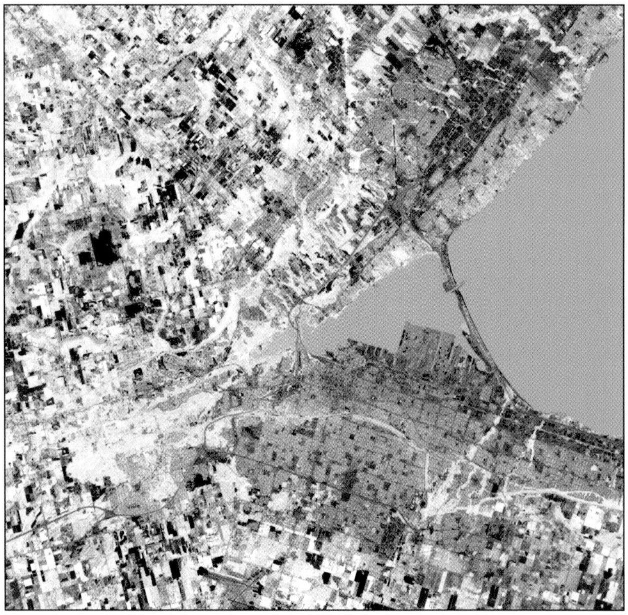

c

Figure 11.34 Spectral mixture analysis endmember fraction images for a Landsat ETM scene of Hamilton, Ontario. The brighter the pixel in each fraction image, the greater the contribution from that endmember. Images a, b, and c represent the soil, concrete, and vegetation fractions, respectively.

Source: © 1999. Government of Canada with permission from Natural Resources Canada.

et al. (1993) suggest that the technique is extensible to any physical attribute that can be measured at each pixel, such as net radiation, radar backscatter, or even elevation, and Piwowar et al. (1997) apply the method to extract temporal signals from long image time series

Artificial neural networks

Conventional computer applications process data following a predefined set of rules. These are typically encoded into procedural computer languages as IF...THEN...ELSE statements. For example, a procedural rule for classifying a multispectral pixel might state:

IF {infrared reflectance is strong}
THEN {assign to class vegetation}
ELSE {assign to class nonvegetated}

One of the difficulties with procedural rules is that they are fixed; once they have been coded into the program, they are not easily modified. After using this classification rule for a while, however, we may *learn* that some nonvegetated surfaces, such as concrete, also have bright infrared reflectances. The particular section of the software containing this rule would have to be recoded to adjust to our new knowledge.

Computer scientists have been studying this limitation for some time and have created a new class of software called *Artificial Neural Networks* (ANNs) that have the capacity to *learn* new rules and self-adapt to unique situations. For a thorough treatment of ANNs see Chen et al. (1995) and Paola and Schowengerdt (1995, 1997).

Like conventional supervised classification methods, neural networks require an initial set of training data to describe the spectral patterns to be learned and the desired class(es) that should be produced from these patterns. However, the training sets for ANN classification are more exacting and extensive than those for traditional classifiers, so this stage is considerably more time consuming.

Of particular interest for remote sensing is the fact that, unlike traditional classification approaches such as maximum likelihood, neural networks do not require an assumption of normality in the data. This means that the classification approach can readily accommodate diverse data sources such as texture and terrain information. Adding additional, nonspectral information to remote sensing analysis has the potential to substantially improve the classification product, although the actual gains using ANNs have not always reached this potential *(figure 11.35)*.

Object-oriented classifiers

The future of digital classification is perhaps best represented by a new breed of object-oriented classifiers. These algorithms build upon the traditional per-pixel classification toolbox using fuzzy boundaries and neural networks to segment images into homogeneous image objects *(figure 11.36)*. They iteratively classify images at different scales and use this combined knowledge to assign a final class to a pixel. The process can be unsupervised, or the analyst can tune the system to look for particular classes. Object-oriented classifiers begin with a traditional per-pixel classification on a full-resolution image. They then aggregate neighboring pixels (image objects) into a new lower-resolution layer and repeat the classification at this level. This is repeated with successively larger pixel aggregates *(figure 11.37)*. At the same time, the classifier traces the classification of each pixel within each layer and uses this semantic information to make a final class assignment. This hierarchical network of image objects enables the image to be examined at multiple scales simultaneously and provides a framework for combining contextual information with the spectral data from each pixel.

For example, a pixel could be classified as a tree using a traditional per-pixel classification method. Although this classification may be correct, it may not be what the user really wanted if the tree was part of a residential neighborhood, golf course, or forest. In essence, the multiscale approach would allow the software to recognize that the tree was part

Figure 11.35   Supervised artificial neural network classification of a Landsat ETM image of Hamilton, Ontario. Compare this result with figure 11.30.

Source:Image created by J. Piwowar. Data provided by © 1999. Government of Canada with permission from Natural Resources Canada.

Figure 11.36   Buildings identified in a cluttered urban scene with Feature Analyst, a feature extraction software by Visual Learning Systems.

Source: Image courtesy of Visual Learning Systems Inc.

of a woodlot that was part of a forest that was part of broader vegetation ecozone. By examining the class assignment of each pixel within the context of its surroundings, object-oriented classifiers have the potential to provide more meaningful results. Preliminary research show positive results when compared with conventional classification methods.

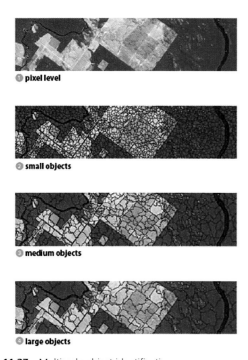

Figure 11.37 Multiscale object identification.

Source: © Definiens Imaging, Germany. www.definiens-imaging.com

Change analysis

Since the launch of the first civilian earth resource satellite, Landsat 1, in 1972, there has been systematic repetitive collection of digital imagery covering most of the globe. Landsat and subsequent satellites have provided more than 30 years of earth observation imagery. Declassification of earlier military satellite reconnaissance photography has extended this historical archive to 1960 for selected coverage areas. Also, government and commercial vendors hold vast archives of aerial photography. One of the most powerful applications of remote sensing has become the identification and analysis of features on the earth that have changed over time.

Due to the raster format of digital imagery, change analysis is easily implemented by an image analysis system. Digital imagery, or even scanned aerial photography from different dates, is first rectified to precisely align the pixel grids of the different images. Changes can then be analyzed by comparing for each pixel location the values for each image date.

For example, Hurricane Hugo struck the U.S. eastern seaboard on September 21, 1989, with winds of over 220 km/h and a 6 m storm surge. Damage was estimated to be $7 billion with a loss of 60 lives, making it the most costly

hurricane in American history. Figure 11.38 shows a pair of Landsat TM images of forested lands in South Carolina acquired before and after Hugo's passing. These images are color infrared composites that show healthy vegetation in red tones. The destruction to the timbered areas is evident in the posthurricane image *(figure 11.38b)* where considerable sections of forest are now light green. Flooding is also

a

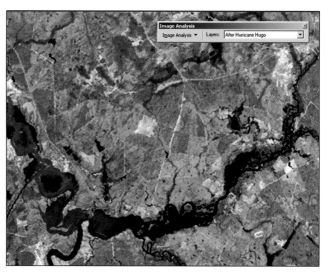

b

Figure 11.38 Landsat TM images of a forested area near Hagan Landing, South Carolina, acquired before (a) and after (b) Hurricane Hugo devastated this area in September of 1989.

Source: Imagery provided by Leica Geosystems GIS and Mapping, LLC.

visible along many of the water courses. To assist foresters in calculating the amount of timber destroyed due to the storm, the before and after images were subtracted, pixel by pixel, to create a *difference* image that was then classified *(figure 11.39 top)*. The computer was subsequently able to tabulate the number of pixels falling into the *destroyed* class showing that more than 14,000 ha of forest had been lost *(figure 11.39 bottom)*.

Although change analyses can be applied to both the built and natural environments, it is in mapping changes arising from unexpected events that the use of satellite imagery is particularly useful. Rarely is adequate data collected before such events occur. Fortunately, the historical remote sensing image archive can fill the gap.

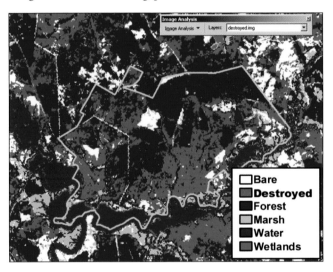

Class	Hectares	% of Scene
Bare	238.77	5.50
Destroyed	1423.73	32.75
Forest	2007.70	46.19
Marsh	199.71	4.60
Water	339.49	7.81
Wetlands	136.63	3.14

Figure 11.39 Classified change image showing the amount of forest destroyed by Hurricane Hugo.

Source: Imagery provided by Leica Geosystems GIS and Mapping, LLC and Positive Systems.

Modeling

The two principal objectives in GIS modeling are to understand and to predict. *Descriptive models* describe what currently exists. They may generalize information and isolate selected features to enhance understanding of the characteristics of a region. A map is in essence a descriptive model of reality. A classified satellite image is a descriptive model of spatial–spectral variation. *Predictive models* assimilate a set of input data to *predict* potential outcomes. Predictive models are useful for building scenarios of conditions that may come to exist at some future time. Spanning the gap between descriptive and predictive models are *temporal models*. Temporal models describe transitional processes that have been observed through time and can form the basis to predict future conditions. Much of the data used to model geospatial data is commonly derived from remotely sensed imagery.

Modeling is in many ways an integral part of remote sensing analysis. The use of topographic information to improve a remote sensing classification is, in effect, distinguishing features by modeling the conditions where they are most likely to occur. As remote sensing data becomes more easily integrated within a GIS environment, it can be more widely used in GIS modeling efforts.

Descriptive modeling: The Southwest Regional Gap Analysis Project

The Southwest Regional Gap Analysis Project[2] covers a large area encompassing the states of Arizona, Colorado, New Mexico, Nevada, and Utah. One of the key products from this effort is a seamless coverage of vegetative land cover modeled from Landsat 7 imagery and other geospatial datasets derived from digital elevation models processed in a GIS environment.

The Gap Analysis Program (GAP) is a descriptive modeling effort with a primary objective to develop and provide regional-scale geographic information on biological diversity to planners, land managers, and policy makers for informed decision making (USGS 2002). One of the traditional goals of *gap analysis* is the identification of spatial gaps in biodiversity protection, i.e., areas of critical wildlife habitat that are not protected from development.

Gaps are determined by overlaying land stewardship maps over potential vertebrate habitat distribution maps, or wild-

2. From Lowry et al. (2002).

life habitat relationship models. A key input to the habitat models is vegetative land cover. This land cover is principally obtained through analysis of multidate remote sensing imagery and landscape characteristics such as slope, aspect, elevation, and landform type. The landscape characteristics significantly improve classification accuracy by providing an independent set of ecological characteristics related to the occurrence of vegetation types. Multidate imagery collected over the growing season provides information on vegetation types and their change over time. Often vegetation classes that cannot be distinguished in one season can be differentiated at a different time.

Rule-based image classification capitalizes on these multiple datasets to more effectively map vegetative land cover. The premise of a rule-based modeling approach is that distinct vegetation communities are associated with different ranges of environmental variables, such as topography, as well as spectral characteristics. Vegetation communities can be characterized by a set of rules that define multiple conditions that must be met for inclusion in a class. The pixels of an image can then be classified not only according to the spectral criteria typically used to classify a remote sensing image, but they can also incorporate landscape characteristics as conditions *(figure 11.40)*.

For example, the Douglas fir vegetation type might be characterized as occurring at an elevation greater than 2,000 m with an NDVI vegetation index on spring imagery (SprNDVI) greater than 0.5 and less than 0.9, and a landform that is moderately moist with a slope of 10 to 35 degrees (here coded as Landform 6). The rule might be coded for the classifier as follows:

IF CONDITION {elevation > 2000 AND SprNDVI > 0.55 and SprNDVI < 0.90 AND landform = 6}

THEN Douglas fir

When landscape characteristics are used in this way, an area is, in effect, stratified first into units with similar physiographic and ecological characteristics. Then remotely sensed imagery is used to further subdivide these units by their spectral characteristics to generate the final land-cover classes. These units have been named *spectro-physiographic areas* or *spectrally consistent classification units* (SCCUs) (Lillesand 1996).

The Southwest Regional Gap Analysis Project used 30 m Landsat TM imagery for three seasons (spring, summer, and fall) and 30 m digital elevation data from the USGS National Elevation Database. From these two sources, several modeling datasets were created for rule generation *(table 11.3)*. Data layers were generated for each mapping zone. Landsat TM imagery was mosaicked and clipped to the mapping zone boundary, and derivative layers such as NDVI (normalized difference vegetation index) and Tasseled-Cap[3] bands (brightness, greenness, and wetness images) were created. DEM derivatives, the topographic relative moisture index (TRMI) and Landform layers were generated in a GIS *(figures 11.41 and 11.42)*. The TRMI is a summed scalar

---

3. The Tassled Cap Transformation is a vegetation index with three components first developed for use with Landsat data. Three image layers, designated brightness, greenness, and wetness are calculated from the multispectral imagery.

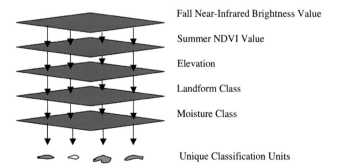

Figure 11.40 Use of modeling for classification. Vegetation classes can often be more accurately classified from remotely sensed imagery when additional biophysical data sources are included.

Source: Image courtesy of J. Lowry, R. Ramsey, and G. Manis. Image in Mapping land cover over large geographic areas: Integrating GIS and remote sensing technologies. In Proceedings of the 2002 ESRI Users Conference. Paper available at gis.esri.com/library/userconf/proc02/pap0331/p0331.htm

Table 11.3 Core modeling datasets derived from TM imagery and DEMs	
Derived from Landsat TM imagery	**Derived from digital elevation model**
Spring NDVI	Slope
Spring brightness	Aspect
Spring greenness	Elevation
Spring wetness	Topographic relative moisture index
Summer NDVI	Landform
Summer brightness	
Summer greenness	
Summer wetness	
Fall NDVI	
Fall brightness	
Fall greenness	
Fall wetness	

Source: Table courtesy of J. Lowry, R. Ramsey, and G. Manir (2002).

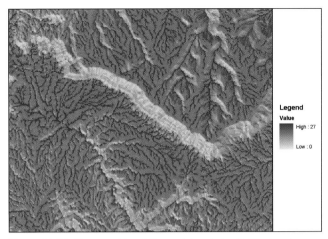

Figure 11.41   Topographic relative moisture index (0-27, dryer to wetter).

Source: Image courtesy of J. Lowry, R. Ramsey, and G. Manis. (2002).

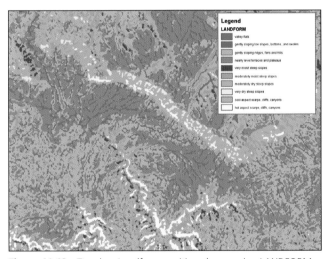

Figure 11.42   Ten class Landform position classes using LANDFORM.

Source: Image courtesy of J. Lowry, R. Ramsey, and G. Manis. (2002).

index of four landscape elements—relative slope position, slope angle, slope shape, and slope aspect—derived from the DEM. The Landform data layer was generated by reclassifying the gridded TRMI data layer according to specified ranges of TRMI index values and slope as derived from the DEM to produce a gridded data layer with 10 Landform position classes. A rule-based classification procedure was then developed to generate the final land-cover classifications for the study area.

Predictive modeling: Gypsy moth damage potential[4]

The gypsy moth is an insect pest that severely weakens and ultimately kills oak and other valuable tree species. In 1997, the Minnesota Department of Natural Resources developed a statewide damage potential model using vegetation parameters derived from AVHRR satellite image data (at 1.1 km resolution) and soils data from existing maps. The model predicted susceptibility to damage (i.e., mortality) based on four factors: cover type and cover density (derived from the AVHRR data), soil type (from the Minnesota Soil Atlas), and environmental stress as measured by moisture shortfall.

The model was designed to assess the damage potential relative to where defoliation is (1) likely to be the heaviest and most sustained and (2) where it has the greatest chance of causing mortality when environmental stress is considered. The different data layers were categorized into classes according to their attractiveness as gypsy moth habitat and their contribution to cumulative stress. Combined, the four weighted variables were considered to effectively predict both infestation potential and subsequent mortality. Two submodels were used. The biological submodel included the cover-type and cover-density components. The environmental submodel included the soil and moisture stress components. The expert opinion of foresters with long experience working in the different regions of Minnesota was used extensively in developing the model ratings and weightings. The components of the gypsy moth prediction model are shown in figure 11.43. Figure 11.44 shows the final damage potential map.

The model was implemented using a raster-based GIS. In 1997, Minnesota did not yet have established gypsy moth populations, so the damage potential model does not predict where infestation is occurring but where mortality is expected once the state is infested. Based on the rate of gypsy moth population spread westward, it was anticipated that the gypsy moth would reach Minnesota in 10 to 20 years. The predictive model is being used to implement management regimes designed to reduce damage using techniques such as thinning stands of susceptible tree species and removing stand components most likely to suffer damage. Using the risk model as advisory data, effort and resources are focused on the areas of greatest risk, thereby obtaining the greatest benefit from limited resources.

---

4. Contributed by S. Aronoff.

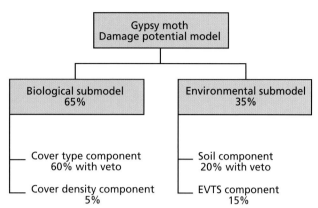

Figure 11.43 Components of the gypsy moth damage potential model.

Source: Image courtesy of T. Eiber of the Minnesota Department of Natural Resources.

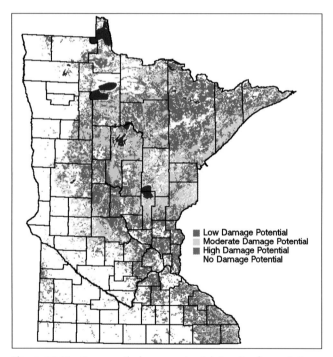

Figure 11.44 Gypsy moth damage potential. Results of a predictive model were classified into four categories ranging from no damage to high damage potential.

Source: Image courtesy of T. Eiber of the Minnesota Department of Natural Resources.

Since this model was generated, a statewide, Landsat-based (30 m) cover-type layer has become available, allowing the model to be regenerated with higher spatial resolution. This new layer is in the final stages of creation and will be used to guide management activities on a more site-specific basis.

In this model, the cover density component was deleted as it was deemed technically inappropriate with the 30 m data. Weighting for the remaining factors is unaltered in the new model. The original model will continue to be used for more generalized planning activity at the landscape level.

Temporal modeling: Trends in Arctic sea ice concentrations

Due to the consistent, repetitive, and numerical nature of satellite images, remote sensing has a key role to play in climate change studies (LeDrew 1992). One part of the earth's geophysical system that is thought to be particularly sensitive to climatically induced changes is the ice cover of the polar oceans (Piwowar and LeDrew 2002). A continuous record of spaceborne observations of the Arctic Ocean extends back to 1978. While it is interesting to compare current images with those from the late 1970s to observe the amount of change, this gives us little insight into what might have occurred between these dates. Temporal modeling attempts to characterize the processes at work within an image time series.

In temporal modeling, a series of remote sensing images acquired consistently throughout a specified period are collected in the form of a temporal image cube or stack *(figure 11.45)*. Then a variety of functions can be applied to the stack in order to examine its inherent temporal structures. In figure 11.46, part of the cube has been cut away to reveal a temporal image face that clearly shows not only the annual seasonality within the data but also some striking anomalies, for example, the significant decrease in ice during the winters immediately following the strong 1983 El Niño event.

Figure 11.46 also shows time series plots at four selected pixels in the scene where the seasonal signals are clearly different. Linear regression trend lines were drawn on each plot, and the slope coefficient from each line was transferred to its corresponding location in an image. When the regression slopes were calculated for every pixel in the image, the trend image in figure 11.47 was produced. This figure describes the spatial distribution of the temporal processes that are present in the Arctic. There are large areas of decreasing ice concentrations (most notably along the eastern coast of North America) but also some regions where there appears to be more ice occurring through time. In general, however, there is an overall decrease in the Arctic ice cover. When these trends were used to forecast future conditions they were found to match current observations quite well.

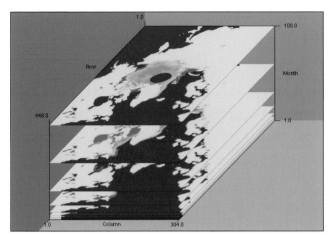

Figure 11.45   A temporal data cube created by stacking 108 monthly (i.e., 9 years of data) time slice images on top of each other. The stack extends from November 1978 at the bottom to October 1987. The data is Northern Hemisphere sea ice concentrations determined from passive microwave imagery.

Source: Image provided by the National Snow and Ice Data Center of Canada.

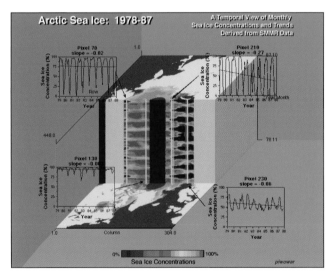

Figure 11.46   A vertical slice was made through the cube in figure 11.45 to expose a time domain data face. This temporal face extends from November 1978 at the bottom to October 1987 at the top and from the Beaufort Sea on the left to the Barents Sea on the right. Moving up the face from the bottom one can see the annual advance and retreat of the sea ice as orange bulges extending from and retreating to the image center. In particular, two winters with significantly less ice, 1984 and 1985, are evident as lesser bulges in the Barents Sea. Four time series of monthly mean ice concentrations from (arbitrary) sample locations indicated by the four vertical lines extending down the temporal face are drawn in the corners.

Source: Image created by J. Piwowar. Data provided by the National Snow and Ice Data Center of Canada.

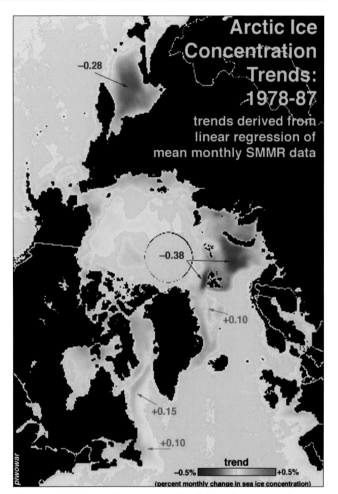

Figure 11.47   A trend image created by fitting a linear regression to the data in figure 11.46 on a per-pixel basis. This image graphically shows the slope of the fit regression lines; pixels with positive slopes are shown in red; pixels with negative slopes are shown in blue. Several regions of statistically significant trends in Arctic sea ice concentrations between 1978 and 1987 are evident, most notably decreasing iciness in the Barents and Okhotsk Seas.

Source: Image created by J. Piwowar. Data provided by the National Snow and Ice Data Center of Canada.

Remote sensing imagery in the GIS environment

The visual and classified products of digital image analyses of remote sensing data are among the most important data sources for many GIS applications. However, the process of incorporating imagery into a GIS can be difficult due to differences in the way GIS data is commonly collected, processed, and stored.

Geographic information systems can generally handle geospatial datasets in both raster and vector formats. In spite of this, most GIS operations still require the input datasets for an operation to all have the same format. Consequently, when both raster and vector data are to be used together in an operation, the program will temporarily (and transparently to the user) convert data from one of the formats to the other in order to satisfy the processing requirements. In these systems, remote sensing imagery and other GIS data can be left in their original formats, increasing efficiency and reducing data loss (Piwowar and LeDrew 1990). Older systems may only accommodate data in one format. In which case, data may have to be specifically converted to a different file structure for import into the GIS.

The raster format is well suited to remote sensing data because the spectral values in this format vary continuously over the image area. However, during image classification, the remotely sensed reflectance values are transformed into a set of discrete features consisting of groups of pixels belonging to the same class around which a distinct boundary can be drawn. Although raster grids can accommodate discrete features, they are more efficiently encoded in a vector environment using points, lines, and areas. So the vector format is generally more suitable for storing classified image data in a GIS. The transfer of remote sensing analysis results, such as a land-cover classification, to a GIS generally requires conversion of the raster-based classification to vector format so that contiguous groupings of pixels with the same class become discrete polygons (*figure 11.48*).

The formats of the other datasets with which the remote sensing imagery will be used also have a decisive impact on how the remote sensing imagery should be ultimately structured. Conversion of spatial data between raster and vector formats always results in some loss of data precision (Piwowar et al. 1990). Consequently, the structure of datasets derived from remote sensing imagery should be selected to match that of other key datasets with which it is to be used.

File exchange formats

One of the first technical issues that must be addressed when transferring a remote sensing image into a GIS is "how to get it there." Some image analysis systems have developed links with one or more GIS to facilitate image exchange, but this is more the exception than the norm. Transferring image data from an image analysis system to a GIS typically requires exporting the data from the imaging system into an intermediate data file format and then importing the data from

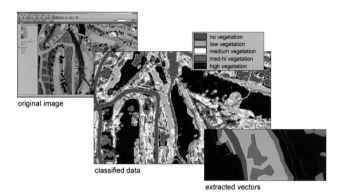

Figure 11.48 Transfer of remote sensing derived data to a vector GIS. In order to import a land-cover classification of remotely sensed data into a vector GIS, the raster image is first classified, and the resulting classification is usually filtered to remove unwanted detail. Conversion to vector format creates polygons in place of contiguous clusters of pixels of the same class, and assigns the class name to the polygon label.

Source: Image courtesy of Positive Systems, Inc.

this format into the GIS. The intermediate format should be one that

- is readable and writable by both the image analysis system and the GIS;
- does not alter the image data;
- retains any georeferencing information; and
- is capable of transferring attribute information.

Standard spatial data file exchange formats have been developed for this purpose. Although some of them are quite effective, so many different ones exist that no single format can truly be considered *standard*. Use of a standard spatial data file format, however, may be the only option if multiple attributes are to be included in the file exchange.

Very often the best solution is to choose a common *graphic* file format, such as the Tagged Image File Format (TIFF) or the structure designed by the Joint Photographic Experts Group (JPEG). The latter format, although capable of significantly reducing the size of a file, employs a data compression algorithm producing noticeable losses in image quality at high compression levels.

The TIFF structure is better suited as a graphics transfer format, not only because it preserves the original image quality and is read and written by most GIS and image analysis systems, but also because it is widely recognized by other graphics display software. In addition, a recent addition to the defined TIFF standard allows for the specification of a *world file* for the exchange of image georeferencing data.

Data volumes

Raster format remote sensing image files tend to be very large compared with geospatial datasets stored in vector format. For example, a 30 m spatial resolution Landsat ETM image of a 10 km by 10 km area contains 111,000 pixels, each of which is represented by seven spectral data values. An image of the same area from the QuickBird sensor would consist of over 260 *million* pixels. Clearly, any GIS operation, such as *select* or *recode*, that needs to examine each point in the data space will require considerably more time to complete with raster imagery than with vector line work. For this reason, image processing systems designed to handle large image files are generally better suited to perform computational operations on imagery.

Class filtering

As discussed earlier, classified images frequently exhibit more category detail than is required or desired for GIS applications. For example, there may be *sand* or *grass* pixels at places where there are tiny islands scattered throughout a river, whereas the GIS data may represent the entire region between the shorelines as a homogeneous class, *river*. An image classification may accurately identify tree classes in an otherwise *urban* area mapped in a GIS as a *multiple-unit residential* zone. The remote sensing data is not incorrect; in fact, it may be *more* accurate than corresponding GIS layers. The image data is simply showing more thematic detail than is required in the GIS.

Thematic detail in a classification may also be the result of class confusion due to the presence of mixed pixels—pixels that represent an area on the ground having more than one land cover, both having strong but different spectral characteristics. In this situation, the pixel reflectance values would not be representative of any of the individual features; instead, they would be an average of the constituent land covers. Such spectrally mixed averages randomly appear as incorrectly assigned pixels in a classified scene.

Regardless of the cause, these small clusters of undesired classification cells can create considerable difficulty when integrating the remote sensing classes into a GIS, particularly if the raster classification is to be converted to vector format for subsequent processing. The sheer number of these cells can quickly overwhelm the capacity of a typical GIS since each one is converted to a tiny polygon *(figure 11.49a)*.

Several approaches have been developed to deal with this issue. A simple method is to recode a cell's class to be the *majority* of a class count between the central cell and its eight neighboring pixels. In the *sieve filter* method, individual pixels or pixel groups of less than a user-specified size are incorporated to the next larger (in spatial extent) adjacent class *(figure 11.49b)*. This *contextual classification* technique reassigns these scattered pixels depending on their relationships with their neighboring cells.

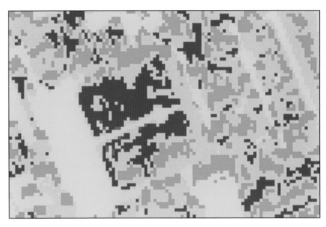

a

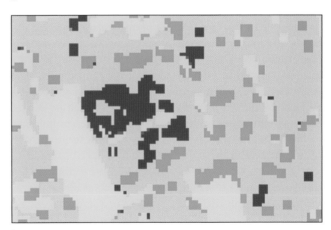

b

Figure 11.49 Postclassification filtering to group isolated pixels into larger contiguous regions. Figure (a) is the original classification; figure (b) is the sieve-filtered data.

Conclusion

Digital image analysis methods have become essential for both visual and digital analysis of remotely sensed data. Image enhancement techniques are used to improve the interpretability of both photographic and digital image products. Geometric rectification and mosaicking of remote sensing imagery to generate images accurately matched to a chosen map projection depend on digital image processing, even if the products are to be interpreted visually.

Human visual interpreters make extensive use of contextual image characteristics such as texture, shape, and associated features. Digital analysis methods can simultaneously analyze many spectral bands and image types. Complex numerical calculations can be used to rapidly generate measurements of environmental parameters collected virtually at the same time over large areas. Only by using remote sensing could global measurements of sea surface temperature, wind patterns, chlorophyll concentration, fire detection, vegetation conditions, and many other environmental factors be collected daily.

Computational methods enable certain types of analyses to be applied to large quantities of image data more rapidly than visual interpretation methods can produce. Automated classifications can generate detailed land-cover maps with summary statistics. The development of segmentation algorithms that incorporate contextual information has improved the correspondence between the maps produced by digital image classification and more traditional map compilation methods that depend on visual image analysis techniques.

With the availability of inexpensive high-speed computing and large data storage capacity, it has become technically feasible to make remote sensing imagery readily available within the GIS environment. The maturing of the digital image analysis field has also resulted in the development of more or less standard image processing functions, such as basic image enhancement and resampling, that are widely used in consumer products. These functions, which make imagery more interpretable, are more easily used by GIS analysts.

Often the specific information provided by imagery is most easily obtained by viewing the image with relevant GIS-based information, such as political or legal boundary data, overlaid. Other functions, such as image classification, while they may be easy to run, must be incorporated within a well-defined study design in order to generate results with the level of accuracy and repeatability suited to practical applications.

Remote sensing analysis depends heavily on the use of other spatial datasets such as digital elevation models and various information sources derived from field observations. Most of these data sources can be stored and managed within a GIS environment. The development of more effective methods of data transfer has facilitated the sharing of information between GIS and image analysis systems. Also, GIS have increasingly incorporated an array of basic image analysis functions sufficient to satisfy basic image enhancement and rectification needs. However, image analysis systems specifically designed to process remote sensing data are still required for more complex processing tasks.

The role of remote sensing and image analysis can be viewed as an important component of a geospatial information environment. Whether the image analysis component is physically networked to an organization's enterprise-wide information system or a stand-alone system, it is, by virtue of its key role as a source of geographic information, always part of a larger geospatial information enterprise. For it is only when incorporated into a larger application framework that remotely sensed data can be used to generate data suited to specific applications and delivered in a timely manner to those prepared to use it.

References

Adams, J. B. 1993. A combined atmospheric and systems calibration from spectral mixture analysis. Proceedings, workshop on the Atmospheric Correction of Landsat Imagery. Geodynamics Corp., June. Torrance, Calif.

Adams, J. B., M. O. Smith, and A. R. Gillespie. 1993. Imaging spectroscopy: Interpretation based on spectral mixture analysis. In *Remote geochemical analysis: Elemental and mineralogical composition*, ed. C. M. Pieters and P. Englert, 145–66. Cambridge: LPI and Cambridge University Press.

Anderson, J. R., E. E. Hardy, J. T. Roach, and R. E. Witmer. 1976. A land use and land cover classification for use with remote sensor data. USGS Professional Paper 964. Washington, D.C.: U.S. Government Printing Office.

Aronoff, S. 1984. An approach to optimized labeling of image classes. *Photogrammetric Engineering and Remote Sensing* 50(6):719–27.

———. 1982a. Classification accuracy: A user approach. *Photogrammetric Engineering and Remote Sensing* 48:1299–1307.

———. 1982b. The map accuracy report: A user's view. *Photogrammetric Engineering and Remote Sensing*. 48:1309–12.

Bachmann, C. M., T. F. Donato, G. M. Lamela, W. J. Rhea, M. H. Bettenhausen, R. A. Fusina, K. R. Du Bois, J. H. Porter, and B. R. Truitt. 2002. Automatic classification of land cover on Smith Island, Va., using HyMAP imagery. *IEEE Transactions on Geoscience and Remote Sensing* 40(10):2313–30.

Ball, G. H., and D. J. Hall. 1967. A clustering technique for summarizing multivariate data. *Behavioral Science* 12:153–55.

Bauer, M. E., T. E. Burk, A. R. Ek, P. R. Coppin, S. D. Lime, T. A. Walsh, D. K. Walters, W. Befort, and D. F. Heinzen. 1994. Satellite inventory of Minnesota forest resources. *Photogrammetric Engineering and Remote Sensing* 60(3):287–98.

Behera, M. D., S. P. S. Kushwaha, and P. S. Roy. 2001. Forest vegetation characterization and mapping using IRS-1C satellite images in the Eastern Himalayan Region. *Geocarto International* 16(3):53–62.

Brodley, C., and M. A. Friedl. 1996. Improving automated land cover mapping by identifying and eliminating mislabeled observations from training data. Proceedings of International Geoscience and Remote Sensing Symposium, (IGARSS'96), May, 27-31, Lincoln, Neb. Piscataway, N.J.: *IEEE* 2:1382–84.

Chen, K. S., Y. C. Tzeng, C. F. Chen, and W. L. Kao. 1995. Land-cover classification of multispectral imagery using a dynamic learning neural network. *Photogrammetric Engineering and Remote Sensing* 61:403–408.

Cihlar J., R. Latifovic, J. M. Chen, and Z. Li. 1999. Testing near real-time detection of contaminated pixels in AVHRR composites. *Canadian Journal of Remote Sensing* 25(2): 160–70.

Congalton, R. G., and K. Green. 1998. *Assessing the accuracy of remotely sensed data: Principles and practices*. Boca Raton, Fla.: Lewis.

Coulombe, A., L. Charbonneau, R. Brochu, and D. Morin. 1991. L'apport de l'analyse texturale dans la définition de l'utilisation du sol en milieu urbain. *Canadian Journal of Remote Sensing* 17(1): 46–55.

Cracknell, A. P., and L. W. B. Hayes. 1991. *Introduction to remote sensing*. London: Taylor and Francis.

Drury, S. A. 1998. *A guide to remote sensing: Interpreting images of the earth*. Oxford: Oxford University Press.

Fung, T., and E. LeDrew. 1987. Application of principal components analysis to change detection. *Photogrammetric Engineering and Remote Sensing* 53(12): 1649–58.

Gloerson, P., D. Cavalieri, W. J. Campbell, and J. Zwally. 1990. *Nimbus-7 SMMR polar radiances and Arctic and Antarctic sea ice concentrations*. Boulder, Colo.: National Snow and Ice Data Center. CD-ROM.

Gong, P., R. Pu, and J. R. Miller. 1992. Correlating leaf area index of Ponderosa pine with hyperspectral CASI data. *Canadian Journal of Remote Sensing* 18(4): 275–82.

Gopal, S., and C. Woodcock. 1994. Theory and methods for accuracy assessment of thematic maps using fuzzy sets. *Photogrammetric Engineering and Remote Sensing* 60:181–88.

Gurney, C. M., and J. R. Townshend. 1983. The use of contextual information in the classification of remotely sensed data. *Photogrammetric Engineering and Remote Sensing* 49:55–64.

Jensen, J. R. 2000. *Remote sensing of the environment*. Upper Saddle River, N.J.: Prentice Hall.

———. 1996. *Introductory digital image processing: A remote sensing perspective*. 2nd ed. Upper Saddle River, N.J.: Prentice Hall.

———. 1978. Digital land cover mapping using layered classification logic and physical composition attributes. *American Cartographer* 5:121–32.

Jensen, J. R., and D. C. Cowen. 1999. Remote sensing of urban/suburban infrastructure and socio-economic attributes. *Photogrammetric Engineering and Remote Sensing* 65:611–22.

LeDrew, E. 1992. The role of remote sensing in the study of atmosphere-Cryosphere interactions in the polar basin. *The Canadian Geographer* 36(4):336–50.

Lillesand, T. M. 1996. A protocol for satellite-based land cover classification in the Upper Midwest. In *Gap analysis: A landscape approach to biodiversity planning*, ed. J. M. Scott, T. H. Tear, and F. W. Davis, 103-118. ASPRS.

Lillesand, T. M., R. W. Kiefer, and J. W. Chipman. 2004. *Remote sensing and image interpretation*. 5th ed. New York: John Wiley and Sons, Inc.

Lowry, J., R. D. Ramsey, and G. Manis. 2002. Mapping land cover over large geographic areas: Integrating GIS and remote sensing technologies. In Proceedings of the 2002 ESRI Users Conference. Redlands, Calif.: ESRI. gis.esri.com/library/userconf/proc02/pap0331/p0331.htm

Lyon, J. G., D. Yuan, R. S. Lunetta, and C. D. Elvidge. 1998. A change detection experiment using vegetation indices. *Photogrammetric Engineering and Remote Sensing* 64(2): 143–50.

Malcolm, N. W., J. M. Piwowar, G. B. Hall, C. Cotlier, and A. Revenna. 2001. An integration of Radarsat and Landsat imagery to identify pockets of urban poverty: A Rosario, Argentina, case study. *Canadian Journal of Remote Sensing* 27(6): 663–68.

Marceau, D., P. J. Howarth, and J. M. Dubois. 1989. Automated texture extraction from high spatial resolution satellite imagery for land-cover classification: Concepts and applications. Proceedings of IGARSS'89/12th Canadian Symposium on Remote Sensing, July. Vancouver Canada.

MicroImages, Inc. 1999. *Getting started: Rectifying images.* Lincoln, Neb.: MicroImages, Inc.

Millward, A. A., and J. M. Piwowar. 2002. Exploring ordination as a method for normalizing disparate datasets: Implications for digital change detection. Proceedings, Joint International Symposium on Geospatial Theory, Processing and Applications, July 8–12. Ottawa Canada.

Paola, J. D., and R. A. Schowengerdt. 1997. The effect of neural-network structure on multispectral land-use/land-cover classification. *Photogrammetric Engineering and Remote Sensing* 63(5):535–44.

———. 1995. A review and analysis of back propogation neural networks for classification of remotely-sensed multispectral imagery. *International Journal of Remote Sensing* 16(6):3033–58.

Peddle, D. R., F. G. Hall, and E. F. LeDrew. 1995. Spectral mixture analysis and geometric-optical reflectance modelling of boreal forest biophysical structure, Superior National Forest, Minnesota. In *Proceedings of 17th Canadian symposium on remote sensing*, 617–22.

Peddle, D. R., E. F. LeDrew, and H. M. Holden. 1995. Spectral mixture analysis of coral reef abundance from satellite imagery and in situ ocean spectra, Savusavu Bay, Fiji. In *Proceedings of third thematic conference on remote sensing for marine and coastal environments*, Vol. II: 563–575.

Piwowar, J. M. and E. F. LeDrew, 2002. ARMA time series modelling of remote sensing imagery: A new approach for climate change studies. *Int. J. Remote Sensing* 23(24): 5225–48.

———. 1995. Climate change and Arctic Sea Ice: Some observations from hypertemporal image analysis. In *Proceedings of third thematic conference on remote sensing for marine and coastal environments*, Vol. I: 625–36.

———. 1990. Integrating spatial data: A user's perspective. *Photogrammetric Engineering and Remote Sensing* 56(11): 1497–1502.

Piwowar, J. M., E. F. LeDrew, and D. J. Dudycha. 1990. Integration of spatial data in vector and raster formats in a geographic information system environment. *International Journal of Geographical Information Systems* 4(4): 429–44.

Piwowar, J. M., and A. A. Millward, 2002. Multitemporal change analysis of multispectral imagery using principal components analysis. In Proceedings of International Geoscience and Remote Sensing Symposium (IGARSS).

Piwowar, J. M., D. R. Peddle, and E. F. LeDrew. 1997. Temporal mixture analysis of Arctic Sea Ice imagery: A new approach for monitoring environmental change. *Remote Sensing of Environment* 63: 195–207.

Richards, J. A., and X. Jia. 1999. *Remote sensing digital image analysis: An introduction.* Berlin: Springer-Verlag.

Singh, A., and A. Harrison. 1985. Standardized principal components. *International Journal of Remote Sensing* 6(6): 883–96.

Taylor, M. M. 1974. Principal components color display of ERTS imagery. In *Proceedings of 2nd Canadian symposium on remote sensing*, 295–313.

TBRS. 2003. Enhanced vegetation index (EVI). Terrestrial Biophysics and Remote Sensing Laboratory.University of Arizona.Tucson, Ariz. tbrs.arizona.edu/project/MODIS/evi.php

Treitz, P. M., P .J. Howarth, O. R. Filho, E. D. Soulis, and N. Kouwen. 1993. Classification of agricultural crops using SAR tone and texture statistics. In *Proceedings of 16th Canadian symposium on remote sensing*, 343–47.

Tso, B., and P. M. Mather. 2001. *Classification methods for remotely sensed data*. London: Taylor and Francis.

USGS. 2002. National GAP Analysis Program. United States Geological Service (USGS). www.gap. uidaho.edu

Vieira, C.A.O., and P. M. Mather. 1999. Assessing the accuracy of classifications using remotely sensed data. In *Proceedings of 4th international airborne remote sensing conference/21st Canadian symposium on remote sensing*, 2:823-30.

Visual Learning Systems, Inc. 2002. The Feature Analyst™ Extension for ArcView® and ArcGIS™. www.featureanalyst.com/downloads/whitePapers/FA_whtpapr.pdf

Wang, F. 1990. Fuzzy supervised classification of remote sensing images. *IEEE Transactions on Geoscience and Remote Sensing* 28:194-201.

Web sites

Vegetation index

An Analysis of Gypsy Moth Damage Potential in Minnesota.
www.dnr.state.mn.us/fid/october97/10309712.html

Global Land AVHRR NDVI Project.
edcdaac.usgs.gov/1KM/1kmhomepage.html

MODIS Land DisciplineWeb site.
modis-land.gsfc.nasa.gov/products

MODIS Vegetation Index Web site.
tbrs.arizona.edu/project/MODIS/index.php

Object-oriented classifiers

e-Cognition.
www.definiens-imaging.com

Feature Analyst.
www.featureanalyst.com

Images and applications

Canada Centre for Remote Sensing.
www.ccrs.nrcan.gc.ca

Earth Observatory (NASA).
earthobservatory.nasa.gov

Earthshots: Satellite Images of Environmental Change (USGS).
edcwww.cr.usgs.gov/earthshots/slow/tableofcontents

Eurimage: Multi-Mission Satellite Data of Europe.
www.eurimage.com

Visible Earth (NASA).
visibleearth.nasa.gov

World Data Center for Remotely Sensed Land Data (USGS).
edc.usgs.gov/wdcguide.html

12 Remote sensing applications

Remote sensing has been applied to a range of applications far broader than can be covered in a single chapter. The selected application areas presented here illustrate diverse analysis techniques ranging from visual interpretation to the integrated use of remotely sensed data and other geospatial information in complex modeling scenarios. A successful remote sensing application depends on matching technology and analysis procedures to information requirements, available skills, and budget. Experience has yielded some more or less standardized analysis procedures. Though there are, of course, technical limitations, remote sensing applications are more commonly compromised by factors related to planning, management, and communications such as failure to identify all of the end users; the content, format, and timeliness of the information they require; or the cost to complete the project. However, the rapid development of remote sensing and related technologies, such as GIS, usually reward innovation with superior results. As such, in presenting the applications discussed in the following sections, listed below, the intention is to stimulate inquiry and encourage experimentation even when employing remote sensing operationally to address practical needs.

Remote sensing in agriculture

Stan Aronoff

The management of food supplies is critical to the health and well-being of every nation. Remote sensing has become an important source of information for the management of agricultural production and assessment of worldwide supply and demand. The scope of such analyses, which integrate remote sensing data at various levels of detail, can be very broad and cover applications ranging from the land parcel to global scale. These analyses will systematically include data on climate; weather; the quality, quantity, and location of arable land; population dynamics; energy production; and environmental quality issues. Cultural factors, such as local farming traditions and differences in national political and social systems, also affect agricultural productivity. Cultural data is becoming increasingly more available in digital formats suitable for integration into agricultural models.

Remote sensing data is most commonly used for crop-type identification and the analysis of crop condition. Specialized image analysis software is used to process the remotely sensed data. The derived information is then typically used within a GIS to perform integrated analyses such as crop-yield forecasting using other agriculture-related data such as weather and climate, soils, field reports, historical planting patterns, and crop yields. The final information products are in the form of maps, reports, and databases that can be interactively queried. Much of the information is time critical and is made available over the Internet or other electronic networks.

The availability of service-bureau-based online data and software enables users to perform sophisticated customized interactive analyses without having to own and manage expensive computer resources. Decision support systems have been developed that integrate current and historical weather and climate data, remote-sensing-derived crop condition assessments, crop yield models, soil characteristics, and market conditions to assist in every phase of crop management, from selecting the crop optimal for each field to the development of treatment plans to improve yields and reduce the risk of disease or pest damage.

Crop-type identification is based on spectral characteristics, image texture, and knowledge of the developmental stages of each crop in the area being assessed. Plant development information is typically summarized in the form of a crop calendar that lists the expected developmental states

and emergence of each crop in the area throughout the year. Crop characteristics change during the growing season, and crops that appear the same at one time may be easily distinguished at another. By using multiple images recorded on dates during the growing season when crop types are most easily discriminated, the accuracy of crop identification and mapping can be substantially improved.

Crop condition assessment evaluates the health and vigor of a crop in relation to some baseline condition, usually an average calculated from several previous years. Remote-sensing-derived crop condition data can be used for crop monitoring for early detection of drought, pests, and disease in time for corrective action to be taken and to optimize irrigation, pest and disease control, and the application of other treatments. Combined with other datasets, crop condition is used to predict crop production shortfalls or surpluses so that measures can be planned to forecast yield and prices; to document crop damage caused by drought, flooding, and other weather related causes; and to verify compliance with regulatory programs.

Remote sensing analyses commonly use indices calculated from digital multispectral image data. The widely used Normalized Difference Vegetation Index (NDVI), discussed in chapter 11, is calculated from two bands of multispectral image data, the visible red band (red) and near-infrared (NIR) bands as follows:

$$NDVI = (NIR - red) / (NIR + red).$$

The NDVI correlates well with biomass. Because healthier plants of a given species tend to have greater biomass, it is a good measure of plant vigor. Since the NDVI values that indicate healthy or stressed conditions vary for different species, the index is best used to compare conditions of the same species at different times of year, in different years, or growing in different areas in the same region or same field. When NDVI values are plotted against time over the growing season, they provide a useful graphic as well as quantitative estimate of yield. The NDVI value for a particular crop has a characteristic curve with NDVI values increasing from when the plants first emerge to when they reach their maximum development, called *peak-of-green* (because it is when the leafy part of the plants are most fully developed and the green reflectance is at a maximum). The NDVI value then declines as the plant matures and goes to seed (see figure 12.2 discussed below). Years when the NDVI curve is lower generally produce lower yields. A shifting of the curve earlier or later in time is indicative of early or late emergence, which affects the time available for the plants to develop and, consequently, the expected yield.

Both airborne and satellite multispectral image data are used for crop identification and condition assessment. Many of the resource satellites currently in operation record imagery in the visible and near-infrared bands most commonly used to produce the NDVI value. Data from high-, medium-, and low-resolution sensors are often used, sometimes in combination, to monitor crop conditions *(table 12.1)*.

The data available at 250 m, 1 km, and 8 km is used extensively worldwide to monitor regional crop condition but are too coarse for detailed crop identification. Global-coverage multispectral data at higher resolutions from Landsat TM (30 m pixels), SPOT (10 m), and the Indian Remote Sensing Satellites IRS-1C and 1D (23 m) are of high enough resolution to be useful in crop identification as well as crop condition assessment. Assessing within-field conditions requires higher resolution multispectral satellite data, like that available from IKONOS (4 m), OrbView (4 m), QuickBird (2.44 m), or airborne multispectral imagery with spatial resolutions as fine as 15 cm.

The within-field data, obtained from high-resolution satellite or airborne imagery, enables variations in soil and moisture conditions, plant vigor, and other variables to be identified and mapped so that planting and treatments can be varied to increase yield and minimize costs—an approach called *precision farming*. This type of within-field optimization has the potential to improve profitability and reduce unwanted environmental impacts where the additional cost in using the technology can be justified.

Regional crop condition monitoring

Global Information and Early Warning System (GIEWS) of the United Nations Food and Agriculture Organization

The Global Information and Early Warning System on Food and Agriculture (GIEWS) operated by the United Nations Food and Agriculture Organization (FAO) monitors famine-prone areas of the globe in order to predict food shortages and possible famine conditions. With sufficient lead time, national governments and relief organizations can mount an effective response to impending food shortages. The GIEWS strives to meet that need by providing detailed analysis and frequent reports on crop conditions and the food supply situation for all countries.

In many drought-prone countries, particularly in sub-Saharan Africa, there is a lack of continuous up-to-date and reliable information on crop conditions. The GIEWS makes use of remotely sensed data, in particular NDVI images derived from the NOAA/AVHRR and SPOT/VEGETATION sensors, to provide seasonal coverage over these large areas. NDVI and other data used for crop monitoring have been collected and archived for several years and are used to assess current conditions and predict agricultural success relative to previous seasons. The NDVI images provide an indication of vegetation vigor and permit GIEWS analysts to monitor vegetation development throughout the crop growing period.

The remotely sensed data is used in conjunction with historical and current crop data, weather data, and field reports from local government and FAO officials to provide an overall

Table 12.1 Sensors commonly used for crop-type and condition assessment			
High spatial resolution	**Panchromatic resolution**	**Multispectral resolution**	**Swath width**
IKONOS	1 m	4 m	11 km
Orbview-3	1 m	4 m	8 km
QuickBird	0.61 m	2.4 m	16.5 km
SPOT 5	2.5 m, 5 m	10 m	60 km[1]
Medium spatial resolution			
SPOT 1 - 4	10 m	20 m	60 km[1]
Landsat 7 enhanced TM	15 m	30 m	185 km
IRS 1-C and 1-D	5.8 m	23 m	70 km
Low spatial resolution			
NOAA AVHRR		1 km, 8 km	2,400 km
SPOT Vegetation Sensor		1.15 km	2,000 km
MODIS sensor on the NASA Terra and Aqua satellites		250 m, 500 m, 1 km	2,330 km

[1] SPOT carries two identical sensors that can be programmed to view adjacent areas giving a combined 117 km swath width in nadir viewing mode (vertically down).

assessment of the crop and food supply situation. Using a geospatial information system, this data is integrated with the digital maps of administrative boundaries, which act as a spatial reference. In addition, recent price and market conditions and information on food stocks and consumption rates are used to predict where prospects for current crops are unfavorable. An interactive Web-based tool is used for information access, evaluation of key indicators, and dissemination of food security analysis at national and global levels.

Figure 12.1 illustrates the use of NDVI images to monitor crop conditions during 1997 in Burkina Faso, a country in the famine-prone Sahel of West Africa. Cloud-free composite imagery is generated from the daily imagery for a 10-day period. The 10 images are registered, and for each pixel, the NDVI values for the 10 days are compared. The pixel with the highest NDVI value (lowest cloud level) is used to produce the composite image. Since clouds appear much brighter than vegetation in both the red and near-infrared bands, pixels depicting clouds are easily identified and not used. To facilitate interpretation, the NDVI values in the images are subdivided into 12 ranges that are color-coded from yellow for low values (indicating bare soil) to

dark green for high NDVI values (indicating well-developed healthy vegetation). It is these classified NDVI images that are used by the crop analysts.

The upper row of three images in figure 12.1 shows the classified NDVI images at the beginning, middle, and end of the growing season (May, August, and October). The bottom row of images shows the difference between the 1997 NDVI values and the 15-year average from 1982 to 1996. The difference image is color-coded by dividing the NDVI difference values into five ranges from red (below average) to green (above average). Viewing these images together, the relationship of growing conditions to crop development can be evaluated.

The 1997 rainy season began in April, earlier than usual, so large areas of the May image appear green. The large areas represented as green classes in the May difference image (lower left image in figure 12.1) indicate that vegetation development was more advanced than average. Low rainfall in the central and eastern areas in June and July caused the early planting to fail and required that crops be replanted in mid-August. The poor conditions in August are clearly indicated with the lower than average NDVI values, as shown in the August difference image. Low rainfall in the later part of September damaged the replanted crops, which resulted in lower than average yields in the central and eastern regions

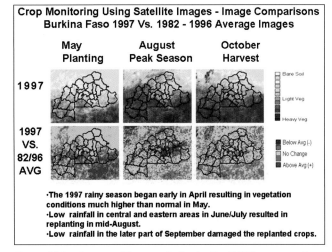

Figure 12.1 Crop condition monitoring by the GIEWS program. The top three images show the progress of crop development in Burkina Faso during 1997. They were produced by calculating an NDVI image from the red and near-infrared bands of 1 km resolution AVHRR image data and then color-coding the NDVI values. In the bottom three images, the NDVI values of the 1997 image were subtracted from the 15-year average NDVI from previous years. The range of difference values were color-coded to produce five classes.

Source: Image courtesy of GIEWS and the United Nations Food and Agriculture Organization.

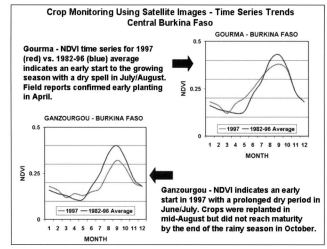

Figure 12.2 Time series comparison of crop development. Crop development in 1997 for two districts within Burkina Faso are compared with the 15-year average (1982–1996) by plotting time series of NDVI values. Higher NDVI values indicate higher plant biomass and greater plant vigor.

Source: Image courtesy of GIEWS and the United Nations Food and Agriculture Organization.

of the country. The October imagery shows this well. Most of these regions are classified below average (red hues) on the difference image. Figure 12.2 shows a plot of the average NDVI values for the province of Ganzourgou located in this region. During months 4 and 5 (April and May), the NDVI curve for 1997 (red line) was above the average (blue line). From June onward, the curve remains below average for the remainder of the year, indicating a poorer crop.

Crop Condition Assessment Program of Statistics Canada

Statistics Canada developed the Crop Condition Assessment Program (CCAP). Since 1989, this program has used AVHRR data from the NOAA satellite to monitor crop and rangeland conditions throughout the Canadian prairies and, from 1993 onward, the northern plains of the United States as well. Assessments are updated weekly throughout the growing season and are provided for the entire region. Reporting of the Canadian data is by province, for each of the 40 Census Agricultural Regions and almost 500 Census Consolidated Subdivisions. Reporting for the United States coverage is by county. Spring wheat yield is forecast weekly, using a linear regression model for the NDVI data and historical yield data published by Statistics Canada. The final remote-sensing-based forecast yield has been within 5.6% or less of the published Statistics Canada estimates (produced from field reports) for 10 of the 13 years from 1989 to 2001 and are available two months earlier than those reports *(table 12.2)*. Discrepancies were greater in 1989, 1993, and 1995 when unusually adverse weather conditions affected results. Spring wheat production is also forecast by multiplying the forecast spring wheat yield by the seeded area estimate produced from other Statistics Canada surveys.

The remote-sensing-based analysis uses seven-day composite NDVI images. The compositing process eliminates most or all cloud effects. Any remaining cloud pixels are automatically identified and removed from calculations. Figure 12.3

is an example of the seven-day color-composite AVHRR image data for Western Canada produced from the visible and near-infrared bands.

Custom software was written to integrate commercially available image analysis and GIS software into a production system that is highly automated, from input of the image data to the generation of crop condition maps, graphs, and tables. Weekly crop condition updates are available to subscribers within 24 hours of the last satellite overpass used to produce the weekly AVHRR composite image. Subscribers are notified by e-mail and can retrieve reports over the Internet as soon as they are available.

Subscribers access the CCAP service over the Internet through a GIS interface. They can interactively choose regions of interest by name or by selecting them from maps at several scales. Several types of image and map products, statistical data, and NDVI curves are produced, all of which are updated weekly. Image products show vegetation conditions on a pixel-by-pixel basis for the entire prairie region of western Canada. In addition, map products derived from the NDVI data illustrate the predominant vegetation condition for the chosen administrative districts *(figure 12.4)*.

Image and map products provide the following comparison analyses:

- comparison of any week of the current growing season with the previous week of the same growing season
- the current week with the same week of the previous growing season
- the current week with the same week of the average
- percent comparison of the current week with the maximum NDVI value from all previous years

In addition, a variety of comparison graphs and tabular summaries can be plotted to show trends for user-selected districts or regions. All products are available to subscribers in a GIS-readable digital format. However, many subscribers prefer to access the data and perform their GIS analyses interactively over the Internet. In this way, they do not have to purchase

Table 12.2 Spring wheat yield forecasts for western Canada based on NOAA AVHRR satellite data													
Western Canada	**1989**	**1990**	**1991**	**1992**	**1993**	**1994**	**1995**	**1996**	**1997**	**1998**	**1999**	**2000**	**2001**
Forecast[1] (bu/ac)	28.8	34.9	31.3	31.0	29.0	29.5	29.4	35.2	33.2	31.8	36.7	34.6	28.8
Statistics Canada Publication[2] (bu/ac)	26.8	33.7	33.1	31.3	32.0	30.7	32.3	36.1	31.5	33.4	37.7	36.1	27.4
Forecast vs Published	7.4%	3.6%	-5.5%	-0.9%	-9.4%	-3.9%	-9.0%	-2.6%	5.6%	-4.8%	-2.7%	-4.2%	-4.8%

[1] Forecasts are viewed as an experimental indicator distributed to CCAP subscribers for evaluation purposes only. The official Statistics Canada crop estimate is based on the Field Crop Reporting Surveys.
[2] Statistics Canada, Field Crop Reporting Series, 1989–2001.

Table courtesy of Statistics Canada.

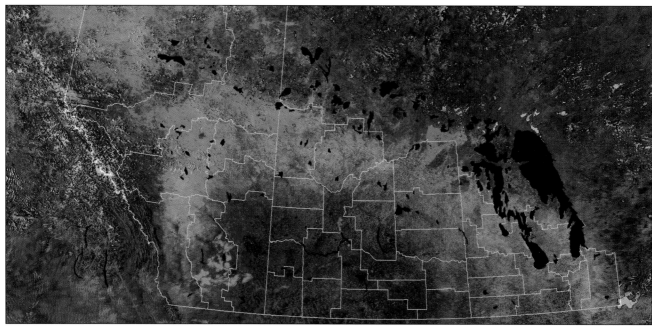

Figure 12.3 Color-composite AVHRR image of Western Canada. This image is a cloud-free composite created from the visible red band and the Normalized Difference Vegetation Index (NDVI) values of seven digital images acquired over a one week period of the summer. The 2,000 km by 1,000 km region extends from central British Columbia east to western Ontario and from the United States border north into the Canadian prairie provinces. Areas where vegetation development is more advanced appear redder. The yellow lines are the crop reporting district boundaries by which crop condition assessment statistics are summarized.

Source: © Natural Resources Canada.

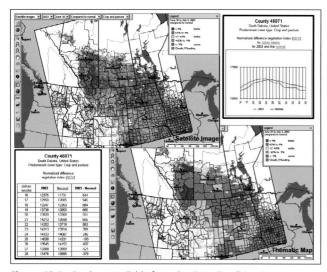

Figure 12.4 Products available from the Crop Condition Assessment Program.

Source: Image courtesy of Statistics Canada.

and maintain GIS software to query the database and generate customized reports (Reichert and Caissy 2002).

Precision farming

Precision farming, also called *precision agriculture* and *site-specific crop management*, refers to a group of techniques, technologies, and management strategies designed to optimize plant growth and farm profitability by adjusting treatments to suit the variable biophysical conditions that occur within an agricultural field instead of applying the same treatment uniformly over the entire area. In addition to improving profitability, precision farming methods can also deliver environmental benefits, such as reduced leaching and runoff.

Implementing precision farming practices requires

- a means to identify variations in biophysical conditions within a field that are relevant to crop production,
- a means to record or map the location and extent of relevant field characteristics to know where different treatments are to be applied, and
- a means to vary treatments in accordance with a plan or prescription designed to match field conditions.

The concept of tailoring farming methods to field conditions is not new. Growers have long sought to match their farm practices to the land. With years of experience farming the same fields, they recognize subtle differences within their fields and, where practical, vary their farming practices accordingly. What has changed is the range of technology that has become available to aid in the assessment of field characteristics, to monitor crop condition, and to automatically control the treatment rates of farm equipment. As a result, precision farming has come to refer to a suite of technologies that enable farming activities such as planting and the application of fertilizers and chemicals to be varied to suit the variable biophysical conditions within a field. These new technologies include remote sensing, GPS satellite positioning, geographic information systems, and yield monitoring and variable-rate-control technology on farm equipment.

The use of remote sensing to monitor crop condition during the growing season and GIS technology to analyze the results has made it possible to identify problems and map their location and extent rapidly enough to take corrective action during the growing season. Also, examination of imagery from past years together with a grower's knowledge of the land's characteristics can be analyzed to optimize the choice of crop and treatments for each field.

Precision farming can be implemented as a comprehensive farm management system using remotely sensed imagery, GPS satellite positioning, field-data collection of soil samples, customized geographic information system software to integrate the data and prepare maps and reports, and specialized farm equipment to apply treatments. The high cost of such systems tends to make them more attractive for large farming operations with high value crops. However, less-expensive analyses of airborne or satellite imagery by an agronomist with remote sensing expertise and the use of less-extensive field-data collection and analysis can provide cost-effective advice for smaller operations and lower value crops.

The role of remote sensing

Remote sensing has long been used to identify and document crop condition within individual fields. Since the early 1930s, soil scientists have relied on large-scale aerial photography to produce regional soil maps. Crop classification and detection of plant stress were among the first civilian applications of color infrared aerial photography when it was declassified after WWII.

Aerial photography, digital airborne and satellite imagery, and, more recently, vehicle-mounted sensors have been used to monitor crop condition. Drought, disease, weeds, water erosion, and nutrient deficiency are plant stresses detectable by remote sensing. Both visual and digital image processing methods are used to analyze the image data to measure and map crop condition, detect change, and identify variable field characteristics such as soils and drainage. Remote sensing, in showing the location and extent of field conditions, provides the spatial information to vary treatments within a field and identify problems and deficiencies such as inadequate irrigation or insufficient nutrients rapidly enough to take corrective action.

Relatively simple and inexpensive methods can provide cost-effective results. For example, figure 12.5a is a color infrared image of a circular corn field watered by center-pivot irrigation. In this image, healthy vegetation appears red. The area where the planter malfunctioned and no corn was planted appears brighter red in this case because it is a different plant species—weeds. An area of sloped ground with lower moisture retention appears darker red, and a poorly drained area is lighter red than normal, indicating lower plant vigor. Figure 12.5b is a classified plant-vigor map derived from the NDVI values of the same digital image. This classified NDVI product more clearly shows the same problem areas noted on the color infrared image. In figures 12.6a and b, the relative moisture level of bare soil areas (gray green in color) are easily discerned. Areas of higher soil moisture appear darker. Satellite imagery with a spatial resolution of 20 m to 30 m is often a lower cost alternative to the higher resolution airborne data and provides sufficient detail for many precision farming applications (*figure 12.7*).

The pattern of plant vigor as represented by a vegetation index image derived from remotely sensed data commonly matches the within-field pattern of soil characteristics, nutrients, and yield. As well, the plant-vigor pattern for a field is commonly consistent for different crops (*figure 12.8*). (Within-field crop-yield data can be collected by harvesting equipment fitted with a yield monitor and GPS.) Where this is the case, the imagery is representing variables related to crop production. The imagery can then be used to monitor and map crop development over the growing season, evaluate field-management effectiveness, such as fertilizer and pesticides treatments, and refine farming prescriptions. Current yields may be improved by early detection and remedial action. Future yields may be improved by changing tilling practices, irrigation, or fertilizer application to better suit the topography, microclimate, and soil characteristics.

Intensive precision agriculture implementations

Precision farming can be implemented using remote sensing in conjunction with geographic information systems, GPS satellite positioning, and specialized controllers and yield monitors on farm equipment. GPS satellite positioning provides accurate encoding of geographic location for field-data collection, remote sensing image rectification, and the controllers and sensors mounted on farm equipment. Crop condition and development are monitored with satellite and

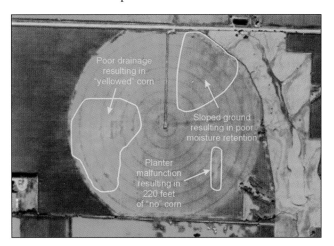

a

b

Figure 12.5 Monitoring center pivot irrigation by remote sensing. Figure (a) is a color infrared image of a center pivot irrigated corn field. The agronomic interpretation annotated on the image highlights problem areas in the field. Figure (b) shows a plant-vigor map derived from an analysis of the visible red and near-infrared bands.

Source: Images courtesy of M. Servilla, EarthScan, Inc. and Photon Research Associates, Inc.

a

b

Figure 12.6 Soil moisture levels on aerial photographs. Bare soil areas appear gray green in color on these color infrared images. The darker areas have a higher soil moisture level. Notice the pattern of drainage tiles appears as linear lighter toned features indicating drier soil conditions. Healthy vegetation appears red.

Source: Images courtesy of Applanix Corporation.

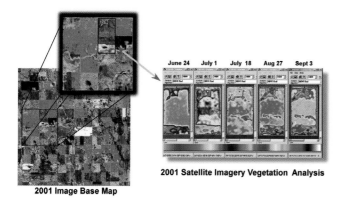

2001 Image Base Map

Figure 12.7   Crop development of a soybean field monitored with satellite imagery. The satellite image on the left is a color-composite image generated from the visible green, visible red, and near-infrared bands with a 30 m spatial resolution that was sharpened using the 15 m panchromatic band. Plant vigor, as measured by the Normalized Difference Vegetation Index (NDVI) (color-coded from green, which indicates high vigor, through red to blue, which indicates low vigor), was generated from the 30 m resolution visible red and near-infrared bands for images acquired during the growing season. The sequence of images shows differences in plant vigor related to different stages of plant development and to differences in growing conditions within the field. Notice the different development cycle of the lower quarter of the field, which is grass that greens up sooner than the soy crop.

Source: Image courtesy of Agri ImaGIS Technologies.

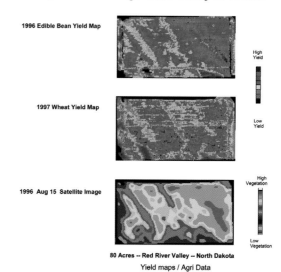

Figure 12.8   Comparison of a plant-vigor map with yield measurements for wheat and beans. The pattern of crop yield within a field for wheat and beans grown in subsequent years was measured by yield monitors on harvesting equipment. The plant-vigor image derived from satellite imagery of the bean crop predicts well the pattern of measured yields for both the wheat and bean crops. The pattern for beans is more distinct, probably due to its greater sensitivity to soil moisture than wheat.

Source: Image courtesy of Agri ImaGIS Technologies.

airborne remote sensing imagery augmented by field observations. Other commonly used data includes current and historical weather data, topography as represented by digital elevation models, historical yield, and field characteristics measured by sampling such as soil texture, pH, nutrients and moisture levels.

Digital imagery is well suited to the time-critical analysis of cropland. Delivered in georeferenced format[1], the data is easily input to a geographic information system as part of a crop management decision-support system. Data integration, implemented within the GIS, can also make use of agronomic models to identify the need for field-management activities, such as irrigation, treatments for pests and disease, and fertilizer application, and also monitor the effectiveness of the treatments that were applied.

Specialized control systems (termed *variable rate technology* or VRT) for farm machinery have been developed that allow the application of herbicide, water, fertilizer, insecticide, and

seed to be continuously varied *(figure 12.9)*. The control system is integrated with a GPS and a GIS to track the location of the machinery and automatically adjust the application rate as it moves through the field in accordance with a computer-based prescription map. (*Figure 12.11b*, shown below, is an example of a spread map used to control VRT machinery applying nitrogen fertilizer.) Crop production data is collected by computer-based yield monitors with integrated GPS satellite positioning. Data such as weight and volume is periodically recorded together with the geographic position of each sample.

Imagery sources

Both satellite and airborne digital imagery are used for precision agriculture. Satellite imagery has been used to monitor variation in crop condition since the launch of Landsat in the mid-1970s. Satellites such as Landsat and the Indian IRS sensor that collect complete imagery for their coverage area are more dependable image sources because imagery will be available from every cloud-free satellite pass. Higher resolution

---

1. Georeferenced imagery is encoded so that an accurate geographic position can be determined for each pixel in the image.

Figure 12.9 Variable rate technology. The variable-rate valve and nozzling system (in white housings) installed on the application equipment adjusts fertilizer application rates as it travels over the field.

Source: Image courtesy of Oklahoma State University.

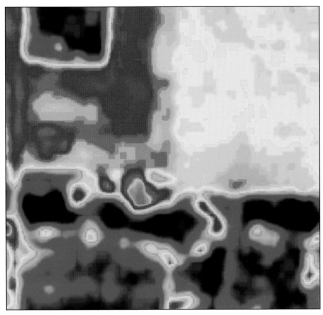

a

sensors, such as SPOT, IKONOS, and QuickBird, while offering spatial resolutions as high as 0.61 m, may not collect imagery for a specific area unless it is ordered in advance so the satellite can be tasked to collect it. Higher resolution imagery generally provides a smaller coverage area per scene *(table 12.1)* and is more costly per unit area. Also, customer tasking of pointable sensors increases image costs.

The spatial resolution of the imagery used should match the required level of detail. Imagery with a higher resolution than needed incurs unnecessary costs, and the finer detail can make analysis more difficult. For example, figure 12.10 shows the plant-vigor images from 3 m airborne and 20 m satellite multispectral imagery. Here the sugar beet plants produce a nearly complete leaf cover over the field, so the coarser 20 m pixels image only the vegetation. The finer detail of the 3 m image highlights linear features and other variations in field conditions that did not significantly improve field management.

Figure 12.11 illustrates the use of satellite imagery for sugar beet management. Thirty meter GSD Landsat 7 imagery was used to generate an NDVI plant-vigor image, shown in figure 12.11a with field-sampled leaf nitrogen values. Since the leaves are left on the field after harvest, areas with plant residues higher in nitrogen require less nitrogen fertilizer the following year. Based on this data, a nitrogen fertilizer spread map was derived *(figure 12.11b)* and used to control subsequent fertilizer application. This management practice reduced nitrogen fertilizer use and increased both the sugar content (and thus the value) of the crop and the consistency within the field as measured by field samples *(figure 12.11c)*.

More recently, small sensors have been designed for use on farm equipment. For example, the GreenSeeker™ line

b

Figure 12.10 Comparison of plant-vigor maps derived from 1 m airborne imagery and 30 m Landsat imagery. For purposes of field management, the patterns visible in the NDVI plant-vigor image generated from Landsat 7 data with a 30 m GSD (a) are essentially the same as those in the higher resolution 1 m GSD airborne scanner image (b). The images were acquired within 30 minutes of each other.

Source: Image courtesy of Agri ImaGIS Technologies.

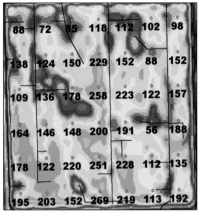
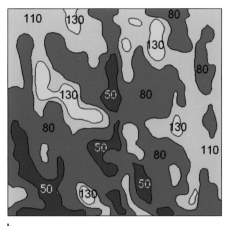
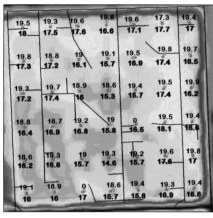

a   b   c

Figure 12.11   Precision farming application to sugar beets. (a) NDVI image generated from 30 m GSD Landsat 7 data with point sample values of measured leaf nitrogen in pounds per acre. (b) Spread map showing pounds per acre of nitrogen fertilizer to be applied.

(c) Comparison of percentage of sugar content of field samples taken in 2001 (upper value) and 1997 (lower value).

Source: Images courtesy of A. W. G. Farms, Inc.

of sensors manufactured by NTech Industries, Inc., collect reflectance measurements in the visible and near-infrared bands. A pulsed light source provides consistent illumination and allows operation day or night. The sensors can be mounted on farm equipment and with a computer-based control system used to vary treatments such as fertilizer application rates as they are applied *(figure 12.12)*. As the sprayer moves forward, a computer control system varies the application rate at each nozzle for each 60 cm by 60 cm square area based on the vegetation index calculated by the sensor and previously collected field calibration data.

The sensors can also be integrated with a GPS to collect geocoded reflectance data from which a plant-vigor (NDVI) digital image and other image products can be generated. The sensor unit is typically mounted in front of farm equipment using a single sensor or multiple units evenly spaced along a boom carried perpendicular to the travel path. As the equipment moves back and forth across the terrain, data is collected with a spatial resolution determined by the spacing of the sensors and the time over which each measurement is collected. Research suggests that collection of data at spatial resolutions as fine as 60 cm by 60 cm (2 ft by 2 ft) samples can significantly improve crop yields and reduce treatment costs. Figure 12.13a shows a single sensor unit mounted on a vehicle with its illumination source, and figure 12.13b is an NDVI plant-vigor image derived from the data.

High-spatial-resolution imagery collected using airborne or vehicle-mounted sensors may be needed to image fields such

Figure 12.12   Close-range remote sensing using tractor/spray boom mounted sensors. Sensors mounted in front of each spray nozzle detect reflectance in two wavelength bands, the visible red and near-infrared (red in the image). The average vegetation index calculated for each 60 cm by 60 cm area is used to control the sprayer application rate for that area as the spray nozzles move over it (yellow). The GreenSeeker™ manufactured by NTech Industries, Inc., uses a pulsed light source that provides consistent illumination of the area detected and allows day or night operation.

Source: Image courtesy of NTech Industries, Inc.

as vineyards that are planted in widely spaced rows separated by large areas of bare soil. To image the vegetation, the spatial resolution must be fine enough for individual pixels to fall completely within the vegetation canopy. Grape vines are grown vertically on trestles. Vehicle-mounted sensors using their own illumination source have been used to collect reflectance measurements of grape plants by aiming the sensors horizontally at the foliage as the equipment moves along the rows *(figure 12.14)*.

a

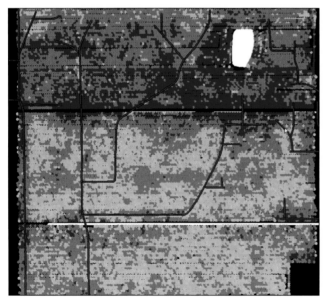

b

Figure 12.13 A single tractor mounted GreenSeeker™ sensor (white unit in figure a) or multiple sensors mounted on a boom can be integrated with a GPS to collect geocoded data as the equipment moves back and forth over the field. The data can then be processed to generate a detailed vegetation index image of the field (b) for use in assessing crop condition and developing treatment prescriptions.

Source: Images courtesy of A.W.G. Farms, Inc.

A potential difficulty in using satellite data for precision agriculture is that in some cases it may be necessary to obtain imagery within a narrow window of a few days in order to record a crop condition at a specific developmental stage in time to take any corrective action that may be needed. With multiple satellites, some with pointable sensors, the chances of obtaining cloud-free imagery within the required time frame are improved but not guaranteed.

Alternatively, airborne imagery can be flown on demand, improving the chance of obtaining cloud-free imagery within the desired time frame. However, the cost of custom flown airborne imagery is generally much higher per unit area than satellite data. Vehicle-mounted equipment gives growers control over data acquisition but requires them to learn the technology and manage the data handling and analysis.

Conclusion

Precision farming is an approach that seeks to optimize farm operations and improve profitability by varying farm practices to more closely match subtle within-field variations in biophysical characteristics. The principles and technology can be applied with varying degrees of intensity, from relatively simple low-cost approaches to more intensive implementations that make use of sophisticated analytical techniques and require more costly equipment and specialized labor. By matching the level of implementation to the resources and economics of the farm, precision farming techniques can be cost effective for a wide range of operations.

Figure 12.14 Horizontal scanning of grape vines with vehicle-mounted sensor.

Source: Image courtesy of NTech Industries, Inc.

References and further reading

Cowan, T. 2000. Precision agriculture and site-specific management: Current status and emerging policy issues. Report RL30630. Resources, Science, and Industry Division. National Council for Science and the Environment. Washington, D.C. www.ncseonline.org/NLE/CRSreports/Agriculture/ag-99.cfm#Other%20Public%20and%20Public-Private%20Roles%20in%20PA

Dixon, J., and M. McCann, eds. 1997. *Precision agriculture in the 21st century*. 1st ed. Washington, D.C.: National Academy Press.

Reichert, G. C., and D. Caissy. 2002. A reliable crop-condition assessment (CCAP) incorporating NOAA AVHRR data, a geographic information system, and the Internet. Proceedings of the 2002 ESRI User Conference. ESRI. Redlands, Calif.

Sevilla, M., and M. Towner. Integrating high-resolution satellite imagery and weather data for improved agricultural management decisions. Proceedings of the 2000 ESRI User Conference. ESRI. Redlands, Calif. gis.esri.com/library/userconf/proc00/professional/abstracts/a601.htm

Web sites

Agri ImaGIS.
www.satshot.com

Alberta Agriculture, Food, and Rural Development.
www1.agric.gov.ab.ca/$department/deptdocs.nsf/all/sag1950

NASA Precision Agriculture Site.
www.ghcc.msfc.nasa.gov/precisionag

NTech Industries, Inc.
www.greenseeker.com/index.html

Oklahoma State University.
www.dasnr.okstate.edu/nitrogen_use

Purdue University.
www.agriculture.purdue.edu/ssmc

South Dakota State University. Precision Farming. College of Agriculture and Biological Sciences.
www.abs.sdstate.edu/abs/precisionfarm

Statistics Canada. Crop Condition Assessment Program (CCAP).
gem.statcan.ca/ccaplogin.asp

United Nations Food and Agriculture Organization (FAO). Global Information and Early Warning (GIEWS) Program.
www.fao.org/giews/english/index.htm

University of Georgia. National Environmentally Sound Production Agriculture Laboratory (NESPAL). College of Agricultural and Environmental Sciences. *nespal.cpes.peachnet.edu/PrecAg*

University of Sydney. Australian Centre for Precision Agriculture. *www.usyd.edu.au/su/agric/acpa*

Remote sensing and GIS in forestry

Michael A. Wulder, Ronald J. Hall, and Steven E. Franklin

Remote sensing and GIS are complementary technologies that, when combined, enable improved monitoring, mapping, and management of forest resources (Franklin 2001).

The information that supports forest management is stored primarily in the form of forest inventory databases within a GIS environment. A forest inventory is a survey of the location, composition, and distribution of forest resources. As one of the principal sources of forest management information, these databases support a wide range of management decisions from harvest plans to the development of long-term strategies.

Historically, forest management inventories were primarily for timber management and focused on capturing area and volume by species. In the past decade, forest management responsibilities have broadened. As a result, inventory data requirements have expanded to include measures of non-harvest related characteristics such as forest structure, wildlife habitat, biodiversity, and forest hydrology.

The entire forest inventory production cycle, from planning to map generation, can take several years. Except for the photo interpretation component, forest inventory production is largely a digital process. *Operational level* inventories, based on both aerial photo interpretation and field-sampled measurements, provide location-specific information required for harvest planning. Forest *management level* inventories meet longer-term forest management planning objectives. Though these levels differ in detail, they both require information fundamentally based on forest inventory data.

A forest management inventory generalizes complex forest resource attributes into mapping units useful for forest management. The types of attributes attached to individual mapping units, or polygons, might include stand species composition, density, height, age, and, more recently, new attributes such as leaf area index (Waring and Running 1998).

Much of the information collected for forest inventory is generated by interpretation of aerial photographs at photo scales of 1:10,000 to 1:20,000, depending on the level of detail required. Other remote sensing sources such as airborne and satellite digital imagery have been valuable in updating forest attributes such as disturbance, habitat, and biodiversity. In providing more frequent information updates, remotely sensed data can improve the quality of forest inventory databases, thereby improving the resource management activities they support.

The quality of photointerpreted data depends on the experience of the interpreters and the use of quality assurance procedures such as interpreter calibration and field verification. Other factors can introduce inconsistencies that compromise the quality of forest inventory data. For example, there may be source data inconsistencies when aerial photography is acquired on different dates or in different weather conditions or inconsistencies in analysis when multiple contractors are used. The quality of the resulting data may vary significantly within a map area. For example, information about disturbances related to fire and insects may be inconsistent within a map area because the aerial photography from which it was interpreted was acquired in different years. Similarly, inconsistencies may occur at the edge of neighboring map sheets because data was collected in different years or was produced by different contractors.

Applications of remote sensing and GIS to forestry

The use of remote sensing by forest managers has steadily increased, promoted in large part by better integration of imagery with GIS technology and databases, as well as implementations of the technology that better suit the information needs of forest managers (Wulder and Franklin 2003). The most important forest information obtained from remotely sensed data can be broadly classified in the following categories:

- detailed forest inventory data (e.g., within-stand attributes)
- broad area monitoring of forest health and natural disturbances
- assessment of forest structure in support of sustainable forest management

Detailed forest inventory data

Forest inventory databases are based primarily on stand boundaries derived from the manual interpretation of aerial photographs. Stand boundaries are vector-based depictions of homogeneous units of forest characteristics. These stand polygons are described by a set of attributes that typically includes species composition, stand height, stand age, and crown closure. Digital remotely sensed data can be used to update the inventory database with change (e.g., harvest) information for quality control, audit, and bias detection. It can also add additional attribute information and identify biases in the forest inventory databases due to vintage, map sheet boundaries, or interpreter preferences.

The objective of managing forests sustainably for multiple timber and nontimber values has required the collection of more detailed tree and stand data, as well as additional data such as gap size and distribution. Detailed within-stand forest inventory information can be obtained from high-spatial-resolution remote sensing data such as large-scale aerial photography and airborne digital imagery. Two methods of obtaining this information are *polygon decomposition* (Wulder and Franklin 2001) and *individual tree crown recognition* (Hill and Leckie 1999).

Polygon decomposition analyzes the multiple pixels representing a forest polygon on a remotely sensed image to generate new information that is then added to the forest inventory database (see Wulder and Franklin 2001). For example, a change detection analysis of multidate Landsat Thematic Mapper satellite images can identify the areal extent and proportion of pixels where conditions have changed.

Individual tree crown recognition is based on analyzing high-spatial-resolution images from which characteristics such as crown area, stand density, and volume may be derived (Hill and Leckie 1999).

Forest health and natural disturbances

Fire, insects, and disease are among the major natural disturbances that alter forested landscapes. Timely update information ensures inventory databases are current enough to support forest management planning and monitoring objectives.

Insect disturbance

Among the insects that cause the most damage to trees are defoliators and bark beetles (Armstrong and Ives 1995). Damage assessment for these insects is typically a two-step process that entails mapping the disturbed area followed by a quantitative assessment of the damage to the trees within the mapped areas.

Aerial sketch-mapping, where human observers manually annotate maps or aerial photographs, has been the most frequently used method for mapping areas damaged by insects (Ciesla 2000). This process is costly, subjective, and spatially imprecise. However, when augmented by ground survey methods and the integrated analysis of remote sensing and GIS, substantial benefits can be realized.

Insect damage causes changes in the morphological and physiological characteristics of trees, which affects their appearance on remotely sensed imagery. Insect defoliation

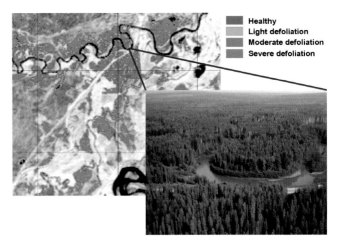

Figure 12.15 Landsat satellite classification for spruce budworm defoliation with field photograph depicting red-colored trees damaged by spruce budworm defoliation. (Location: Junction of Troy Lockhart Kledo Creek and Alaska Highway, Fort Nelson, B.C.).

Source: © Her Majesty the Queen in the right of Canada, Natural Resources Canada.

causes loss of foliage that results in predictable color alterations. For example, residual foliage after attack by spruce budworm will turn the tree a reddish color *(figure 12.15)*. The mountain pine beetle is a bark beetle that bores through the bark and creates a network of galleries that girdle the tree and cause the foliage to become a reddish-brown color. These foliage loss and color changes often occur during a short time period—this is the optimal time for detection by remote sensing. Knowing the characteristics of a particular damage agent, the most appropriate sensor characteristics and acquisition times can be selected (see example by Hall below).

Integrated remote sensing and GIS analyses that support insect damage monitoring and mitigation include:

- detecting and mapping insect outbreak and damage areas
- characterizing patterns of disturbance relative to mapped stand attributes
- modeling and predicting outbreak patterns through risk and hazard rating systems
- providing data to GIS-based pest management decision support systems.

Fire

Fire is an ecological process that governs the composition, distribution, and successional dynamics of vegetation in the landscape (Johnson 1992). Knowledge of fire disturbance is necessary to do the following:

- understand fire impacts on timber and nontimber values
- define salvage logging opportunities
- understand the effect of climate change and feedback processes on forest fire occurrence
- quantify the influence of fire on regional, national, and global carbon budgets (Kasischke and Stocks 2000)

To address this range of issues, foresters employ a multitude of field, global positioning system (GPS), and remote sensing (airborne and satellite) methods and data sources. Integrated remote sensing and GIS fire support systems are used in real-time, near real-time, and postfire applications. For example, infrared and thermal infrared cameras with integrated GPS/INS (inertial navigation system) technologies can observe fire hot-spots, active fires, and fire perimeters in real-time. Data on fire location and size is sent from the aircraft to field-based systems from which precise directions can be given to water-bombers and firefighting crews. Near real time remote sensing and GIS systems are generally based on daily observations from coarse-resolution satellites such as the AVHRR (1 km pixel) and MODIS (250 m to 1 km pixel) satellites. Daily hot-spot information identifies the occurrence of fire activity over large areas and helps to target locations to collect more detailed information. Postfire applications largely entail mapping the extent of burned areas from aerial photographs or satellite imagery and assessing fire damage to vegetation.

The Canadian Wildland Fire Information System (CWFIS) and the Fire Monitoring, Mapping, and Modeling System (Fire M3) are integrated remote-sensing- and GIS-based systems providing nationwide coverage to support fire management and global change research. NOAA AVHRR and SPOT VEGETATION remote sensing products can be used to monitor actively burning large fires in near real time (*figure 12.16*) to estimate burned areas and model fire behavior, biomass consumption, and carbon emissions (Fraser et al. 2000, Lee et al. 2002).

The rapid fire detection and response system implemented by the U.S. Forest Service Remote Sensing Applications Center, in cooperation with NASA and the University of Maryland, uses MODIS satellite imagery to identify hot spots throughout the United States. MODIS Active Fire Map products are compiled daily at 3:00 AM and 3:00 PM, mountain time, and are available over the Internet approximately two hours later. In addition to forest fire detection, the center provides image data from several different sensor sources in support of fire response and postfire assessment

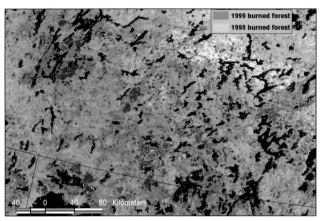

Figure 12.16   Sample of Canada-wide burn area mapping from Fire M3 depicting an area in the Northwest Territories.

Source: © Natural Resources Canada.

activities (Quayle et al. 2002, Orlemann et al. 2002). An example of MODIS data for forest fire detection is shown in figure 7.17 in chapter 7.

Landscape ecology, habitat, and biodiversity

Sustainable forest management requires that landscape ecological characteristics related to habitat and biodiversity be included in forest inventory and certification procedures (Vogt et al. 1999). The characteristics of interest are (1) spatial patterns within the landscape, (2) specific habitat-related forest conditions, and (3) the ecological processes that link spatial pattern, habitat, and ecosystem functioning.

Land-cover information is one example of spatial patterns readily obtainable by classifying remotely sensed data. Other useful datasets include forest canopy information (e.g., crown closure or leaf area estimates), understory information (Hall et al. 2000), and measures of the distribution and boundaries of landscape units such as forest fragmentation (Debinski et al. 1999). Remote sensing can provide repeatable and consistent methods to develop these data layers such that changes over time can be monitored and habitat models can be developed and validated for individual species.

Habitat assessment is typically GIS-based; it involves selecting data layers likely to be of value in developing predictive models for the occurrence and distribution of individual species or species assemblages, as well as the identification of species useful as indicators of ecological condition (see example by Franklin below). The use of remotely sensed data, together with other spatial datasets integrated within a GIS environment, has greatly enhanced the habitat assessment process.

Improvements in forest management also depend on increased understanding of ecological processes within the carbon, nutrient, and hydrological cycles. Remotely sensed data provides key inputs to models of carbon flux, nutrient uptake and the influence of fertilization, and drought and water stress indicators (Lucas and Curran 1999).

Future directions of remote sensing in forestry

A key development in remote sensing has been the increased availability of high-spatial- and high-spectral-resolution remotely sensed data from a wide range of sensors and platforms including photographic and digital cameras, video capture, and airborne and spaceborne multispectral sensors. Hyperspectral imagery promises to provide improved discrimination of forest cover and physiological attributes. Radar applications are being developed that penetrate the forest canopy to reveal characteristics of the forest floor (discussed in chapter 8). New technologies such as lidar can provide estimates of forest biomass, height, and the vertical distribution of forest structure with unprecedented accuracy (Lim et al. 2003). The use of advanced digital analysis methods and selective use of complementary data have provided more detailed information about forest structure, function, and ecosystem processes than ever before (Culvenor 2003, Hill and Leckie 1999).

As the availability of multiresolution remotely sensed imagery and multisource data increases, so will the capability to generate timely and accurate maps of forest composition and structure. Operational capabilities continue to improve forest attribute mapping with a precision commensurate with forest management scales. This, in turn, will contribute to efforts aimed at assessing the sustainability of our forests through better forest practices and improved decision making in forest management.

References

Armstrong, J. A., and W. G. H. Ives, eds. 1995. Forest insect pests in Canada. Natural Resources Canada, Canadian Forest Service, Science and Sustainable Development Directorate, Ottawa, Ont.

Ciesla, W. M. 2000. Remote sensing in forest health protection. United States Department of Agriculture, Forest Service, Forest Health Technology Enterprise Team, Remote Sensing Applications Center, Salt Lake City, Utah. Report, No. 00–03.

Culvenor, D. 2003. Extracting individual tree information: A survey of techniques for high spatial resolution imagery. In *Remote sensing of forest environments: Concepts and case studies*, ed. M. Wulder and S. Franklin, 255-77. Dordrecht: Kluwer Academic Publishers.

Debinski, D. M., K. Kindscher, and M. E. Jakubauskas. 1999. A remote sensing and GIS based model of habitat and biodiversity in the Greater Yellowstone Ecosystem. *International Journal of Remote Sensing* 20: 3281–91.

Franklin, S. E. 2001. *Remote sensing for sustainable forest management*. Boca Raton, Florida: CRC Press.

Fraser, R. H., Z. Li, and J. Cihlar. 2000. Hotspot and NDVI differencing synergy HANDS): A new technique for burned area mapping over boreal forest. *Remote Sensing of Environment* 74: 362–76.

Hall, R. J., D. R. Peddle, and D. L. Klita. 2000. Mapping conifer under story within boreal mixed woods from Landsat TM satellite imagery and forest inventory information. *The Forestry Chronicle* 76: 887–902.

Hill, D. A., and D. G. Leckie, eds. 1999. International forum: Automated interpretation of high spatial resolution digital imagery for forestry. Proceedings of a workshop, February 10–12, 1998. Natural Resources Canada, Canadian Forest Service, Victoria, B.C.

Johnson, E. A. 1992. *Fire and vegetation dynamics: Studies from the North American boreal forest*. Cambridge: Cambridge University Press.

Kasischke, E. S., and B. J. Stocks, eds. 2000. *Fire, climate change, and carbon cycling in the boreal forest*. New York: Springer-Verlag.

Lee, B. S., M. E. Alexander, B. C. Hawkes, T. J. Lynham, B. J. Stocks, and P. Englefield. 2002. Information systems in support of wildland fire management decision making in Canada. *Computers and Electronics in Agriculture* 37: 185–98.

Lim, K., P. Treitz, M. Wulder, B. St-Onge, and M. Flood. 2003. LiDAR remote sensing of forest structure. *Progress in Physical Geography* 27(1): 88–106.

Lucas, N. S., and P. J. Curran. 1999. Forest ecosystem simulation models: The role of remote sensing. *Progress in Physical Geography* 23: 391–423.

Orlemann, A., M. Saurer, A. Parsons, and B. Jarvis. 2002. Rapid delivery of satellite imagery for burned area emergency response (Baer). In *Proceedings of the 9th biennial forest service remote sensing applications conference*, ed. J. D. Greer. Bethesda, Md.: American Society of Photogrammetry and Remote Sensing.

Parsons, A., and A. Orlemann. 2002. Mapping post-wildfire burn severity using remote sensing and GIS. In Proceedings 2002 ESRI User Conference. ESRI. Redlands, Calif. gis.esri.com/library/userconf/proc02/pap0431/p0431.htm

Quayle, B., M. Finco, K. Lannom, R. Sohlberg, J. Descloitres, M. Carroll, J. Norton, and D. Kostyuchenko. 2002. USDA Forest Service MODIS active fire mapping and delivery.

Proceedings of RS2002: Ninth Biennial Forest Service Remote Sensing Applications Conference, San Diego.

Vogt, K. A., B. C. Larson, J. C. Gordon, D. J. Vogy, and A. Fanzeres. 1999. *Forest certification: Roots, issues, challenges, benefits.* Boca Raton, Fla.: CRC Press.

Waring, R., and S. Running. 1998. *Forest ecosystems: Analysis at multiple scales.* 2nd ed. San Diego, Calif.: Academic Press.

Wulder, M., and S. E. Franklin. 2001. Polygon decomposition with remotely sensed data: Rationale, methods, and applications. *Geomatica* 55(1):11–21.

Wulder, M., and S. Franklin, eds. 2003. *Remote sensing of forest environments: Concepts and case studies.* Dordrecht: Kluwer Academic Publishers.

Web sites

Canadian Wildland Fire Information System.
cwfis.cfs.nrcan.gc.ca/en/index_e.php

Parsons, A., and A. Orlemann. 2002. Mapping post-wildfire burn severity using remote sensing and GIS. In Proceedings 2002 ESRI user conference. ESRI. Redlands, Calif.
gis.esri.com/library/userconf/proc02/pap0431/p0431.htm

U.S. Forest Service Remote Sensing Applications Center Rapid Response Web site.
activefiremaps.fs.fed.us

Acknowledgment

We would like to thank Mark Gillis of the Canadian Forest Service in Victoria, B.C., for valuable comments and suggestions.

Case study: GIS, remote sensing, and jack pine budworm defoliation

Ronald J. Hall

There are many insects that ravage the forests of North America. Among the important insect defoliators are the spruce budworm, jack pine budworm, and forest tent caterpillar. Jack pine budworm is a major defoliator of jack pine forests in Canada and the Great Lakes States of the United States (Mallett and Volney 1990). Severe defoliation reduces tree growth, causes top kill (i.e., dead tree tops), mortality, and may predispose trees to attack by other destructive agents (Gross 1992). Methods to map and assess defoliation damage are of interest to those concerned with the health and sustainability of the jack pine timber resource.

Forest stand structure varies due to differences in species composition, crown closure, height, and age distribution on various sites. These characteristics strongly influence the number of budworms inhabiting a stand and the consequent growth losses and morphological damage that may be sustained from budworm defoliation (Wulf and Cates 1987). Stand structure information is typically captured during an inventory and represented on forest inventory maps that are digitally stored in a geographic information system (GIS). These maps describe cover types by nominal (e.g., species composition) and ordinal (e.g., age, height, crown closure) forest classification systems. The interpretation of stand structure information represented in a forest inventory is an indicator of general stand health (Luther et al. 1997). Spatial associations between stand structure attributes and insect damage can be assessed statistically if a map of insect damage produced from remote sensing could be compared with the corresponding forest inventory map. Strong associations would suggest characteristics of stands that are vulnerable to jack pine budworm damage. A remote-sensing-GIS integration exercise was employed to identify the composition and structure of stands that sustained the greatest damage from defoliation that, in turn, could be used in hazard rating (Hall et al. 1995; 1998) and input to the jack pine budworm decision support system (Power and Gillis 1995).

Analysis procedures

The data used for this study consisted of color and color-infrared aerial photographs, digital forest inventory and site quality data residing in GIS databases, and field observations.

The color-infrared aerial photographs at 1:5,000 and 1:25,000 scales were acquired and interpreted for severity of defoliation using a classification system that rated damage as nil, light, moderate, and severe (Hall et al. 1995, 1998) *(figure 12.17)*. The classification system was based on discrete levels that appeared separable, given that field data over selected stands and 1:900 large-scale photographs acquired for a previous study (Hall et al. 1993) were also available for the study area. Operationally, this process entailed the collection of ancillary information to be used in support of the interpretation exercise that was no different than for any other digital or analog image interpretation exercise.

Digital forest inventory maps of the study area were combined into a mosaic and then reclassified to produce separate map coverages of species, stand height, crown closure, and stand age attributes. Stand age was originally represented in 10 age classes that were subsequently reduced to three jack pine stand maturity classes: young (25 to 35 years), mature (45 to 65 years), and over-mature (85 to 125 years). A site quality map was also produced by interpreting landform and vegetative patterns on the 1:25,000 scale color-infrared aerial photographs. This map was supplemented by field data

Figure 12.17 Interpreted aerial photo depicting light (L), moderate (M) and severely (S) damaged forest stands caused by jack pine budworm defoliation.

Source: R. Hall. © Natural Resources Canada.

and ordination analysis from which a site quality class (poor, poor-moderate, moderate, moderate-good) was assigned.

The Cramer's V statistical test was employed to determine the association between insect defoliation and the stand and site characteristics. Cramer's V is the preferred means of determining associations when analyzing ordinal data (Foody 1994) and is suitable for use with rectangular contingency tables (i.e., when the number of classes between two maps being compared is not equal) (Fosnight and Fowler 1996). The computed coefficient in Cramer's V ranges from zero to one. The coefficient equals zero when the variables are not correlated. The significance of Cramer's V was tested with the x^2 statistic at the 0.05 level of significance. Random point sampling procedures were implemented to avoid the influence of spatial autocorrelation that would exist with map data that is not sampled randomly (Dale et al. 1991). Although Cramer's V is considered the most appropriate measure of the strength of the association for the types of ordinal data in this study, its value is always positive, which provides no information on the direction of the relationship. Spearman's nonparametric rank correlation coefficients (r_s), which range from −1 to +1 (Conover 1980) were therefore computed to further amplify the association measures obtained with Cramer's V.

Products, time frame, and costs

The products generated from this study were empirical relationships between the stand and site attributes and defoliation damage. Site quality and stand maturity were more highly associated with budworm damage than stand height and crown closure. Jack pine areas that sustained moderate and severe damage were those on poor sites that were over-mature (> 85 years), 15 to 20 m tall with 30% to 55% crown closure (Hall et al. 1998). The remote-sensing-GIS integration approach proved an effective tool for identifying vulnerable stands. The project entailed the acquisition

of aerial photographs to see the state of defoliation damage (two days); field work to assess stand condition, budworm damage, and site quality (three weeks); interpretation and digital translation (two weeks); and processing and statistical analysis of GIS forest inventory and site quality datasets (four weeks). The results of this study enable jack pine budworm hazard rating maps to be produced for other forest areas in the region by GIS-based analyses using existing forest inventory data without the need for further remote sensing input (*figure 12.18*). These maps can then be used to guide forest management planning that would help to ensure the sustainability of the jack pine timber resource.

Figure 12.18 Hazard rating map produced from GIS depicting forest stands that were rated as light (blue), moderate (green) and high (red) hazard to damage from jack pine budworm defoliation.

Source: R. Hall. © Natural Resources Canada.

References   Conover, W. J. 1980. *Practical nonparametric statistics.* 2nd ed. New York: John Wiley and Sons, Inc.

Dale, M. R. T., D. J. Blundon, D. A. MacIsaac, and A. G. Thomas. 1991. Multiple species effects and spatial autocorrelation in detecting species associations. *Journal of Vegetation Science* 2:635–42.

Doliner, L. H., and J. H. Borden. 1984. Pesterms: A glossary of forest pest management terms. B.C. Ministry of Forests, Victoria, B.C. Pest Management Report Number 3.

Foody, G. M. 1994. Ordinal-level classification of sub-pixel tropical forest cover. *Photogrammetric Engineering and Remote Sensing* 60(1):61–65.

Fosnight, E .A., and G. W. Fowler. 1996. Measures of association and agreement for describing land cover characterization classes. In *Proceedings of spatial accuracy assessment in natural resources and environmental sciences: 2nd international symposium*, tech. cords. H. T. Mowrer, R. L. Czaplewski, and R. H. Hamre, 425–33. Gen. Tech. Rep. RM-GTR-277.

Gross, H. L. 1992. Impact analysis for a jack pine budworm infestation in Ontario. *Canadian Journal of Forest Research* 22:818–31.

Hall, R. J., S. J. Titus, and W. J. A. Volney. 1993. Estimating top-kill volumes with large-scale photos on trees defoliated by the jack pine budworm. *Canadian Journal of Forest Research* 23(7):1337–46.

Hall, R. J., W. J. A. Volney, and K. Knowles. 1995. Hazard rating and stand vulnerability to jack pine budworm defoliation using GIS. In *Jack pine budworm biology and management: Proceedings of the jack pine budworm symposium*, ed. W. J. A. Volney, V. G. Nealis, G. M. Howse, A. R. Westwood, D. G. McCullough, and B. L. Laishley, 121–32. Information Report NOR-X-342.

Hall, R. J., W. J. A. Volney, and Y. Wang. 1998. Using a geographic information system (GIS) to associate forest stand characteristics with top kill due to defoliation by the jack pine budworm. *Canadian Journal of Forest Research* 28(9):1317–27.

Luther, J. E., S. E. Franklin, J. Hudak, and J. P. Meades. 1997. Forecasting the susceptibility and vulnerability of balsam fir stands to insect defoliation with Landsat Thematic Mapper data. *Remote Sensing of Environment* 59:77–91.

Mallett, K. I., and W. J. A. Volney. 1990. Relationships among jack pine budworm damage, selected tree characteristics, and Armillaria root rot in jack pine. *Canadian Journal of Forest Research.* 20:1791–95.

Murtha, P. A. 1972. A guide to air photo interpretation of forest damage in Canada. Canadian Forest Service, Forest Management Institute, Ottawa, Ontario. Publication 1292.

Power, M., and T. Gillis. 1995. Decision support tools for jack pine budworm management - progress and report. In *Jack pine budworm biology and management: Proceedings of the jack pine budworm symposium*, ed. W. J. A. Volney, V. G. Nealis, G. M. Howse, A. R. Westwood, D. G. McCullough, and B. L. Laishley, 133–41. Information Report NOR-X-342.

Wulf, N. W., and R. G. Cates. 1987. Site and stand characteristics. In *Western spruce budworm, tech. coords.* M. H. Brookes, R. W. Campbell, J. J. Colbert, R. G. Mitchell, and R. W. Stark, 89–115. U.S. Department of Agriculture, Forestry Service, Washington, D.C., Tech. Bull. 1694.

Case study: Satellite-based grizzly bear habitat mapping and models

Steven E. Franklin

Mapping large areas for the purpose of understanding habitat and providing improved management for wide-ranging wilderness species with universal environmental requirements can only feasibly be accomplished with the aid of satellite remote sensing technology. For example, Mackinnon and de Wulf (1994) state that "two time series of satellite images showed the reduction and rapid fragmentation of the giant panda's habitat in China. More than any other factor, it was this perspective provided by satellite imagery that changed the. . . manager's view about the main threats to panda survival" (p.130). Like the panda in China, grizzly bears in Alberta have been classed as a species of special concern and may require protection and a recovery plan. A key component of this plan is that resource managers must identify critical habitat areas where resource conflicts may occur and bears may be threatened. Satellite remote sensing imagery, together with GIS and DEM data layers, were used to map habitat; such maps were necessary to develop the appropriate context within which bear population viability analysis and mitigation strategies could be conducted using data from GPS-collared bears and models based on resource selection functions (Nielsen et al. 2002). First this geospatial data and imagery were used to map land cover (Franklin et al. 2001); second, these land-cover classes were related to bear habitat characteristics, including landscape structure (Popplewell et al. 2003, Linke et al. 2004); and third, models of resource selection by bears were developed to allow prediction of bear habitat use and movement across the landscape (Nielsen et al. 2002).

Data used

The data used includes Landsat TM and ETM+ multitemporal imagery, a digital elevation model (DEM), GIS databases, field observations, and bear GPS-collar locations.

Analysis procedures

The analysis procedure is outlined in figure 12.19. Satellite imagery was radiometrically corrected using standard atmosphere models and coregistered with the available geospatial data before using the large-area mosaicking tools within the PCI image analysis system. We used a combination of unsupervised and supervised classification techniques on the image

transformed into brightness/greenness/wetness indexes and elevation, slope, and incidence value data extracted from the DEM. Clustering procedures separated the nonvegetation classes (e.g., rock, ice, and water) from the vegetation classes, which in turn were trained individually using field and air photo data for classification with a maximum likelihood decision rule. GIS data was used to separate some spectrally-similar habitat classes (e.g., alpine shrub and riparian shrub based on a distance-to-stream feature rule). Slope rules were used to separate mountain shadows and lakes. Where available, standard forest inventory GIS data was used to attribute additional characteristics to image classes (e.g., age and closure of coniferous forest stands).

The habitat layer was processed for landscape metrics (using the FRAGSTATS software system) and was a necessary input to the development of resource selection functions based on GPS-collar data. Landscape metrics are computations that describe total landscape area, diversity of patch types within the landscape, and various aspects of the patches themselves, such as their size, shape, variability in size and shape, distance between patches of the same type, and the degree to

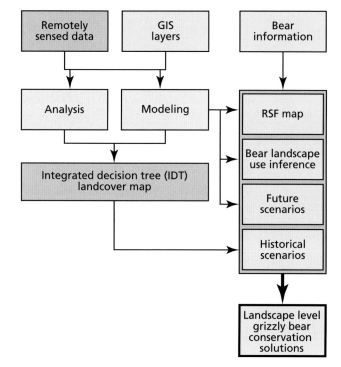

Figure 12.19 Flowchart of the grizzly bear habitat analysis procedure.

Source: S. Franklin.

which patch types are isolated (Forman and Godron 1986, McGarigal and Marks 1995). Resource selection functions are generalized linear models for populations or individuals (Boyce and McDonald 1999). These functions can be used to explain small-scale selection and avoidance for patch or habitat types and therefore can provide critical and timely information to forest and wildlife managers.

Products, time frame, and costs

Production of a large-area habitat map from land cover classified on satellite imagery mosaics was accomplished in approximately four months from time of image reception in the lab to delivery of the multiple data layers (e.g., greenness maps, land cover, habitat). Training and accuracy assessment data collection for the image classification was accomplished in a few weeks by two well-trained field crews supported

with the appropriate field resources (e.g., vehicles, air photos, mensuration equipment); preliminary image analysis and GIS data preparation took several weeks; and the final image classification product (*figure 12.20*) was generated following several iterations and accuracy assessment runs in less than one month. Costs were distributed approximately as shown in the table below:

Task	Number of personnel	Time (week)	Proportional cost (%)	Range ($1,000)
Data preparation	2	2	25	5-15
Field data collection	4	2	50	10-20
Image classification	1	3	20	5-10
Map/data layer generation	1	1	5	1-3

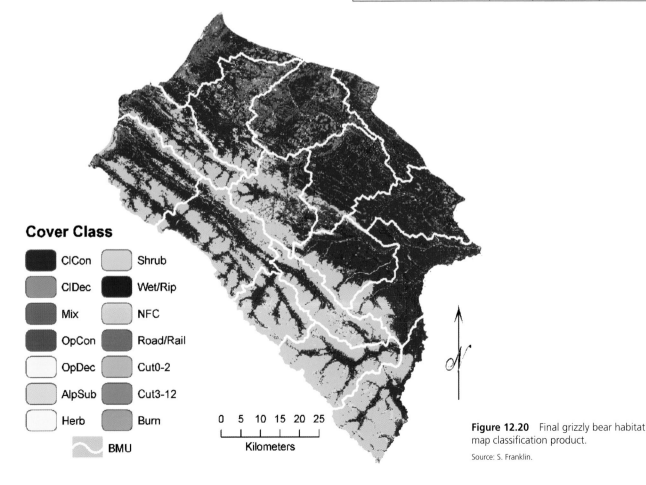

Figure 12.20 Final grizzly bear habitat map classification product.

Source: S. Franklin.

References and further reading

Boyce, M. S., and L. L. McDonald. 1999. Relating populations to habitats using resource selection functions. *Trends in Ecology and Evolution* 14:268–72.

Franklin, S. E., M. J. Hansen, and G. B. Stenhouse. 2002. Quantifying landscape structure with vegetation inventory maps and remote sensing. *Forestry Chronicle* 78(6): 866–75.

Franklin, S. E., M. B. Lavigne, M. A. Wulder, and G. B. Stenhouse. 2002. Change detection and landscape structure mapping using remote sensing. *Forestry Chronicle* 78(5): 618–625.

Franklin, S. E., D. R. Peddle, J. A. Dechka, and G. B. Stenhouse. 2002. Evidential reasoning using Landsat TM, DEM, and GIS data in support of grizzly bear habitat analysis. *International Journal of Remote Sensing* 23(21): 4633–52.

Franklin, S. E., G. B. Stenhouse, M. J. Hansen, C. C. Popplewell, J. A. Dechka, and D. R. Peddle. 2001. Integrated Decision Tree Approach (IDTA) to classification of land cover in support of grizzly bear habitat analysis in the Alberta Yellowhead Ecosystem. *Canadian Journal of Remote Sensing* 27(6):579–93.

Franklin, S. E., and M. A. Wulder. 2002. Remote sensing methods in large-area land cover classification using satellite data. *Progress in Physical Geography* 26(2): 173–205.

Forman, R. T. T., and M. Godron. 1986. *Landscape ecology*. New York: John Wiley and Sons, Inc.

Linke, J., S. E. Franklin, and G. B. Stenhouse. 2005. Seismic lines, grizzly bear landscape use and habitat structure. *Landscape Ecology* 15: in press.

Mackinnon, J., and R. de Wulf. 1994. Designing protected areas for giant pandas in China. In *Mapping the diversity of nature*, ed. R. I. Miller, 127–42. London: Chapman and Hall.

McGarigal, K., and B. J. Marks. 1995. FRAGSTATS: Spatial pattern analysis program for quantifying landscape structure. USDA Forest Service General Technical Report PNW-GTR-351. Corvallis, Ore.

Nielsen, S. E., M. S. Boyce, G. B. Stenhouse, and R. H. M. Munro. 2002. Development and testing of phenologically driven grizzly bear habitat models. *Ecoscience* 10(1): 1–10.

Popplewell, C., S. E. Franklin, G. B. Stenhouse, and M. Hall-Beyer. 2003. Using landscape structure to classify grizzly bear density in Alberta Yellowhead Ecosystem bear management units. *Ursus* 14:27–34.

Grizzly bear habitat Web sites

Wildlife Status Reports: Grizzly Bear.
www3.gov.ab.ca/srd/fw/status/reports/grizzly/index.html

Alberta Species at Risk Program.
www3.gov.ab.ca/srd/fw/riskspecies/index.html

Remote sensing in geology

Zeev Berger and Danny Fortin

The debut of satellite imaging systems on board Landsat 1 in 1972 was a technological advance of considerable interest to earth scientists. Satellite images gave geologists a unique opportunity to observe the complex interaction of large-scale geological structures that make up earth's landscape. Further, digital satellite data could be manipulated and enhanced in order to accentuate the surface expressions of certain geological features. The availability of digital data also allowed integration of satellite imagery with traditional exploration datasets such as geological maps, digital elevation models, and borehole information. Today, satellite images are further integrated with potential field data such as aeromagnetics and gravity.

This section presents a general understanding of geological applications of satellite imagery. Key geological structures commonly mapped with satellites are introduced, and how GIS can be used to integrate multiple datasets to refine interpretation is illustrated. Textbooks providing more comprehensive coverage of geological remote sensing include Berger (1994), Drury (1987), Prost (1997), Sabins (1978), and Vincent (1997).

Key geological structures imaged with remote sensing data

The ability to recognize and map geological structures with remote sensing imagery is highly dependent on the level of bedrock exposure of the features under investigation. As a result, geologists typically encounter two main categories of structures: exposed and obscured (or buried). Both are equally important, providing practical information about the geological setting of underlying bedrock, which is useful in reconstructing the geological history of an area.

Exposed geological structures

Exposed structures are recognized and analyzed from remote sensing data by the surface expressions of their inclined bedrock strata and fault-line traces. Folds, fault and fracture systems, and lithologies (rock units) are readily identified in satellite data. The identification of folds is important since synclinal structures frequently host groundwater supplies, while anticlines form traps for oil and gas *(figure 12.21)*.

Mapping folds with remotely sensed imagery requires the identification of the dip and strike of inclined bedrock.

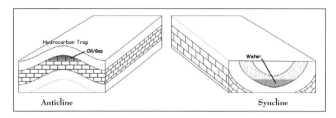

Figure 12.21 Fold structures. Anticlines are geological structures where rock beds on both sides of the axis dip away from the axis. Synclines are structures where the beds dip towards the axis. Anticlines commonly host hydrocarbon deposits, while water commonly accumulates within synclines.

Source: Courtesy of Image Interpretation Technologies. Calgary, Alberta. © IITECH 2003.

Folded strata generally have diagnostic surface features that can be used to recognize the orientation (dip and strike) of their exposed limbs. The power of using remote sensing data for mapping exposed geological structures is illustrated by the Landsat image for the San Rafael Swell in Utah *(figure 12.22)*. The data provides geologists with a clear view of the size and shape of this large domal feature, as well as detailed information on the types of rock that are exposed around the rims of the structure. The figure also illustrates how geologists can use different indicative features to estimate the inclination of the rock units, which in turn can be used to determine the possible shape of the structure at depth.

The recognition of fault and fracture systems is essential during the analysis of the tectonic history of an area. In addition, faults can form trapping structures for hydrocarbons and are localities where migrating fluids can precipitate mineral deposits. Exposed faults typically produce topographic scarps easily recognized in remote sensing data. An excellent example of a fault's surface expression is shown in the Landsat imagery of the Altyn Tagh fault in western China *(figure 12.23)*. The fault-line traces associated with this structure manifest clearly at surface and can be used to assess the geological characteristics of the fault and determine which segments are likely to form prospective areas for exploration.

The identification of lithological units in remote sensing data requires a good knowledge of the interaction between geology and landforms, as well as familiarity with different rock type response to weathering and erosion. Texture, pattern, tone, and geological context are some of the most useful attributes to distinguish lithologies in remotely sensed imagery. Using physical attributes to identify rocks is often difficult since variable climatic and erosion conditions cause different appearances. Other characteristics can be considered.

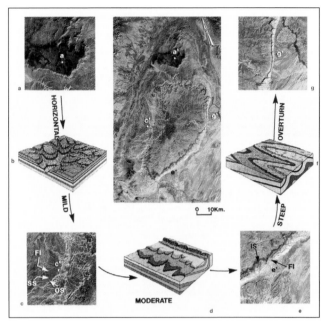

Figure 12.22 Landsat imagery over the San Rafael Swell, Utah. The image illustrates the surface expressions of exposed bedrock units at various magnitudes of inclination. Starting in the northwest corner, and proceeding counter-clockwise, the exposed rims of the swell increase in inclination. (a) gently dipping strata; (c, e) moderately inclined rock units; (g) steeply dipping and overturned beds. The accompanying block diagrams (b, d, f) are from Hamblin and Howard (1989). FI flatirons; SS subsequent streams; OS obsequent streams; IS interrupted slopes.

Source: Berger, Z. *Satellite hydrocarbon exploration*. © Springer-Verlag.

Figure 12.23 Expression of a fault-line trace (FLT). (a) Landsat imagery over the Altyn Tagh fault in western China. (b) Geological interpretation of the Landsat imagery. The images illustrate that a linear negative topographic feature is associated with the fault. CSP compressional splays; NF negative FLT.

Source: Berger, Z. *Satellite hydrocarbon exploration*. © Springer-Verlag.

The internal chemical properties of minerals result in distinct spectral absorption features. This characteristic allows minerals to be recognized by measuring the wavelength of light reflected or scattered from different rock units in several spectral bands. Figure 12.24 illustrates how geologists can manipulate different spectral bands of data to create false-color images designed to enhance specific lithological units of interest. The example shown is from the Wind River Basin in Wyoming, which is characterized by the presence of red beds along the exposed margins of the basin.

While conventional multispectral sensors collect data in only a few bands, advances in computer and sensor technology permit new hyperspectral imaging systems to acquire data in several hundreds of bands. Whereas broadband (multispectral) sensors can simply discriminate between materials, high-resolution hyperspectral platforms allow for the identification of specific materials.

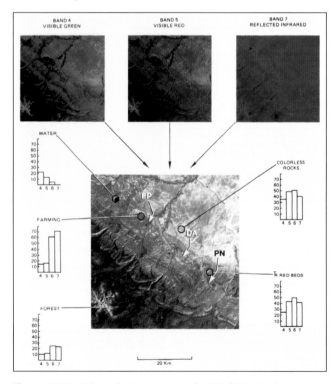

Figure 12.24 False-color imagery over the Wind River Basin, Wyoming. The bar graphs illustrate the response of four separate Landsat bands as the total number of digital data counts (out of a total of 256 possible digital numbers) recorded for specific surface materials. FP floodplain; DA Dallas anticline; PN plunging nose. (After Townsend and Dodge 1983).

Source: Berger, Z. *Satellite hydrocarbon exploration*. © Springer-Verlag.

Obscured or buried geological structures

Geological structures with outcropping units either partially obscured by soil and vegetation or completely buried under sediments are usually indirectly recognized in remote sensing images by their subtle influence on the regional topography and related drainage, vegetation, and soil moisture patterns. Figure 12.25 illustrates the use of satellite imagery to identify buried and obscured structures. This Landsat image captures the subtle topographic expressions of buried salt anticlines in the Colville Hills area of the Northwest Territories. The elongated salt ridges are recognized by their bright spectral response at surface, which is attributed to dry moisture conditions.

Although obscured or buried structures can be identified in surface imagery alone, their proper recognition often requires the integration of subsurface data such as seismic, magnetics, gravity, and borehole information.

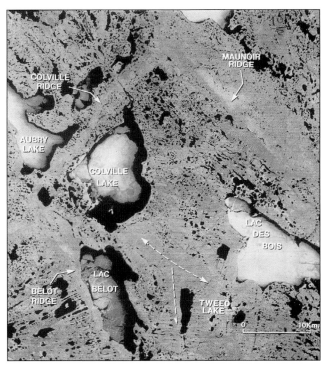

Figure 12.25   Surface expression of buried salt anticlines. Landsat imagery and interpretation over the Colville Hills area, Northwest Territories, Canada. The crests of the salt ridges are recognized in the imagery by their bright spectral anomalies, which are attributed to dry conditions.

Source: Berger, Z. *Satellite hydrocarbon exploration.* © Springer-Verlag.

Direct detection of hydrocarbons

Though satellite imagery can be used to recognize surface evidence of geological structures that may form hydrocarbon traps in the subsurface, some oil fields were discovered through direct evidence of petroleum at surface. Oil and gas recognition at surface is referred to as *direct detection* of hydrocarbons. When hydrocarbons seep to the surface, they interact with rocks, soil, and vegetation to produce anomalous conditions that can often be recognized in remote sensing images.

Onshore seepage can cause alteration of minerals, such as bleaching of red beds, and the conversion of feldspars into clay minerals. Hydrocarbon seepage can also lead to the formation of minerals such as calcite, pyrite, and certain magnetic iron oxides. These minerals have diagnostic spectral features readily detected in remote sensing data such as hyperspectral imaging. Vegetation is also used as an indicator of hydrocarbon seepage. In some cases, hydrocarbons are detrimental to plant communities and tend to stress vegetation. On the other hand, changes in soil chemistry can cause anomalous plant species to develop.

Waterborne oil seeps can be detected in remotely sensed data since presence of hydrocarbons will affect the spectral behavior of water. Compared to clean water, oil on water has stronger absorption in the visible and higher reflectivity in the near-infrared portions of the electromagnetic spectrum.

Data integration

Data integration involves merging several datasets in order to extract useful information for the interpretation. The availability of a wide range of data in digital form allows the exploration geologist to integrate various datasets in order to assist with the interpretation and analysis of a specific area. In the early days of remote sensing, when the only source of data was aerial photography, data integration was very limited, but with the advent of GIS systems, the merging of information gathered from different datasets has become common practice.

Effective exploration concepts are developed through the integration of geological and geophysical datasets. These datasets include satellite imagery, potential fields (magnetics and gravity), seismic data, digital elevation models (DEMs), borehole information, and structural maps on key horizons. Figure 12.26 illustrates the way different geological datasets are integrated to analyze the complex three-dimensional relationships between surface structures observed with remote sensing data and subsurface structures mapped from other datasets.

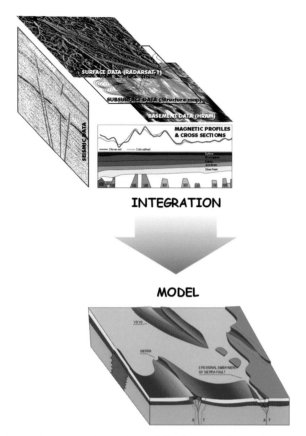

Figure 12.26 Data integration and development of an exploration concept. This diagram illustrates that effective development of new exploration concepts involves the integration of multiple datasets, including seismic, potential field and satellite remote sensing data, as well as structural maps and magnetic models.

Source: Courtesy of Image Interpretation Technologies, Calgary, Alberta. © IITECH 2003.

By integrating various datasets that are accurately registered, geologists can observe the spatial relationships between structures at different depths and their possible interaction through time and space. Such integrated analyses assist in the development of geological concepts that can guide exploration for hydrocarbons, mineral deposits, and water.

Case study: Integrated geological study of the Gabon Sedimentary Basin

High-resolution aeromagnetic (HRAM) data, Radarsat-1 imagery, and regional seismic data were used to investigate the hydrocarbon potential of deep-seated structures in the Dianongo Trough area of Gabon. The objectives of the study were to improve the understanding of the area's structural style and to relate structures to potential hydrocarbon accumulations. Previous exploration efforts in the area, based mainly on seismic data, proved difficult due to the presence of a thick salt-bearing horizon in the subsurface. Seismic data in the area was generally of poor quality as a result of low acoustic impedance contrast between the salt and underlying formations. In addition, thick vegetation cover over much of the area impeded standard exploration techniques. It was therefore considered that a remote sensing study could be conducted at reduced costs and provide greater impact on the exploration potential of the area than conventional exploration techniques. This study showed that the integration of HRAM and Radarsat-1 data is a particularly valuable resource for structure mapping in frontier hydrocarbon-bearing basins. The integration of seismic and well log data allowed for an even more effective use of the HRAM and satellite data. Finally, the study demonstrated that two hydrocarbon-trapping mechanisms potentially exist in the Dianongo Trough area.

Data used

The HRAM data was collected using a fixed-wing airplane over an area 45 km wide and 140 km long. The survey was flown with north–south traverse lines at 600 m line spacing with control lines flown approximately perpendicular at 1800 m line separation. The position of the aircraft was maintained using a differential GPS navigation system.

Two Radarsat-1 scenes were used as part of the study. These scenes were collected using the satellite's Standard mode (Beam 7) in descending orbit. The incidence angle of the radar beam provided a nominal ground resolution of 28 m. The coverage of the radar imagery included nearly all of the HRAM survey area. Radarsat-1 imagery was chosen because it has proven to be a valuable exploration tool in tropical regions since it can penetrate the near constant cloud cover and collect images of the topographic expression of geological structures covered by thick vegetation.

Additional surface information came from published and unpublished geological maps of Gabon, which were used to outline the main lithological units present in the study area. Subsurface information came primarily from structural maps derived from the sparse well log data. Seismic data was also used as part of the analysis, although it was only useful in defining structures present above the salt horizon.

Analysis procedures

A mosaic of the Radarsat-1 imagery was first used to map and evaluate the surface expression of basement structures in the Dianongo Trough, as well as to analyze the exposed basement structures east of the trough. The analysis of the radar imagery identified the dominant faults and fractures system present in the area and the presence of a profound topographic escarpment near the western edge of the exposed basement, which forms the eastern margin of the trough.

The HRAM data was then used to further analyze the overall geometry of the trough. Images derived from the Total Magnetic Intensity (TMI) data were used to map and evaluate the basement structures and geometry of the trough. The high-frequency signature of the data enabled the approximate shape of the Dianongo Trough within the study area to be discerned, while several generations of basement faults within the trough were identified within the lower frequency data.

The usefulness of the integration of HRAM and Radarsat-1 data is shown in figure 12.27. The figure illustrates how the two datasets were used to define the outline of the trough and identify the major faults and fractures system present in the study area.

The regional seismic and well log data available within the study area was used to constrain the analysis of the HRAM and Radarsat-1 data and establish possible links among basement structures, sedimentary structures, and topographic structures.

The information extracted from the various datasets was then combined to establish an overall tectonic model for the area. The model shows how pre existing basement faulting led to the development of the Dianongo Trough. The tectonic model also illustrates the structural components of the trough and the hydrocarbon potential of the study area.

Time frame

This project began with the acquisition of the HRAM survey, which lasted about three months. During this period, we began to prepare for the integrated interpretation of the remote sensing data by (1) selecting the satellite imagery best suited for the investigation (in this case the two Radarsat-1 Standard mode scenes); (2) processing the remote sensing data; (3) collecting all available geological information for the region; and (4) converting the datasets into a common digital format. The interpretation of all datasets began soon after the completion of the HRAM survey. The complete interpretation procedure lasted an additional two months.

Cost

The cost of the whole study was approximately $240,000 CAD. The most expensive portion of the project was the collection of the HRAM data, which is estimated at $200,000 CAD. The cost of the two Radarsat-1 scenes was about $10,000 CAD, while the interpretation and integration of all datasets was approximately $30,000 CAD. The seismic data used as part of the study was proprietary data valued at several millions of dollars.

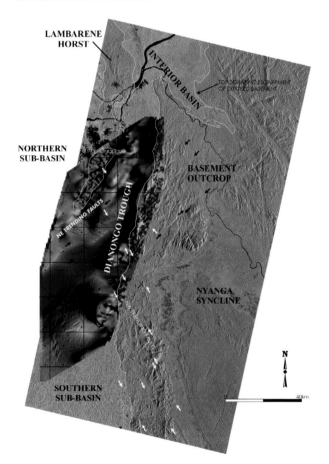

Figure 12.27   HRAM and Radarsat-1 data integration over the Dianongo Trough, Gabon. The imagery has been overlain with a color transparency to indicate the main geological outcrops in the area. The white arrows indicate the surface expression of northwest-trending rift faults, while the black arrows indicate the surface expression of the basement structural fabric. The high-frequency HRAM signature of the basement allows the outline of the Dianongo Trough to be defined.

Source: Courtesy of Image Interpretation Technologies, Calgary, Alberta. © IITECH 2003.

Products

The client for this project received a series of images and maps that illustrated the results of the integrated geological study. This included merged images of HRAM and Radarsat-1 data, as well as maps showing the location of prospective geological structures. Three-dimensional diagrams illustrating the relationship between geological structures observed in seismic data and their surface expressions were also delivered. All products were provided in both hard-copy and digital formats.

References

Berger, Z. 1994. *Satellite hydrocarbon exploration*. New York: Springer-Verlag.

Billinsley, F. C. 1983. Data processing and reprocessing. In *Manual of remote sensing*. 2nd ed., ed. R. N. Collwell. Falls Church, Va.: American Society of Photogrammetry.

Drury, S. A. 1987. *Image interpretation in geology*. London and Boston: Allen and Unwin Publishers.

Hamblin, W. K. and J. D. Howard. 1989. *Exercises in physical geology*. 5th ed. Minneapolis: Macmillan.

Morley, C. K., R. A. Nelson, T. L. Patton, and S. G. Munn. 1990. Transfer zones in the east African rift system and their relevance to hydrocarbon exploration in rifts. *American Association of Petroleum Geologists Bulletin* 74(8): 1234–53.

Prost, G. L. 1997. *Remote sensing for geologists: A guide to image interpretation*. Amsterdam: Gordon and Breach Science Publishers.

Sabins, F. F. 1978. *Remote sensing: Principles and interpretation*. New York: W. H. Freeman and Co.

Townsend, T. E. and R. L. Dodge. 1983. Techniques for geologic mapping based on the spectral component of Landsat imagery. Exxon Production Research Company, Internal Report.

Vincent, R. K. 1997. *Fundamentals of geological and environmental remote sensing*. Upper Saddle River, N.J.: Prentice Hall.

Remote sensing in meteorology, oceanography, and climatology

Kevin Gallo

The world's first meteorological satellite was launched on April 1, 1960. The Television and InfraRed Observation Satellite, TIROS-1, although experimental, demonstrated the capability of acquiring images of the earth's cloud cover. The usefulness of this data has resulted in a nearly continuous flow of meteorological observations from space. The United States currently operates three series of operational meteorological satellites: the Geostationary Operational Environmental Satellites (GOES), the Polar-Orbiting Operational Environmental Satellites (POES), and the Defense Meteorological Satellite Program (DMSP) series of satellites.

The GOES satellites are located approximately 36,000 km above the earth's surface and positioned such that the satellite is stationary above a given earth location. This is accomplished through the orbital characteristics of the satellite. The satellite orbital plane is the same as the earth's equator, and the orbital velocity of the satellite matches the rate at which the earth turns on it axis. Thus, the satellite appears stationary above a specific location on the earth as it follows this location as the earth rotates on a daily basis. Thus, the GOES instruments can observe a given location in North or South America on a nearly continuous basis throughout the day and night.

The POES and DMSP satellites are in a polar orbit about 850 km above the earth's surface, and their orbital characteristics are such that the earth rotates beneath the satellites. This results in the POES and DMSP having the capability to make observations of most of the earth's surface twice a day.

The operational satellites of the civilian and defense agencies of the United States were primarily designed for near real-time meteorological and oceanographic uses. New climatological applications of the data acquired by these satellites have become possible with the greater availability of historical data. The objective of this overview is to give a brief introduction to some of the instruments on board, and products available from, these operational satellites.

Remote sensing instruments

The remote sensing instruments on board the current operational meteorological satellites include three primary sensor types: imagers, sounders, and microwave sensors.

Imagers are designed to measure energy that is reflected or emitted in the visible and infrared wavelengths. Typical uses of data from the visible wavelengths (nominally 0.4–0.7 μm) include cloud detection and monitoring of water and land surface features. Infrared (IR) uses of the data include monitoring vegetation, aerosol properties, and clouds (near-IR: 0.7–1.0 μm); cirrus cloud and snow-versus-cloud detection (reflective and mid-IR: 1.0–3.0 μm); and monitoring atmospheric water vapor and aerosols, land, and ocean temperatures (thermal IR: 3.0–15 μm).

The sounder instruments are used to acquire vertical profiles of temperature and moisture data in the atmosphere, similar to that acquired by balloon-carried radiosondes.[2] Atmospheric sounder instruments observe radiation emitted by atmospheric gases. The radiation emitted by atmospheric gases at a specific wavelength is associated with a specific layer (altitude) within the atmosphere. By observing at specific wavelengths, the sounder can measure atmospheric properties at various altitudes and thus provide a vertical profile of these properties much more frequently than is available with radiosondes.

The sounder instrument on the current GOES series of satellites includes 19 channels with wavelengths that range from the visible (0.70 μm) to thermal IR (14.71 μm), although 18 of the 19 channels are in the thermal IR region. Products derived from the sounder include total precipitable water, atmospheric winds, and several indices related to atmospheric stability.

The microwave instruments, similar to the sounder instruments, acquire vertical profiles of temperature and moisture in the atmosphere. The instruments on the current POES series of satellites include 20 channels that range from 23 to 183 GHz. The data acquired in these channels can also be used to monitor the land surface. Current products from the microwave instruments include temperature and humidity profiles, total precipitable water, and snow cover. Products under development include land surface temperature and surface wetness.

In the following sections, examples of current operational and developmental products are presented and briefly discussed.

Atmospheric products

Atmospheric products are generated from the data collected by all three of the instruments previously discussed. These products include atmospheric soundings (measurement of

---

2. A radiosonde contains instruments that make direct measurements of atmospheric temperature, humidity, and pressure at different altitudes of the atmosphere.

temperatures at several levels in the atmosphere), high-density winds, aerosol concentrations, fog, and other aviation-related products. Examples of these products include total precipitable water *(figure 12.28)* derived from data measured by the GOES and used in weather forecast models; data on the ozone present at the North Pole *(figure 12.29)* and South Pole derived from POES, which is used for climate monitoring; and aerosol concentrations, which are also derived from POES data and used for climate monitoring *(figure 12.30)*.

Additional products include indexes related to atmospheric stability (useful for short-term severe weather forecasts) and total column ozone (useful for climate monitoring), which are available from the Cooperative Institute for Meteorological Satellite Studies, University of Wisconsin-Madison. A review of several quantitative products derived from GOES imager and sounder data can be found at *www.orbit.nesdis. noaa.gov/smcd/opdb/goes/soundings/#informational*.

Oceanographic products

Examples of oceanographic uses and products include sea surface temperature derived from POES data *(figure 12.31)*, which is useful for climate monitoring and seasonal climate predictions (e.g., El Niño events); the ocean surface winds product from DMSP data *(figure 12.32)*; and monitoring of coral reef conditions with POES-derived data *(figure 12.33)*. Additional examples include GOES and POES sea surface temperatures available from the NOAA CoastWatch Program *(coastwatch.noaa.gov/index.html)*.

Land products

Land-related products include the snow cover during winter product *(figure 12.34),* derived from a combination of GOES, POES, and DMSP data, and the seasonal changes in

green vegetation product (fractional green vegetation cover) *(figure 12.35)* derived from POES data. Both of these products are useful as data sources for weather forecast models and for climate monitoring. Nighttime lights emitted by surface features such as urban areas, fires, fishing vessels, and gas flares provide a unique perspective of the land surface *(figure 12.36)*. This data has been used to analyze the influence of urbanization on local air temperatures (the urban heat-island effect).

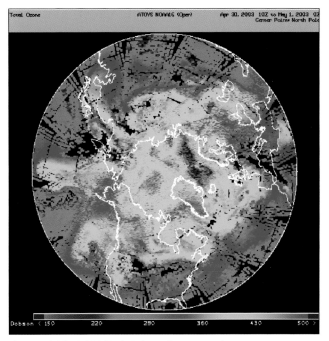

Figure 12.29 POES North Pole total ozone product.

Source: Image courtesy of NOAA.

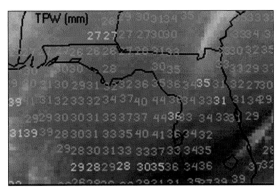

Figure 12.28 GOES total precipitable water product.

Source: Image courtesy of NOAA.

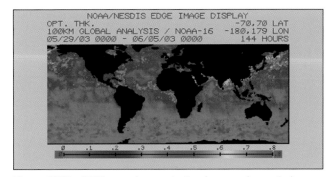

Figure 12.30 POES aerosol concentration (measured as optical thickness) product.

Source: Image courtesy of NOAA.

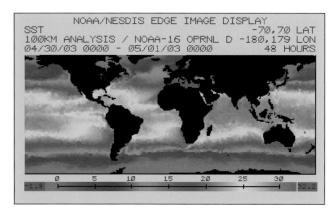

Figure 12.31 POES sea surface temperature analysis.

Source: Image courtesy of NOAA.

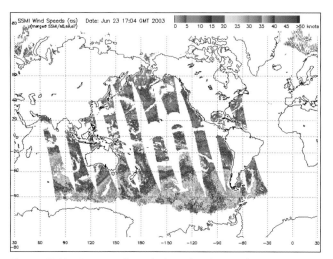

Figure 12.32 Ocean surface winds product from Defense Meteorological Satellite Program.

Source: Image courtesy of NOAA.

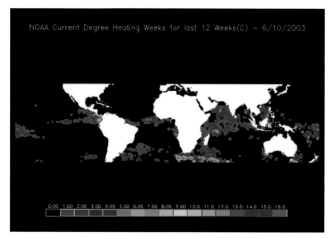

Figure 12.33 Coral reef monitoring product.

Source: Image courtesy of NOAA.

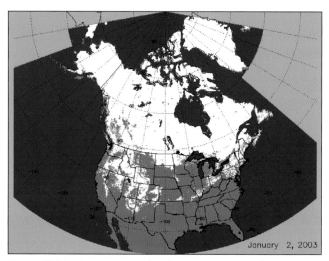

Figure 12.34 Snow cover product for January 2, 2003, for a portion of North America.

Source: Image courtesy of NOAA.

The availability of relatively long term vegetation information derived from satellite data has led to climatological datasets that include the Fractional Green Vegetation products, as well as drought-related products (e.g., the vegetation health product derived from POES data *(figure 12.37)* and products that monitor changes in global terrestrial net primary vegetation production *(figure 12.38, derived from POES data)*. With increased forest fire concerns, satellite derived forest fire products have also evolved *(figure 12.39)*.

The availability of satellite-derived datasets and continuing development of ecosystem models has enabled systems to be developed that integrate remote sensing and other spatial datasets to monitor global ecosystem dynamics. For example,

the Terrestrial Observation and Prediction System (TOPS), is a computer-based geospatial information system designed to model, monitor, and forecast the effects of climate variability on ecosystems *(figure 12.40)*. Developed by the Numerical Terradynamic Simulation Group at the School of Forestry, University of Montana, Missoula, the system integrates weather and climate forecasting, ecosystem modeling, and satellite remote sensing to support management decisions related to floods, droughts, forest fires, human health, and the productivity of crop, range, and forest lands.

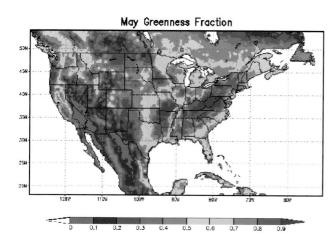

Figure 12.35 Seasonal fractional green vegetation for May and July.

Source: Image courtesy of NOAA.

Figure 12.36 Nighttime lights of a portion of North America derived from DMSP data.

Source: Image courtesy of NOAA and NASA.

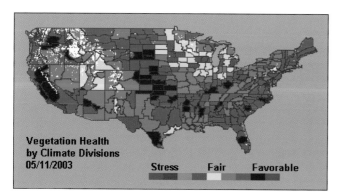

Figure 12.37 Vegetation Health product for the conterminous United States, 11 May 2003.

Source: Image courtesy of NOAA.

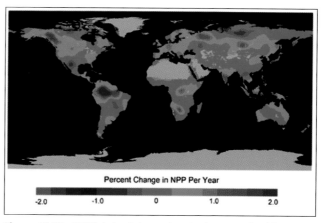

Figure 12.38 POES-derived change in net primary productivity between 1982 and 1999.

Source: Image courtesy of R. R. Nemani of NASA Ames Research Center.

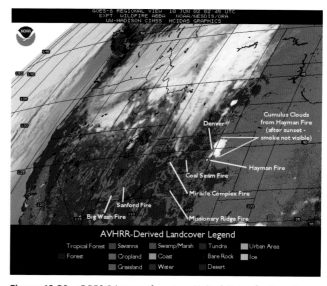

Figure 12.39 GOES 8 image of western United States for June 9, 2002, with Wildfire Automated Biomass Burning Algorithm applied.

Source: GOES Biomass Burning Monitoring Team, UW-Madison, Cooperative Institute for Meteorological Satellite Studies (CIMSS).

Conclusion

The applications of both the Geostationary (GOES) and Polar-Orbiting (POES and DMSP) operational satellites are quite diverse, a result of the different instruments on board and the creativity of the scientists who have analyzed the data in the past and present. There are numerous satellites operated by other countries, commercial enterprises, and other agencies of the U.S. government (e.g., the NASA experimental Earth Observing System, and USGS Landsat, satellites) that provide information also considered valuable for meteorological, oceanographic and climatological applications. The Web sites provided below are a good place to search for more information.

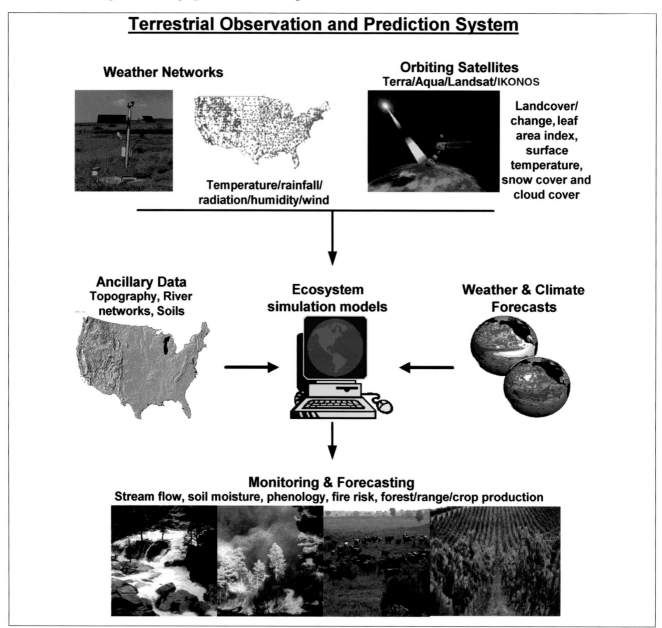

Figure 12.40   Overview of the Terrestrial Observation and Prediction System (TOPS).

Source: Image courtesy of R. R. Nemani of NASA Ames Research Center.

Web sites

CIMSS GOES Sounder and Imager Page.
cimss.ssec.wisc.edu/goes/goes.html

NASA Earth Observing System.
eospso.gsfc.nasa.gov/index.php

NASA Earth Observatory Nighttime Lights earthobservatory.
nasa.gov/Newsroom/NewImages/Images/earth_lights_lrg.jpg

NOAA/ National Weather Service, National Centers for Environmental Prediction Seasonal Fractional Green Vegetation for May and July.
www.emc.ncep.noaa.gov/mmb/gcp/sfcimg/gfrac/index.html

NOAA/ NESDIS Center for Satellite Applications and Research.
Monitoring Vegetation Condition From NOAA Operational Polar-Orbiting Satellites
www.orbit.nesdis.noaa.gov/smcd/emb/vci/index.html

Oceanic Research and Applications Division.
www.orbit.nesdis.noaa.gov/sod/orad/sod

Snow cover product for a portion of N. America.
www.orbit.nesdis.noaa.gov/smcd/emb/snow/HTML/snow.htm

Temperature and Moisture Soundings Fields.
www.orbit.nesdis.noaa.gov/smcd/opdb/goes/soundings/html/fields.html

NOAA/NESDIS Office of Satellite Data Processing and Distribution Aerosol Products.
www.osdpd.noaa.gov/PSB/EPS/Aerosol/Aerosol.html

Sea Surface Temperature Images.
www.osdpd.noaa.gov/PSB/EPS/SST/sst_anal_fields.html

Tropical Ocean Coral Bleaching Indices.
www.osdpd.noaa.gov/PSB/EPS/CB_indices/coral_bleaching_indices.html

NOAA/NESDIS Office of Satellite Operations.
www.oso.noaa.gov

NPOESS-National Polar-orbiting Operational Environmental Satellite System.
ipo.noaa.gov

USGS National Center for Earth Resources Observation and Science.
edc.usgs.gov

Remote sensing in archaeology

Scott Madry

Archaeology is the scientific study of the physical evidence of past human societies. Archaeologists not only attempt to discover and describe past cultures, but they also formulate explanations for the development of human culture. Aerial photography and remote sensing have been widely used by archaeologists around the world for over 50 years for a variety of purposes, including site discovery, regional analysis, understanding the environmental context of sites, determining the location of road and settlement patterns, developing site preservation plans, and for integration with other data using GIS. The development of predictive models and management plans for the protection of our cultural resources relies increasingly on remote sensing, GPS, and GIS data and analysis (Farley et al. 1990).

This section presents a general overview of the uses of remote sensing for archaeology. The details of aerial archaeological methods are presented in a variety of volumes, including Parrington (1983), Ebert (1984), and El-Baz (1997). The theory and practice of low-level aerial archaeology are outlined in detail by Agache (1970), Chevallier (1964), Dassie (1976), and Kunow et al. (1995), among others.

History

The application of the aerial perspective to archaeology dates to the earliest days of aerial photography (Deuel 1969). Stonehenge was photographed from a tethered balloon as early as 1906. Major advances were developed in World War I in Europe, when British, French, and German archaeologists, serving in their respective air services, discovered many strange patterns and anomalies from the air that they guessed were vestiges of ancient landscapes. These were the origins of aerial archaeology in Europe and the Middle East. Charles Lindbergh searched from the air for Mayan sites in Mexico and the southwestern United States in the 1920s. World War II created many new technologies that were quickly adopted, including infrared film, and a vast amount of surplus aerial cameras and aircraft were for sale. Much of the earth was photographed from the air for the first time.

The later development of satellite remote sensing was quickly adopted but was initially limited to general environmental analysis by poor spatial resolution (Lyons and Ebert 1978). As the resolution of data improved over the 1980s and 1990s, archaeologists also adopted these new systems

(Madry and Crumley 1990). The recent development of ultra-high-spatial-resolution space systems has finally provided researchers with the equivalent of aerial photography from space. These systems provide the ability to locate individual sites (Fowler 1996), although at a cost that is often beyond the means of archaeologists. Airborne and space radar, hyperspectral sensors, lidar, and related tools are quickly being adopted by the archaeological community, although cost remains an issue.

Archaeological applications with remote sensing data

Archaeological site discovery

The discovery of the vestiges of ancient landscapes is significantly aided by the use of the aerial perspective. The aerial view provides a systematic overview of an area and permits the discovery of both individual sites as well as their contextual relationships such as road networks. These can be discovered from the air even though the actual structures are buried below the modern surface. Faint lines and color differences visible in imagery are often not visible or are unrecognizable when viewed from the ground. Visible straight lines, circles, squares, etc., are often indications of archaeological sites and ancient landscapes, but these are also often the result of natural processes or modern human activities. Care must be taken to verify such features on the ground.

Several features mark archaeological remains that are visible in this way. The primary features are crop marks, soil marks, and shadow marks (figure 12.41). Crop marks result from visible differences in crop or vegetation height or color that can be caused by the effects of buried features. Buried stone walls or foundations can cause decreased crop vigor, especially in times of drought, that creates clear outlines of the structure from the air that are often not visible at all on the ground. These are known as negative crop marks. The European drought of 1976 permitted the discovery of thousands of new sites from the air (Dassie 1976). Positive crop marks are caused by buried ditches or post holes that provide additional vegetation vigor that can be seen as patterns of increased crop growth from the air. Soil marks are caused by variations in surface soil type or color. Soil marks can be a reflection of past activities such as trenches, field divisions, ramparts, or roads. Slight variations in the modern topography that are vestiges of ancient landscapes can often be seen in imagery as linear shadow marks (when low sun elevation and other conditions are right). Snow can be piled up against small microtopography and show patterns from the air, and snow melting patterns can also be indicative of buried structures. All of these faint patterns can be identified by trained specialists in archaeological image interpretation.

Regional archaeological environmental analysis

Human cultures exist within a complex web of interaction with their environment, natural resources, and other factors. Remote sensing provides an excellent source of data to understand the overall environmental context of archaeological sites and settlement patterns. Data derived from remote sensing, such as modern vegetation and land-use patterns, hydrology, elevation, and roads are all useful to archaeologists for a variety of purposes. Archaeological field campaigns can be complex and expensive activities that are usually conducted on a very limited budget. Satellite imagery provides important information on the current environment of the study area, providing data on the best routes of access into remote areas and for logistical planning. Cultural features such as large sites, road networks, and resource extraction sites are often visible on moderate-resolution imagery and

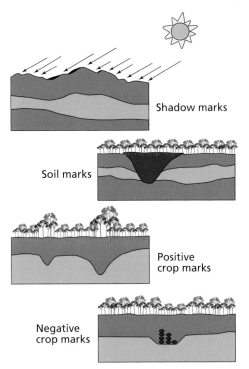

Figure 12.41 Drawings showing the processes causing the visibility of archaeological sites from the air.

Source: S. Madry.

can be used to plan fieldwork and to determine new areas of interest. Much of the world is still very poorly mapped, and archaeologists frequently have to create their own maps of remote areas using imagery. Acquiring new imagery periodically also provides vital change detection data that is useful for a variety of purposes, including monitoring site looting, natural disaster impacts on archaeological sites, and modern human impacts such as deforestation and the establishment of new settlements. A recent example comes from China, where archaeologists report finding over 28,000 sites using space remote sensing data for the massive Three Rivers Gorge Dam project (*People's Daily* May 10, 2000). Such large-area activities are only feasible using remote sensing data.

Geographic information systems

GIS has become a standard tool for archaeology and the management of cultural resources (Allen, Green, and Zubrow 1990). Satellite imagery often provides the basis for GIS data layers (land use and land cover, elevation, slope, aspect, modern roads, hydrology, and modern settlement patterns) that are often not commercially available for remote areas outside the developed world. GIS provides a powerful tool for modeling of settlement patterns and creating predictive models of site locations. It also provides the ability for archaeologists to analyze their data between field seasons and, perhaps most importantly, to quantitatively test hypotheses and theories using real data.

Data integration

Early aerial archaeologists worked primarily with uncontrolled aerial photos, a compass, and an out-of-date topographic map. It was very difficult to accurately locate features on the ground that were discovered through aerial survey and prospecting. Archaeologists now use a fully integrated suite of spatial analysis tools including GPS, laser theodolites, digital image processing of imagery, and geophysical prospecting tools such as ground penetrating radar, resistivity, and magnetometry. This disparate data is integrated and synthesized in the GIS environment, where data from multiple sources is georeferenced and combined with other data such as historic maps, soil data, property ownership, and other environmental information. Sophisticated modeling of site location patterns, including predictive models of settlement patterns, road networks, and site intervisibility patterns can be created using GIS for a variety of research and management applications. Cultural resource management using these tools by governmental entities is playing an increasingly important role in protecting our irreplaceable cultural resources around the world. This is indicated by the recent adoption of remote sensing technologies in support of the World Heritage Convention and for monitoring of UNESCO World Heritage sites (Hernandez 2003).

Future direction of remote sensing and GIS in archaeology

The continuing development and integration of spatial analysis tools provide archaeologists and cultural resource managers with a powerful capability to locate, study, and protect our cultural resources. The spatial, temporal, and radiometric resolution of imagery has continued to improve and provides improved information for decision makers. New ultra-high-resolution space systems (less than 1 meter spatial resolution) provide the ability to locate and map individual sites (Beck 2003) in areas where aerial photography is not available. Hyperspectral systems offer the promise of locating individual features such as shell middens and resource extraction sites. Space radar systems such as SIR-C have demonstrated the ability to penetrate dry sand and locate landforms that contain archaeological sites (McCauley et al. 1986, McHough et al. 1988). New higher resolution radar systems will be extremely useful in forested areas such as Central America and areas that are often covered by clouds. lidar provides microtopography data that can show subtle terrain changes caused by buried structures, roads, and field patterns. Advanced digital image processing techniques continue to allow the extraction of more information from the imagery that we acquire. GIS modeling of site locations, road networks, and human interactions continue to be developed (Maschner 1996, Lock and Stancic 1995). The GIS environment is also being expanded to include simulation and visualization techniques to reconstruct structures and landscapes (Ogleby 2001).

The primary future direction is the *integration* of data from multiple sources in an overall geomatics context. Archaeologists now integrate data into a GIS environment from geophysical data, field survey data, historic maps, aerial photography, and space imagery. The ability to routinely integrate such disparate data over large areas provides for powerful analysis and a more complete understanding of the complexity of the archaeological record. Several very large-area, GIS-based predictive modeling projects are either underway or have been recently completed for comprehensive cultural resource management in Ontario (Dalla Bona 1994), Minnesota (Hudak et al. 2002), and North Carolina

(Madry et al. 2004). The development of quantitative models of archaeological site patterns and the changes in these over time are of significant value for scientific research and cultural resource management. Our irreplaceable cultural heritage deserves no less.

Case study: Integrated application of remote sensing and GIS for regional archaeological analysis in Burgundy, France

A research program has been conducted over 20 years in the application and integration of remote sensing and GIS for regional archaeological analysis in the Burgundy region of France. The objectives of the project are to determine the utility of various geomatics tools in understanding the changing nature of settlement patterns and land use over a period of 2000 years, from the Iron Age through the Gallo-Roman, medieval, and into the modern era. Our goal is to understand the changing patterns of interrelation between different cultures and their environment at various scales over time.

Data used

A multidisciplinary program using a wide variety of data sources has been developed for this project. Archaeological ground survey, historical documentation, cadastral records, aerial prospecting and survey, ethnographic interviews, analysis of historical maps, airborne thermal data, and various satellite imagery have been combined into a GIS database for analysis. A French airborne thermal scanner was flown over the area in order to locate vestiges of ancient landscapes. Extensive aerial prospecting has been conducted over the years from low-level aircraft, with sites and road segments located and field verified. A detailed GIS database has been constructed, which now includes a variety of modern environmental and cultural parameters such as geology, vegetation at different time periods, road networks, hydrology, and modern settlement data. Much emphasis has been placed on the analysis and integration of historic cartographic data, and maps from the seventeenth, eighteenth, and ninteenth centuries have been scanned and georeferenced, and features have been extracted. Cassini maps from 1759 have significant detail and have provided much information. Remote sensing data from early Landsat 1 (1973), SPOT (1986), Radarsat (1999), and Landsat 7 ETM (2002) have been processed and integrated. A complete set of 1945 aerial photographs taken by the U.S. Army Air Forces in World War II has been acquired and analyzed in stereo, with numerous sites and road sections located. These are compared with historic maps, site survey data, and GIS models in an interactive manner.

Analysis procedures

This project has been a part of a larger interdisciplinary program of research by a group of archaeologists, historians, ecologists, medievalists, and other scientists. The goals of the geomatics research program were to investigate the evolving potential of these tools for the larger study. We have also provided an integrative and quantitative database that is available to all researchers and which provides a long-term repository for project data. As new data is acquired, such as new field survey data using GPS, it is integrated into the overall database. Analysis is conducted to support specific research questions as they develop. Theories about settlement and land-use patterns are tested in a quantitative manner.

Time frame

This project has been ongoing since the mid-1970s. It is our belief that, in order to understand the complexities of the interaction between different cultures and the environment over time, it is necessary to engage in a long-term multidisciplinary analysis. This is unusual. Most archaeological projects are conducted for only a few years as funding is acquired. We have taken a different approach, dedicating significant time and resources over decades in order to attempt to understand the complexity of the systems involved while testing the utility of new technologies as they are developed.

Cost

Most of the cost of this project has been born by the researchers themselves. Funding and priorities for such a long-term and integrative program fall outside the boundaries of most established funding programs. We have received small support over the years from several sources, but the research team has met the majority of the cost of data acquisition, fieldwork, and analysis.

Results

This project has demonstrated the utility and effectiveness of such an integrated, long-term, multidisciplinary analysis (Crumley and Marquardt 1987). Numerous archaeological sites have been discovered, and we have placed them into their environmental context as that has changed over time. We have mapped the changing nature of settlement and

transportation networks over time. The functional integration of numerous disparate sources of data has provided a variety of results. Features located on 1945 aerial photos have been cross-checked against field surveys, maps from the 1750s, cadastral records, and modern imagery. Significant emphasis has been placed on understanding the changing nature of the road networks, and GIS techniques of cost surface analysis were developed to model the environmental and cultural factors involved in Celtic road networks between hill forts (Madry and Rakos 1996). Figure 12.42 shows the location of a Gallo-Roman era road segment on a 1759 map. This segment was then relocated on aerial photographs and mapped using GPS. Figure 12.43 shows an ancient road segment visible on 1945 aerial photography that was mapped in the same way. Numerous cultural resources in the region have been destroyed by a series of gravel mines in the river valley, including a large, intact Gallo-Roman villa complex. This villa was located in the aerial and field survey program and was the first such villa complex ever discovered in the area. It was destroyed before it could be excavated, and its destruction was the impetus for the development of archaeological predictive models for the region, in order to locate other sites before they were destroyed.

Figure 12.44 shows an aerial photograph of the villa complex; it clearly shows the outline of the structure's walls as negative crop marks (areas of reduced crop vigor). A positive crop mark (areas of increased crop vigor), possibly an outbuilding or animal enclosure, is visible at the top right. Figure 12.45 shows a ground view of the villa complex taken

Figure 12.42 Casinni map of France of 1759 showing the location of an ancient Roman road segment marked "Ancien Chemin des Romans."

Source: S. Madry.

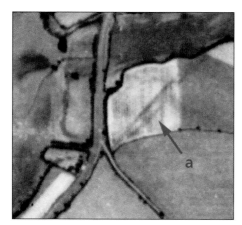

Figure 12.43 Segment of an ancient roadway (a) visible as a soil mark in a 1945 aerial reconnaissance photo taken by the U.S. Army Air Forces in France.

Source: S. Madry.

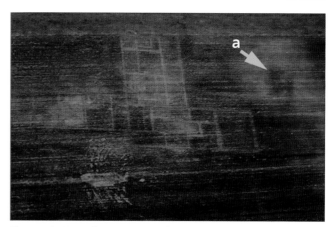

Figure 12.44 Gallo-Roman era villa complex in France, discovered from the air in 1979, with negative crop marks (lighter lines) showing the outline of the walls. A positive crop mark in green in the upper right corner (a) indicates an enclosure or other structure. Curved lines at bottom are modern plow scars.

Source: S. Madry.

Figure 12.45 Gallo-Roman villa complex shown on the ground. The lighter colored straight lines are the outlines of the walls shown in figure 12.44.

Source: S. Madry.

the same day; it shows how difficult it is to recognize such features on the surface. GIS predictive modeling of site location patterns for the Iron Age, Gallo-Roman, and medieval periods were developed (Madry and Crumley 1993), and these have been used to focus ground, geophysical, and aerial surveys in the areas threatened by mining operations. Figure 12.46 shows the area around the destroyed villa complex with the predictive model. The villa is shown as the diamond in the center of the red area. Figure 12.47 shows the predictive model overlaid on a SPOT land-cover image. Our work in the region continues with continuing refinements of the models, new aerial surveys, cartographic analysis, and visualization approaches.

Case study: Development of an archaeological predictive model and GIS database for St. Johns County, Florida

St. Johns County, Florida, sought an archaeological predictive model for use in long-term planning and conservation of the cultural resources in the county. The project was initiated by the St. Johns County Growth Management Services Department, Planning Division, which sought archaeological services to assist in preserving below-ground cultural resources. The scope of work called for "the development and delivery of a reconnaissance-level archaeological survey of St. Johns County, including a predictive model to determine areas of the unincorporated county with high-, medium-, and low-probability of archaeological resources." A contract was awarded to Environmental Services, Inc., of

Jacksonville, Florida, in 2001 to conduct this work. St. Johns County lies in the northeastern part of the Florida peninsula. The total area is some 609 square miles (389,760 acres). The archaeological record shows a continuous 8,000 year human occupation of the area.

Data used

The project used a GIS analysis approach incorporating remote sensing data. A total of 230 prehistoric archaeological sites, including burial mounds, shell middens, habitations, and debris scatters are recorded within the county. There are also 62 historic sites recorded, ranging from the late sixteenth

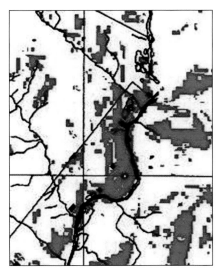

Figure 12.46   GIS predictive model of Gallo-Roman occupation in the study area in France. It is highly likely that the red areas were occupied during the Roman era. The blue dot at center is the site of the destroyed villa complex.

Source: S. Madry.

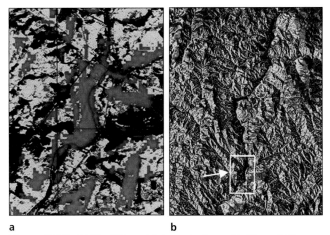

a   **b**

Figure 12.47   (a) The GIS model shown in figure 12.46 overlaid on the SPOT satellite land-use/land-cover image. Yellow areas are fields, green areas are forest. The Arroux River in blue runs top to bottom. (b) A SPOT 20 meter resolution land-use/land-cover classification of the Arroux River valley in France showing location of figure 12.47a. Pasture areas are shown in yellow, forests in green, water in blue, and towns and cities in orange. The river runs top to bottom along the center of the image. This image has been further enhanced with a shading algorithm using a slope map derived from a digital elevation model.

Source: S. Madry.

century through modern times. The archaeological data was acquired in relational database format, including map coordinates, cultural affiliation, size, and other parameters. New sites were recorded through field survey using GPS. A GIS environmental database was created using existing data sources.

Analysis procedures

Five general tasks were undertaken.

1.   A literature review was conducted to locate known sites and gather information to facilitate the development of the GIS database.
2.   Field investigations and verification of known sites and limited field testing was conducted to identify previously unrecorded sites and revisit known sites of interest.
3.   An inventory of known archaeological sites and new sites was developed for the county's GIS system.
4.   Three predictive models in GIS format were developed.
5.   A final report of the GIS data was prepared and delivered.

A GIS-based predictive model for prehistoric and historic archaeological sites in St. Johns County, Florida, was developed using a total of 274 known archaeological sites recorded in the Florida Master Site File (FMSF) database. The GIS database was constructed and univariate statistical analysis was conducted to determine which categories of the environmental data were positive predictors of site presence, neutral, or predictors of site absence. This statistical analysis revealed that five significant variables, modern land cover (derived from remote sensing sources), U.S. Dept. of Agriculture Natural Resource Conservation Service STATSGO digital soil map units, Federal Emergency Management Administration (FEMA) flood zones, distance to perennial streams, and distance to major hydrologic sources were significant predictors for the presence or absence of archaeological sites.

Models of high-, medium-, and low-potential for prehistoric, historic, and all cultural resources were developed *(figure 12.48)*. The technique used was a weighted ranking approach, as presented by Dalla Bona (1994) and Madry and Crumley (1993). The final model accounted for 65.6% of all known sites in 26% of the county considered to be the high-probability zone, medium-probability areas accounted for 30.8% of all sites in 47% of the county, with low-probability accounting for 3.6% of sites in an area of 27% of the county.

The results of these models were then intersected with current (1996) and future (2015) land-use maps to locate and

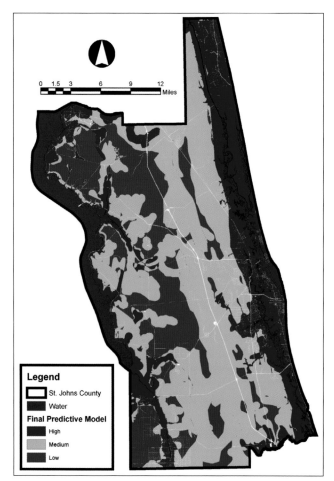

Figure 12.48 GIS predictive model of archaeological site potential in St. Johns County, Florida.

Source: S. Madry.

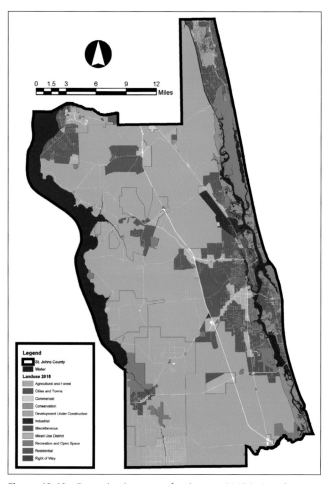

Figure 12.49 Future land-use map for the year 2015 in St. Johns County, Florida.

Source: S. Madry.

predict the potential for undiscovered cultural resources in areas destined for future development. The future land-use map was created by recategorizing the 2015 zoning map into current land-use categories consistent with the 1996 imagery data *(figure 12.49)*. Analysis of this data shows that half of the developing areas will be in the high-probability zone for the presence of archaeological sites and that this area also contains three-fourths of all recorded sites in the county *(figure 12.50)*. These data will be useful for general planning and growth management in the county and will assist in preserving and protecting the county's cultural resources and heritage.

Time frame

This project began with the acquisition of the development of the GIS database, which was created from various available data. This process took one month. New field surveys were conducted for this project, which located nine new sites and revisited five known sites. The interpretation of all datasets began soon after the completion of the GIS database and the incorporation of the archaeological site data. The complete project lasted a total of six months.

Cost

The project was funded by a grant from the state of Florida. Administration of the project was by St. Johns County. The project was financed in part with historic preservation grant

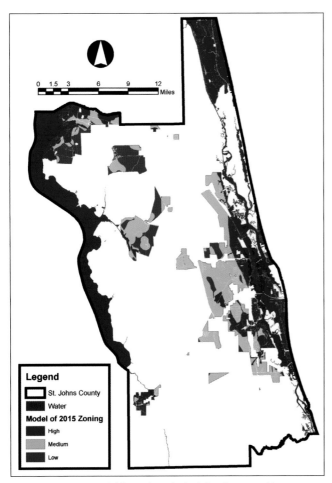

Figure 12.50 Potential fpr archaeological sites in area subject to development by 2015. This map was derived from the archaelogical predictive model and from land-use and zoning data.

Source: S. Madry.

assistance from the Bureau of Historic Preservation Division of Historical Resources, Florida Department of State. The cost of the entire project was approximately $70,000. This also included a GPS survey of all historic structures in the county.

Products

A final report was delivered to the county, as well as all GIS and imagery data used in the project. The three final predictive models that were developed (prehistoric, historic, and all cultural resources) were included, as well as new data showing the location of currently known sites and areas of high probability with relation to current zoning and land use and future development patterns out to the year 2015.

References

Agache, R. 1970. Détection aérienne de vestiges protohistoriques gallo-romans et medievaux dans le bassin de la Somme et ses abordes. *Bulletin de la Société de préhistoire*, special issue, 7 Amiens: Musée de d'Amiens.

Allen K. M. S., S. W. Green, and E. B. W. Zubrow. 1990. *Interpreting space: GIS and archaeology*. London: Taylor and Francis.

Beck. 2003. Satellite applications and landscape archaeology: A case study in the Homs region of Syria. Paper presented at the workshop on the Contribution of Remote Sensing to the Management of Cultural and Natural Heritage Sites. Beruit, Lebanon, 15–17 Dec. 2003.

Chevalier R. 1964. *L'aviation à la decouverte du passé*. Paris : Artheme Fayard.

Crumley, C., and W. Marquardt, eds. 1987. *Regional dynamics: Southern Burgundy from the Iron Age to the present*. San Diego, Calif.: Academic Press.

Dalla Bona, L. 1994. Volume 3: Methodological considerations. A report prepared for the Ontario Ministry of Natural Resources. Thunder Bay, Ontario: Lakehead University, Center for Archaeological Resource Prediction.

Dassie, J. 1976. *Manuael d'archéologie aeriénne*. Paris: Edition Technip.

Deuel, L. 1969. *Flights into yesterday: The story of aerial archaeology*. New York: St. Martins Press.

Ebert, J. I. 1984. Remote sensing applications in archaeology. *Adv. Arch. Methods and Theory* 7:293–362.

El-Baz, F. 1997. Space age archaeology. *Scientific American* 277(2): 40–45.

Farley J. A., W. F. Limp, and J. Lockhart. 1990. The archaeologist's workbench: Integrating remote sensing, EDA and database. In *Interpreting space: GIS and archaeology*, ed. K. M. S. Allen, S. W. Green, and E. B. W. Zubrow, 141–64. London: Taylor and Francis.

Fowler, M. J. F. 1996. High-resolution satellite imagery in archaeological application: A Russian satellite photograph of the Stonehenge region. *Antiquity* 269:667–71.

Gaffney, V., and Z. Stancic. 1991. GIS approaches to regional analysis: A case study of the island of Hvar. Znanstveni institut Filozofske fakultete, Ljubljana.

Hernandez, M. 2003. Conference on the contribution of remote sensing to the management of cultural and natural heritage sites. December 15-17, 2003. Beirut, Lebanon.

Hudak, G., E. Hobbs, A. Brooks, C. Sersland, and C. Phillips. 2002. Mn/Model final report: A predictive model of precontact archaeological site location for the state of Minnesota. Minnesota DOT.

Kunow, J., and G. Joseph, eds. 1995. *Luftbildarchaeologie in Eos-und Mittelwuropa: Aerial Archaeology in Eastern and Central Europe*. Brandenburgishes Landesmuseum Fur Ur-Und Fruhgeschichte. Brandenburg, Germany.

Lock, G., and Z. Stancic, eds. 1995. *Archaeology and geographical information systems*. London: Taylor and Francis.

Lyons, T. R., and J. I. Ebert, eds. 1978. Remote sensing and non-destructive archaeology: Remote Sensing Division, Southwest Cultural Resources Center, National Park Service, Publication No. 36.

Madry, S. 1987. A multiscalar approach to remote sensing. In *Regional dynamics: Burgundian landscapes in historical perspective*, ed. C. Crumley and W. Marquardt. San Diego, Calif.: Academic Press.

Madry, S., and C. Crumley. 1993. Integrating advanced satellite and airborne remote sensing and geographic information systems technologies for regional archaeological settlement pattern analysis in the Arroux River Valley, Burgundy, France. Report submitted to the National Geographic Society.

Madry, S., and C. Crumley. 1990. An application of remote sensing and GIS in a regional archaeological settlement pattern analysis: The Arroux River valley, Burgundy, France. In *Interpreting space: GIS and archaeology*, ed. K. M. S. Allen, S. W. Green, and E. B. W. Zubrow, 364–80. London: Taylor and Francis.

Madry, S., and L. Rakos. 1996. Line-of-sight and cost surface analysis for regional research in the Arroux River Valley. In *New methods, old problems: Geographic information systems in modern archaeological research*, ed. H. D. G. Maschner, 104–126. Center for Archaeological Investigations, Occasional Paper No. 23.

Madry, S., G. Smith, and C. Whitehill. 2001. Archaeological survey and development of a GIS-based site probability model: St. Johns County, Florida. Report submitted to St. Johns County, Florida by Environmental Services, Inc. (ESI Report #249).

Madry, S., G. Smith, S. Seibel, B. Resnick, M. Cole, and S. Gould. 2004. Development of a state-wide archaeological predictive model for the North Carolina Department of Transportation and computerized archaeological database for the NC Office of State Archaeology. Forthcoming in *Analyzing space in time: New directions for GIS in archaeology*, ed. S. Branting. Oxford: Archaeopress.

Maschner, H. 1996. *New methods, old problems: Geographic information systems in modern archaeological research*. Center for Archaeological Investigations, Occasional Paper No. 23.

McCauley J. F., et al. 1986. Palaeodrainages of the Eastern Sahara-The radar rivers revisited (SIR-A/B implications for a mid-tertiary trans-African drainage system). *IEEE Trans Geoscience & Remote Sensing* GE-24:624–48.

McHugh W. P., C. S. Breed, G. G. Schaber, J. F. McCauley, and B. J. Szabo. 1988. Acheulian sites along the "Radar rivers," southern Egyptian Sahara. *Journal of Field Archaeology*. 15: 361–79.

Ogleby, C. L. 2001. The ancient city of Ayutthaya: Explorations in virtual reality and multimedia. *Asian Journal of Geoinformatics* 2(1): 3–9.

Parrington, M. 1983. Remote sensing. *Annual Review of Anthropology* 12: 105–24.

People's daily. Accessed on May 10, 2000. *fpeng.peopledaily.com.cn/200005/10/eng20000510_40473. html*

Web sites

Aerial archaeology work group.
aarg.univie.ac.at

Satellite remote sensing and archaeology.
ourworld.compuserve.com/homepages/mjfff/homepage.htm

Archaeological computing Web site.
www.serve.com/archaeology/comp.html

ESRI archaeology interest group Web site.
www.esri.com/industries/archaeology

GIS and remote sensing for archaeology in Burgundy, France.
www.informatics.org/france/france.html

Minnesota statewide archaeological predictive model (MnModel).
www.mnmodel.dot.state.mn.us

Archaeological predictive modeling: An assessment.
srmwww.gov.bc.ca/risc/o_docs/culture/016/index.htm

Remote sensing in the military

Stan Aronoff and David Swann

Remote sensing has been used for military reconnaissance since the 1860s when cameras were carried aloft by hot air balloons to photograph enemy positions during the American Civil War. In World War I, the use of aerial photography from aircraft was first viewed with skepticism by military authorities but by the end of the War had become an indispensable intelligence source. Photointerpretation expertise developed for wartime was later applied to civilian endeavors, particularly for mapping and resource inventory. But much of the military photointerpretation expertise was lost during the years between the world wars. At the outbreak of World War II, Germany had a well-developed airphoto acquisition and photointerpretation capability that was superior to that of Allied forces. During the war, rapid advances were made by Allied forces in the use of aerial photography for such applications as mapping, analysis of terrain trafficability and water depth in shallow coastal waters, bomb damage assessment, assessment of the position and capabilities of enemy forces, and the identification of facilities and materiel related to an adversary's fighting capability.

After World War II, the development of nuclear weapons and intercontinental ballistic missiles ushered in the cold war period. The openness of Western societies made it much easier to gather useful intelligence about them from open sources than from the tightly controlled Soviet bloc countries. The United States and its allies depended heavily on the rapid development of ever more capable remote sensing systems and new technologies to meet critical intelligence needs, especially the assessment of Soviet nuclear capability and military intentions. During the 1950s and 1960s, radar, thermal-infrared imaging, and satellite photography were developed in secret. In the interests of national security, military remote sensing was isolated from the civilian remote sensing world.

Ever more sophisticated remote sensing systems continue to be developed for military applications and their capabilities remain classified. However, with the end of the cold war and the introduction of commercial satellites offering spatial resolutions better than 2 m, publicly available satellite imagery became an important resource for military planners. As well, the high cost of specialized airborne and satellite military reconnaissance resources has made the use of lower cost commercial satellite imagery wherever suitable an important

means of controlling escalating budgets. (A more detailed discussion of the historical context of military applications of remote sensing can be found in chapter 2.)

The nature of military intelligence

Governments engage in a broad range of intelligence gathering activities that are withheld from public view. They range from information collected to guide the formulation of international policy to detailed information on a nation's military capabilities or the disposition of forces on an active battlefield. Much of the information is obtained by the systematic analysis of open sources such as foreign newspapers, military and scientific journals, radio and television broadcasts, statistical compilations, publicly available maps, and other geospatial data including airborne and satellite imagery. Critical information unobtainable by other means may be obtained covertly through the use of informants or undercover agents to directly infiltrate an organization or by technical means such as the interception of communications or the collection and analysis of remotely sensed imagery.

The efficient and effective use of intelligence data and the setting of priorities for further data collection depend on understanding the interrelationships and interdependence of the information obtained from independent sources as much if not more than correctly assessing individual items of information. To the extent that these analyses provide information on a nation's military activities, capabilities, and intentions, they can provide information valuable in anticipating, planning, and executing military operations.

Remotely sensed imagery is used to collect a broad range of strategic, operational, and tactical[3] information of value in anticipating the actions of potential adversaries and in the planning and execution of military operations. They range from current information on battlefield activities during a conflict (called *situation awareness*), information collected during peacetime to monitor potentially hostile forces and to plan future military actions, evidence gathered to monitor

---

3. Military activity is commonly classified into three hierarchical levels that differ in scope and time frame. *Strategic planning* involves the formulation of broad policy and military plans at the national and international levels. In *operational planning*, a series of specific military objectives are defined that will contribute to one or more strategic goals. At the *tactical level*, the specific details of an operation are defined such as the specific area, objectives, plan of action, and the units to be involved. For example, in wartime, a strategic objective might be to cut off an enemy's outside supply of critical resources. An operational plan might identify disruption of oil imports by destroying pipelines and port facilities as one of the means to advance that goal. A tactical level plan might involve specific actions to destroy the oil handling facilities of a specific port.

Table 12.3 Military information commonly derived from the analysis of remotely sensed data	
Application	**Specific uses**
Mapping, charting, and geodesy	Production of orthorectified image maps Production of digital elevation data Terrain characterization and trafficability analysis[1] Map production Change detection
Broad area search	Automated change detection Cueing support
Disaster support	Natural-disaster assessment Man-made disaster assessment
Strategic industries and resources	Monitoring of sites and facilities Natural resource exploration/mining POL facilities Industrial material process flow Seaport usage Power supply Underground facilities Chemical and biological weapons production Assessment of transportation and communication networks
Contingency planning support	Intelligence preparation of the battlefield Beach and landing zone analysis[2] Water depth and characteristics[3] Amphibious operations planning Airfield analysis Noncombatant evacuation operations Environmental hazards
Mission planning and rehearsal	Large area orientation Operations planning Mission rehearsal fly-through Mission assessment
Current operations support	Order of battle analyses (naval, ground, air-defenses) Detection of chemical and biological weapons Theater surveillance
Targeting support	Target detection Target identification and tracking Target vulnerability characterization Target penetration analysis Bomb damage assessment Camouflage, concealment, and deception detection Cruise missile targeting
Counter-narcotics	Identification of the growing, processing, and shipment of narcotics
Treaty monitoring	START (Strategic Arms Reduction Treaty) Detection of chemical and biological weapons Environmental treaty monitoring Nuclear weapons proliferation
Counter-terrorism	Counter-terrorism operations support

[1] Terrain analysis—Assessment of a geographic area to determine the effect of natural and manmade features on military operations, including cover and concealment, obstacles, terrain composition and moisture characteristics, avenues of approach, and trafficability. Trafficability—Determination of the type and characteristics of land and water features over which personnel and equipment will travel.
[2] Beach and landing zone analysis—Determination of terrain composition, slope, soil, and foliage in support of aircraft operations, ground equipment use, personnel/equipment movement, and amphibious operations.
[3] Water depth and characteristics—Determination of the bathymetric, thermal, salinity, turbidity or turbulence characteristics of a body of water.

Source: Adapted from Anderson and Malila et al. 1994, and information from the Topographic Information Center, U.S. Army.

adherence to arms limitation treaties, as well as broad strategic information to guide policy such as evaluations of a nation's strategic resources including food crops, energy, transportation, and communication networks. Much of this broader level of intelligence gathering makes extensive use of geographic analysis, and it is not surprising that intelligence organizations make extensive use of geographers as analysts. Table 1 lists the types of military intelligence information commonly obtained from the analysis of remotely sensed data.

Current military remote sensing systems

Details of the specialized sensor systems developed for military intelligence applications remain closely guarded secrets. With the end of the cold war, the declassification and sale to the public of early military intelligence imagery collected by the United States and the former Soviet Union, and the availability of commercial high-resolution satellite imagery (1 m GSD spatial resolution and better), the level of secrecy surrounding military remote sensing has been somewhat relaxed. Information about these systems has made its way into the public domain through leaked or captured information. Degraded imagery from these sensors (often referred to as information obtained by "national technical means") is sometimes released to support military or political objectives such as to illustrate breaches of arms control agreements or

to report on the progress of military interventions *(figure 12.51)*. (In fact, such classified intelligence satellites were key to enabling arms control agreements to be negotiated during this time. Without on-site inspections, it was essential for the United States to have confidence in its ability to monitor compliance with such agreements using "national technical means.") In addition, manufacturers of remote sensing systems used by the military provide limited information about their products. From these sources and knowledge of the remote sensing technologies involved, estimates of the capabilities of these sensors have been made and published in open literature. There are even Web sites devoted to the collection and public dissemination of information about military resources including remote sensing systems.

Military remote sensing systems are typically operated from satellites, specialized aircraft designed to fly at high altitudes such as the U-2[4] *(figure 12.52)*, modified fighter aircraft that fly at high speed and low altitude to evade radar detection and interception, and unmanned aerial vehicles (UAVs), which are operated remotely and can transmit the imagery they acquire in real time. For example, the Predator is an armed mid-range UAV that can operate at altitudes as high as 7.6 km (25,000 ft) for 40 hours (figure 8.40 in chapter 8) and the Global Hawk, an unarmed long-range UAV capable of 22,000 m (65,000 ft) altitudes and a range of 2,200 km *(figure 12.53)*.

Virtually every type of remote sensing system has been used for military applications. Digital and film cameras, multispectral scanners, thermal-infrared scanners and radar are

---

4. The U-2 aircraft in current use were built in the 1980s and upgraded in the 1990s. They are some 40% larger than the original version developed in the late 1950s. They can carry electro-optical/infrared scanners, radar, and film cameras, as well as the electronics to manage high bandwidth data links and collect radio signal traffic.

Figure 12.51   Degraded image from a U.S. KH-11 military satellite. This degraded image acquired by an advanced KH-11 satellite was used in a Pentagon press briefing on December 19, 1998. The arrows show two areas where the Secretariat Presidential was damaged by air strikes during Operation Desert Fox in Iraq.

Source: Image courtesy of the U.S. Department of Defense.

Figure 12.52   American U-2 high-altitude reconnaissance aircraft. The fleet of over 30 U-2 aircraft in current use was built in the 1980s and upgraded in the 1990s.

Source: Image courtesy of USAF Air Combat Command.

Figure 12.53 The Global Hawk is a jet-powered UAV capable of operating at altitudes of about 22,000 m (65,000 ft) with a range of 2,200 km and on-station endurance of 24 hours. It can be fitted with optical, infrared, and radar sensors that relay its imagery to ground receiving stations directly or via satellite relay.

Source: U.S. Air Force photo by G. Rolhmaller.

the systems most widely reported. Examples of these systems have been discussed previously in chapter 2 and chapter 8. Most military applications require imagery with high spatial resolutions on the order of 1m GSD or better. However, coarser resolution imagery can be valuable in detecting changes over large areas and selecting locations that merit more detailed imaging. Also, broad strategic-level analyses can make use of lower resolution imagery for such analyses as the assessment of a nation's projected food production.

The value of imagery for military reconnaissance depends not just on its spatial resolution. It also depends on such factors as the timeliness of the data, the availability of alternate sources, and the reliability of the information that can be derived.

Sensors differ in the type of information they can provide, how soon they can collect images of a target area, and the conditions under which they can operate. For example, radar imagery can be acquired day or night and in most kinds of weather, but it does not provide imagery with as high spatial resolution as an optical system flown at the same altitude. Optical imagery generally requires cloud-free daylight conditions but can provide higher spatial resolutions and spectral information useful in target identification. Satellites provide a more stable platform than airborne systems that are buffeted by air currents. As a result, they can produce sharp images at greater magnifications but must operate at higher altitudes. Satellites can over-fly hostile territories unchallenged and acquire large quantities of imagery rapidly. They

can repeatedly monitor an area at relatively low cost. But a satellite's orbital path limits the times when it can acquire imagery of a specific location, and it may not be able to provide urgently needed imagery soon enough to allow an effective response.

For example, during the Iraq conflict in 2003, commercial satellite imagery of the area was widely used in the press. However, the satellites that provided this imagery can only collect data over Iraq once during a day for 10 to 15 minutes as the satellite passes over the country in daylight hours. The next pass does not occur for several days. Military satellites, while capable of providing higher resolution imagery, are still restricted by their orbital paths.

Airborne sensors can be dispatched rapidly and repeatedly to acquire imagery of a target area, if they are within range, but they may run the risk of being intercepted and destroyed. During the Iraq conflict, air defenses were systematically destroyed at the start of the conflict, which allowed airborne reconnaissance relatively safe access to the entire country.

Coalition forces obtained frequent if not continuous coverage of the country using a combination of high-altitude U-2 aircraft, UAV, and low-flying American and British tactical reconnaissance aircraft to generate the imagery needed for surveillance, intelligence gathering, and damage assessment. During the day, imagery was produced mainly by electro-optical sensors (scanners) supplemented by high-resolution photography generated by film cameras. At night, lower resolution radar and thermal-infrared sensors carried by aircraft and UAV provided image coverage.

Imagery could be transmitted from reconnaissance aircraft to ground stations by direct line-of-sight data links at shorter ranges and via communications satellites at longer ranges. Commanders at headquarters and in the field were provided with a continuous flow of airborne imagery in support of military operations. Change detection and analysis was of particular value in detecting the movement of vehicles and military formations to target and attack these military assets, often using precision guided munitions.

The war in Iraq saw the first large-scale deployment of the Joint Surveillance and Target Radar system (J_STARS) *(figure 12.54)*. Mounted in a remanufactured Boeing 707-300 commercial aircraft, the surveillance radar carried by these systems can be operated as an imaging SAR system or as a moving target indicator. Powerful computers handle the remote sensing system, as well as a command and control center. The surveillance radar is capable of determining the

Figure 12.54 J-STAR E-8. The Joint Surveillance and Target Radar System (J-STARS) consists of a command and control system and a surveillance radar system mounted on a remanufactured Boeing 707-300 commercial aircraft. It can operate at altitudes of over 12,000 m (40,000 ft) with an endurance of 11 hours without a refuel or 20 hours with in-flight refueling.

Source: Image courtesy of the USAF 116th Air Control Wing.

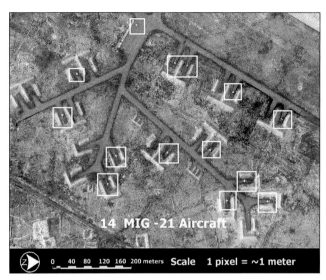

Figure 12.55 IKONOS commercial satellite imagery of Bagram Airbase, Afghanistan. This 1m GSD pan-sharpened image, acquired August 13, 2001, illustrates the value of high-resolution commercial satellite imagery for military intelligence. A skilled interpreter can readily identify the 14 aircraft visible in this image as Mig-21 fighters.

Source: Image courtesy of Space Imaging Corporation.

location, direction, speed, and activity of vehicles moving on the ground at ranges up to 240 km.

Satellite imagery undoubtedly supplemented imagery acquired from airborne platforms. However, satellite imagery is of more limited use in a battlefield situation. There were relatively few satellites in operation: six U.S. military satellites (three radar and three optical imagers operated by the U.S. National Reconnaissance Office) and the three commercial satellites (IKONOS, EROS-A1, and QuickBird). These polar-orbiting satellites will over-fly the country of Iraq in about two minutes once in daylight and once in darkness every two or three days. With this restricted acquisition capability, they could not provide the flexibility or continuous image coverage that the coalition's fleet of reconnaissance aircraft could provide, but the high-resolution satellite imagery was probably a useful addition.

Virtually every military organization with access to aircraft has some form of airborne image reconnaissance capability. However, until the mid-1980s with the launch of the SPOT satellite by France, only the United States and the Soviet Union possessed the technology and had the means to launch satellites and were the only sources of satellite imagery for military intelligence. The technology has since become widely available, and satellite launch services are available from several competing sources. With the availability of high-resolution satellite imagery from commercial operators, the distinction between military and civilian satellite imaging has blurred. Considerable military intelligence information can be obtained from commercial satellite imagery with spatial resolutions of about 1m GSD or finer *(figure 12.55)*.

For many nations, high resolution commercial satellite imagery is the highest resolution imagery available to them to meet their national security intelligence needs. Even for those nations that operate classified military systems, the use of commercial high-resolution satellite imagery wherever possible is a cost-effective alternative to building additional (very costly) military satellite systems.

In addition, the availability of multiple satellites increases the likelihood that one will be in the right position to acquire imagery of a specific location on short notice. By negotiating preferential access to commercial satellite imagery, government organizations can increase their information gathering capability in times of national emergency at considerably lower cost than operating additional satellites. It is not surprising that government military and intelligence organizations are major customers for the high-resolution commercial satellite industry

High-resolution satellite imagery is available commercially from several sources. IKONOS-2 and OrbView-3 produce 1m GSD panchromatic and 4 m GSD multispectral imagery. QuickBird-2 produces imagery with nominal spatial resolutions of 0.7 m panchromatic and 2.8 m multispectral and a follow-on satellite under development will provide imagery with spatial resolutions as fine as 0.5 m GSD.

The EROS-A1, providing panchromatic imagery with 1.8 m or 0.9 m GSD, is operated by ImageSat International. This commercial operator is unique in offering clients licenses that grant periods of exclusive and confidential control of the image acquisition area and direct download of the data. In effect, a client is able to rent the satellite for specified time periods.

Many countries operate or are in the process of developing classified intelligence gathering satellite programs. They include China, France, Germany, India, Israel, Italy, Japan, Russia, South Korea, Taiwan, the United Kingdom, and the United States. The United States has operated military reconnaissance satellites since 1960. The Keyhole satellite series, bearing the KH designation, carries optical sensors. The KH-11 and KH-12 satellites are fitted with very long focal length lenses, similar to that carried by the Hubble space telescope but pointing to earth instead of the stars. These sensors are purported to achieve spatial resolutions as high as 10 cm GSD. The Lacrosse satellite series carry synthetic aperture radar (SAR) sensors providing all-weather day or night imagery with spatial resolutions reportedly as high as 1 m GSD.

The former Soviet Union and now Russia have flown military intelligence satellites since the 1960s, but there is little reliable published information on its current military intelligence satellite programs.

Information technology and military intelligence imaging

Acquisition of suitable imagery in a timely manner, though essential, is but one part of the process by which useful intelligence information is generated. Military reconnaissance is an art as much as it is a science. It typically involves deducing specific types of information using a variety of image and nonimage sources within a short enough time frame for the data to still be of value. That is, it is fundamentally a process of integrating multiple geospatial datasets as rapidly, efficiently, and accurately as possible. In addition to providing the appropriate imagery, for imagery to be of intelligence value, it is essential that there be personnel with the technical expertise to analyze the data, efficient access to other relevant data sources, the tools to integrate the information, and the means to rapidly disseminate the results. Information technology has revolutionized these processes.

The introduction of information technology has precipitated a revolution in military affairs. Defense structures, processes, and organizations are being reengineered to make ever more extensive use of information technologies, an approach that has been termed *network-centric warfare* (NCW). Advances in computer processing power, data storage, and network-data transfer capacities (i.e., network bandwidth)

have provided the means to collect, analyze, and disseminate information, including remotely sensed imagery and information products, more rapidly than ever before. The development of geospatial data handling technology has enabled the development of common data infrastructures capable of storing and manipulating multiple geospatial data formats including raster image and vector data.

The use of remote sensing for military reconnaissance and intelligence gathering has traditionally been mission oriented. Specific information needs were identified by units within the defense organization, image requirements were defined, and the necessary mission activities were carried out to obtain and analyze the imagery. The resulting information was then delivered to those users designated to have the need for the information and the necessary level of security clearance to be entitled to access it.

Image analysis procedures tended to be narrowly focused, in part limited by the nature of the supporting technology (initially analog devices such as stereo viewing instruments and then early information technology that made the integrated analysis of multiple datasets difficult). As a result, analysis procedures tended towards a linear production model where image frames would be input and a focused analysis of the images would be output. The critical decision-support function that this operation was to support was one step removed from the analysis process—the analysis process came to be regarded as the goal rather than as a step that supports the real need—the decision.

The paradigm of an image as a stand-alone picture tends to focus resources on the *collection* process rather than on the dissemination and analysis processes critical to support decision making. This was the basis of the collection-centric criticism the U.S. intelligence community garnered by a commission tasked to review the U.S. National Imagery and Mapping Agency (now the U.S. National Geospatial-Intelligence Agency or NGA) (Marino 2000). The commission found the intelligence community was focused on the development and operation of sophisticated technical collection systems such as reconnaissance satellites, and "only as an afterthought," preparing to properly task the systems and to process, exploit, and disseminate the collected products[5].

---

5. Page viii of *The information edge: Imagery intelligence and geospatial information in an evolving national security environment.* Report of the Independent Commission on the National Imagery and Mapping Agency, December 2000. Washington, D.C.

Image analysis has always made extensive use of information from other sources. However, information technology such as geographic information systems makes possible the integrated use of image analysis and its information products in a larger context—that of an overall decision support information system in which many other sources of useful geospatial information can be accessed and included in the image analysis. Conversely, integrated geospatial analyses, as undertaken using a GIS, can include image data. In order to develop such a decision-support information system, the imagery data must itself first be organized and encoded in a consistent and cartographically accurate form. It must be structured as an imagery infrastructure.

Imagery infrastructure

In information technology terms, an image without georeferencing (i.e., an image not encoded so as to obtain an accurate location coordinate for each pixel) is a document. Early image exploitation systems could transform an individual pixel from internal to external coordinate space by applying ephemeris, camera model, and other orientation information, but that was still clever manipulation of a document. The procedure was useful in calculating the geographic position of specific points within the image—useful for such applications as targeting—but was not actually georeferencing every location in the image to a defined level of accuracy.

With an image infrastructure, each and every pixel is georeferenced to a consistent, continuous global spatial reference system. As a result, users can not only retrieve a geographic coordinate for any pixel location, but they can also obtain information from any other georeferenced datasets accessible to the information system such as elevation, soil type, terrain trafficability by vehicle type, load capacities of bridges, or current rainfall. The ability to query multiple geospatial datasets and use them together in an integrated analysis is a well-developed function of geographic information systems. Users, be they image analysts or decision makers, can assess an image within a broader context of corroborating data to improve the decision-making process

In addition, imagery of different types or with different spatial resolutions for the same area can be retrieved and viewed together for comparison. A series of images with progressively higher spatial resolutions can be structured to allow interactive zoom with smooth transitions to higher resolution images as magnification is increased *(figure 12.56)*.

The high position accuracy of orthorectified image products has led to their extensive use, with or without vector overlays,

as location maps, such as in street maps. Where imagery can be provided in near-real time, the image data layer may be visually interpreted directly by the user to monitor the status of rapidly changing conditions.

By choosing to implement image information within a consistent infrastructure that conforms to information technology infrastructure standards, off-the-shelf database management system (DBMS) technology can be used, and data can be disseminated using Web services over standard public or secure (private) networks. The data can then be exploited using a range of server technologies to make different levels of retrieval and analysis capabilities available to individual users, be they strategic planners or commanders in the field.

Improvements being sought for military intelligence applications of remotely sensed imagery include the following:

- Rapid image acquisition. During the cold war, the acquisition, delivery, and analysis of satellite or airborne reconnaissance imagery could take days or weeks. Current remote sensing systems can provide much more rapid turnaround on the order of hours or minutes. Some systems are capable of providing real-time imaging to remote receiving locations.

- Broadened availability. Traditional military reconnaissance operations placed imagery in the control of small specialized groups operating under tight security. Emphasis is now being placed on making imagery a more widely disseminated information source available to multiple levels of personnel over secure computer networks and multiple security access levels. The concept of tightly connecting *sensor to shooter* is a key deliverable of network-centric warfare, with a goal of less than 10 minutes from sensing to effect.

- Advanced image exploitation tools. More intelligent analysis tools are being designed to enable decision-support staff to derive useful information from remotely sensed imagery, though sophisticated analyses will still require the abilities of expert image analysts. (An illustration of systematic image-analysis procedures for military and business intelligence is illustrated in the subsequent section of this chapter, *Remote sensing in geospatial intelligence analysis*.)

These changes are recasting the role of advanced intelligence gathering sensor systems and the expert image analyst to serve wider mandates within military organizations. Not only can end-users be provided with imagery, but they can also be provided the means to task or directly control remotely sensed image acquisition. Machine-to-machine

Figure 12.56   Imagery held within a georeferenced image infrastructure can be interactively zoomed to show increasing levels of detail and introduce additional data sources. This sequence of screenshots (running counterclockwise from top right) illustrates a continuous zoom from global-scale imagery to high-resolution aerial imagery overlain with extruded vector data.

Source: (top left) Image courtesy of I-Cubed; (top center) Image courtesy of I-Cubed and Earth-Sat Corporation; (bottom row) Data courtesy of the City and County of Honolulu.

interfaces can assist in providing this extended control and reduce human error.

Integrated control systems for military intelligence data collection

The more current and complete a commander's knowledge of a theater of operation (i.e., situational awareness), the more effectively a commander can carry out the mission. Rapid communication of information needs and delivery of information products can offer significant military advantage. The process may involve tasking a database to retrieve existing imagery and related data or tasking a sensor to acquire new imagery. The process by which the information is provided is fundamentally irrelevant to the user. Only the speed of delivery, suitability for use, and the quality of the content are of concern. Integrated military information control systems are being developed to provide end-users with greater control of the acquisition, analysis, and dissemination of image intelligence.

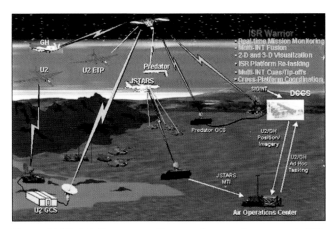

Figure 12.57　Intelligence surveillance and reconnaissance assets (ISR) are independent systems that can be more effectively used under a central control system.

Source: Image courtesy of Raytheon Company.

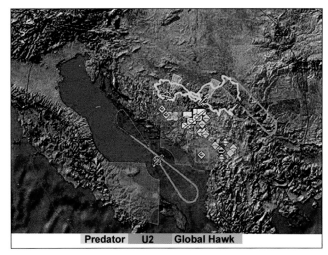

Figure 12.58　Intelligence assets control system. Battle space display from the ISR Warrior control system.

Source: Image courtesy of Raytheon Company.

Figure 12.57 illustrates some of the intelligence gathering resources that might be used to support military action: UAV such as the Predator and Global Hawk and manned reconnaissance aircraft such as the high-altitude U-2. Joint Surveillance and Target Attack Radar Systems (JSTARS) are equipped with radar systems for surveillance and target acquisition and a command and control center. Also, commercial satellites can be tasked to acquire imagery of selected areas to be used directly or to guide the tasking of other airborne and ground-based sensors.

To make effective use of such diverse information collection resources requires some form of central control (i.e., horizontal integration) to coordinate information requirements, data collection, analysis, and dissemination. The Distributed Common Ground System (DCGS) being developed for the Air Force is an example of one system being developed to fill this need.

Figure 12.58 is a pictorial representation of a simulated battle space being managed by a central control system designed to provide real-time monitoring and visuals of data collection missions and analysis tools for the operator to rapidly exploit the intelligence information being generated. The system can provide a consolidated picture of the theater of operation, order of planned military actions, information collection plans, and planned targets along with tip-off information from Signals Intelligence (SIGNIT), and Moving Target Indicator (MTI) sources. Perspective views can be generated to assist the operator in developing an integrated understanding of such information as weather, terrain limitations, and threat domes (three-dimensional zones of vulnerability to enemy attack).

In Iraq, the Theater Battle Management Core System (TBM-CS) is a Web-based system used for planning, managing, and executing the air war. Some 50 computer programs track information on targets, weapons, fuel loads, weather, and navigation. The system makes use of data from manned and unmanned surveillance aircraft to share tactical information and deploy offensive and defensive forces and provide targeting information to precision guided munitions. Such systems enable the very high volumes of geographic data generated by airborne and satellite-based sensors to be processed and acted upon quickly.

These technologies are being applied both for homeland security as well as for battlefield situation awareness and planning. With broadband communications technology enabling imagery to be transmitted anywhere in the world in real time, remotely sensed imagery has evolved from being an independent source of information to being an integral part of a continuously updated decision-support system[6].

6. Current technology makes possible the rapid collection and management of large volumes of data. Yet the concept is not as new as it seems. The Soviet General Staff in the late 1970s reportedly thought of such an information capability, which they designated "a reconnaissance-strike complex."

References and further reading

Anderson, J. R., W. Malila, R. Maxwell, and L. Reed. 1994. Military utility of multispectral and hyperspectral sensors. Unclassified ERIM Report No. 246890-3-F. Ann Arbor, Michigan: General Dynamics Advanced Information Systems Ann Arbor Research and Development Center. www.iriacenter.org/iriaplib.nsf/0/266297acf6e954cd85256c320067df87?OpenDocument

Anderson, J. R., E. E. Hardy, J. T. Roach, and R. E. Witmer. 1994. A land use and land cover classification system for use with remote sensor data. Professional paper no. P0964. Reston Va.: U.S. Geological Survey.

Avery, T. E. 1977. *Interpretation of aerial photographs*. Minneapolis: Minn.: Burgess Publishing Company.

Burrows, W. E. 1988. *Deep black*. New York: Berkley Publishing Group

Clark, E. 2001. Military reconnaissance satellites (IMINT). Center for Defense Information. Washington, D.C. www.cdi.org/terrorism/satellites-pr.cfm

Colwell, R. N. 1997. History and place of photographic interpretation. In *Manual of photographic interpretation*. 2nd ed., ed. W. R. Philipson. Bethesda, Md.: American Society for Photogrammetry and Remote Sensing.

Colwell, R. N., ed. 1960. *Manual of photographic interpretation*. Falls Church, Va.: American Society of Photogrammetry.

Hays, P. 2001. Transparency, stability, and deception: Military implications of commercial high-resolution imaging satellites in theory and practice. Presentation at the International Studies Convention, Chicago.

Jones, A., and P. Marino. 2000. Report of the defense science board task force on national imagery and mapping agency. Defense Science Board.

Marino, Peter. 2000. The information edge: Imagery intelligence and geospatial information in an evolving national security environment. Report of the Independent Commission on the National Imagery and Mapping Agency, December 2000). Washington, D.C. www.nima.mil/StaticFiles/Acquisition/nima_commission.pdf

Owens, B. 2001. *Lifting the fog of war*. Baltimore: The John Hopkins University Press.

Petrie, G. 2004. High-resolution imaging from space a worldwide survey. *GeoInformatics* Part I–North America 7(1): 22–27. Part II–Asia 7(2): 22–27. Part III–Europe 7(3): 38–43. web.geog.gla.ac.uk/~gpetrie/pubs.htm

———. 2003. Iraq: Winning the war; reconstructing the peace. *GI News* 3(4): 35–40. web.geog.gla.ac.uk/~gpetrie/pubs.htm

———. 2003. Monitoring Iraq: Imagery options for monitoring and gathering intelligence. *GI News* 3(1): 28–33. web.geog.gla.ac.uk/~gpetrie/pubs.htm

———. 2001. Imagery for surveillance and intelligence over Afghanistan. *GI News* 2(3): 28–34. web.geog.gla.ac.uk/~gpetrie/pubs.htm

Richelson, J. T. 1999. *U.S. satellite imagery, 1960–1999*. National Security Archive Electronic Briefing Book No. 13. Washington, D.C.: George Washington University. www.gwu.edu/~nsarchiv/NSAEBB/NSAEBB13

Van Creveld, M. 1991. *The transformation of war*. New York: The Free Press.

Web sites

Federation of American Scientists. Imagery Intelligence.
www.fas.org/irp/imint/index.html

Global Security.Org.
www.globalsecurity.com

Intelligence Operations. Air University Library and Press. Maxwell Air Force Base.
www.au.af.mil/au/aul/school/acsc/intel.htm

National Security Archive. George Washington University. Washington, D.C.
www.gwu.edu/~nsarchiv

Spy Satellites.
danshistory.com/spysats.shtml

U.S. Army Topographic Engineering Center.
www.tec.army.mil/tio

U.S. National Geospatial Intelligence Agency. Historical Imagery Declassification (HID): Historical Background. Washington, D.C.
www.nima.mil/portal/site/nga01

Remote sensing in geospatial intelligence analysis

Tom Last

Prior to World War II, in 1938, the Chief of the German General Staff, General Werner Von Fritsch said, "The nation with the best photoreconnaissance will win the next war." These prophetic words are as true today as they were during World War II. The collection and effective use of remote sensing data for military operational purposes has been and will continue to be the difference between victory and defeat.

Since the mid-nineteenth century, there have been countless examples of how remote sensing data has been effectively used within a combat environment. During World War I, Allied and German forces realized the value of photographic reconnaissance. Initially used to identify troop deployment and supply depots, aerial photography later served to identify enemy artillery placements and more accurately direct artillery fire to destroy them. Air photo reconnaissance further expanded, both in collection ability and interpretation skills, during World War II *(figure 12.59)*. It was during this time that strategic imagery reconnaissance took place at a very large scale. Not only were enemy military forces observed and detected, but an enemy nation's complete industrial ability to wage war was also interpreted and analyzed using remote sensing imagery in the form of aerial photographs. Aerial photography became indispensable for planning and executing military operations and assessing results. Since that time, the range of sensor systems and analysis tools has greatly expanded to include radar, thermal infrared, multispectral and hyperspectral scanners, and sophisticated analytical tools have been developed using state-of-the-art image processing, photogrammetry, and GIS. Yet visual interpretation skills remain essential to extract useful military and industrial intelligence from remotely sensed imagery.

Military intelligence

There are various definitions of intelligence in the military context. For the purpose of this section, the following intelligence definition is used:

Intelligence is the product resulting from the processing of information concerning foreign nations, hostile or potentially hostile forces or elements, or areas of actual or potential operations. The term is also applied to the activity which results in the product and to the organizations engaged in such activity.

Figure 12.59   World War II air photo reconnaissance. Elements of the Seventh Canadian Infantry Brigade moving off Juno Beach on the coast of France during the D-Day invasion, June 6, 1944.

Source: Analysis © ImStrat Corporation 2003.

The principal goal of military intelligence is to determine the status and intention of an adversary through the use of various information collection methods, remote sensing being one of them. There are normally three levels of intelligence responsibility, each having a unique scope and point of view. They are

- strategic intelligence, which is primarily concerned with supporting defense planning at the national level;
- operational intelligence, which is intelligence of a regional nature in support of regionally deployed military units; and
- tactical (combat) intelligence, which is intelligence to support a specific active military operation.

Intelligence reports generally address all three levels of intelligence, and remotely sensed imagery is analyzed to extract information at all three levels as well.

Critical to the value of intelligence products is the timely delivery of the right information to the right people. The effectiveness of intelligence operations is thus as dependent on the organization and expertise of the personnel as it is on the capabilities of the intelligence gathering technologies.

The overall objective of the intelligence process is to provide military or political decision makers the most accurate and complete understanding of a crisis or combat situation, termed *situational awareness*, so they can make well-informed decisions.

For example, knowing and understanding the intention, geolocation, equipment types, and capabilities of an adversary, a commander can better choose the types of combat forces, munitions, and other resources to achieve the maximum desired effect using the least resources. Situational awareness of the battlefield is a key component of capturing and maintaining battlefield dominance.

The intelligence cycle

The intelligence cycle is a systematic procedure designed to provide a continuous flow of intelligence products to maintain situational awareness of a battlefield or other theater of activity. It is a sequence of activities where information is obtained, assembled, converted into intelligence, and made available to users *(figure 12.60)*. The cycle is commonly divided into four phases, each comprising a detailed set of activities:

1. Direction: To meet specific intelligence requests, information requirements are defined and sources identified, including remote sensing resources.
2. Collection: Information is gathered and delivered to the appropriate units.
3. Processing: The diverse information collected is sorted, evaluated, and grouped for interpretation and integrated analysis. Products are generated to satisfy the intelligence requests.
4. Dissemination: Intelligence products are distributed in a useful form and by a suitable means to all those who require them.

While a specific request is processed sequentially, in generating a constant flow of intelligence reports these phases operate concurrently. As new information is being collected, existing information is being processed, and intelligence products are being disseminated, resulting in the identification of additional information needs, which lead to further direction.

Intelligence is a perishable commodity. Intelligence products delivered too late, to the wrong personnel, or in a form that is misunderstood, are of little value or, at worst, lead to incorrect decisions. Intelligence reporting is designed to be

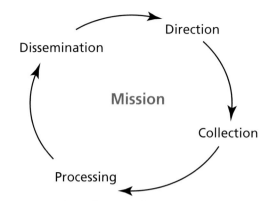

Figure 12.60   The intelligence cycle.

Source: © ImStrat Corporation 2003.

clear and concise in a standardized format, regularly updated and rapidly disseminated in a secure manner to all those who should use it.

Remote sensing data and intelligence collection

Military image intelligence analysts use imagery from virtually every type of airborne and satellite remote sensing system. Sensors used by the military are typically classified, although commercially available imagery, particularly high-resolution satellite imagery, is used extensively. Radar, thermal infrared, multispectral and hyperspectral, and conventional aerial photography are the most commonly used sensors. Information extracted by visual interpretation, as well as digital image analysis, is integrated or *fused*[7] with other sources of information to produce imagery intelligence products.

The principal military applications of remotely sensed data are the assessment of adversary activity and topographic analysis. Combat operations are rapidly changing fluid activities demanding rapid processing and delivery of intelligence. Image analysts require the fundamental skills of visual interpretation, experience, and a comprehensive knowledge of military equipment, tactics, and the organizational structure of military forces.

A skilled analyst derives much of the militarily valuable information in an image by inferences made about the features identified. For example, an analyst may identify 10 battle

---

7. *Data fusion* in a military context refers to the compilation and integration of information about a particular target or area of interest. It does not refer to the electronic merging of remote sensing data.

tanks parked along a tree line next to a road. Being able to recognize equipment types, the analyst can establish whether it is a friendly or adversary force. With knowledge of organizational structure, it can be inferred that the group is a tank company or squadron. If tank tracks were visible over an area used for maneuvers, analysis of the track patterns and knowledge of the tactics used by different military forces could be used to infer the type of training the tank crews had received and the tactics they would be likely to use in combat. The following example uses satellite imagery of a military training exercise to illustrate the process.

Case study: Military intelligence scenarios

The following image analysis of a military exercise at the Wainwright military training base, Alberta, Canada, illustrates the type of military information that can be derived from commercially available 1 m spatial resolution satellite imagery, in this case from IKONOS. Gathering intelligence on a military exercise might be done to verify and assess the training activity and project future activities and intentions.

An intelligence unit might be directed to address the following tasks:

- Confirm or deny that military training is occurring at the location
- Identify the military size of the training exercise, e.g., battalion, brigade group, or division
- Identify the military equipment in use
- Determine what activity, if any, is occurring which can further assist in determining what future activity could take place, (i.e., intention)

Visual image interpretation can identify key military equipment and military parking and convoy formations. Other sources of information on how military formations operate can then assist in determining what activity is occurring and infer possible intentions.

In figure 12.61 three areas of military vehicle concentration were identified on the extreme western edge of the training area near the Battle River. They are

- vehicle convoy and combat engineer support unit,
- armor combat team,
- parked soft-skin vehicles (trucks).

Figure 12.62 is an enlargement of the northwest corner of the imagery. A convoy of military pattern trucks and possible wheeled armored vehicles were identified on a primary gravel road, facing south. Parallel to the primary gravel road was a combat engineer support unit in a convoy formation. The key indicators of this being a combat engineer support unit

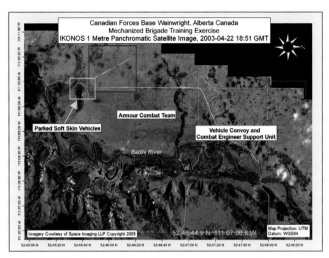

Figure 12.61 Overview IKONOS satellite image of a military training area at Canadian Forces Base, Wainwright, Alberta Canada.

Source: Analysis © ImStrat Corporation 2003. Data provided by Space Imaging.

Figure 12.62 Enlargement of the vehicle convoy and combat engineer support unit shown in figure 12.61.

Source: Analysis © ImStrat Corporation 2003. Data provided by Space Imaging.

are the two Beaver armored bridge layers, along with other support vehicles and possible tracked armored vehicles.

Under the Canadian organizational army structure, armored bridge layers are classified as a brigade-level asset. Therefore, the identification of these two bridge layers, along with supporting wheeled and other armored vehicles, suggests that a Canadian brigade is operating within the training area. In addition to this, the identified presence of the bridge layers also suggests that future training activity could include river or other obstacle crossing activity.

Figure 12.63 is an enlargement of an area approximately 2 km south of the vehicle convoy and combat engineer support unit. A large armored combat team is parked in a leaguer formation in an open field. A leaguer formation is a typical tactical armored parking pattern used in areas of minimum natural cover. It is designed to ensure all around defensive protection, while being supplied with ammunition, fuel, and rations. This formation consists of Leopard main battle tanks (MBT), possible wheeled light armored vehicles, 10-ton trucks and other soft-skin support vehicles. The parking pattern of the support vehicles within the confines of the parked Leopard tanks suggest the combat team is in the process of being refueled and resupplied.

Continuing south on the primary gravel road, approximately 1 kilometer from the armored combat team, a series of soft-skin vehicle trucks are parked next to a road T-junction (figure 12.64). The reason for the presence of these vehicles at this location is unknown. However, vehicle tracking leading into the adjacent forested area suggests additional vehicles could be located within the treed area. Further analysis of radar-based or high-resolution multispectral imagery could assist in determining whether additional vehicles and their types are located in this area.

No evidence of occupied defensive positions was identified west of the Battle River. However, east of the Battle River unoccupied vehicle revetted positions were located in various areas.

Nonetheless, they appeared not to be recently used, as no fresh tracks were identified leading into the defensive positions.

It can be inferred that the armored heavy brigade group could be preparing for a large offensive type exercise, which would include heavy tank support and river crossing procedures along the Battle River.

Sometimes other types of imagery such as radar and multispectral imagery may be required to extract additional intelligence data. More sophisticated analytical techniques may be used such as change detection using multidate imagery and multisensor imagery to detect features uniquely identifiable with different sensors. In some cases, imagery from multiple sensors is used because it is the only coverage available for the required acquisition time.

Collateral information input

In order for an imagery intelligence analyst to be highly effective and as accurate as possible, other sources of information, called collateral information, must be available during the analysis.

Collateral information complements and corroborates the feature and activity information extracted from the remotely sensed imagery. It may include information gathered by monitoring communications, field operations, and analysis of open source materials such as newspapers and news broadcasts.

Geographic information systems have become widely used in the military as a decision support tool, providing visualizations

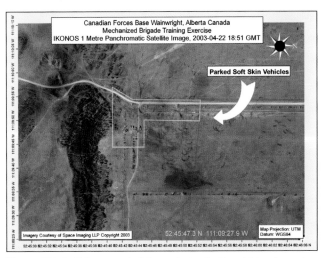

Figure 12.63 Enlargement of the armor combat team shown in figure 12.61.

Source: Analysis © ImStrat Corporation 2003. Data provided by Space Imaging.

Figure 12.64 Enlargement of the parked soft-skin vehicles shown in figure 12.61.

Source: Analysis © ImStrat Corporation 2003. Data provided by Space Imaging.

of intelligence data, tactical modeling such as predicting the actions of enemy forces, and operational- and strategic-level planning support. *Geospatial intelligence analysis* refers to the integrated analysis of imagery and digital geographic information within a GIS environment. A valuable feature of the GIS environment is that it provides an effective means for the intelligence analyst to store detailed information about objects of military interest and their appearance on different types of imagery. This knowledge base is readily made available to other analysts, thereby improving the effectiveness of the unit.

Imagery intelligence and competitive intelligence

Publicly available information on corporate activities and business status in press releases, financial data, and other open sources often prove to be ambiguous or misleading. The same techniques used for military intelligence gathering are readily adapted to the collection of business intelligence. The use of remotely sensed imagery is an extension of other forms of competitive analysis standard within the corporate world.

Imagery is particularly valuable in assessing resource extraction industries and heavy industry such as vehicle manufacturing enterprises that have large outdoor facilities to store assembled product. Imagery can be used to estimate a competitor's number of employees or customers, changes to facilities, and production flow rates.

In many countries, aerial photography of a commercial competitor's facilities is illegal, but the use of satellite imagery is not. The commercial availability of high-resolution imagery from satellites such as IKONOS, QuickBird, and OrbView-3, with spatial resolutions of 0.6 to 1 m, has become a valuable new source for competitive intelligence analysis. Sending observers to a site may be an option, but access may be restricted. Satellite imagery avoids such restrictions and can be a more cost-effective means of information gathering.

Competitive intelligence analyses are generally confidential. The following examples are hypothetical but are representative of current applications of satellite imagery to assess business activities. These simplified examples illustrate the type of information that can be derived from images collected periodically of an enterprise's facilities.

Scenario 1

A container pier 2,000 miles from your corporate headquarters is your competitor's international shipment point *(figure 12.65)*. The competitor publicly claims that its new product is selling beyond expectations, but imagery of its shipping

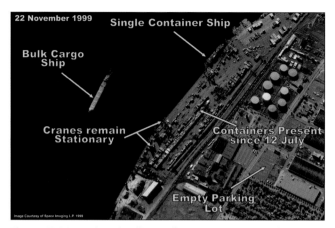

Figure 12.65   Business intelligence from container port analysis. An absence of activity at a company shipyard over several months may indicate deteriorating business conditions. Shipping activity can be monitored by analyzing multiple satellite images (1 m spatial resolution) collected over several months.

Source: Analysis © ImStrat Corporation 2003. Data provided by Space Imaging.

Figure 12.66   Vehicle production monitoring from satellite imagery. Analysis of IKONOS satellite images with 1 m spatial resolution collected periodically can provide valuable business intelligence information on the level of production at a vehicle manufacturing plant.

Source: Analysis © ImStrat Corporation 2003. Data provided by Space Imaging.

area tells another story. A bulk cargo ship is observed near the facility, but it isn't associated with shipping activity at this particular port. Periodic analysis shows that only one container ship is docked at the pier, and many of the containers haven't moved for several weeks. The employee parking lot is empty, and the gantry cranes haven't been moved since the first of the month. Despite public statements to the contrary, your competitor probably isn't doing as well as claimed.

Scenario 2

Your competitor claims that vehicle production is down and its second research and development (R & D) facility will be shut down, leaving it with only one R & D facility *(figure 12.66)*. The facility that's supposed to be shut down is heavily secured, and access is limited to employees only. In addition, high walls surround the facility, precluding ground observation. However, imagery of the R & D facility shows that four production buildings have been built, and the foundation for a fifth building is in place. The test track has been improved, and numerous vehicles await shipment. No signs of facility down sizing are apparent. Contrary to public statements, evidence suggests that the company is ramping up production capacity.

Conclusion

Remote sensing imagery has been a valued means of intelligence gathering since the early days of photography. Increasingly sophisticated sensors and computing resources have greatly expanded the scope of intelligence gathering capabilities, and geographic information systems have enabled more diverse information sources to be integrated in the image analysis process. Yet the effectiveness of image intelligence efforts ultimately depends on the abilities of the human image analyst, and visual photo interpretation remains a critical skill.

References

Aronoff, S. 1989. *Geographic information systems: A management perspective*. Ottawa: WDL Publications.

Canadian Forces. 2000. Land force information operations, Intelligence field manual, B-GL-357-001/FP-001. Ottawa, Canada.

Mohr, D. 2003. Transparent battle situation awareness: Ramping up the looking glass. PowerPoint presentation.

Richards, J. A., and X. Jia. 1986. *Remote sensing digital image analysis: An introduction*. 3rd ed. Berlin: Springer-Verlag.

Remote sensing in urban infrastructure and business geographics

James D. Hipple and Timothy L. Haithcoat

The use of remotely sensed data, either from satellite-based sources or more traditional sources such as aerial platforms, in traditional urban applications has increased within the past decade. A number of advances have enabled this. Computing hardware offers increased processing speed and storage capacity at decreasing cost. Geographic information systems (GIS) software can better handle and integrate vector and raster data layers. Commercial satellite imagery has become available with spatial resolutions of 1 m and better from Space Imaging (IKONOS satellite), DigitalGlobe (QuickBird), and ORBIMAGE (OrbView-3), as well as digital airborne imagery from scanned aerial photography and digital camera and scanner systems.

High-quality image products are available from many sources now. Also in the United States, the National Aerial Photography Program acquires complete 1:40,000 scale aerial photography coverage of the coterminus states on a seven-year cycle. This imagery is also used to generate the 1:24,000 scale Digital Orthophoto Quadrangle (DOQ) and 1:12,000 scale Digital Orthophoto Quarter Quadrangle (DOQQ) products available from the USGS. State and local governments also acquire aerial photography or airborne digital imagery on a regular basis. Existing imagery from government sources such as these are generally available at relatively low cost.

The use of remotely sensed data for urban infrastructure is rapidly increasing. Urban infrastructure uses include the application of GIS and remote sensing data for regional and municipal planning, public works, and transportation. GIS is used as a data management and analysis tool, and remotely sensed data (from satellite or aerial imagery sources) provides a quick, spatially accurate data source.

These technologies are also being used in *business geographics* applications. Business geographics refers to the analysis of specialized business problems through the use of a geographic information system. Business geographics problems include market planning, site selection, target marketing, and sales district management. GIS software make analyses of these tasks simpler by allowing large quantities of spatially referenced customer and site data to be viewed spatially and analyzed quickly by the lay practitioner. The theories used in business geographics come from the academic tradi-

tions of location theory, geography, demography, economics, and regional science. Traditional models such as Proximal Area Method, Reilly's Law of Retail Gravitation, Converse's Breaking Point Theory, and the Huff Model—extensions of the Gravity Model (Cadwallader 1996)—have been incorporated in GIS software, thereby allowing the user to integrate them easily into their information systems, while allowing managers to apply them effectively in business decision making. These analyses would benefit from the integration of remotely sensed data.

Urban infrastructure uses

Remotely sensed imagery has many uses in mapping and analyzing urban infrastructure such as planimetric base mapping, the generation of detailed topography, mapping land cover and inferred land use, urban forestry assessment, infrastructure condition assessment, development monitoring, and emergency response and disaster management. Many business geographics applications can also benefit from remotely sensed data, including economic development and business site development, market analysis and market planning, and risk analysis and damage assessment.

Planimetric base mapping

Planimetric base mapping applications using remote sensing data are quite numerous. The planimetric base, a layer that presents only the horizontal positions for features represented such as cadastral, often serves as a foundation for other urban infrastructure applications, including urban infrastructure condition assessment, utility line mapping, and precisely locating utility poles, signs, fire hydrants, and other infrastructure elements. The spatial resolution required for these applications are often quite fine. For general road centerline mapping and utility line mapping, spatial resolutions need to approach 1 meter, and for mapping rights-of-way, utility poles, and signage, a spatial resolution of 0.25 to 0.62 meters is needed. The panchromatic band of the commercial high-resolution satellites (IKONOS, QuickBird, OrbView) is useful for road centerline mapping. Imagery at spatial resolutions of 0.25 meters to 0.60 meters is limited to imagery taken by airborne sensors (both digital and traditional analog aerial photography). Temporally, these images are acquired every one to five years, often in conjunction with the decadal U.S. census, and then again mid-decade. For an image to be used as a base map in most GIS applications, it must be orthorectified (a process in which camera and terrain displacements have been removed from the image, including all sources of scale variation in the original photograph or image). Because many vector-based data layers, including street centerlines, parcel boundaries, and even sewer lines, are routinely superimposed upon the image base, these images must have a high degree of horizontal resolution and planar accuracy *(see table 12.4)*.

Detailed topography

New sources for terrain and digital elevation data (DEMs) include airborne Interferometric Synthetic Aperture Radar (IfSAR) and lidar (Light Detection And Ranging) data. These sensors serve to provide close to the horizontal and vertical resolution typically only extracted through stereoscopic analysis of aerial photography. High-resolution DEMs can be derived from 1:40,000 scale or larger scale aerial photographs (e.g., 1:24,000; 1:12,000; or even 1:6,000) with sufficient horizontal resolution and vertical accuracy to be useful for many urban applications, including U.S. Environmental Protection Agency (USEPA) Phase II Storm Water Regulation requirements for National Pollutant Discharge Elimination System (NPDES). For example, stereoscopic large-scale metric aerial photography at a scale of 1:86,000 scanned at 15 m can produce a pixel size of 1.2 meters and a z-accuracy of 3 to 5 meters (Tapley 2000). Larger scale imagery can provide higher accuracies.

There are several commonly used satellite data sources for high-resolution DEMs. Satellite sources of DEMs (produced through stereo correlation processing of panchromatic imagery, e.g., SPOT satellite) typically have a 20 meter horizontal (x,y) resolution and a 7 to 11 meter relative RMS vertical (z) accuracy (Spot Image 2000). Similarly, DEMs produced from IfSAR image processing (e.g., ERS-1/2) have yielded 25 m x,y resolutions and 10 to 30 m RMS z-accuracies (Davis and Wang 2002).

The most widely available high-resolution DEMs in the United States are the 7.5-minute Level 1 DEMs available from the United States Geological Survey (USGS). These DEMs are produced by automated stereocorrelation from 1:40,000 scale panchromatic aerial photographs acquired by the National Aerial Photography Program (NAPP) (Light 1993). The USGS 7.5-minute DEMs have a 30 m x,y resolution and a 7 to15 m RMS z-accuracy (USGS 1997).

The horizontal resolution and vertical accuracy of all the DEMs described above are not suitable for the vast majority of urban applications. Recently, this has prompted development of higher-resolution and higher-accuracy DEM data products from commercial remote sensing data providers using airborne IfSAR and lidar. The Intermap STAR 3i X-band IfSAR system provides digital elevation data with a 5 meter x,y posting and a 3 meter RMS z-accuracy (Intermap 2000). Commercial lidar systems can provide upwards of 1.5 meter x,y resolution and 0.4 m vertical accuracy *(see table 12.5)*.

Unless significant changes occur, DEMs only need to be acquired once every 5 to 10 years in urban areas. This is beneficial because these datasets are often costly to acquire.

Land cover and inferred land use

With regard to urban infrastructure uses of remote sensing, land-cover and land-use data is important for monitoring growth and change within a region. While land cover can

Table 12.4 Planimetric base mapping			
Source	Spatial resolution	Temporal	Uses
Airborne (digital or analog aerial photography)	0.25-0.60 meters	Frequency of one to five years	Right-of-way mapping, utility poles, and signage
Satellite (panchromatic)	1 meter	Frequency of one to five years	Road centerline mapping and utility line mapping

Table 12.5 Detailed topography			
Source	X, Y, and Z	Temporal	Uses
Photography derived 1:86,000 and larger scale	Better than 1.2 m x,y; 3 to 5 m z RMS	Frequency of 5 to 10 years	Building height, 2 ft. contours, storm water modeling, cut and fill analysis for transportation, DEM for orthorectification
SPOT stereo imagery	20 m x,y; 7 to-11 m z RMS	Frequency of 5 to 10 years	Runoff and slope analysis, DEM for orthorectification
Airborne IfSAR	5 m x,y; 3 m z RMS	Frequency of 5 to 10 years	Building height, storm water modeling, DEM for orthorectification
lidar	1.5 m x,y; 0.4 m z RMS	Frequency of 5 to 10 years	Building height, 2 ft. contours, storm water modeling, cut and fill analysis for transportation
7.5-minute Level 1 DEM	30 m x,y; 7.5 to 10 m z RMS	Frequency of 5 to 10 years	Runoff and slope analysis, DEM for orthorectification

be interpreted from an image directly, land use can only be inferred from remotely sensed data. Remote sensing has long been used to map urban growth, sometimes referred to as mapping urban morphology—implying the mapping of *form*, *land use*, and *density*, each having their associated shape, configuration, structure, pattern, and organization of land use. Sometimes simply mapping an urban or non-urban dichotomy is important; sometimes detailed morphologic mapping is needed where the positions of buildings and roads or the extraction of the three-dimensional and topographical aspects of urban areas are needed.

Conventions for the classification of land use or land cover vary by application. A number of very different systems have been devised for the mapping of urban land cover. Probably the most widely used system, the *Anderson Land Cover Classification* (Anderson et al. 1976), was developed by the USGS for use with remotely sensed data. Many states in the United States have their own classification systems; these are often derived from the Anderson Classification.

The Anderson Land Cover Classification is both a land-cover and land-use classification system designed for use with remotely sensed data. It is a hierarchical system with two pre-defined broad levels *(table 11.1 in the chapter 11)*. The Level I class includes urban or built-up land, agricultural land, rangeland, forest land, water, wetland, barren land, tundra, and perennial snow or ice. The remote sensing system best suited for inventorying these features at this broad scale include Landsat MSS (79 m spatial resolution), Landsat TM and ETM+ (30 m spatial resolution), and SPOT HRV XS (20 m spatial resolution) (Jensen and Cowen 1999).

For the USGS Level II class, higher resolution sensors with a minimum spatial resolution of 5 to 15 m (Jensen and Cowen 1999) are needed. At this level, the USGS classification system starts to include specific types of urban and human settlement structures. The Level II classification for *urban or built-up land* includes classes such as residential, commercial and services, industrial, and industrial and commercial complexes. In addition to including more specific urban classes, the USGS system switches predominantly to a land-use-based classification from a land-cover-based scheme. Higher resolution datasets are needed to identify these classes, including SPOT panchromatic (10 m spatial resolution) or Landsat 7 panchromatic (15 m spatial resolution).

The USGS Level III class (which is defined by the particular problem at hand) is best inventoried using sensors with spatial resolutions of about 1 to 5 meters (Cowen and Jensen 1998, Jensen 2000). These include the EROS-A1 0.9 or 1.8

m spatial resolution panchromatic imagery (ImageSat International, Inc.), IKONOS 1 m panchromatic and 4 m multispectral imagery (Space Imaging, Inc.), IRS-1C and -1D and ResourceSat-1 5.8 m resolution panchromatic and 23 m multispectral imagery (Indian National Remote Sensing Agency), QuickBird 0.62 m panchromatic and 2.8 m multispectral imagery (DigitalGlobe, Inc.), OrbView-3 1 m panchromatic and 4 m multispectral imagery (ORBIMAGE, Inc.), SPOT-5 2.5 m or 5 m panchromatic and 10 m multispectral imagery (SpotImage, Inc.). Aerial photography (0.25 m to 1 m spatial resolution) and digital multispectral and hyperspectral cameras (0.25 to 5 m spatial resolution) can also aid in the derivation of items in the Level III class. For the highest level of detail, a spatial resolution of 0.25 to 0.5 m is desirable (Cowen and Jensen 1998). For many local government uses (particularly in public works, utilities management, and transportation engineering), airborne photographic or digital imagery are preferred for Level III and finer classification detail.

In rapidly growing urban areas where significant changes are occurring, land use and land cover only need to be acquired once every five years at the coarsest of levels (USGS Level I and II). Municipalities can improve the currency of their image data and realize substantial savings by selectively acquiring imagery of rapidly growing urban areas every five years while flying complete image coverage, including the fairly static core areas of the city, less frequently.

Urban forests

The assessment of urban vegetation, particularly urban forests and their contribution to the quality of urban life and other benefits, ties greatly into land-use and land-cover mapping. The characteristics of urban vegetation (high species diversity, low homogeneity, and high fragmentation) often hinder the process of delineating or mapping vegetative land cover automatically. Moreover, the ecological relationships present in natural settings are not present in urban settings due, in part, to human interference and modification.

One example of an application developed specifically for the mapping and delineation of urban forest resources and urban vegetation is American Forests and their CITYgreen program (*www.americanforests.org*). The CITYgreen system integrates remote sensing and GIS to provide analysis of storm-water runoff, air quality, summer energy savings, carbon storage and avoidance, and tree growth. Urban forest management cuts across divisions, integrating planning, public works, transportation, and parks interests. Data collected by public works includes high-resolution planimetric orthophotography, Landsat 7 data for monitoring local and regional growth, transportation's need for high resolution digital elevation data, and ecological monitoring and inventory data *(table 12.6)*.

Urban infrastructure condition assessment

Remote sensing can play a valuable role in assessing the condition of urban infrastructure. For example, the condition of road surfaces can be assessed from large-scale aerial photography with spatial resolutions of 0.5 m or finer. Less-expensive videography methods using true-color or color-infrared video systems with integrated GPS satellite positioning system can be vehicle mounted and driven over the streets to be assessed. Though the imagery is less expensive to acquire, it can be more time consuming and difficult to analyze.

High-resolution aerial photography (0.5-meter resolution or finer) is generally used to collect building infrastructure information such as a building footprint (an outline of the base of a building). Some of this information can be extracted from the highest resolution satellite data but with a somewhat lower level of positional accuracy due both to the lower positional accuracy of the image data and the additional degradation of digitizing errors. Building footprints can be digitized off orthorectified aerial imagery displayed in a GIS. Building height data can be extracted from lidar and IfSAR data and added to the GIS database. Other data such as roof type and building types can be interpreted visually from the imagery *(figure 12.67a,b,c)*.

Development monitoring

Monitoring of existing construction sites within the limits of a city is essential for a city government. Monitoring helps control construction site permits issued by a city, delineate locations where natural ground cover or preexisting structures were removed and exposed bare soil (a potential source of sediments that lower water quality in local watersheds), and monitor construction progress at those sites. By monitoring

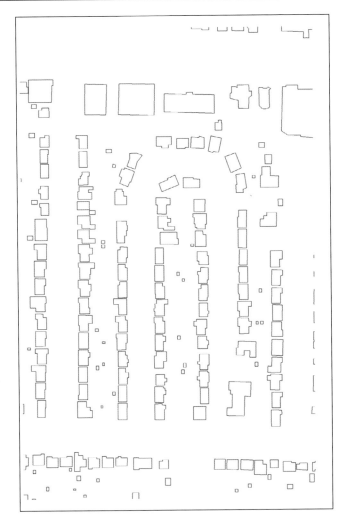

Figure 12.67a Planimetric base mapping digitized from orthophotography using heads-up digitizing, i.e., interactively drawing outlines on a displayed image. Ground control points, a digital elevation model, and GPS satellite positioning are used to rectify the imagery and verify positional accuracy.

Source: J. Hipple. Underlying DOQ courtesy of USGS.

urban development, a city can also track compliance with EPA water quality regulations for urban watersheds.

The spatial and spectral resolution needs for monitoring construction are dependent on the size of the areas to be monitored. If the sites are large (20 acres or larger), imagery with spatial resolutions as coarse as 10 m may be suitable, but for smaller areas, resolutions of 1 to 5 meters or better are generally needed, which requires the use of high-resolution satellite or airborne imagery. Ideally, acquisition of this imagery

Table 12.6 Urban forestry			
Source	**Spatial resolution**	**Temporal**	**Uses**
Airborne (digital or analog aerial photography)	0.25 to 0.60 meters	Frequency of 1 to 5 years	Tree canopy delineation and species identification
Satellite (multispectral)	10 to 30 meters	Frequency of 5 to 10 years	General land use and land cover; vegetation map (NDVI)

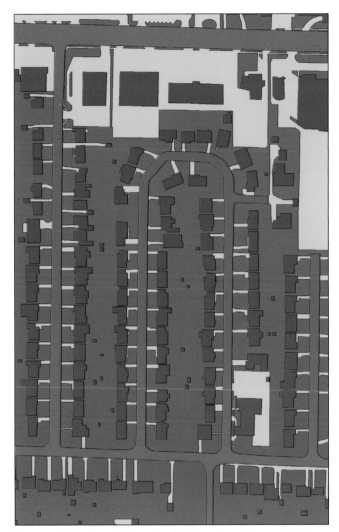

Figure 12.67b   A preliminary classification (from photography) of land-cover type. Red represents road (10.05%); purple represents buildings and built structures (16.96%); yellow represents parking lot, sidewalk, and driveway (22.51%); green represents vegetation (50.44%); impervious surfaces (the total minus the vegetation class) represents about 49.56% of the area.

Source: J. Hipple.

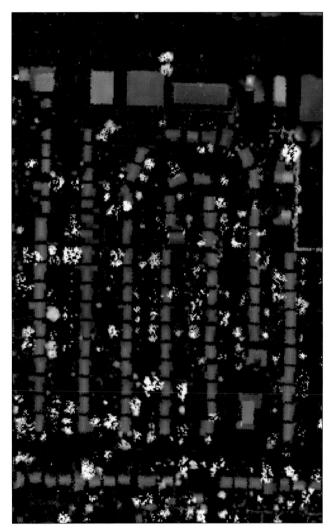

Figure 12.67c   A bald earth digital elevation model (i.e., terrain surface without surface features such as buildings and trees) generated from lidar data was subtracted from the lidar first surface profile to generate an image of above surface features. Notice the tree signatures and the building signatures.

Source: J. Hipple.

would occur at least every year. In rapidly developing places, imagery might be needed quarterly, especially for tracking residential development. Figure 12.68a is a USGS Digital Orthophoto Quad (DOQ) acquired in 1995 of Ashland, Missouri. Figure 12.68b is an image acquired by the IKONOS satellite in 2000. The zoning map of the same area *(figure 12.68c)* is the local government record of development, which is based upon the plats submitted by the devel-

oper and the building permits issued. This map gives a false sense of development in the region.

Emergency response and disaster management

A very basic use of remote sensing for emergency response is the integration of orthophotos or high-resolution satellite imagery into the E911 emergency response system. The photos allow for the pinpointing of structures (as opposed

Figure 12.68a USGS DOQ of Ashland, Missouri, acquired in 1995.

Source: Image courtesy of USGS.

Figure 12.68b Image acquired by Space Imaging IKONOS satellite in 2000.

Source: Image courtesy of Space Imaging Corporation.

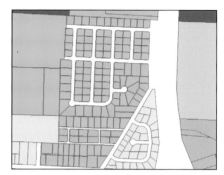

Figure 12.68c Local government record of development based upon the plats submitted by the developer and the building permits issued.

Source: J. Hipple.

to the general geocoding of an address), giving emergency responders a better idea of the environment and location to which they are responding.

Other applications for emergency response and disaster management use a wide variety of remotely sensed data collected from various airborne and satellite sensors, digital orthophotos, and digital elevation models. Additional data related to the disaster response area (e.g., buildings, roads, land use and land cover) can also be integrated. The source of this data is often a GIS in use by the local administrative services.

In responding to acts of God (floods, hurricanes, tornadoes) or human-caused disasters (chemical spills, toxic waste, urban violence, terrorism), imagery must be acquired soon after the event. Satellite imagery with low spatial resolutions on the order of 1 km to 250 m provides almost daily global coverage. Medium-resolution satellites without pointable sensors, such as Landsat with its 16-day revisit cycle, could have significant delays in image acquisition. Satellites with pointable sensors that enable off-nadir acquisition, such as the medium-resolution SPOT sensors and the high-resolution

commercial satellites, can shorten the revisit period to 1 to 4 days depending on the satellite and location. (Locations at higher latitudes can be imaged more frequently.) Airborne sensor systems can usually acquire imagery within hours of a disaster event if the location is within the range of suitably equipped aircraft *(table 12.7)*.

Business geographics applications of remotely sensed data

Within the area of business geographics, remote sensing and image data can be a valuable ancillary data source. Three of the main areas where imagery has become key in business geographics are economic and business site development, market analysis, and risk analysis and damage assessment. Analysis of infrastructure also plays a key role in the use of imagery for business geographics, including trade area analysis and urban change detection.

Business geographics use imagery for such applications as base mapping, site selection, market analysis, planning, change detection, and emergency response. These applications tend to require two levels of spatial resolution. Lower spatial resolutions of 30 to 250 m can be used to analyze large areas (e.g., state or multistate regions) to provide an overview for strategic planning and regional analyses. The capability to distinguish relatively general classes such as urban areas, forest, and agricultural land is usually sufficient. Currency of the imagery also tends to be less important. A wide range of exisiting satellite and airborne imagery are suited to these applications.

Larger scale imagery is needed to provide the detail for such applications as site selection, market planning, studies of competition, and logistical relationships. These applications

Table 12.7 Emergency response and disaster management			
Source	**Spatial resolution**	**Temporal**	**Uses**
Airborne (digital or analog aerial photography)	0.25 to 2 meters	Within hours to 1 day of event	Damage assessment, assessing impacts to infrastructure, search and rescue, fire control
Satellite (multispectral)	1 to 30 meters	Next available acquisition	General land-use and land-cover impacts; regional damage assessment

require the identification of detailed land-use patterns and areas of rapid growth, as well as visualizations of an area's appearance after development. Imagery with spatial resolutions of 1 m and better acquired within 1 year or less is typically required. High-resolution satellite imagery with 1 m spatial resolution or better and airborne data are typically used. In some cases, airborne imagery may be flown specifically for a project.

Economic development and business site development

One of the most effective economic development tools for a community is a GIS. As a community tries to attract new businesses as part of their economic development plan, the GIS provides a quick and easy way for planners to map the needs of the firm with available sites within the community. In this case, imagery integrated within the GIS provides the icing on the cake. An image of potential sites—and overlay of various GIS data layers, including transportation, utilities, and zoning—allows for the quick visual assessment of the benefits of each proposed site. These industrial sites atlases use high-resolution orthorectified satellite imagery (1 meter or better panchromatic or pan-sharpened multispectral imagery) or airborne imagery as an underlay over which GIS data layers are displayed, giving potential developers valuable site information.

Market analysis and market planning

A GIS used for market analysis and market planning allows a company to include spatial analysis in evaluating its competitive success. A business can make better marketing plans when it knows where its target group actually lives. A GIS can aid in finding target markets and in market analysis and planning. This allows a company to develop spatially specific target-marketing campaign by enabling market segmentation by age group, income, or even community growth characteristics. Similar to the economic development and business site development approaches, a company can use this data to analyze the available locations for new business locations and new markets within a region. To that extent, mapping of customers by street address or ZIP code, and the integration of demographic information within the analysis can be greatly enhanced with imagery. These demographics are often estimates, and the vector GIS data reveals only part of what is going on. With imagery as a backdrop, an analyst can easily obtain additional information such as the residential-versus-nonresidential-area breakdown of a U.S. census

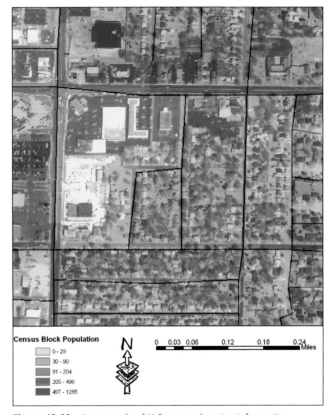

Census Block Population
- 0 - 29
- 30 - 90
- 91 - 204
- 205 - 496
- 497 - 1265

0 0.03 0.06 0.12 0.18 0.24 Miles

Figure 12.69 An example of U.S. census housing information overlayed on an IKONOS 1 meter panchromatic image. The image shows that a significant portion of the area within the U.S. census block is non residential.

Source: J. Hipple.

block group or customer locations by identifying actual households on the imagery *(figure 12.69)*.

Risk analysis and damage assessment

Risk management and damage assessment applications can be considered both an urban infrastructure and business geographics application. Rapid evaluation of the impact of a storm event or other natural disaster such as fire or flooding can aid planners and businesses in recovery operations and insurance companies to process claims more quickly. One example of using remotely sensed data integrated within a GIS is the analysis of development within the flood-impacted areas from the 1993 flooding of the Mississippi and Missouri rivers. Since the flood events of 1993, millions of dollars of new development have poured into the flood-impacted areas. There has been development in many areas impacted by the 1993 floods in the past 10 years, but nowhere as notable as in the St. Louis metropolitan region. Figure 12.70 illustrates some of the development that is occurring behind the Monarch Levee in the Chesterfield Valley area of St. Charles County, Missouri. Image (a) shows a portion of a Landsat 5 TM image acquired October 17, 1991, illustrating *Pre-Flood* development; image (b) is a portion of a Landsat 7 ETM+ image acquired April 1, 2003, showing development that occurred since the 1993 flooding *(Post-Flood)*; and image (c)

is a portion of a USGS digital orthophotograph acquired in March 1996 for comparison. The Monarch Levee broke July 30, 1993. The height of the new levee is being raised to that of the 500-year flood level (Carey et al. 2003). Since the 1993 flooding, the Chesterfield Valley (located behind this levee) has been the recipient of over \$400 million in new development (Heisler 2003). Such imagery, integrated within a GIS, enables the tracking and monitoring of wide areas for development that could be at risk from future floods. The levee boundaries are shown as white lines. It is important to note that development is continuing in the area protected by the Monarch Levee.

The integration of remote sensing with GIS

The area of remote sensing integration with geographic information systems is poised for rapid growth. Many new tools have been developed; software has been enhanced to improve the extraction of information from remotely sensed imagery. Improved and cheaper computer hardware (in terms of storage space, network speed, and processor speed) has made the processing technology available to a wider range of users. In addition, the spatial, temporal, and radiometric resolution of space-based imagery is continually improving. High-resolution sensor systems, such as IKONOS, Quick-Bird, and OrbView-3, provide orthorectified imagery with a quick turnaround time unmatched by traditional aerial camera systems. As more high-resolution satellites become operational near real-time delivery of imagery may become available to nonmilitary users. As with many other application areas, the primary change will be in the development of improved methods to integrate remotely sensed data within the GIS environment. The integrated remotely sensed data will allow for rapid updating of other GIS data layers, integration of field survey data, rapid assessment of land-use and land-cover change and community growth—all leading to better assessment of the urban infrastructure and better data for business planning.

a

b

c

Figure 12.70 Development behind the Monarch Levee in the Chesterfield Valley area of St. Charles County, Missouri. Image (a) is a portion of a Landsat TM image acquired October 17, 1991 showing *Pre-Flood* development. Image (b) is a portion of a Landsat 7 ETM+ image aqcuired April 1, 2003 showing development *(Post-Flood)* since the 1993 flooding. Image (c) is a portion of a USGS digital orthophoto acquired in March 1996 for comparison (leevee boundaries are illustrated as white lines).

Source: J. Hipple. Imagery courtesy of USGS.

References

Anderson, J. R., E. E. Hardy, J. T. Roach, and R. E. Witmer. 1976. A land use and land cover classification system for use with remote sensor data. Geological Survey Professional Paper 964. landcover.usgs.gov/pdf/anderson.pdf

Cadwallader, M. T. 1996. *Urban geography: An analytical approach.* 2nd ed. Upper Saddle River, N.J.: Prentice Hall.

Carey, C., et al., 2003. A flood of development: Building booms in the flood plain. *Saint Louis Post-Dispatch*, 27 July.

Cowen, D. and J. Jensen. 1998. Extraction and modeling of urban attributes using remote sensing technology. In *People and pixels: Linking remote sensing and social science.* Washington, D.C.: National Academy Press.

Davis, C. H., and X. Wang. 2002. Planimetric accuracy of IKONOS 1-m panchromatic orthoimage products and their utility for local government GIS base-map applications. *International Journal of Remote Sensing* 24(22):4267–88.

Haithcoat, T. L., L. Warnecke, and Z. Nedovic-Budic. 2001. Geographic information technology in local government: Experience and issues. In *International City/County Management Association: Municipal Yearbook 2001,* 47–57.

Heisler, E. 2003. A flood of development: $400 million in investment flows into Chesterfield Valley. *Saint Louis Post-Dispatch* 28 July.

Intermap, Inc. 2000. www/digitalglobe.com

Jensen, J. 2000. *Remote sensing of the environment: An earth resource perspective.* Saddle River, N.J.: Prentice Hall.

Jensen, J. and C. Cowen. 1999. Remote sensing of urban/suburban infrastructure and socio-economic attributes. *Photogrammetric Engineering & Remote Sensing* 65(5):611–22.

Light, D. L. 1993. The National Aerial Photography Program as a geographic information system resource. *Photogrammetric Engineering and Remote Sensing* 59(1):61–5.

Sugumaran, R., D. Zerr, and T. Prato. 2001. Multi-temporal IKONOS images for local government planning and management. *Canadian Journal of Remote Sensing* 28(1): 90–95.

Tapley, I. J. 2000. Advances in digital elevation datasets for exploration and mapping. In *Improving the chances of exploration success: Australian Mineral Foundation.* Perth, Western Australia.

United States Geological Survey. 1997. *Standards for digital elevation models. Part 1: General. Part 2: Specifications. Part 3: Quality control.* Washington, D.C.: Department of the Interior.

13 Remote sensing and the organization

James W. Merchant

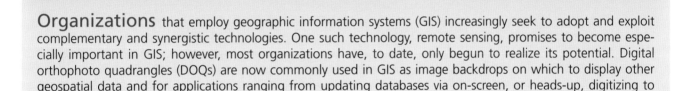

Organizations that employ geographic information systems (GIS) increasingly seek to adopt and exploit complementary and synergistic technologies. One such technology, remote sensing, promises to become especially important in GIS; however, most organizations have, to date, only begun to realize its potential. Digital orthophoto quadrangles (DOQs) are now commonly used in GIS as image backdrops on which to display other geospatial data and for applications ranging from updating databases via on-screen, or heads-up, digitizing to georeferencing.

Major GIS software packages provide capabilities for display and tools to manipulate DOQs and other remotely sensed images, and GIS analysts are comfortable with such uses. However, remote sensing can clearly contribute much more.

During the next decade, organizations using GIS will experience growing demands to more fully take advantage of the opportunities afforded by remote sensing. Some organizations will endeavor to build internal capabilities for using remote sensing. Others will outsource their remote sensing and seek assistance from private-sector corporations or public-sector organizations, including universities, technical and community colleges, state mapping agencies, and not-for-profit corporations offering training and support. This chapter is intended to provide guidance for those considering such options.

Implementing remote sensing in the organization

In many important respects, the issues confronted when considering the place of remote sensing in an organization are not fundamentally different than those dealt with when implementing GIS. The literature on the latter is substantial (see Aronoff 1989, Huxhold and Levinsohn 1995, National Academy of Public Administration 1998, Obermeyer and Pinto 1994, Tomlinson 2003) and should be consulted on general guidelines and considerations pertinent to establishing geospatial information technologies in an organization.

Recently, studies have begun to identify issues specifically related to remote sensing (see Steering Committee on Space Applications and Commercialization, National Research Council 2003 and Transportation Research Board, National Research Council 2001). These studies suggest that organizations considering implementation of remote sensing must address a number of matters, including determining needs, estimating costs and benefits, identifying human resources requirements, and identifying the appropriate institutional model for conducting remote sensing projects.

Assessing needs

The initial steps in considering the role of remote sensing in an organization must involve (1) a systematic and realistic appraisal of what remote sensing has to offer, and (2) a detailed evaluation of the organization's requirements for data and information that might, at least in part, be provided through remote sensing. Each assessment should include a

properly designed cost–benefit analysis. These assessments can be conducted simultaneously and may sometimes benefit from the involvement of a consultant skilled in such work.

Potential benefits of remote sensing

As noted above, most organizations currently using GIS understand the benefits of using DOQs. Several organizations have begun to use high-spatial-resolution images like those from commercial systems such as the IKONOS or QuickBird satellites in a similar fashion. However, only a few routinely go beyond image display to undertake the more sophisticated applications of remote sensing involving information extraction using procedures such as image classification and thematic mapping.

Many opportunities to learn about the state of the art in remote sensing exist through professional conferences, periodicals and books, short courses, and the Internet. However, organizations must be judicious in separating experimental successes from operational applications. They must consider factors such as the accuracy achievable, the level of user technical skill, and software capabilities needed for specific applications. Because so many potential applications of remote sensing exist, it is virtually impossible to generalize about success in this rapidly changing field. Instead, organizations must focus on their specific information requirements in order to gauge the value of individual techniques in their operations.

Organization information requirements

No organization should adopt remote sensing until there is a clear indication that remote sensing can profitably contribute to the organization's ongoing responsibilities, mandated activities, and mission. An organizational needs assessment should explore the types of products required (e.g., thematic maps such as land use, water temperature, and turbidity), and the spatial (e.g., land use *at a resolution of one hectare*), categorical (e.g., water temperature *at a resolution of 1 °C*), and temporal (e.g., turbidity measured *hourly*) resolution required. In addition, the availability of staff and software and the costs of data must be considered. Information extraction (e.g., image classification) typically requires greater software capabilities, user technical skill, and data cost than the simple use of images such as DOQs in a visualization mode.

In assessing the potential for remote sensing, it is extremely important to avoid unrealistic expectations. Like any instrument or process used for data collection, remote sensing systems are capable of generating errors. Errors in data collection are often artifacts of sensor design (i.e., the level of spatial, spectral, and radiometric resolution), but errors may also be introduced during data analysis and information extraction.

In remote sensing, these error rates vary greatly depending on objectives, imagery used (e.g., spatial and spectral resolution, time and season of data acquisition), and the skills of the analysts. As in a GIS, error rates must be within the levels of tolerance acceptable to the organization's needs. Certain categories such as wetlands are, in many types of terrain, frequently difficult to map and characterize accurately using satellite imagery. In these cases, the uncertainty of the error rate may render remote sensing unsuitable for the particular project.

Crops can often be mapped effectively if imagery (typically two or three scenes per growing season) is available at critical times during the growing season (e.g., May, July, September). Accuracies will decrease if fewer dates are available or if images are collected at times that are not optimal for capturing the phenology and land management indicative of certain crops. Low-risk pilot or demonstration projects are often helpful for ascertaining error rates, real costs, staff, and technical skills required, as well as the probable benefits. Assessments of the potential benefits of remote sensing must be carried out with due understanding of the advantages, as well as the limitations, of the technology.

After assessing the potential contributions of remote sensing and the potential value of the products derived, an organization must evaluate the types, scope, and frequency of the work to be accomplished. How much staff training will be required? How often will remote sensing analyses need to be conducted, and as a consequence, how often will information need to be updated? The size of the area with which an organization deals must also be taken into account. In general, organizations that manage or otherwise require information for large areas on a frequent basis will be those most likely to establish in-house capabilities for remote sensing analyses. Organizations that require information derived from remote sensing infrequently or that deal with relatively small areas may wish to contract for remote sensing work. Likewise, it may pay an organization to hire or train a staff expert in sophisticated analysis methods only when that staff expert is likely to be fully employed in remote sensing work.

Human resources

The implementation of remote sensing will be most likely to succeed when the organization has at least one strong advocate for the technology (Steering Committee on Space Applications and Commercialization, National Research Council 2003). This person must be well acquainted with current remote sensing technology, have a fundamental technical understanding of remote sensing, and must be familiar with the information needs of the organization. It is essential that the advocate be an articulate spokesperson to ensure that the merits (and limitations) of remote sensing are well represented to, and understood by, the management of the organization.

In addition, most organizations will require, at the very least, one person who has significant technical expertise in remote sensing. Even if the organization chooses to contract for most remote sensing services, a person having substantial technical skills will be needed to negotiate and monitor contracts (Steering Committee on Space Applications and Commercialization, National Research Council 2003 and Transportation Research Board, National Research Council 2001). Of course, technical expertise in remote sensing can only be maintained if put to regular use, and the expert must be periodically reeducated in order to keep pace with new developments. With each passing year there are new sensors, new modes of data analysis, innovations in image analysis software, changing costs for data, and more sources from which to acquire imagery. For example, in many organizations there is currently a great deal of interest in lidar for mapping terrain elevations with high precision. However, since operational applications of lidar are recent, relatively few individuals currently possess the skills needed to understand analyses of lidar data. Thus, it is critical that funds be provided for technical staff to participate in training courses and professional conferences to maintain and update their skills.

Beyond what is identified above, the specific amount of staff and technical skill in remote sensing required in an organization will vary greatly with the size of the organization and the scope and nature of the remote sensing work to be accomplished. It should be noted, however, that staff will rarely be able to stay current with the state-of-the-art in remote sensing technology or apply advanced methods of data analysis if they devote only a few hours of effort now and then. If the organization has determined that an investment in remote sensing is warranted to achieve its goals, then a commitment of some full-time staff will almost certainly be required.

Implementing remote sensing

The preliminary investigative work outlined above should culminate in a written plan that reflects organizational goals and information needs, potential opportunities for using remote sensing, funding, staffing, expected benefits, management, and a timetable for implementation. This plan will serve to guide adoption of remote sensing initially but should be revised periodically to reflect new developments in remote sensing and changes within the organization.

The organization's plan should include assessment of options for carrying out remote sensing. Several models may be used. Unlike GIS, in which the technology and operation is often distributed throughout an organization, many organizations will likely consider implementing remote sensing in a centralized mode, recognizing that specialized skills and tools will be required to conduct remote sensing applications. Some organizations will then endeavor to conduct virtually all remote sensing in-house, others will join with partners to form consortia, and still others will choose to contract with private firms or public institutions for remote sensing projects.

Conducting remote sensing in-house

Organizations may wish to carry out all or most remote sensing work within the organization. This option works well when the organization has sufficient demand for remote sensing that a well-trained, full-time staff and appropriate hardware and software facilities can be justified. Such justification may be based upon the frequency with which remote sensing projects must be conducted, the size of the area over which the organization has interests, or both. The in-house model may not be economically viable when remote sensing work is to be carried out infrequently or when highly specialized technical skills are likely to be needed on a short-term basis.

External contracting

Some organizations may wish to contract for virtually all their remote sensing work. This option will be especially attractive to organizations that are small, that are just beginning to evaluate the technology, that will use remote sensing infrequently, or that will be conducting remote sensing work that is highly specialized for which in-house technical skills are not available. As noted above, even when work is contracted, the organization will still need to retain someone technically proficient in remote sensing to monitor contractual activities. At present, this is the model most often employed by organizations using GIS.

Partnerships and consortia

In some instances, organizations may wish to join together to develop shared remote sensing capabilities or to enter into cooperative contractual agreements with vendors. Such partnerships will work best when organizations have overlapping geographical jurisdictions, similar requirements for information, and compatible interests that portend potential long-term cooperative relationships.

For example, it is common in many states and provinces for several agencies to require information on land use and land cover. Often this is done for different applications (e.g., wildlife habitat assessment, land-use planning, water-resources modeling). In these circumstances, organizations should clearly seek to collaborate on development of common products that optimize spatial, temporal, and categorical requirements of all partners. This provides an efficient way to share costs of data, image analysis, and product generation. Such work may entail development of shared in-house capacity, external contracting, common data holdings, or some combination of these approaches.

The three models outlined above (in-house operations, external contracting, and partnerships and consortia) are not mutually exclusive. Consequently, organizations may employ various modes of operation for different types of projects. In addition, within a particular organization the mix may well change over time as the organization becomes better versed in remote sensing and integrates the technology into routine activities.

The scope of remote sensing within an organization

Clearly, there are several alternative approaches an organization may use to successfully integrate remote sensing into its GIS operations. However, it is important to note some of the common remote-sensing-related tasks any organization will need to accommodate in some fashion.

Whatever approach to implementation is adopted, the organization must ensure that it is provided the most current information on remote sensing technology. A major task of those responsible for providing oversight of remote sensing will be educating themselves (and perhaps others) about new developments in sensors, applications, costs, modes of analysis, and so forth. Judicious selection and acquisition of imagery (including the costs, advantages, and disadvantages of using airborne or satellite imagery; archival or new data;

and copyrighted or public-domain data) is critical to most applications, and may require a substantial investment of time and effort.

The organization must also ensure that image analysis procedures employed in information generation are up-to-date. This requires key oversight staff to be familiar with state-of-the-art software options. Also, oversight staff must be familiar with design and implementation of data analysis strategies that involve many modes of image classification. Depending on the application, analysis may require use of ancillary data, ground calibration, sophisticated numerical techniques, or other procedures.

It is critical that such staff be conversant with quality assurance and accuracy assessment methods in order to document successes and shortcomings of analyses. There should be a well-defined process by which information generated from remote sensing analyses is documented, evaluated, and then either rejected until the information quality is improved or accepted into the organization's information infrastructure. The importance of rigorous quality assurance of remote-sensing-derived information before it is made widely available to users within an organization should not be underestimated. It should be assumed the information will be used by those unfamiliar with the remote sensing analysis procedures employed to generate it. Thus the assessment of its suitability for use should be made in advance by those with the technical remote sensing expertise to better judge the consequences of inadvertent misapplication.

Finally, remote sensing staff must be well acquainted with procedures to integrate products derived from remote sensing with GIS. A successful remote sensing program should not be isolated in any regard from other components of the organization's geospatial information technology infrastructure.

Conclusion

Remote sensing is a potentially powerful complement to GIS technology. Skilled analysts can, in many instances, employ remote sensing to provide current information on, among other things, land use and land cover, water resources, terrain elevation, biophysical characteristics of the earth's surface, and landscape change. Organizations seeking to integrate remote sensing into the GIS processing stream should begin by carefully assessing the potential of the technology for meeting data requirements.

Subsequently, a plan should be developed to guide implementation of remote sensing through (1) building in-house capability, (2) external contracting, (3) developing partnerships and consortia, or some combination of these. The specific approach adopted by an organization will be largely dependent on the size of the area dealt with routinely, the type of information analysis that needs to be conducted, and the frequency with which remote sensing must be used. In any event, organizations should not underestimate the importance of having at least one full-time remote sensing expert on staff.

References

Aronoff, S. 1989. *Geographic information systems: A management perspective.* Ottawa: WDL Publications.

Huxhold, W. E., and A. G. Levinsohn. 1995. *Managing geographic information system projects.* New York: Oxford University Press.

National Academy of Public Administration. 1998. *Geographic information for the 21st century: Building a strategy for the nation.* Washington, D.C.: National Academy of Public Administration. www.napawash.org/pc_management_studies/GIFullReport.pdf

Obermeyer, N. J., and J. K. Pinto. 1994. *Managing geographic information systems.* New York: The Guilford Press.

Steering Committee on Space Applications and Commercialization, National Research Council. 2003. *Using remote sensing in state and local government: Information for management and decision making.* Washington, D.C.: The National Academies Press. www.nap.edu/books/0309088631/ html

Transportation Research Board, National Research Council. 2001. *Remote sensing for transportation: Report of a conference, December 2000.* Washington, D.C.: The National Academies Press. gulliver. trb.org/publications/conf/reports/remote_sensing_1.pdf

Tomlinson, R. 2003. *Thinking about GIS: Geographic information system planning for managers.* Redlands, Calif.: ESRI Press.

Appendix A Rectification and georeferencing of optical imagery

Gordon Petrie

Data acquired by airborne and spaceborne imagers (imaging systems) often forms an invaluable source of information that needs to be input to a GIS system. The data may take the form of the actual continuous tone image itself, or it may take the form of line and polygon data or raster data that has been extracted from the image data. Whichever form it takes, it will contain geometric displacements and scale variations, which means that this data will not fit the geospatial data that resides within the GIS system. The process of removing these geometric displacements and ensuring that the image has a consistent scale is called *rectification*.

Closely associated with this rectification process is the process of transforming the image data from its own arbitrary image coordinate system into either geographic coordinates of the latitude/longitude system or into the specific map projection system that is being used as the reference system in the GIS—a procedure known as *georeferencing*. Since the two processes—rectification and georeferencing—are very closely linked with one another, they are often implemented as a single combined procedure in the systems used to carry out the processing of remotely sensed imagery. This appendix examines both processes under three headings:

- Image geometries
- Rectification and georeferencing based on the use of two-dimensional transformations
- Image rectification and georeferencing based on the use of three-dimensional transformations

A. Image geometries

Fundamental to all image rectification and georeferencing is a knowledge of the geometry of the particular type of remote sensing imagery from which the GIS data is to be generated or extracted. In fact, imagers operate on different principles, have different constructions and geometries, and generate a wide variety of image and map products from which GIS data can be derived. Thus there is no single generic type of remotely sensed image to which a single universal image rectification and georeferencing procedure can be applied. However, looking at image rectification from a completely general point of view, three main image geometries can be identified:

- *frame camera image geometry*
- *line scanner image geometry*
- *radar image geometry*

All three types of image can be acquired from both airborne and spaceborne platforms. In general terms, the rectification techniques and procedures developed for use with imagery having one of these three basic geometries can be used irrespective of the platform used. In this particular account, only the rectification procedures used with the first two types—frame cameras and line scanners—producing *optical imagery* will be covered.

A.I Frame camera images

Frame camera images are acquired by photographic and digital cameras. The images are generated in the form of

individual frames rather than the continuous strip images of the ground that are generated by line scanners or radar imagers. From the point of view of image rectification, frame camera images can be divided into two main types: those produced as *planar (flat) frame images* and those produced on a *cylindrical imaging surface*.

A.I (a) Planar frame images

The best known and most commonly encountered type of frame image is that produced by the classical metric film cameras or their modern digital camera equivalents. Each of these cameras produces a planar (flat) frame image that is acquired from a *single exposure station* either in the air or in space. This ensures the *simultaneous exposure* of the whole area covered by the frame. With regards to geometry, this arrangement produces a *central projection* with a *fixed set of tilt values* that apply to the whole of the planar frame image (*figure A.1a*). These very simple and stable geometric characteristics make the rectification of frame images a much easier and more straightforward task as compared with the procedures needed to rectify line scanner or radar images.

Tilt displacement

If a frame image is exposed in a camera with its optical axis pointing in a truly vertical direction in the air or in space, then a set of squares marked out on a truly flat piece of ground will be imaged as squares on the frame image. If, however, the camera has been tilted, then the image of the squares will be deformed (*figure A.1b*). In geometric terms, these deformations manifest themselves as displacements of the corners of the squares either radially inwards or outwards from the center of the frame image. These are called *tilt displacements*. Associated with these displacements are variations in the scale of the image across the frame. Obviously these effects need to be removed if the image or any data extracted from the image is to be used in a GIS system.

Relief displacement

A further set of geometric displacements occurs in frame images due to the relief that is present in the terrain. These are called *relief displacements*. Again they take place radially from the center of the image (*figure A.1c*), though in a quite different pattern to the tilt displacements since they are related to the variations in relief that are present on the ground. These relief displacements also cause a further set of scale variations in the image. The removal of these tilt and relief displacements and their associated scale changes across the whole of the frame image is the objective of the *image rectification* process. If this can be carried out successfully, then the rectified image will have the geometry of a map, making it easy to use data extracted from a rectified image in a GIS system.

Vertical and near-vertical frame images

Most frame images of the terrain are acquired with the optical axis of the camera pointing vertically downward from the airborne or spaceborne platform on which it is mounted. In practice, due to the effects of atmospheric turbulence, the vertical, or nadir, pointing frame images acquired from an airborne platform will always deviate a little from a truly vertical orientation. Thus, the resulting images are better described as being near-vertical frame images. The small tilts

(less than 3°) that are present will cause small but important tilt displacements and scale changes on the images that still need to be eliminated through a prior image rectification if data is to be extracted and used in a GIS system.

Oblique frame images

In order to increase the area of the terrain covered by a single frame image, sometimes the camera's optical axis will be tilted deliberately through a substantial angle from the vertical. Images taken at these angles are called *oblique images*. The main distinction is between *high-oblique images*, which will include an image of the earth's horizon (*figure A.2*), and *low-oblique images*, which do not include the horizon. In each case, tilt displacement will cause the scale of the image to change very substantially across the frame, resulting in

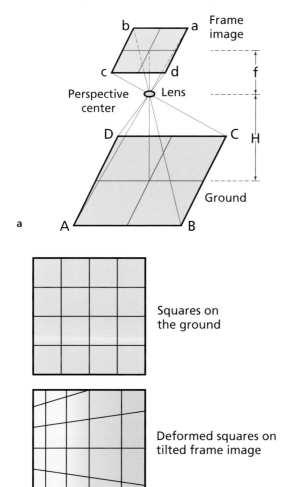

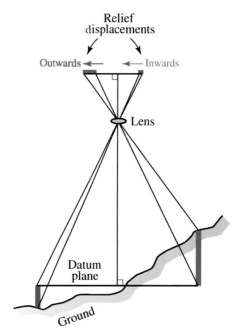

Figure A.1   (a) Geometry of a planar frame image produced by a photographic or digital frame camera with a simultaneous exposure of the entire ground area from a single point (the perspective center) in the air or from space. The scale of a vertical frame image of flat terrain is the ratio of the focal length of the lens (f) to the flying height above the terrain (H). (b) If the optical axis of the camera is truly vertical, then the squares on flat terrain would appear as squares on the image. If the camera is tilted, areas further from the camera will be imaged at a smaller scale, and the squares will appear deformed in the image. (c) Further displacements on a frame image will result if the ground is not flat. These are called relief displacements and result in further irregular changes in scale across the frame image.

Source: G. Petrie and M. Shand.

large shifts in the positions of terrain objects recorded on the image due to tilt displacement.

A special form of low-oblique frame imagery acquired from airborne platforms flying at high altitudes is *Long-Range Oblique Photography* (LOROP). This involves the use of a camera equipped with a very narrow angle lens having a long focal length (1 to 2 m) that can produce high-resolution oblique imagery of a distant target area from a standoff position over a very considerable range, on the order of 30 km *(figure A.3)*. Both film and electro-optical (EO) types of LOROP cameras are used in large numbers aboard military reconnaissance aircraft worldwide.

Overlapping frame images

Almost always, near-vertical frame images are taken in such a way that they overlap by 60% or more in the along-track direction to provide *stereocoverage* of the imaged area from a single flight line *(figure A.4)*. The resulting overlapping frame images can then be used to form *3D stereomodels* of the terrain from which orthoimages (having the same geometry as a map) and digital elevation models (DEMs) can be extracted

directly, often using highly automated techniques. Increasingly, the photographic images acquired by metric film cameras are being converted to digital form using high-precision film scanners prior to their rectification in image processing systems or digital photogrammetric workstations (DPWs).

Forward motion compensation (FMC)

For the acquisition of high-resolution imagery, provision has to be made within the frame camera for *forward motion compensation* (FMC). A device is needed to compensate for the

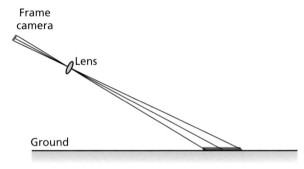

Figure A.3 A LOROP (Long Range Oblique Photographic) frame image is acquired from a reconnaissance aircraft over great distances (as much as 30 km) using a camera equipped with a long focal length lens that is operated in a tilted (low-oblique) attitude.

Source: G. Petrie and M. Shand.

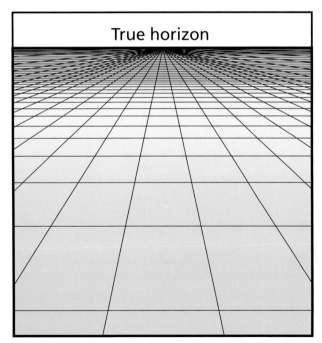

Figure A.2 High-oblique frame images, where the camera is tilted so that the earth's horizon appears within the image, have large tilt displacements and scale changes.

Source: G. Petrie and M. Shand.

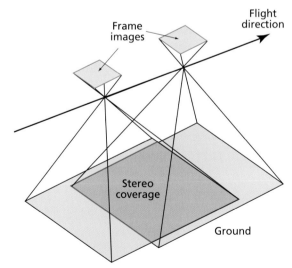

Figure A.4 Stereocoverage is produced with a frame camera by capturing a series of overlapping images along a flight line such that each frame contains a portion (usually 60%) of the terrain imaged in the preceding frame.

Source: G. Petrie and M. Shand.

forward movement of the airborne or spaceborne platform while the image is being exposed or captured to eliminate blurring of the image and to ensure its maximum sharpness. In modern airborne frame cameras, this is achieved by a controlled mechanical movement of the imaging surface within the camera's focal plane *(figure A.5)* while the camera's shutter is open and the image is being exposed and recorded.

FMC is even more important with images acquired from the frame cameras being operated in space, since the satellite is moving over the earth's terrain at the very high speed of 6 to 7 km/sec Thus, during an exposure time of 2 milliseconds (1/500th sec), the satellite will have moved 10 m over the ground. If not compensated for, the image of a high-resolution frame image acquired from the satellite would be quite blurred. FMC can be applied either through the movement of the whole frame within the focal plane of the space camera at a suitable rate or by rocking the camera as a whole through the required angular change *(figure A.6)* to compensate for the satellite motion during the time interval when the shutter is open. From the point of view of the GIS user, this important matter is being taken care of mechanically in the frame camera, so it does not become an issue during the process of image rectification.

Auxiliary instruments and data

In order to keep the image displacements resulting from camera tilts to a minimum with near-vertical imagery, many of the modern metric film cameras being operated from airborne platforms are placed on *gyro-controlled mounts* which keep the camera pointing in a near-vertical direction *(figure A.7)*. These units reduce the tilt values to about ±1° from the vertical, which is a favorable situation from the point of view of image rectification. Furthermore, the positions and flying height of the frame camera in the air, together with the residual tilt values, can be measured in-flight using an *inertial measurement unit* (IMU) coupled with a *differential GPS* (DGPS) *(figure A.8)*. In those cases where accuracy requirements are not too demanding, these measured values can be used to carry out the direct rectification or orthorectification of the images. However, more commonly, when high-accuracy map or GIS data is required, the in-flight measurements from the IMU/DGPS combination are not of a high enough quality to permit such a direct rectification to be undertaken. Instead, the IMU/DGPS measurements are used mainly as auxiliary data in procedures such as aerial triangulation that provide the network of ground control points that are needed for image rectification. Arising from the high cost of gyro-controlled mounts and of IMU/DGPS units, it must be said that many airborne film and digital cameras are being operated quite successfully without these aids. In which case, rectification and orthorectification, together with the associated georeferencing, can still

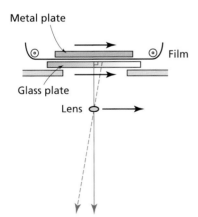

Figure A.5   Forward motion compensation. The forward motion of an airborne or spaceborne photographic frame camera during the exposure of its film needs to be compensated for, otherwise the image will be blurred. This is achieved by an appropriate small forward motion of the film sandwiched between two plates while the camera shutter is open.

Source: G. Petrie and M. Shand.

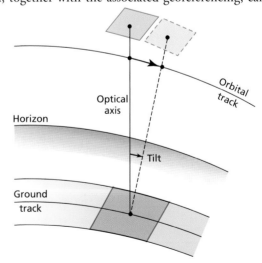

Figure A.6   Forward motion compensation for frame cameras on earth orbiting satellites. Another technique used to prevent the blurring of the frame image that would occur due to the very high speed (6 to 7 km per sec) of the satellite is to rock (i.e., tilt) the camera during the short time period when the shutter is open so that the optical axis continues to point toward the same point on the ground.

Source: G. Petrie and M. Shand.

be carried out successfully using well-established techniques and procedures based on the use of ground control points (GCPs).

Airborne film frame imagers

Typical of the *airborne film frame cameras* used for the acquisition of GIS data are the well-known RC series from Wild (now Leica) *(figure 5.3 in chapter 5)* and the RMK and LMK series from Zeiss (now Z/I Imaging, a division of Intergraph). All of these metric cameras produce planar frame images of an excellent quality having a standard square format of 23 by 23 cm (9 by 9 in). Each metric camera is calibrated in a specialist laboratory to provide accurate data about its geometric characteristics. This data comprises the exact position of the principal point (center) of the camera frame, the precise value of the focal length of the camera lens, and the pattern of geometric distortion generated by the camera lens. This calibration data is sometimes referred to as the *camera model.*

Many military reconnaissance cameras produce image data of a similar format size that has excellent interpretative qualities. However, often these reconnaissance cameras exhibit certain deficiencies in terms of their geometric qualities—for example, they are usually uncalibrated, and they may have substantial geometric lens distortions that need to be eliminated during image rectification.

Smaller-format frame cameras have also been used extensively for the acquisition of airborne remote sensing data from light aircraft. These include 35 mm film frame cameras

and the 70 mm film frame cameras (with 6 by 6 cm format) from Hasselblad, Vinten, and others. Since most of these small-format cameras have not been designed to generate data for mapping or GIS purposes, again they often exhibit undesirable geometric characteristics. These include large lens distortions or, in some cases, the use of focal plane shutters that distort images when taken from a moving platform. These effects need to be removed or compensated for during rectification.

At the other end of the format size range are the *large-format* 23 x 46 cm (9 by 18 in) frame cameras. These cameras have been used extensively by U.S. defense mapping and intelligence agencies, by NASA on its ER-2 aircraft, and by the National Ocean Survey (NOS) for its extensive mapping of coastal areas.

Spaceborne film frame imagers

Modified *aerial film cameras,* such as the ESA Metric Camera (MC) with its 23 by 23 cm (9 by 9 in) format and the NASA Large Format Camera (LFC) with its 23 by 46 cm (9 by 18 in) format, were operated from the space shuttle for short periods during the early 1980s. Only limited terrain coverage was achieved during these missions, and the resulting imagery with its wide angular coverage, small image scales (1:820,000 for MC and 1:740,000 for LFC), and moderate

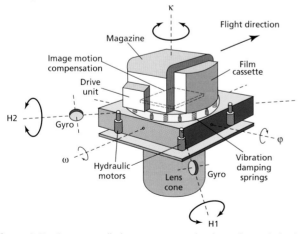

Figure A.7 Gyro-controlled mounts are used to keep the optical axis of a frame camera in a near-vertical position; the resulting images will then have the smallest possible tilts.

Source: G. Petrie and M. Shand.

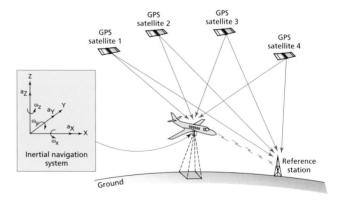

Figure A.8 The position and height of the airborne camera when the frame image is being exposed is measured by a GPS receiver linked to the camera. Small differential corrections to these position and height values are provided continuously by the signals generated at a reference or base station on the ground that are transmitted to the aircraft. The IMU provides continuous measurements of the tilt or attitude values of the camera and also provides intermediate values of the camera's position and height between those given by the DGPS system (if required).

Source: G. Petrie and M. Shand.

ground resolutions (18 m for MC and 10 m for LFC) is now mainly of historic interest *(figure A.9)*.

However, Russian agencies have continued to use large-format film frame cameras extensively from space to the present day. Of particular importance are the KFA-1000 with its 30 by 30 cm (12 by 12 in) format and the large-format TK-350 camera with its 30 by 45 cm (12 by 18 in) format. These cameras have been used to cover large areas of the earth for reconnaissance and mapping purposes, and the resulting images are now available through several outlets in the United States and Western Europe. The TK-350 is a metric camera that has been used to carry out the mapping of large parts of Europe, the Middle East, and Africa. The KFA-1000 is a reconnaissance camera that is often operated in a twin-camera, split-vertical configuration with each individual camera tilted by 8.5° to point obliquely to each side of the flight line *(figure A.10a)*. This low-oblique configuration (which has also been used with airborne reconnaissance cameras) is designed

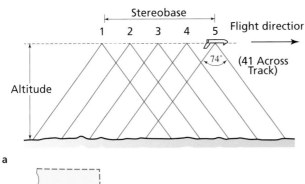

a

b

Figure A.9 (a) The use of a Large Format Camera (LFC) from the space shuttle taken with 80% forward overlap. (b) The ground coverage of a single LFC space image acquired from the space shuttle over California.

Source: G. Petrie and M. Shand.

to increase the coverage of the terrain during a single flight *(figure A.10b)*. Obviously the substantial tilts present in these images need to be eliminated through the use of suitable rectification procedures. Tests carried out by scientists in Germany have shown that the image geometry of the KFA-1000 is not too stable. Furthermore, the images acquired by this camera exhibit quite large lens distortions. These have to be taken into account during their rectification.

Airborne digital frame imagers

With airborne *digital cameras* and *video cameras*, the resulting image data is invariably of a much smaller size or format as compared with that produced by metric film cameras—the use of Kodak digital cameras equipped with a 4,000 by 4,000 pixel CCD areal array being typical. A modern example having such an array is the DSS digital frame camera *(figure 5.36 in chapter 5)*.

Larger sizes of digital cameras that give larger area coverage of the ground are now in prospect, for example, the 7,000 by 9,000 CCD aerial array developed by Lockheed (now BAE Systems). However, in general terms, the same image rectification or orthorectification procedures can be employed with frame images irrespective of their format size or whether they were recorded photographically on film or digitally on a CCD areal array.

Multiple airborne digital frame imagers

Another approach to achieving greater area coverage of the ground to overcome the limited format size of current digital cameras is to synthesize images from multiple digital cameras into a single large-format image. This particular approach has been adopted, for example, by Z/I Imaging with its Digital Mapping Camera (DMC). This system comprises four individual digital cameras, each producing a panchromatic (black-and-white) image *(figure A.11a)*. These are arranged (tilted outward) in a star-type arrangement *(figure A.11b)*. The resulting four individual tilted images, which overlap slightly, are exposed simultaneously. They are then processed (i.e., rectified) to form a single (virtual) perspective frame image *(figure A.11c)* that can be treated in the same way as a frame image from an individual film or digital camera.

Spaceborne digital frame imagers

Besides the large-format film cameras mentioned above, small-format *digital frame cameras* have also been deployed and operated from space on the several micro- and mini-satellites that have been built in the United Kingdom by SSTL and

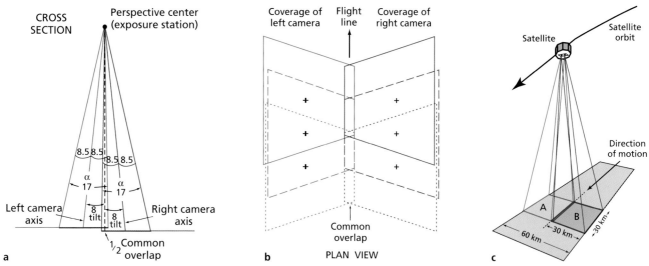

Figure A.10 (a) A cross section of split-vertical photography, comprising a pair of low-oblique (tilted) frame images. These images are acquired by a pair of frame cameras, one tilted to the left of the flight line, the other to the right of the line. The two images overlap slightly along the flight line. The diagram shows the configuration for the twin KFA-1000 cameras used in Russian spacecraft.
(b) A plan view showing the ground coverage produced by the twin-camera configuration.
(c) The split-vertical images may be acquired either from airborne or spaceborne platforms.

Source: G. Petrie and M. Shand.

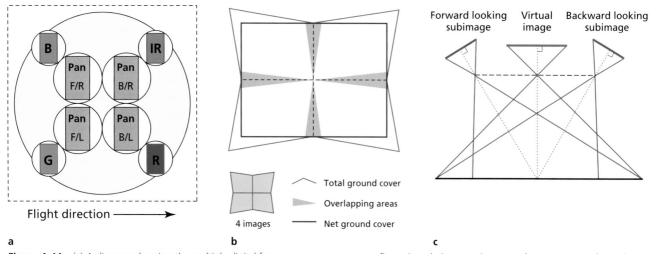

Figure A.11 (a) A diagram showing the multiple digital frame cameras making up the Z/I Imaging Digital Modular Camera (DMC) imaging system. (b) The four pan frame cameras forming the central part of the DMC system are all tilted outwards in a star-type configuration; their respective ground coverages are shown in this diagram. (c) The individual tilted DMC pan frame images need rectification to form a single virtual near-vertical image.

Source: G. Petrie and M. Shand.

launched on behalf of various countries—Korea, China, Thailand, Malaysia, Turkey, Algeria, Nigeria, Portugal, and Chile—wishing to acquire a national space imaging capability. With all of these planar frame images, more or less the same image rectification procedures can be applied as those used with the corresponding airborne images—though, of course, certain matters such as earth curvature will be more important with the images acquired from spaceborne platforms. Some of these spaceborne digital cameras have been operated in the same twin-camera, split-vertical configuration described above for the KFA-1000 film camera *(figure A.10c)*. Once again, the effects of the large tilts need to be removed during the image rectification process.

A.I (b)   Cylindrical frame images

Panoramic frame cameras employ a sequential line-by-line exposure of the frame image that uses either a moving slit traveling across the photographic film or a scanning mirror that rotates in front of the camera lens. This produces a cylindrical imaging surface *(figure A.12)*.

Airborne panoramic cameras

Film panoramic cameras have been used in many military reconnaissance aircraft, whereas only one or two digital panoramic cameras have appeared until now. While panoramic film cameras produce wide angular coverage and high-resolution frame images of an excellent interpretative quality, from the geometric point of view, the image scale will vary systematically across the field of view of each image *(figure A.12c)*. Besides these regular changes of scale that occur across the image format, further random and less predictable changes are produced by the forward movement and the variations in the tilts of the airborne or spaceborne platform that occur during the time that it takes for the exposure slit to scan across the cylindrical imaging surface of the camera. In summary, the frame images acquired by panoramic cameras exhibit quite a complex geometry with a new exposure station for each successive line making up the image. The large and somewhat unpredictable image displacements and scale changes lead to a quite complicated image rectification procedure being required with cylindrical frame images.

Spaceborne panoramic cameras

As noted above, military reconnaissance aircraft are often fitted with panoramic frame cameras. However, in recent years, this type of camera has achieved further prominence within the remote sensing and GIS communities with the availability of cylindrical frame images taken from space platforms. In particular, the vast library of Corona images acquired by U.S. reconnaissance satellites up until 1972, were released into the civilian domain in 1995. These form a unique historic record

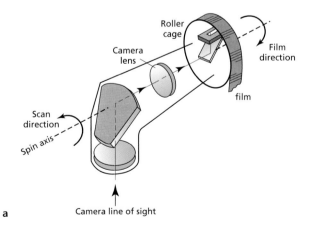

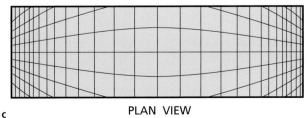

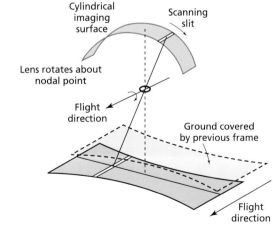

Figure A.12   (a) A panoramic frame camera using a rotating optical bar to scan the terrain. The film occupies a cylindrical image surface. (b) This image shows the cylindrical imaging surface and the corresponding ground coverage of a panoramic frame image. (c) The scale of a panoramic frame image will vary greatly across the image and requires rectification.

Source: G. Petrie and M. Shand.

and are now beginning to be used quite widely, especially for the detection and tracking of the changes that have taken place on the ground over a substantial period of time. The high-resolution Corona frame images were taken by a pair of panoramic frame cameras usually in a tilted twin-convergent configuration to produce overlapping coverage in the along-track direction *(figure A.13)*. The resulting stereoimagery allowed the stereoscopic interpretation and 3D measurement of ground objects and even the generation of DEMs. The other major source of high-resolution panoramic frame images that have been taken from space are those acquired by Russian agencies using the KVR-1000 panoramic camera. The resulting high-resolution images (with 1 to 2m ground pixel size) are now available for many areas outside Russia. The same rectification procedures that are used with airborne panoramic images will apply to the corresponding images obtained from spaceborne platforms.

A.II   Line scanner images

All scanner images are built up sequentially line-by-line using the platform motion over the ground to ensure coverage of the appropriate piece of terrain for each successive line of the image. This process results in the exposure of a *continuous strip image* of the terrain instead of the frame-type images discussed above. With scanner imagery, each line of the image has its own individual and different perspective center instead of the single perspective center that exists for a complete frame image acquired by an aerial or space photographic or digital frame camera. Thus, a line scanner image,

such as those acquired from SPOT satellites, in effect, comprises 6,000 *individual perspective images* that have butted together to form a single continuous strip image. The *perspective projection* for each individual line of the image is constrained to occur only in the direction of the scanned line, i.e., at right angles to the flight direction *(figure A.14)*.

Need for auxiliary instruments

Besides the need to establish the 3D coordinates of each of the 6,000 perspective centers corresponding to each individual line of the scanner image for rectification and georeferencing purposes, it is also necessary to take into account the changing attitude or tilt of the airborne or spaceborne platform and its imager over the time period during which the image was acquired. Usually the preliminary values needed to carry out the required modeling of the flight path and the orientation of the platform and its imager required for image rectification purposes are derived from the data generated by the *auxiliary instruments* operating onboard the platform. The IMU/DGPS combination often used in the case of airborne scanner imagery and the DORIS/stellar tracker combination used onboard the SPOT satellites are examples of these auxiliary instruments. Later the measured values obtained from the auxiliary instruments will be refined through the use of accurate ground control point (GCP) data, e.g., GCPs measured by DGPS.

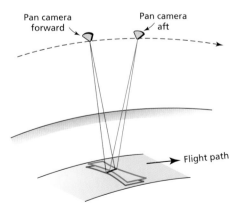

Figure A.13   The stereocoverage produced by Corona space images were acquired by U.S. reconnaissance satellites equipped with twin convergent panoramic cameras pointing in the forward and backward directions along-track.

Source: G. Petrie and M. Shand.

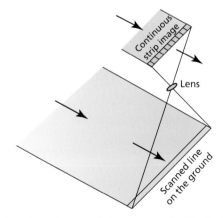

Figure A.14   A continuous strip image is acquired line-by-line using a line scanner. Each individual line of the image has a different perspective center, while the perspective projection occurs only in the across-track direction.

Source: G. Petrie and M. Shand.

Types of line-scanner images

From the point of view of image rectification, line-scanner images can be classified into three main types:

- those produced by *optical-mechanical scanning of the terrain*
- those produced by *linear arrays*
- those generated by *conical scanners*

Each has a different type of geometry that must be taken into account during image rectification.

A.II (a)   Optical-mechanical scanner images

Optical-mechanical scanning has been used widely in all the imaging scanners operated on board the satellites in the Landsat series. These include the Multi-Spectral Scanner (MSS) used in Landsats 1, 2, and 3; the Thematic Mapper (TM) used in Landsat 4, and 5; and the Enhanced Thematic Mapper (ETM+) used in Landsat 7. In each case, an oscillating plane mirror is used to scan the terrain in the direction perpendicular to the flight line *(figure A.15)*. The radiation from the ground picked up by the scanning mirror then passes through the main optical system and is measured by detectors. The satellite will have moved forward over the ground during the time it takes for the scanning mirror to complete its scan of a complete line. To correct for this movement, an electro-mechanical *scan line corrector* (SLC) is inserted between the lens and the detectors *(figure A.16)*. Similarly with the NOAA AVHRR scanners, a flat mirror is used to carry out the scanning of the terrain at right angles to the flight direction. However, in this case, the scan mirror performs a continuous 360° rotation, with the speed of rotation selected so that the adjacent scan lines are contiguous at the point on the terrain directly beneath the satellite (nadir). Similar arrangements using scanning mirrors and detectors have also been employed in early airborne scanners such as the Daedalus visible/infrared (VIS/IR) line scanners and various types of thermal infrared line scanners.

Geometric aspects

From the geometric point of view, the resulting continuous strip image produced by an optical-mechanical scanner comprises a series of very thin line images, each of which has been recorded on a *cylindrical surface* and abuts exactly against the two adjacent line images on either side of it *(figure A.17)*. Each individual line has its own (slightly different) perspective center and, at least potentially, a slightly different but slowly changing tilt value in the case of the spaceborne optical-mechanical scanners. With airborne scanners of this

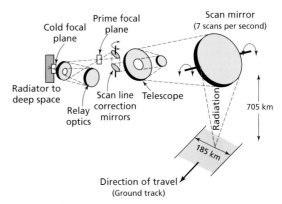

Figure A.15 With Landsat line scanner images, optical scanning is carried out only in the across-track direction using a rapidly oscillating mirror.

Source: G. Petrie and M. Shand.

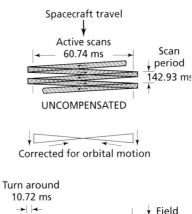

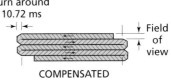

Figure A.16 An electro-mechanical scan line corrector (SLC) is used to compensate for the forward motion of the Landsat satellite while the scanning of the ground is taking place to form the image.

Source: G. Petrie and M. Shand.

type, the tilt values will be much larger and so will the resulting *tilt displacements*. Moreover, the tilts may change rapidly in different and quite arbitrary directions and amounts between one line and another due to the effects of atmospheric turbulence.

Relief displacements can only take place in the scan line direction, i.e., at right angles to the flight line. With the higher-resolution TM and ETM+ scanners used in the later Landsat series, the effects of terrain relief can quite definitely be seen and felt. These relief displacements need to be compensated for during image rectification through the use of a digital elevation model obtained from other sources. With the much coarser resolution AVHRR images (with 1km ground pixel size), there is generally very little relief displacement. A very substantial mountain range must be present before any relief effects can be felt. At the other end of the scale and resolution range, relief displacements will be large when an optical-mechanical scanner is operated from an airborne platform at low altitudes. Furthermore, in built-up urban areas, many dead areas will occur (i.e., no image will be recorded) due to the shadowing effects of tall buildings.

A.II (b)   Linear array (pushbroom) scanner images

In its basic configuration, this type of scanner contains a linear array of multiple CCD detectors located in the focal plane of the instrument's main lens or mirror optical system. This eliminates the need for a scanning mirror. Moreover, it ensures that all the image data being sensed by this linear array is being detected, measured, and recorded simultaneously. Hence there is no need for a scan line corrector device to be incorporated within the instrument. As with all scanners, the forward motion of the airborne or spaceborne platform over the ground provides the change in position required for the next line of the image to be exposed—hence the common description of instruments equipped with CCD linear arrays as being pushbroom scanners. The geometry of each complete line is that of a central projection constrained to lie within the plane containing the scan line and the instrument's lens or projection center. The projection of each line onto the terrain forms a very narrow rectangle on the terrain (*figure A.18*). If tilt was not present, then the continuous

series of exposed lines butted together would form a planar flat surface (*figure A.19*). This distinguishes it geometrically from the cylindrical shape of the imaging surface produced by the optical-mechanical scanner.

Twin-coupled linear arrays

Although, in principle, the basic geometry of the pushbroom scanner equipped with a single CCD linear array is relatively simple, in practice, matters can become complex. In turn, this affects the image rectification process being carried out later. First, in many modern pushbroom scanners, two linear arrays are butted together parallel with the additional array shifted horizontally with respect to the first by half-a-pixel.

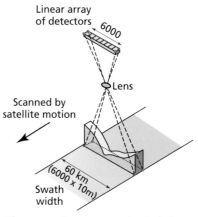

Figure A.18   The perspective geometry of a single line on a pushbroom scanner, such as that used in the SPOT satellite. Relief displacements on the image take place in the cross-track direction only.

Source: G. Petrie and M. Shand.

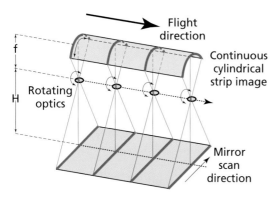

Figure A.17   The continuous strip image produced by an optical-mechanical line scanner (as used in Landsat) is recorded on a cylindrical surface.

Source: G. Petrie and M. Shand.

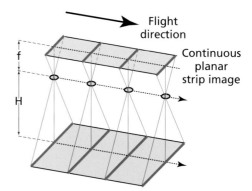

Figure A.19   The continuous strip image produced by a pushbroom scanner equipped with a linear array of CCDs is recorded on a flat (planar) surface.

Source: G. Petrie and M. Shand.

The purpose of this arrangement is to provide an improved radiometric and geometric resolution of the image by averaging the values measured by the two linear arrays. Thus, the two image datasets need to be superimposed, interlaced, and averaged during subsequent data processing *(figure A.20)*. This particular type of arrangement is used in the SPOT-5 HRG pushbroom scanner (called Supermode), in the Quick-Bird pushbroom scanner deployed in space, and in the Leica ADS40 airborne pushbroom scanners.

Multiple linear arrays

A second important matter concerns the limitations in the length of past and present CCD linear arrays. In the past, they were limited to around 4,000 detectors; in 2003, the longest CCD linear array contained around 12,000 to 13,000 detectors. In order to ensure greater angular coverage or swath width over the terrain, some pushbroom scanners have multiple linear arrays that are set in the across-track direction within the focal plane of the imager. Examples are the imagers on the earlier IRS-1C and -1D satellites, which used three linear arrays, each containing 4,096 individual detectors to give a total length of more than 12,000 pixels. More recently, the QuickBird imager also featured three linear arrays, each with around 9,000 individual detectors, to provide an across-track coverage of 27,000 pixels *(figure A.21)*. These arrangements all produce shifts or offsets between the resulting three strip images. Adjustments need to be made for these offsets during image rectification, after which the three strip images are stitched together to produce the final composite (strip) image. Sometimes, as with the IRS-1C and -1D images, small discontinuities or mismatches will exist between the component strip images. These also need to be removed during image rectification.

Multispectral linear array images

A further complication from the geometric and image rectification point of view occurs with the *multispectral imagery* acquired by pushbroom scanners. If, for example, the imager is generating four images in the blue, green, red, and near-infrared parts of the spectrum, then each of the four bands will be imaged using a separate linear array equipped with a suitable filter *(figure A.22)*. This is the case, for example, with the QuickBird multispectral scanner/imager. Since these four linear arrays cannot all occupy the same position within the focal plane of the imager, they are offset relative to one another. This offset or shift has to be allowed for or corrected during processing to produce a single (ideal) final line. In some cases, as for example, with the Leica ADS40 airborne pushbroom scanner, only the central high-resolution panchromatic image is acquired in the vertical (nadir) position *(figure A.23)*. The other linear arrays that produce the ADS40's multispectral images are offset within the focal plane from the vertical (nadir) position. Thus, they occupy a forward-pointing position having tilt values of 14.2° from the vertical. Obviously, this particular geometric arrangement needs to be taken care of during the subsequent image rectification procedure.

Tilted linear array images

A further important consideration with pushbroom scanners concerns the method used to point the imager at the ground in a deliberately tilted (i.e., oblique) position. This has particular

Multiple linear arrays

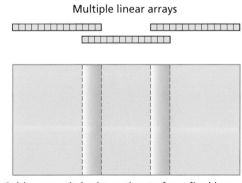

Subimages stitched together to form final image

Figure A.21   Multiple linear arrays are often used to provide a wider angular coverage in a pushbroom scanner, producing a greater swath width over the ground. The subimages produced by the individual arrays need to be stitched together during the subsequent image data processing.

Source: G. Petrie and M. Shand.

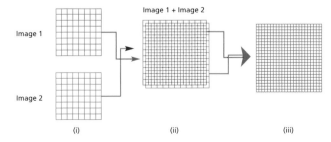

Figure A.20   Twin-coupled linear CCD arrays use two linear arrays shifted by half-a-pixel relative to the other. To generate image products, the two datasets must first be superimposed, interlaced, and averaged.

Source: G. Petrie and M. Shand.

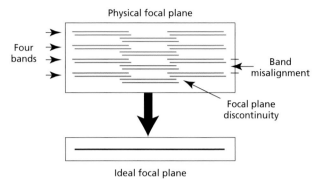

Figure A.22 For the production of multispectral imagery using linear CCD arrays, numerous arrays need to be accommodated in the focal plane of the pushbroom scanner imager. The resulting offsets and shifts need to be corrected later to produce image products with the different spectral band images in correct registration.

Source: G. Petrie and M. Shand.

relevance to the acquisition of stereocoverage from airborne and spaceborne platforms and, in the case of the spaceborne imagers, to the reduction of the time required to revisit an area and acquire new coverage of the area. Needless to say, the images produced from these tilted configurations will require substantial image rectification if they are to be used in a GIS system.

(i) Across-track configuration

This particular configuration is used in certain satellites such as the SPOT and IRS-1C/D series. It makes use of a mirror mounted in front of the pushbroom line scanner that can rotate in the across-track direction to provide off-nadir imaging capability. Under command from a ground station, the mirror can be rotated in steps to either side of the flight line or orbital track *(figure A.24)* to allow images to be acquired to the side of the satellite's ground track. Stereocoverage is produced from two separate orbits flown on adjacent ground tracks using the overlapping images acquired with the linear array sensor pointing in opposite directions during the two orbits *(figure A.25)*. The scanner image systems on the SPOT satellites have been used extensively in this way to produce stereocoverage from which DEMs can be generated. The IRS-1C and -1D satellites have also been operated in a similar manner.

(ii) Along-track configuration

This arrangement involves the use of two linear arrays of detectors, each tilted in a fixed orientation: one pointing in the forward direction down the flight line and the other

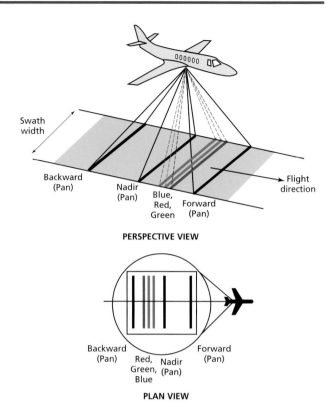

Figure A.23 The diagram shows a three-line pushbroom scanner with the individual linear CCD arrays pointing in the forward, nadir, and backward positions to produce overlapping pan (black-and-white) images. The linear arrays that produce the multispectral images then have to be placed in an intermediate position between the individual arrays that produce the pan images. Thus, in this case, they point forward in a tilted configuration.

Source: G. Petrie and M. Shand.

pointing in the backward direction along the same line *(figure A.26a)*. These sensors produce overlapping images of the same piece of terrain from different positions in the air or in space during a single flight. The combination of the two images provides overlapping stereocoverage of the swath of the earth's surface recorded by the two linear arrays. The Leica ADS40 and DLR HRSC-A airborne scanner imagers both operate in this mode. From space, this particular configuration was demonstrated first with the JERS-1 OPS and MOMS-02 image systems and has now been adopted in the form of the dedicated HRS (high-resolution stereo) imager mounted on the SPOT-5 satellite *(figure A.26b)*. Other arrangements, such as the combination of a vertical-pointing image and a forward-pointing image used by the ASTER imager on board the Terra satellite, are also in use.

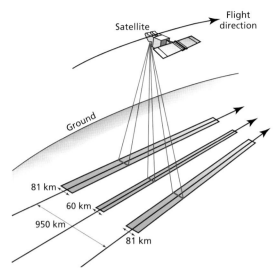

Figure A.24   Off-nadir images are produced by tilting a mirror placed in front of the pushbroom linear array scanner in the across-track direction, as is done with the SPOT satellites. The resulting tilted images require rectification.

Source: G. Petrie and M. Shand.

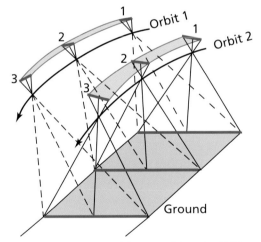

Figure A.25   Stereocoverage of the ground from the SPOT-1 to SPOT-4 satellites and the Indian IRS-1C and IRS-1D satellites is produced from two separate orbits flown on different dates with the linear array scanners pointing inward in the across-track direction to give overlapping images of the ground. Each of the overlapping images needs to be rectified to fit the ground coordinate system.

Source: G. Petrie and M. Shand.

(iii) Flexible pointing configuration

To implement this type of configuration, either gimballed mirrors or a whole body movement of the spacecraft is used to point the line scanner in any required direction off-nadir.

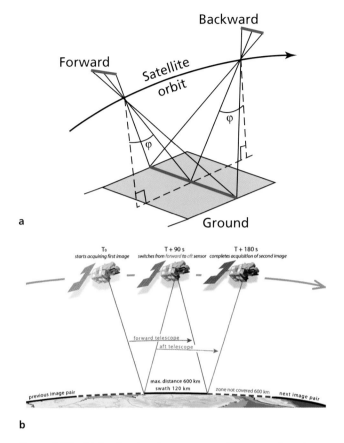

Figure A.26   (a) Overlapping stereocoverage of the ground can be produced along-track during a single orbit from certain spaceborne platforms (e.g., MOMS-02, SPOT-5, ASTER) and from airborne platforms (e.g., Leica ADS40, DLR HRSC-A) using pushbroom scanners equipped with twin linear CCD arrays: one pointing in the forward direction; the other in the backward direction.
(b) The overlapping stereocoverage of a 600 km segment of the terrain surface as acquired by the HRS along-track scanner image system mounted on the SPOT-5 satellite.

Source: (a) G. Petrie and M. Shand. (b) © CNES 2004/Courtesy of SPOT Image Corporation

Whichever method is used, the sensor can be commanded to point in any direction at viewing angles of up to 45° from the vertical. This allows the acquisition of image data tilted in both the across-track and along-track directions or in any other intermediate direction *(figure A.27)*. Stereocoverage of an area of terrain can be acquired either from a single orbital flight or from adjacent orbits. The high-resolution imagers equipped with linear CCD arrays mounted on the IKONOS, QuickBird, OrbView-3, and EROS satellites provide this type of operating mode.

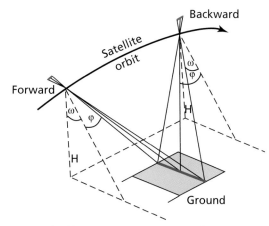

Figure A.27 Flexible pointing towards an area of interest lying off nadir on either side of the satellite ground track combines both across-track and along-track tilting of the images to form an overlapping stereopair.

Source: G. Petrie and M. Shand.

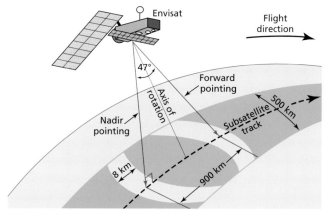

Figure A.28 The conical scanning action of the ATSR and AATSR imagers mounted on the ERS-1, ERS-2, and Envisat satellites produces two scans of each patch of the ocean or land surface, one from the nadir position, the other from an off-nadir tilted position along-track.

Source: G. Petrie and M. Shand.

A.II (c) Conical scanner images

The imagery generated by this type of scanner is perhaps less likely to be encountered by GIS users than the two previous types. However, for those engaged in environmental applications such as global monitoring, it can be important. An early example of this type of imager was the S-192 conical multispectral scanner operated from *Skylab* in the early 1970s. The S-192 collected data in 13 spectral bands in the visible and near-infrared (VIS/NIR) parts of the spectrum. However, better known and more widely used are the along-track scanning radiometer (ATSR) imagers (ATSR-1, ATSR-2, and AATSR), which have been operated continuously on board the ERS-1, ERS-2 and Envisat satellites respectively since 1991. All of these imagers employ a unique conical scanning arrangement that results in images having a dual-scan viewing geometry *(figure A.28)* for the same piece of ground or ocean. During conical scanning, one image is acquired in the nadir direction, and the other is acquired at an angle of 53° to the vertical. This allows two sets of radiometric measurements to be made of the same piece of land or ocean from two different positions, at two different viewing angles, and over two different path lengths. This allows the measured radiation to be corrected for the effects of atmospheric absorption, so ensuring greater radiometric accuracy. These conical scanner images need substantial rectification before the image can be related to the ground reference system used in maps and GIS.

B. Image rectification and geo-referencing using two-dimensional transformations

The use of 2D transformations to carry out the rectification and georeferencing of remotely sensed imagery is quite common within the remote sensing community. In this particular approach to image rectification, it is assumed that a quite simple mathematical relationship exists between the locations of all the points recorded on a remotely sensed image and the corresponding locations of these points in the ground coordinate system being used either for mapping or as the reference system within the GIS. Normally this mathematical relationship or model comprises a 2D *transformation* based on the use of *polynomial equations* containing various terms of differing degrees. Thus, the mathematical relationship that has been adopted for the image rectification and georeferencing simply attempts to relate the image space to the object space (i.e., the ground) as well as possible using this type of 2D transformation. No attempt is made to model the geometric relationship between the sensor position and orientation and the terrain at the time the image was acquired. Nor is there any attempt to model and deal with the effects of the terrain relief present in the area covered by the image that is being rectified.

Use of polynomial equations

Two-dimensional transformations based on polynomial equations have been used extensively by remote sensing practitioners. This is particularly the case where the main interest lies in the interpretation of the features present in the image and in the thematic mapping of vegetation, forests, soils, geology, etc.—where the demands for the positional accuracy of the final data are often quite moderate. Thus, virtually every remote sensing image processing software package provides a 2D transformation capability for the purposes of image rectification and georeferencing. While these 2D transformations are almost always based on polynomial equations, the actual terms that are used will vary from one package to another. Image data that has been processed in this way will often pass into the hands of GIS users. Some understanding of the basis of these 2D transformations is useful for GIS users to assess the suitability of the geometric characteristics of the image data rectified in this way for their specific application.

B.I   Two-dimensional transformations using polynomial equations

The polynomial equations that form the basis of most of the 2D transformations used for image rectification in remote sensing have the following general form:

$$
\begin{aligned}
X = {}& a_0 && \text{(a constant term)} \\
& + a_1x + a_2y && \text{(linear [1st order] terms)} \\
& + a_3xy + a_4x^2 + a_5y^2 && \text{(quadratic [2nd order] terms)} \\
& + a_6x^2y + a_7xy^2 + a_8x^3 + a_9y^3 && \text{(cubic [3rd order] terms)} \\
& + a_{10}x^3y + a_{11}xy^3 + a_{12}x^4 + a_{13}y^4 + a_{14}x^2y^2 \\
& && \text{(quartic [4th order] terms)} \\
& + a_{15}x^3y^2 + a_{16}x^2y^3 + a_{17}x^5 + a_{18}y^5 + a_{19}x^4y + a_{20}xy^4 \\
& && \text{(quintic [5th order] terms)} \\
& + \ldots\ldots\text{[still higher order terms].}
\end{aligned}
$$

$$
\begin{aligned}
Y = {}& b_0 && \text{(a constant term)} \\
& + b_1x + b_2y && \text{(linear [1st order] terms)} \\
& + b_3xy + b_4x^2 + b_5y^2 && \text{(quadratic [2nd order] terms)} \\
& + \ldots\ldots && \{\text{as above for X, but with "b" substituted for "a"}\}
\end{aligned}
$$

Where

X and Y are the *ground coordinates* of one of the ground control points (GCPs)

x and y are the corresponding *image coordinates* of the same point as measured on the image

a and b (i = 1 to n) are the *transformation parameters* or coefficients connecting the ground and the image coordinate systems

The use of sets of these equations allow the transformation of coordinate data measured on the image (in x,y coordinates) into the ground or GIS reference system (in X,Y coordinates), or the reverse operation.

Selection of the terms used in the 2D transformation

The pair of basic equations that have been set out above have been stated in a very general form. While some software packages may include all of these terms in the image rectification and georeferencing procedure that they provide, others will only provide a *subset* of these terms—those that the designers of the software package have deemed to be most useful for the particular application for which the software is to be used. Furthermore, even when all of the various terms set out in the equations above are available in the 2D transformation engine included in the software package, it is often possible for the user to select only a subset of them. In this situation, it is helpful for the user to understand the effects of each of the terms being used or offered by the 2D transformation procedure and the pattern of distortion or displacement that is being modeled or corrected through the inclusion of a particular term. These are shown in graphical form for each term in figure A.29. The user can then select which set of terms will best carry out the image rectification, having regard to the image geometry and the effects of

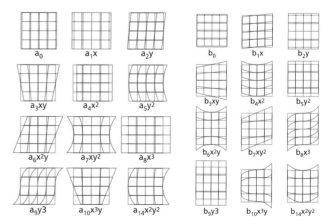

Figure A.29   Graphical representation of the individual **a** and **b** terms in the 2D polynomial equations.

Source: G. Petrie and M. Shand.

tilt and relief displacement on the particular type of image that is being handled. Some of the more commonly used 2D transformations are described below.

(i) Affine transformation

Starting with a simple example, an *affine transformation* will comprise only the first three terms of each equation given in the general set of equations set out above, as follows:

$$\left.\begin{array}{l} X = a_0 + a_1x + a_2y \\ Y = b_0 + b_1x + b_2y \end{array}\right\} \text{(i)}$$

This particular transformation can carry out the corrections needed for the scaling, rotation, and shearing of the image during the rectification process and convert the measured image data (in x,y coordinates) into the GIS reference system (in X,Y coordinates).

The addition of second-order or other higher-order terms to this initial subset will allow more complex distortions or displacements in the original image to be modeled and removed using the 2D transformation procedure (refer to figure A.29 to see the effects of the addition of these terms). The following sections discuss typical transformations using higher-order terms that can be found in remote sensing image processing packages.

(ii) Bilinear transformation

$$\left.\begin{array}{l} X = a_0 + a_1x + a_2y + a_3xy \\ Y = b_0 + b_1x + b_2y + b_3xy \end{array}\right\} \text{(ii)}$$

The polynomial equation used in the bilinear transformation comprises the constant term (a_0 or b_0) and the two first-order terms (a_1x, a_2y or b_1x, b_2y) from each of the two equations contained in the general polynomial expression given above, together with the product (a_3xy or b_3xy) of these two linear terms—hence the use of the term bilinear. Essentially it comprises the affine transformation plus the two (x,y) product terms.

(iii) Quadratic transformation

If all the first- and second-order terms in the general equation are used together with the constant term, this gives rise to the following quadratic transformation:

$$\left.\begin{array}{l} X = a_0 + a_1x + a_2y + a_3xy + a_4x^2 + a_5y^2 \\ Y = b_0 + b_1x + b_2y + b_3xy + b_4x^2 + b_5y^2 \end{array}\right\} \text{(iii)}$$

(iv) Bi-quadratic transformation

$$\left.\begin{array}{l} X = a_0 + a_1x + a_2y + a_3xy + a_4x^2 + a_5y^2 + a_6x^2y + \\ \quad a_7xy^2 + a_8x^2y^2 \\ Y = b_0 + b_1x + b_2y + b_3xy + b_4x^2 + b_5y^2 + b_6x^2y + \\ \quad b_7xy^2 + b_8x^2y^2 \end{array}\right\} \text{(iv)}$$

Again it can be seen that the individual terms included in the polynomial equations used in this transformation comprise all six of those up to second order occurring in the general polynomial equations together with the additional terms containing their products (xy^2; x^2y; x^2y^2)—hence the use of the term *bi-quadratic* to describe this transformation.

(v) Cubic transformation

The cubic transformation includes the first-, second-, and third-order terms together with the constant term, this gives rise to the following form of equation in the cubic transformation:

$$\left.\begin{array}{l} X = a_0 + a_1x + a_2y + a_3xy + a_4x^2 + a_5y^2 + a_6x^2y + \\ \quad a_7xy^2 + a_8x^3 + a_9y^3 \\ Y = b_0 + b_1x + b_2y + b_3xy + b_4x^2 + b_5y^2 + b_6x^2y + \\ \quad b_7xy^2 + b_8x^3 + b_9y^3 \end{array}\right\} \text{(v)}$$

(vi) Bi-cubic transformation

$$\left.\begin{array}{l} X = a_0 + a_1x + a_2y + a_3xy + a_4x^2 + a_5y^2 + a_6x^2y + \\ \quad a_7xy^2 + a_8x^2y^2 + a_9x^3 + a_{10}y^3 + a_{11}x^3y + a_{12}xy^3 \\ \quad + a_{13}x^3y^3 + a_{14}x^3y^2 + a_{15}x^2y^3 \\ Y = b_0 + b_1x + b_2y + b_3xy + b_4x^2 + b_5y^2 + b_6x^2y + \\ \quad b_7xy^2 + b_8x^2y^2 + b_9x^3 + b_{10}y^3 + b_{11}x^3y + b_{12}xy^3 \\ \quad + b_{13}x^3y^3 + b_{14}x^3y^2 + b_{15}x^2y^3 \end{array}\right\} \text{(vi)}$$

The polynomial equations used in this particular transformation comprise the ten terms up to the third order given in the general polynomial equations set out above. However, it will also include the specific fourth-order (e.g., x^2y^2, x^3y, xy^3), fifth-order (e.g., x^3y^2, x^2y^3), and sixth-order (x^3, y^3) terms formed by the products containing the second- and third-order terms —hence the use of the name *bi-cubic*.

Requirements for ground control points (GCPs) for image rectification and georeferencing

Ground control points (GCPs) are those points located on the earth's surface on which the image rectification procedure will be based. Such points should be very well defined and very clearly and easily identified and measured on the image on which they are to be used. Typical GCPs are features such as road and stream intersections, wall and fence intersections or corners, bridges, and very distinctive topographic points such as a sharply defined headland projecting into a lake.

The matter of obtaining accurate positional coordinate values for these GCPs is also an important consideration. Very often, remote sensing practitioners will measure the coordinate values on a vector- or raster-based topographic map, either manually with a scale or using a tablet digitizer. This has an immediate effect on the image rectification. In the first place, there is the limitation produced by the *scaling/measuring process* itself. Even if the X,Y coordinates could be measured off the map to an accuracy of ± 0.1 mm, this limits the accuracy to 5 m in the case of a 1:50,000 scale map (where 1 mm on the map ≡ 50 m on the ground) and 2.5 m in the case of a map at 1:25,000 scale. In many cases, the accuracy actually achieved in the measuring process is substantially lower—±0.2 to 0.3 mm is not uncommon.

Accuracy of ground control points

Furthermore, the *actual features* shown on the map may not be sufficiently accurate for image rectification. In this respect, the U.S. National Map Accuracy Standards (NMAS) specify that 90% of the well-defined features shown on a topographic map should lie within 1/50th inch (0.5 mm) of their correct position—equivalent to a root mean square error (RMSE) of ±0.3 mm. At 1:50,000 scale, this amounts to ±15 m; at 1:25,000 scale, this will be ±7.5 m. So there are limitations in the accuracy of the data shown on the map. Thus, if accurate image rectification is to be achieved, and especially if orthorectification is required, then the ground coordinates need to be obtained using differential GPS (DGPS) or some other comparable ground survey method, e.g., the use of total stations in the field. Using such methods, the accuracy of the GCP coordinate data will then be in the submeter class; the exact requirement with regard to the accuracy of the coordinates will depend on the scale and resolution of the image being rectified.

Number of GCPs required for image rectification and georeferencing

If an *affine transformation* has been selected to carry out the image rectification and georeferencing, it will be seen that there will be six unknown terms—a_0, a_1, a_2 and b_0, b_1, b_2—in the equations. In this case, the use of three suitably placed ground control points (GCPs) will generate six equations (two for each GCP). This will allow the numerical values of all six terms to be determined through the solution of the six simultaneous equations. In practice, more GCPs will

be provided than this minimum in order to provide a check against errors in the identification of the GCPs either on the map or on the image and to provide redundancy in the measurements. Typically five or six GCPs will be provided when an affine transformation is being implemented.

If the two second-order (x,y) product terms have been selected, i.e., added by the user, to the original six terms of the affine transformation, then, as explained above, it produces the *bilinear transformation*. In this case, there are eight unknown terms—a_0 . . . a_3 and b_0 . . . b_3—whose values need to be determined. Eight equations need to be formed to enable a solution for the values of these eight terms; this will be provided through the provision of four ground control points (GCPs) with known X and Y coordinate values in the ground reference system. In practice, more GCPs will be provided—the use of eight points is typical in this particular case.

With the simple *quadratic transformation,* there will be twelve unknown polynomial terms—a_0 . . . a_5; b_0 . . . b_5—whose values need to be determined. Thus, twelve equations are needed for a solution. In turn, this means that a minimum of six GCPs with known ground coordinate values must be provided, although 10 to 12 are typically used when a quadratic transformation is employed as the basis for the image rectification.

The GCP requirements for the *bi-quadratic, cubic,* and *bi-cubic transformations* can be established in a similar manner. If, as is usual, more ground control points (GCPs) are available than the minimum number needed to solve the set of simultaneous equations and determine the values of the unknowns (a_0 . . ., b_0 . . .), then a *least-squares solution*[1] will be implemented to give the most probable value for each of these terms.

Transformation procedure

(i) As for the actual image rectification and georeferencing procedure, the two equations for a particular point are set up, using the known ground coordinates (X,Y) of a ground control point (GCP) in the ground reference system and the measured coordinates (in x,y pixel values) of the corresponding point on the remotely sensed frame or scanner image. The equations for each of the other GCPs available are also set up in a similar manner. In this way, the required set of simultaneous equations is created.

---

1. *The least-squares criterion* states that the sum of the squares of the residuals or corrections to the set of measurements must be a minimum (Mikhail, Bethel, and McGlone 2001, p. 390).

(ii) The set of simultaneous equations can then be solved mathematically so that the previously unknown numerical values of the parameters (a,b) of the 2D transformation can be obtained by the computation.

(iii) Next the now known numerical values of the parameters (a,b) that have just been determined are substituted back into the equations of the transformation model. They can then be used to compute the 2D positions (X,Y) in the ground coordinate system of all or any other points whose coordinate positions have been measured as x,y values on the image.

(iv) To carry out the *complete rectification* of the whole image, for every (x,y) pixel in the image, the corresponding ground (X,Y) position will be calculated. In this way, rectification and georeferencing will have been carried out for the whole of the image. This then allows the reverse operation to be implemented—as may be required in the GIS system. Thus, if the X,Y ground coordinates of a point are input to the system, then the corresponding point on the rectified image (with x,y coordinates) will be located.

Image resampling

Once the 2D transformation has been completed, the rectified image has to undergo a final resampling procedure. This involves the assignment of a brightness value for each individual pixel in the rectified image. These values are based on the corresponding brightness values that were present in the original (unrectified) image. The three alternative techniques (*figure A.30*) commonly used in this resampling operation are called nearest neighbor, bilinear resampling, and cubic convolution respectively. With the *nearest neighbor* procedure, the brightness value of the nearest input pixel on the unrectified image is assigned to the corresponding output pixel on the rectified image. Using the *bilinear* procedure, the distance-weighted average of the brightness values of the four closest input pixels is calculated and assigned to the output pixel on the rectified image. Using the *cubic convolution* procedure, the brightness values of the 16 nearest pixels are used to determine the value for the output pixel on the rectified image. Frequently this is carried out by fitting a second-order polynomial surface through the brightness values for the relevant four by four pixel patch in the unrectified image and then calculating the brightness value at the center of the output pixel.

The use of patchwise polynomials

The use of polynomial equations to carry out the image rectification may result in the image fitting well at the ground

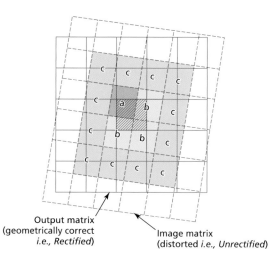

Figure A.30 Image resampling. After the rectification and georeferencing of an image has been carried out, a brightness value has to be assigned to each pixel in the rectified image. These values are based on the corresponding brightness values in the original unrectified image. In the nearest neighbor technique, the nearest value (a) is adopted. In the bilinear procedure, the distance-weighted average of the brightness values of the four nearest pixels (the a and b values) are used. In the cubic convolution procedure, the value for the output pixel on the rectified image is derived from the brightness values of the 16 nearest pixels (the a, b, and c values) in the unrectified image.

Source: G. Petrie and M. Shand.

control points (GCPs), whereas elsewhere in the image, the fit of the rectified image to the map or GIS data may be much less good. Often this will result from the unpredictable effects of the higher-order terms of the polynomial on the rectified image, especially in the areas lying between the ground control points (GCPs) on which the rectification and georeferencing has been based. Thus, in many cases, remote sensing practitioners will use only the lower (first- and second-) order terms in the 2D transformation procedure being used to implement the image rectification. To try to cut down the effects of using higher-order terms, an alternative approach is to employ a *patchwise procedure* using low-order polynomials within a limited area or patch of the image. These lower-order polynomials are often described as *patchwise polynomials* or functions. Of course, this does mean that a certain minimum number of ground control points (GCPs) need to be present in each patch. If the image has been subdivided into a large number of patches, than the requirements for GCPs will rise accordingly. Other more complex types of 2D polynomial, such as the *multiquadric approach*, have also been used for image rectification.

B.II Two-dimensional projective transformations

Besides the 2D polynomial transformations commonly used for image rectification and georeferencing in remote sensing image processing software, a few packages will also provide 2D projective transformations. Essentially these can be regarded as simplified versions of the 3D projective transformations that will be discussed in more detail later in this appendix.

Frame images

For frame images, the image plane is projected through a single projective center directly on to the object plane, i.e., the ground. This relationship is shown graphically in figure A.31. In essence, this particular transformation is based on the simple projection of an image as carried out in an optical projector or enlarger. Expressed in numerical form, the transformation can take the following form:

$$X = \frac{a_1x + a_2y + a_3}{c_1x + c_2y + 1}$$

$$Y = \frac{b_1x + b_2y + b_3}{c_1x + c_2y + 1}$$

Where

X and Y are the *ground coordinates* of one of the ground control points (GCPs)

x and y are the corresponding *image coordinates* of the same point

a, b, and c (i = 1 to n) are the *projective transformation parameters* connecting the ground and the image coordinate systems

Since there are eight projective transformation parameters—a_1, a_2, a_3, b_1, b_2, b_3, c_1, c_2—that have unknown values, these can be found using a minimum of four ground control points (GCPs) lying in the ground plane. As with the 2D polynomial transformation, a set of eight simultaneous equations can then be set up and solved to provide the numerical values of the projective transformation parameters.

Pushbroom scanner images

As described above, the linear array of a pushbroom scanner forms a very narrow rectangle with its long side oriented at right angles to the forward motion of the platform on which it has been mounted. Each scan line has its own individual perspective or projection center. These perspective centers form a

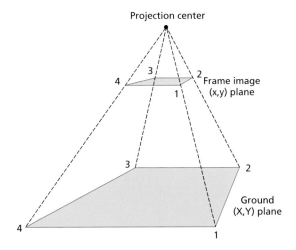

Figure A.31 The 2D projective transformation of a frame (x,y) image into the ground plane and its (X,Y) coordinate system.

Source: G. Petrie and M. Shand.

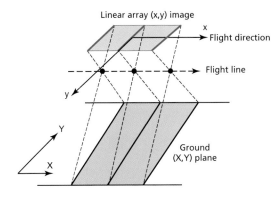

Figure A.32 The 2D projective transformation of a linear array scanner (x,y) image into the ground plane and its (X,Y) coordinate system.

Source: G. Petrie and M. Shand.

nearly straight line as shown graphically in figure A.32. The corresponding transformation can take the following form:

$$X = \frac{a_1x + a_2y + a_3}{1}$$

$$Y = \frac{b_1x + b_2y + b_3}{c_1x + c_2y + 1}$$

Given the large area of the ground that will be covered by a line scanner image of this type from space—60 x 60 km in the case of SPOT images—the effects of earth curvature need to be considered. In practice, these effects will be removed before the 2D projective transformation is applied for the rectification and georeferencing of the pushbroom linear array imagery with the ground then being considered as a plane surface.

Limitations of the 2D transformation approach

In using the various 2D transformations described above, no attempt is being made to take account of the geometries of the many different types of remotely sensed image data described in the first section of this appendix. Nor is the data from the geometric calibration or camera model being used. Moreover, with this 2D transformation approach, no attempt is being made to derive or make use of the *orientation* or *tilt values* that were present when the image data was being acquired. It will be apparent to the reader that this does not accord with the actual situation that existed in the air or in space during the time when the image data was being captured—tilts are always present. Furthermore, almost invariably, it is assumed that the ground is flat and that *no relief displacements* are present in the image. Nor does this assumption agree with the situation that exists on the ground there will almost always be some terrain relief. In hilly or mountainous regions, this can easily amount to hundreds or even thousands of meters within the area being covered by a single image, so quite large residual tilt and relief displacements together with the associated scale variations can still be present in an image rectified in this way.

Actual use of the 2D transformation approach

In spite of their limitations, 2D transformations are widely used in remote sensing practice. The 3D transformations are based on a much more rigorous geometric analysis of the imagery but are more complex and often regarded as too difficult to understand. As well, 2D transformations can be implemented with less information about the imager. For a 3D transformation the imager's *internal calibration data* and the *external data* giving its tilt/orientation values at exposure are needed. If a fully map accurate rectification (i.e., orthorectification) is to be achieved, a *digital terrain model* (DTM) is also required. As well, the analyst needs to be well-versed in photogrammetric procedures to ensure a high quality result.

By contrast, the 2D transformation does not require these datasets, is simpler to implement, and makes fewer demands on the analyst. But the results are demonstrably much less accurate than those achievable when a full 3D approach is adopted, especially if there is substantial terrain relief. However, the wide use of 2D transformations suggests that the resulting limitations in positional or planimetric accuracy are acceptable to many remote sensing practitioners. GIS managers and users should assess the limitations in the planimetric or positional accuracy of remotely sensed imagery rectified by 2D transformation or the spatial data products derived from such imagery and judge their suitability for the GIS applications for which they would be used.

C. Image rectification and geo-referencing using 3D transformations

The use of 3D transformations for image rectification takes into account the *spatial relationships* that exist between the *image* as it was acquired in the air or in space and the *ground*, including the terrain relief. However, an important point that has to be made at the outset of any discussion about 3D transformations is that it is impossible to construct an accurate 3D representation of the terrain and to obtain accurate 3D data for GIS purposes from a single image only. To obtain accurate planimetric and height data for use in a GIS, a second overlapping image is required. This allows a *stereomodel* to be formed that accurately represents the terrain and allows both planimetric (positional) coordinate data and height (elevation) data to be obtained through precise measurements of this 3D model of the terrain. The digital 3D coordinate data that is obtained from measuring the stereomodel can take the form of a vector line map (including contours) or it can be delivered to the GIS users in the form of an image (an orthophoto or orthoimage) having the geometry of a map. In the latter case, this orthoimage will often be supplemented by a digital terrain model (DTM) or digital elevation model (DEM) that represents the varying heights of the terrain surface. Again, this will often have been extracted directly from the stereomodel. If, however, a DTM or DEM already exists for a particular area and is available from another source, then a 3D transformation can be carried out using a single image to create an orthophoto or orthoimage of the area that it covers. Such an operation is called *orthorectification*.

C.I   Basic 3D transformations for frame imagery

The following basic equations for a frame image relate the position of any point (i) on the image to the perspective center (O) of the frame imager and to the equivalent point

(I) on the ground *(figure A.33)*. These equations take the following form:

$$x_i = -f. \frac{a_1(X_I - X_O) + a_2(Y_I - Y_O) + a_3(Z_I - Z_O)}{c_1(X_I - X_O) + c_2(Y_I - Y_O) + c_3(Z_I - Z_O)}$$

$$y_i = -f. \frac{b_1(X_I - X_O) + b_2(Y_I - Y_O) + b_3(Z_I - Z_O)}{c_1(X_I - X_O) + c_2(Y_I - Y_O) + c_3(Z_I - Z_O)}$$

Where

f is the exact focal length of the camera exposing the frame image obtained through a prior laboratory calibration

X_O, Y_O, and Z_O are the 3D ground coordinates of the perspective center (O) of the imager

X_I, Y_I, and Z_I are the 3D ground coordinates of the ground point (I)

x_i and y_i are the image coordinates of the corresponding point (i) on the frame image

a_1, a_2, a_3; b_1, b_2, b_3; c_1, c_2, c_3 are the transformation parameters.

[Note: These last nine terms—a_1...c_3—are directly related to and are functions of the rotations (ω, φ, κ) around the three (X,Y,Z) coordinate axes that have their origin in the perspective center (O)—which is the position that the frame imager occupied in the air or in space when the image was being acquired.]

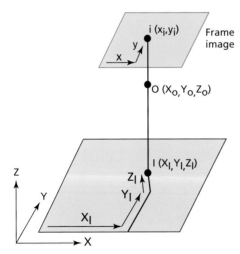

Figure A.33 The collinearity condition for a frame image showing that the image point (x_i,y_i), the perspective center (X_O, Y_O, Z_O) and the corresponding ground point (X_I, Y_I, Z_I) are collinear, i.e., they lie on a straight line. This forms the basis of both the space resection and space intersection procedures.

Source: G. Petrie and M. Shand.

The above pair of equations are called the *collinearity equations* since they state, in mathematical terms, that the image point (i), the perspective center (O), and the corresponding ground point (I) lie on a straight line, i.e., they are collinear.

Space resection (SR)

A typical use of these collinearity equations is in the procedure known as *space resection*. This procedure allows the coordinates (X_O, Y_O, Z_O) of the perspective center of the frame imager to be determined together with the values of the tilts (ω, φ, κ) of the frame imager when the image was being acquired. Essentially, this is the first step in the procedure used to carry out the rectification and georeferencing of a frame image using a 3D transformation. It is used extensively with frame images, irrespective of whether they were acquired from airborne or spaceborne platforms.

To implement this procedure, a minimum of three ground control points (GCPs) are required. Thus, the accurate values of the 3D ground coordinates (X_I,Y_I, Z_I) of each of these three points must be available. The corresponding image coordinate values (x_i, y_i) of each of these three points are then measured very accurately on the image. For each of these three well-defined image points, the two equations set out at the beginning of this section can then be formed. Having the minimum of three GCPs available means that a total of six equations will be created. These allow the values of the six unknowns—X_O, Y_O, Z_O and ω, φ, κ—to be determined. In this way, the position and elevation of the frame imager in the X, Y, Z ground coordinate system, together with its orientation at the time of its exposure, will be obtained. As with the 2D transformations, usually more points than the minimum three GCPs will be used in order to provide redundancy, to detect any errors in the identification or measurement of the points occurring during the procedure, and to provide a check on the accuracy of the whole space resection procedure. In this case, a least-squares solution will be implemented to produce the most probable value of each of the six unknown parameters.

Space intersection (SI)

Once the values of the six parameters have been determined for each of the two frame images that form an *overlapping stereopair* using the space resection technique, it is then possible to complete the 3D transformation and determine the positions and elevation values—X_I, Y_I, Z_I—of all the points that occur within the area of the ground covered by the overlapping pair of frame images. Again, this can be implemented on the basis of the collinearity equations and results in the formation of an exact stereomodel of the terrain from which high-accuracy map and GIS data can be derived. This second procedure, which normally follows directly from space resection, is called space intersection *(figure A.34)*. It should also be noted that, once this space intersection procedure has been completed, the algorithm can be used in reverse. Thus, starting with a ground point with known positional (X,Y) and elevation (Z) coordinates in the stereomodel, together with the known values of the projection centers and tilts of the two overlapping frame images obtained from the prior space resection operation, the corresponding image positions (x,y) of that point on both images can be determined. In fact, this reverse procedure is used extensively both in digital photogrammetric workstations (DPWs) and in certain orthorectification operations.

Comparison with 2D transformations

Comparing the 3D image transformation produced by combining these two procedures (SR + SI) with the 2D image transformation procedures described earlier in this appendix, it will be seen that the final result of the 3D transformation generates an accurate value of the elevation (Z) of each point appearing in the overlapping images, as well as its correct planimetric (X,Y) position. Furthermore, the final result will have a much higher accuracy than that produced by image rectification using a 2D transformation since the combined SR/SI procedure models, as precisely as possible, the position, orientation, and attitude of the imager in the air or in space and its relationship with the ground. Thus it does not contain the approximations and the limitations in accuracy inherent in the 2D transformations used in the simpler image rectification procedures described earlier.

Orthogonal projection

From the purely geometric point of view, by employing the SR/SI procedure, the two overlapping frame images that constitute a pair of perspective projections have been used to form an exact 3D model of the ground, including its relief. Once this 3D model has been created, it is then a simple matter to project any or all of the data contained in this model orthogonally into the ground reference plane where it will have exactly the same geometry as that of a map *(figure A.35)*.

Applications of the SR/SI technique for feature extraction and map compilation

The use of the combined SR/SI procedure to implement a full 3D transformation and form a stereomodel of the ground allows the conversion of the frame image data into accurate 3D digital maps and GIS data. The use of this particular procedure is widespread throughout the mapping industry. A typical application is to produce 3D digital map data from the stereopair of images using a *digital photogrammetric workstation* (DPW). With this device, the exact stereo-model of

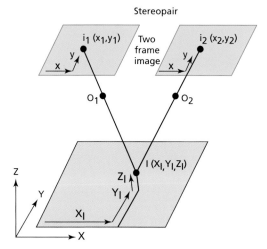

Figure A.34 Space intersection allows the 3D coordinates (X_I, Y_I, Z_I) of a ground point to be determined from the measurements x_1, y_1 and x_2, y_2 of the positions of that point on each of the overlapping frame images making up the stereopair. The tilt values ω, φ and κ, of each of the frame images were determined via a prior space resection operation.

Source: G. Petrie and M. Shand.

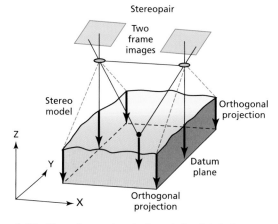

Figure A.35 The orthogonal projection of all points in the stereomodel of the ground surface of the terrain into the data plane of the map.

Source: G. Petrie and M. Shand.

the terrain formed by the SR/SI procedure can be viewed stereoscopically in 3D on the monitor screen of the DPW. The actual 3D stereoscopic viewing and measurement of this exact model of the terrain can be carried out on the screen using any one of several alternative stereoviewing techniques. Examples are the active glasses employing high-speed alternating shutters, as used in the CrystalEyes® system from the StereoGraphics Corporation *(figure A.36a)*, or passive glasses using polarizing filters, such as the NuVision™ system from MacNaughton, Inc. *(figure A.36b)*. The extraction of the features needed for mapping or GIS purposes from this stereomodel is then carried out in 3D stereo by an operator using a measuring cursor (or floating mark). This cursor can be set continuously up and down to the correct height in the stereomodel by the operator for stereomeasurement and to map the required features.

Feature extraction and contouring

Using this technique, the operator uses the DPW to measure the whole of the stereomodel in a systematic manner, measuring as precisely as possible the exact positions of the edges or center lines of roads, tracks, railways and streams; the edges

or boundaries of fields, forests, lakes, ponds, etc.; or the corners or edges of buildings, bridges, dams, and other man-made constructions—as required by the map or GIS data specifications. Besides the correctly georeferenced positional (X,Y) data, the *elevation (Z) values* for each measured point or line feature will be recorded simultaneously. In addition, if required, the measuring cursor can be set to a predetermined height value, and the operator can then trace individual contour lines by keeping the cursor (or floating mark) continuously in contact with the ground as it is being viewed in 3D while moving the cursor within the stereomodel. In this way, the successive planimetric (X,Y) positions of the contour line for that particular height (Z) value will have been measured continuously across the whole of the stereomodel. This contouring operation can then be repeated for the next required contour line once the measuring cursor has been reset to the appropriate height value in the DPW. During all of these various operations, the 3D coordinates of every point being measured in the stereomodel are being determined very accurately in real time by the DPW system using the space intersection computation.

Application of the SR/SI technique—DEM generation

The purely manual extraction and compilation of planimetric detail and its associated elevation and contour data for mapping or GIS purposes is still practiced widely, especially in areas of complex terrain. These include urban areas with numerous high buildings and rural areas with numerous scattered trees and woodland. In such areas, attempts to use automated image matching techniques result in large numbers of false or erroneous elevation values being generated for these features caused by mismatches. These errors need extensive and time-consuming correction and editing. However, in areas of open ground, the use of automated image matching techniques for the generation of DEMs is widespread. Once again, in such an operation, the initial steps of the SR/SI procedure are implemented as before with the formation of the 3D stereomodel that has been fitted (i.e., georeferenced) to the available ground control points (GCPs). Once the stereomodel has been formed, instead of the manual measurement and extraction of data, an automated image correlation procedure is implemented. This matches the image data present on one image with the corresponding data on the second overlapping image using an automated procedure, usually on a patch-by-patch basis, until the whole of the stereomodel has been covered. This image-

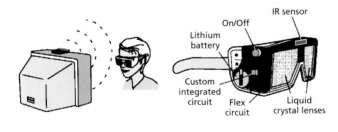

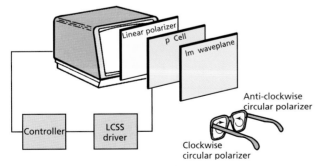

Figure A.36   The 3D stereo-viewing of overlapping images on a monitor screen as used in the extraction of features from stereomodels for inclusion in a GIS, which employ (a) alternating shutters forming a pair of active glasses, or (b) passive polarizing spectacles.

Source: G. Petrie and M. Shand.

matching operation actually generates x-parallax values which can then be converted directly into the corresponding terrain elevation values using a very simple computational procedure. Once again, the computation of the 3D (X,Y,Z) coordinate values is carried out continuously during this operation using the space intersection (SI) technique. Once the automated generation of the height values has been completed, the resulting elevation data is then edited manually to remove the mismatches, spikes, etc., that will inevitably be generated in this type of automated operation.

Orthorectification and production of orthophotos

The production of orthophotos from frame images is a widespread and very popular activity at the present time. Formerly this type of rectified image was produced directly in hard-copy photographic form through the profile scanning of a stereomodel in a specially constructed or modified stereo-plotting instrument. However, nowadays orthophotos are produced in digital form from frame images using *differential rectification* on the basis of an existing DEM *(figure A.37)*. This DEM may have been generated using any one of a variety of methods, including the following:

- through the *automatic image correlation* of stereopairs of photographs in digital photogrammetric workstations (DPWs) as described above
- directly from *airborne laser scanning or interferometric SAR* (InSAR) data

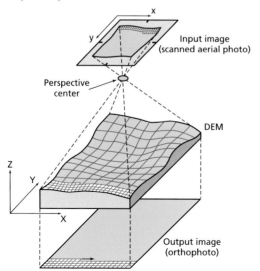

Figure A.37 Differential rectification of a single frame image using a DEM to produce an orthoimage.

Source: G. Petrie and M. Shand.

- by interpolation from *digitized contours* taken from an existing map

The actual orthoimage rectification procedure is carried out using as its starting point the known values of the perspective center coordinates (X_O, Y_O, Z_O) and tilts (ω, φ, κ) of one of the two overlapping frame images of the stereopair, as determined through space resection. Thereafter, for each successive ground position given by the DEM (with coordinates X_I, Y_I, and Z_I), using the collinearity equations, the corresponding position on the frame image (with coordinates x_i and y_i) can be found and its brightness value recorded. The final orthorectified image is then generated and stored in a computer as a matrix of brightness values in digital form from which hard-copy images may be produced on a film or on a laser or inkjet plotter. If needed by a GIS user, the extraction and compilation of the vector data can then take place using the digital orthophoto data as the base. However, this procedure will not have the benefits of interpreting the features stereoscopically in 3D that are present when carrying out the feature extraction in a DPW. Features can be viewed in greater detail and their position and dimensions more accurately measured when viewed in 3D on a DPW instrument.

Orthoimage mosaics

Once the individual adjacent orthoimages have been generated either directly from the stereopair of frame images or from the individual frame images through the use of a DEM from another source, they can be combined to form an orthoimage mosaic. This allows a much larger geographic area to be covered and viewed as a single composite image. The georeferencing of the individual orthorectified images must have been carried out first, using a single consistent ground coordinate system in order to produce a seamless orthoimage mosaic. Otherwise, mismatches or discontinuities will occur along the junctions or overlaps between the individual orthoimages when they are stitched together.

Orientation procedure

While the space resection/space intersection (SR/SI) procedure described above has been incorporated into many software packages, an alternative procedure, called *orientation*, is also in widespread use with frame images, particularly in DPWs. With this procedure, the setting-up of the stereomodel is divided into three stages. The first is called *inner orientation*, which involves the establishment of the exact positions of each of the two frame images in relation to their respective perspective centers using data given by a prior camera calibration. (Note: This preliminary procedure must also be carried out with the space resection/space intersection procedure described above.) The second stage is called *relative orientation*, which uses the y-parallax values between the two images as measured at specific locations in the overlap between them in conjunction with a special geometric condition (the coplanarity condition) to establish the ω, φ, κ tilt values of the two images with respect to one another. This relative orientation operation also results in the formation of the 3D stereomodel. The third stage is *absolute orientation*. This involves the scaling and rotation of the stereomodel as a whole to fit the ground control points (GCPs)—i.e., it provides the required georeferencing. Thereafter, the operations for the extraction of features or the generation of DEMs from these stereomodels are exactly the same as those described above using the SR/SI technique. It should be emphasized that, when this orientation procedure is being implemented, the requirements for control points (a minimum of three GCPs) and the geometric accuracy of the final results will be exactly the same as those achieved using the SR/SI technique. Essentially, the orientation procedure is the classical method used to set up stereomodels in analog stereoplotting instruments, but it can be used equally well in DPWs.

Calibration of frame imagers

With all the frame imagers that have been designed specifically for mapping purposes—such as the metric film and digital cameras discussed above—a prior calibration of the imager will have been carried out in an approved laboratory. This calibration procedure establishes the *principal point* (or fiducial center) of the frame image, the exact *focal length* of the imager's lens, and the extent and pattern of any *geometric distortion* that may be present in the lens of the imager. This calibration data establishes the inner orientation of the image system, which is necessary to generate very high-precision coordinate data not only for mapping and GIS purposes but also for engineering work where the demands for very accurate data for design and construction work can be very high. However, there is always pressure from GIS users to employ nonmetric (i.e., uncalibrated) airborne cameras rather than purpose-built calibrated metric cameras for the acquisition of GIS data—usually on grounds of cost or availability. Unfortunately, although these nonmetric cameras are usually optimized to produce images with good picture quality, often their lenses will feature large geometric distortions.

Additionally, they will not have the calibrated position of the center of the image, nor will the focal length be known to the required high degree of precision. All of which leads to results that, in terms of the accuracy of the resulting ground coordinate data, are substantially poorer than those produced from images acquired using a calibrated frame imager or camera. So the data from nonmetric cameras will not meet the normal requirements for positional and elevation accuracy as specified for mapping and GIS purposes.

C.II   Basic 3D transformations for line scanner imagery

As discussed previously in this appendix, the images acquired by line scanners are collected on a line-by-line basis with the forward motion of the airborne or spaceborne platform being used to generate a single continuous strip image. Thus, each line in the image will have a different perspective center and tilt value. Considering the strip image as a whole, it therefore comprises multiple lines with multiple projection centers and contains tilt values that vary along the length of the image in the flight direction. These values need to be determined, while keeping in mind the additional complication that the collinearity equations will need to be modified since they are constrained to lie in a single plane defined by the individual line and its corresponding projection center. The exterior orientation of each individual line is given by the usual parameters—i.e., the three positional values of the perspective center and three rotation or tilt values associated with it. But unlike frame images, where once the single set of these parameters has been determined, they apply to the whole image, with line scanner imagery, they will be different for every scan line since the projection centers and tilt values are continuously changing. Thus, arising from the dynamic nature of the geometry and the changing positional and attitude parameters of line scan imagery, they have to be regarded and treated as being time dependent.

However, with a line scan imager mounted on a spaceborne platform, the next scan line will only have very small and fairly predictable changes in the position of its projection center (since the imager is moving in a well-defined orbit) and in its tilt value (since the platform is operating in the near vacuum of space and the tilt values change very slowly). These relatively small changes can be modeled to a sufficient level of accuracy using quite *low-order polynomials*, usually based on the line number or the time interval between successive lines. On the basis of this additional modeling, sets of collinearity equations can be set up that apply to the whole of the strip image. With most spaceborne line scan imagers, the relevant calibration data (focal length, lens distortion, etc.) is made available to users by the satellite operators. After measuring the positions of the image points on the strip image corresponding to the ground control points (GCPs), the values of the unknown parameters (i.e., the projection centers, tilt values, and polynomial terms) can be determined by space resection through the solution of the appropriate number of simultaneous equations based on the relevant collinearity equations.

Use of auxiliary data with space line scanner imagery

With most of the software packages that are geared towards the processing of *spaceborne line scanner images*, the position and attitude information produced by the auxiliary instruments (GPS, DORIS, star trackers, inertial systems, etc.) mounted onboard the satellite or spacecraft is incorporated in the solution. This additional information is usually presented to the user in the form of header information that accompanies the actual image data file. The incorporation of this additional information usually has the effect of reducing the numbers of GCPs needed to produce a solution for the whole of the strip image. However, it should be noted that the minimum number of GCPs required to implement a solution will vary according to the number of parameters that have been used in modeling the position and attitude of the imager in space, each of which will need to be determined. Thus, with different software packages from different suppliers, the minimum number of GCPs required to achieve a solution will often vary quite substantially from one package to another. As with all image-rectification procedures, whether based on 2D or 3D transformations, it is normal to use more than the minimum number of GCPs needed for the solution, in which case, a least-squares solution will be applied.

Use of auxiliary data with aerial line scanner imagery

In the case of airborne line scanner images, even if a gyro-controlled mount is being used, atmospheric turbulence causes larger and more abrupt and unpredictable changes in the position and attitude of the line scanner imager. In this case, the use of highly accurate auxiliary instruments is necessary to obtain a workable solution. Thus, a combined IMU/DGPS unit is an integral part of an airborne line scan imager. The DGPS unit supplies very accurate data on the positions

(X,Y) and elevations (Z) of certain perspective centers in the air at time intervals of around one second, whereas the IMU supplies positional and attitude data of a lesser quality every few milliseconds. By combining the two sets of DGPS and IMU data, the perspective centers and attitude (tilt) data for each line in the image can be derived for the purposes of the strip resection and subsequent rectification of the image.

Space intersection (SI)

Once the successive projection centers and tilt values have been determined for each of the two overlapping line scanner images, a space intersection technique can then be used to calculate the ground coordinates of any point in the terrain. This is based on the measured coordinate values of the corresponding image points on the two overlapping strip images. Where the overlap is in the *across-track position*, as used with SPOT and IRS stereopairs, the situation will be as shown in figure A.38. The specific orientation parameters relating to each of the two corresponding lines containing the relevant image points must first be computed before the actual space intersection can be calculated. With *along-track stereoimagery*, as acquired from the JERS-1 OPS, ASTER, and SPOT-5 HRS imagers, similar considerations and procedures will be applied with first a space resection of each of the overlapping strip images being implemented, followed by a space intersection using the measured points on the two overlapping strip images. In some cases, the software used is basically a modification of the software originally developed for use with the SPOT and IRS across-track imagery. Obviously, with the across-track case, the large tilt angles (which are not fixed values) will occur around the along-track axis to produce the required across-track overlap, whereas with

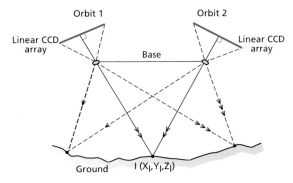

Figure A.38 Space intersection is applied to across-track scanner imagery to form a stereomodel, shown here in cross section.

Source: G. Petrie and M. Shand.

an along-track scanner, imager such as the JERS-1 OPS, ASTER, and SPOT-5 HRS imagers, the along-track viewing angles are fixed with the rotations being set around the across-track axis, as shown in figure 28a above—i.e., at right angles as compared with the across-track situation (shown previously in figure 27). As for the airborne pushbroom scanner imagers operating in the same along-track mode, the same basic procedures will be followed as with their spaceborne equivalents.

DEM and orthoimage generation—across-track imagery

In general, the main use of overlapping spaceborne line-scanner imagery acquired *across-track* by the SPOT and IRS line scanner imagers has been the generation of DEMs, followed by the production of orthorectified images (orthoimages). While stereomodels are indeed formed in the course of these operations, there has only been a very limited use of stereomeasurement, feature extraction, and compilation of vector line maps and GIS data from SPOT and IRS data in the way that is popular with stereopairs of frame images. Instead, the DEMs generated from overlapping SPOT and IRS images are produced almost invariably using automatic image matching procedures. However, some special difficulties can arise with these image matching operations. In particular, the overlapping across-track images taken from two separate orbits may have been acquired several months apart to obtain cloud-free coverage and prevent gaps appearing in the DEM and in the orthoimage. However, this will often result in the very different appearance of the same objects in the two overlapping strip images. In particular, the vegetation, cultivated areas, and water features may have an entirely different appearance in each image due to seasonal weather changes between summer and winter, wet and dry seasons, etc. In this case, it will be impossible to carry out automatic image matching operations for DEM and orthoimage production. However, these difficulties are less likely to occur in desert and semiarid areas—for example, in the Middle East and North Africa—where much greater use has been made of the procedures outlined above.

DEM and orthoimage generation—along-track imagery

In the case of *along-track* stereoscopic cover, the overlapping images of an area will be acquired within about a minute during a single orbital pass, so the same object will appear the same in both images. Indeed, for this reason, most recently-built spaceborne optical line scanners have been designed

to operate in an along-track mode for the acquisition of overlapping stereocoverage. With regard to *orthoimage generation*, once an accurate DEM is available from whatever source, the orthoimage can again be generated on the basis of the position and attitude data obtained from the prior space resection of one of the strip images and the use of the appropriate collinearity equations to determine the specific pixel position i (with x_i, y_i coordinates) corresponding to the position of the ground point I (with X_I, Y_I, Z_I coordinates) in the DEM.

High-resolution scanner images

With the new generation of high-resolution line scanner imagers mounted on the IKONOS, QuickBird, OrbView-3, and EROS-A1 satellites, the possibility exists to acquire image data both in the across-track and along-track directions and in any intermediate direction, i.e., with flexible pointing. In the case of the QuickBird imagery, the calibration data relating to its imager has been made available by DigitalGlobe to software suppliers and users in the same way it has been done with SPOT, IRS, and other earlier satellites. Reportedly, the same open policy will be applied by ORBIMAGE with respect to the imagery acquired by its newly launched OrbView-3 satellite. However, the situation regarding IKONOS imagery is somewhat different. In particular, Space Imaging has adopted a policy of nondisclosure of the calibration data relating to the IKONOS imager and to not supply the position and attitude data collected in-flight. In addition, Space Imaging also placed restrictions on the supply of stereoimagery to nongovernment users—although this has since been relaxed. The reasons for these various restrictions appear to be commercial. Through this combination of measures, Space Imaging has attempted to ensure that, instead of selling either raw imagery or imagery with very basic processing direct to users or to mapping or remote sensing companies—which then carried out further value-added processing such as DEM or orthoimage production—these processing activities could only be carried out by Space Imaging and its affiliates. Needless to say, this matter has become highly controversial and has provoked a variety of responses from the mapping and remote sensing communities on the one hand and from the Space Imaging company on the other.

Space Imaging production of rectified images

Space Imaging offers a variety of products called Geo, Reference, Pro, Precision, and PrecisionPlus respectively, which have undergone different levels of rectification of the IKONOS images. The Geo product is derived solely from the company's in-house knowledge of the scanner imager's calibration data, the in-flight positions recorded by the satellite's GPS receivers, and the tilts derived from the star tracker and gyro data measured onboard the IKONOS satellite. No GCPs are used in the image rectification carried out for the Geo product, for which the absolute positional accuracy in terms of RMSE is stated (by Space Imaging) to be ±25 m without including terrain effects, e.g., relief displacements. At the other end of the accuracy spectrum are the *Precision* and *PrecisionPlus* products with stated RMSE positional values of ±2 m and ±1 m, respectively. The generation of these higher-end products is based on the use of high-quality GCPs and DEMs besides the company's own calibration and exterior orientation data. Obviously, the cost of these higher accuracy products is much greater—in the order of five to ten times the cost of the corresponding Geo product, which itself is not inexpensive (Fraser et al. 2002). The nonavailability of the calibration data and the measured in-flight orientation data was not too well received by commercial mapping companies, the software suppliers, and the potential users of remote sensing data, since these restrictions prevented them from using IKONOS imagery in the manner they were accustomed to with aerial photography and with SPOT and IRS satellite image data. To meet the various criticisms, Space Imaging introduced its so-called Rational Polynomial Camera (RPC) model, which is based on the use of rational polynomial equations or functions—though, of course, IKONOS is a line scanner and not a camera! These functions had already come into use with the imagery used by the U.S. intelligence community.

Rational polynomials

The adjective *rational* is normally associated with the noun *reason*. However, in the context of this present discussion, it is, in fact, associated with the word *ratio*. A rational polynomial is one where the ratio between two polynomial equations or functions has been defined. With the RPC model used by Space Imaging, these rational polynomials are used to express the relationship between the ground points (with X,Y,Z coordinates) and their corresponding image points (with x,y coordinates). They take the following form with 20 terms (up to and including third-order terms) in both the numerators and denominators (Grodecki 2001):

$$x = \frac{a_0 + a_1X + a_2Y + a_3Z + a_4XY + a_5XZ + a_6YZ + \ldots\ldots + a_{19}Z_3}{b_0 + b_1X + b_2Y + b_3Z + b_4XY + b_5XZ + b_6YZ + \ldots\ldots + b_{19}Z_3}$$

$$y = \frac{c_0 + c_1X + c_2Y + c_3Z + c_4XY + c_5XZ + c_6YZ + \ldots\ldots + c_{19}Z_3}{d_0 + d_1X + d_2Y + d_3Z + d_4XY + d_5XZ + d_6YZ + \ldots\ldots + d_{19}Z_3}$$

Where

X and Y are the *ground coordinates* of one of the ground control points (GCPs)

x and y are the corresponding *image coordinates* of the same point

a, b, c, and d (i = 0 to 19) are the *rational polynomial coefficients* (RPCs)

Image rectification using rational polynomials

The RPC model coefficients $(a_0 \ldots a_{19}; b_0 \ldots b_{19}; c_0 \ldots c_{19}; d_0 \ldots d_{19})$ are determined from a three-dimensional grid of object (ground) points that is generated using a rigorous solution based on the IKONOS camera model—including the imager's focal length, principal point location, lens distortion, etc.—which is known only to Space Imaging. Using this so-called camera model, the 3D grid of object points is generated by intersection of the rays from a 2D grid of image points with a number of elevation planes *(figure A.39)*. The 80 RPC coefficients (4 equations with 20 coefficients per equation) generated are consistent with the specific type of image product (Geo, Precision, etc.) that is being supplied to the user. If overlapping IKONOS stereoimagery has been supplied, it is also possible to generate the 3D (X,Y,Z) coordinates of ground points based on the RPC values that have been supplied for each of the two overlapping images (Grodecki 2001) as illustrated in figure A.40. The big question for users is whether the results that can be obtained using rational polynomials are comparable to those that would be obtained using an exact solution based on the imager's calibration data. In their commercial literature and published papers, Space Imaging assures users that there is only a very small loss of accuracy. However, it is impossible for this to be confirmed independently by outside agencies or researchers when the calibration data is being withheld by Space Imaging.

Further developments

Understandably, the introduction by Space Imaging of this approach of withholding the calibration data of the scanner imager and the in-flight data and supplying instead the rational polynomial coefficients to users has provoked a strong

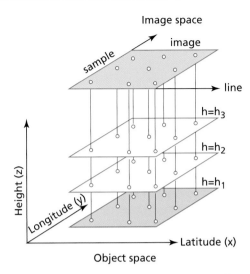

Figure A.39   Generation of the rational polynomial coefficients (RPC) with the rays from a 2D grid of image points intersecting a number of constant elevation planes (h_1, h_2, h_3).

Source: Reproduced with permission of the American Society for Photogrammetry and Remote Sensing. J. Grodecki. IKONOS stereo feature extraction. Proceedings ASPRS Annual Conference. Bethesda: ASPRS, 2001.

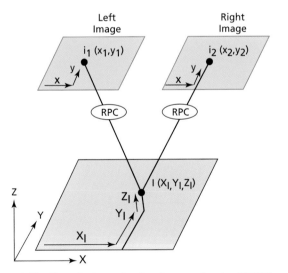

Figure A.40   Stereofeature extraction from overlapping IKONOS images based on the respective sets of RPC coefficients from each of the two overlapping images.

Source: Reproduced with permission of the American Society for Photogrammetry and Remote Sensing. J. Grodecki. IKONOS stereo feature extraction. Proceedings ASPRS Annual Conference. Bethesda: ASPRS, 2001.

reaction among the mapping and remote sensing communities. Several system suppliers (such as ERDAS, PCI, and Z/I Imaging) have added modules to their software packages to allow users of their software to use the RPC data supplied with their IKONOS images. Also, a number of the leading analytical photogrammetrists from universities and research institutes in different countries have also become involved in this particular matter in various ways. One development has been that some photogrammetrists used a reverse engineering approach to construct a camera model for IKONOS that appears to be very close to the correct values. Using this derived camera model, photogrammetrists have implemented a fully rigorous rectification solution in the same way as they have been able to do for other types of spaceborne scanner imagery. Results from accuracy tests suggest that this independently derived model does indeed give results that, in terms of accuracy, are similar to those that could be expected from an exact solution where the calibration parameters are known. Other recent developments in this area have been the investigations into and the use of other *alternative models*—e. g., an affine projection model, a relief-corrected affine transformation, and a variant of the Direct Linear Transformation (DLT) approach, which is normally used with nonmetric cameras in close-range photogrammetry (Fraser, Baltsavias, and Gruen 2002). All of these alternative approaches have been used on an experimental basis with IKONOS image data for which the calibration and in-flight auxiliary data have been withheld. All of these trials have been conducted with the least-expensive Geo data and appear to give results that, in terms of geometric accuracy, are those normally associated with the more expensive Precision and Precision Plus products from Space Imaging. However, shortcomings in the radiometric quality of the IKONOS images that were tested in these investigations "gave rise to accuracy and interpretability concerns in both manual and computer mensuration" (Fraser, Baltsavias, and Gruen 2002).

Conclusion

As this account has attempted to show, it is vital for GIS managers and users to take into account (1) the geometry of the specific type of remotely sensed image data being used, and (2) the rectification and georeferencing procedures that have been applied to the images. This assessment should be carried out before attempting to incorporate the data into a GIS system. As a general rule, one can say that the rectification and georeferencing of image data using 2D transformations will not provide the level of geometric accuracy that will result from the proper application of the appropriate 3D transformations that take into account the tilts present in the imagery and the terrain relief that exists on the ground. Indeed, some of the 2D rectification and georeferencing procedures being used by the remote sensing community are oversimplified and are not compatible with the requirements of a GIS, to say the very least! While these latter procedures may well satisfy the requirements of the remote sensing specialists concerned, the resulting limitations in the positional accuracy of the image data will then come to the fore when attempts are made to merge this data with the map-accurate data that is held in the GIS. In this respect, it is important for GIS managers and users to take into account the spatial accuracies being used within their own GIS environment in order to judge the geometric accuracy that is required or is being offered by the rectified imagery or the data that has been derived from it.

References   Fraser, C. S., E. Baltsavias, and A. Gruen. 2002. Processing of IKONOS imagery for submetre 3D positioning and building extraction. *ISPRS Journal of Photogrammetry and Remote Sensing* 56(3): 177–94.

Grodecki, J. 2001. IKONOS stereo feature extraction. Proceedings ASPRS Annual Conference. Washington, D.C.

Mikhail, E. M., J. S. Bethel, and J. C. McGlone. 2001. *Introduction to modern photogrammetry.* New York: John Wiley and Sons, Inc.

Appendix B Characteristics of selected satellite sensors

The following tables present important characteristics of satellite sensors widely used for earth resource remote sensing. This appendix can be used to compare the spatial, spectral, and temporal resolution of these sensors.

Table B.1	Sensors with low spatial resolution						
Platform/ Altitude	**Sensor**	**Spectral resolution**		**Spatial resolution (GSD) in km**	**Swath width**	**Revisit period/ Orbit**	**Common applications**
		Channel	Waveband (μm)				
GOES 8–11 36,000 km	Imager	1 2 3 4 5	0.52–0.71 VIS 3.73–4.07 SWIR 5.80–7.30 SWIR 10.2–11.2 TIR 11.50–12.50 TIR	1 4 4 4 8	100° of longitude	Full earth disc acquired every 26 minutes or less. Subareas can be acquired in less time/ geosynchronous	Daytime cloud and surface mapping Nighttime cloud cover Water vapor Earth and cloud images, sea surface temp, water vapor, fires Night cloud and sea surface temperature, fires
Defense Meteorological Satellite 830 km	OLS	1 2	0.40–1.1 VIS/NIR 10.0–13.4 TIR	2.8 km	3000 km	12 hours/ sun-synchronous	Clouds, low illumination nighttime imagery Night cloud and sea surface temperature
NOAA 11,12,14 [a,b] 15,16,17 [a] 15,16,17[a] 833 km	AVHRR	1 2 3 3A 3B[a] 4 5	0.58–0.68 VIS 0.725–1.10 NIR 3.55–3.93 SWIR 1.58–1.64 SWIR 3.55–3.93 SWIR 10.3–11.3 TIR 11.5–12.5 TIR	1.1 1.1 1.1 1.1 1.1 1.1 1.1	2400 km	12 hours/ sun-synchronous	Daytime cloud and surface mapping Land–water boundaries Night cloud and sea surface temperature Snow and ice detection, distinguishing snow from clouds Night cloud and sea surface temperature Night cloud and sea surface temperature Sea surface temperature
OrbView-2 or SeaStar	SeaWiFS	1 2 3 4 5 6 7 8	0.402–0.422 VIS 0.433–0.453 VIS 0.480–0.500 VIS 0.500–0.520 VIS 0.545–0.565 VIS 0.660–0.680 VIS 0.745–0.785 NIR 0.845–0.885 NIR	1 1 1 1 1 1 1 1	2800 km	2 days/ sun-synchronous	Visible and near-infrared bands used for the assessment of chlorophyll concentration, water depth, turbidity, currents, bottom characteristics, and other oceanographic information. Also useful for land and atmospheric studies and the identification, monitoring, and damage assessment of extreme natural events such as forest fires, volcanic eruptions, floods, and hurricanes.

[a] NOAA-15 and later satellites included an additional SWIR band designated channel 3A, and the former channel 3 was designated channel 3B.
[b] NOAA-13 failed
Abbreviations: NIR: near-infrared, SWIR: shortwave infrared, TIR: thermal infrared, VIS: visible

Table B.2	Characteristics of Selected European Low Spatial Resolution Sensors						
Platform	Sensor/ Altitude	Spectral resolution		Spatial resolution (GSD) in km	Swath width	Revisit period/ Orbit	Common applications
		Channel	Waveband (µm)				
Meteosat 1–7	Imager 36,000km	1 2 3	0.5–0.9 VIS/NIR 5.7–7.1 MWIR 10.5–12.5 TIR	2.5 km 5km 5km	100 degrees of longitude	Geosynchronous Images generated every 30 minutes for all three channels	1. Daytime cloud and surface mapping. 2. Water Vapor 3. Cloud and Sea Surface Temperature (SST) Meteosats 5–7 are all still operational.
Meteosat 8	SEVERI (Spinning Enhanced Visible and Infra Red Imager) 36,000km	1 2 3 4 5 6 7 8 9 10 11 12	0.56 –0.71 VIS 0.74 –0.88 NIR 1.50–1.78 SWIR 3.48–4.36 MWIR 8.30–9.10 TIR 9.80–11.80 TIR 11.0–13.0 TIR 5.35–7.15 MWIR 6.85–7.85 MWIR 9.38–9.94 TIR 12.40–14.40 TIR 0.5–0.9 VIS/NIR	1 km in VIS and NIR channels 3 km in IR and WV channels	100 degrees of longitude	Geosynchronous Images generated every 15 minutes	Channels 1 to 4 and Channels 6 and 7 are similar to those on AVHRR. Channel 5 is a new TIR channel. Channels 8 and 9 are medium wave infrared channels for Water Vapor analysis similar to that on Meteosats 1 to 7. Channel 10 is an ozone absorption channel as on HIRS. Channel 11 is a CO2 absorption channel as on the GOES-VAS sounder. Channel 12 is a broadband visible channel like the VIS/NIR channel on Meteosats 1 to 7. N.B. Originally this satellite was called Meteosat Second Generation (MSG). After launch, it was re-named Meteosat-8.
ERS-1	ATSR-1 785km	1 2 3 4	1.6 SWIR 3.7 MWIR 10.2–11.2 TIR 11.5–12.5 TIR	1 km	500 km	Polar Orbiting Sun-synchronous	Designed to measure global Sea Surface Temperature (SST) Radiometric resolution—0.1 K Launched July 1991; Failed March 2000.
ERS-2	ATSR-2 785km	1–4 5 6 7	as ATSR-1, plus 0.53–0.57 VIS 0.65–0.67 VIS 0.855–0.875 NIR	1 km	500 km	Polar Orbiting Sun-synchronous	Main mission (like ATSR-1) to measure global SST. Also 3 additional VIS/NIR channels to develop land applications— e.g. monitoring of vegetation biomass, vegetation health and growth stage. Launched April 1995; Still operational.
Envisat	AATSR 800 km	1–7	7 bands, as on ATSR-2, but fully digital output	1 km	500 km	Polar Orbiting Sun-synchronous	Radiometric resolution—0.1°K Sea Surface Temperature (SST) accuracy better than 0.5°K Continuity of the ATSR-1 and -2 datasets used for climate research and global vegetation monitoring Launched March 2002; Still operational.
Envisat	MERIS 800km	1 2 3 4 5 6 7 8 9 10 11 12 13 14 15	0.4125[a] 0.4425[a] 0.490[a] 0.510[a] 0.560[a] 0.620[a] 0.665[a] 0.68125[a] 0.705[a] 0.754[a] 0.760[a] 0.775[a] 0.865[a] 0.890[a] 0.900[a]	300m (Full Resolution) or 1.2 km (Reduced Resolution)	1,150 km	Polar Orbiting Sun-synchronous 3 day repeat cycle	1. Yellow substance and turbidity 2. Chlorophyll absorption maximum 3. Chlorophyll and other pigments 4. Turbidity, suspended sediment and red tides 5. Chlorophyll, suspended sediment 6. Suspended sediment 7. Chlorophyll absorption 8. Chlorophyll fluorescence, red edge 9. Aerosol, red edge transition 10. Oxygen absorption reference band, vegetation 11. O2 absorption R-branch 12. Aerosol, vegetation 13. Aerosol 14. Water vapor, vegetation 15. Water vapor

Abbreviations: NIR—near infrared; SWIR—shortwave infrared; MWIR—medium wave infrared; TIR—thermal infrared; VIS—visible
[a]For the MERIS sensor band centers are given because the spectral bandwidth is variable between 0.00125 µm and 0.03 µm depending on the application.

Table B.3   MODIS (Moderate Resolution Imaging Spectro-Radiometer) sensor

Platforms: NASA Terra and Aqua satellites
Swath width: 2,330 km
Orbit: 705 km, sun-synchronous, near-polar, circular orbit
10:30 AM descending node (Terra) or 1:30 PM ascending node (Aqua)
Quantization: 12 bits

Channel	Band	Wavelength (μm)	Spatial resolution	Primary use
1	VIS–Red	620–670	250 m	Land/Cloud/Aerosols Boundaries
2	NIR	841–876	250 m	
3	VIS–Blue	459–479	500 m	Land/Cloud/Aerosols Properties
4	VIS–Green	545–565	500 m	
5	SWIR	1230–1250	500 m	
6	SWIR	1628–1652	500 m	
7	SWIR	2105–2155	500 m	
8	VIS–Blue	405–420	1 km	Ocean color/ Phytoplankton/ Biogeochemistry
9	VIS–Blue	438–448	1 km	
10	VIS–Blue	483–493	1 km	
11	VIS–Green	526–536	1 km	
12	VIS–Green	546–556	1 km	
13	VIS–Red	662–672	1 km	
14	VIS–Red	673–683	1 km	
15	NIR	743–753	1 km	
16	NIR	862–877	1 km	
17	NIR	890–920	1 km	Atmospheric Water vapor
18	NIR	931–941	1 km	
19	NIR	915–965	1 km	
20	TIR	3.660–3.840	1 km	Surface/Cloud Temperature
21	TIR	3.929–3.989	1 km	
22	TIR	3.929–3.989	1 km	
23	TIR	4.020–4.080	1 km	
24	TIR	4.433–4.498	1 km	Atmospheric Temperature
25	TIR	4.482–4.549	1 km	
26	SWIR	1.360–1.390	1 km	Cirrus clouds Water vapor
27	TIR	6.535–6.895	1 km	
28	TIR	7.175–7.475	1 km	
29	TIR	8.400–8.700	1 km	Cloud properties
30	TIR	9.580–9.880	1 km	Ozone
31	TIR	10.780–11.280	1 km	Surface/Cloud Temperature
32	TIR	11.770–12.270	1 km	
33	TIR	13.185–13.485	1 km	Cloud top Altitude
34	TIR	13.485–13.785	1 km	
35	TIR	13.785–14.085	1 km	
36	TIR	14.085–14.385	1 km	

Abbreviations: NIR: near-infrared, SWIR: shortwave infrared, TIR: thermal infrared, VIS: visible

Table B.4   Landsat sensors

Platforms: Landsat satellites
Orbit: Sun-synchronous, near-polar, circular orbits crossing the equator on the southward portion of each orbit at about 9:45 AM

Satellite	Launched	Decommissioned/status	Sensors
Landsat 1	July 23, 1972	January 6, 1978	MSS and RBV*
Landsat 2	January 22, 1975	February 25, 1982	MSS and RBV*
Landsat 3	March 5, 1978	March 31, 1983	MSS and RBV*
Landsat 4	July 16, 1982	June 30, 2001	MSS and TM
Landsat 5	March 1, 1984	(Operational)	MSS and TM
Landsat 6	October 5, 1983	(Did not achieve orbit)	ETM[1]
Landsat 7	April 15, 1999	Degraded operation	ETM+[1]

* The Return Beam Vidicon (RBV) cameras had technical difficulties and little imagery was collected.
[1] The sensor onboard Landsat 6 was called the enhanced thematic mapper (ETM).
Landsat 7 carries the enhanced thematic mapper plus (ETM+).

Platform/Altitude	Sensor	Spectral resolution			Spatial resolution (m)	Swath width (km)	Revisit period (days)
		Channel	Band	Waveband (µm)			
Landsat 1,2,3/900 km	MSS	4	Green	0.50–0.60	79	185	18
		5	Red	0.60–0.70	79		
		6	NIR	0.70–0.80	79		
		7	NIR	0.80–1.10	79		
	(Landsat 3 only)	8	TIR	10.4–12.5	240		
Landsat 4,5/705 km	MSS	1	Green	0.50–0.60	82	185	16
		2	Red	0.60–0.70	82		
		3	NIR	0.70–0.80	82		
		4	NIR	0.80–1.10	82		
	TM	1	Blue	0.45–0.515	30	185	16
		2	Green	0.525–0.605	30		
		3	Red	0.63–0.69	30		
		4	NIR	0.75–0.90	30		
		5	SWIR	1.55–1.75	30		
		6	TIR	10.4–12.5	120		
		7	SWIR	2.09–2.35	30		
Landsat 7/705 km	ETM+	1	Blue	0.45–0.515	30	185	16
		2	Green	0.525–0.605	30		
		3	Red	0.63–0.69	30		
		4	NIR	0.75–0.90	30		
		5	SWIR	1.55–1.75	30		
		6	TIR	10.4–12.5	60		
		7	SWIR	2.08–2.35	30		
		8	PAN	0.52–0.90	15		

Abbreviations: NIR: near-infrared, SWIR: shortwave infrared, TIR: thermal infrared, VIS: visible, PAN: panchromatic band

Table B.5	SPOT sensors			
Satellite	**Launched**	**Decommissioned**	**Sensors**	
SPOT 1	February 1986	Decommissioned 2001	HRV and RBV*	
SPOT 2	January 1990	Operational	HRV and RBV*	
SPOT 3	September 1993	Failed November 14, 1996	HRV and RBV*	
SPOT 4	March 1998	Operational	HRVIR and vegetation instrument	
SPOT 5	May 2002	Operational	HRG, HRS, and vegetation instrument	
* The Return Beam Vidicon (RBV) cameras had technical difficulties and little imagery was collected.				

Satellite/Altitude	Sensor	Wavebands (µm)	Spatial resolution (GSD[2])	Swath width in km (vertical viewing)
SPOT 1, 2, 3/ 832 km	HRV[1] panchromatic (PAN) mode HRV[1] multispectral (XS) mode	0.51–0.73 (VIS-PAN) 0.50–0.59 (GREEN) 0.61–0.68 (RED) 0.79 - 0.89 (NIR)	10 m 20 m	60 (each HRV) 60 (each HRV)
SPOT 4	HRVIR[1] in monospectral mode HRVIR[1] in multispectral mode Vegetation instrument	0.61–0.68 (RED) 0.50–0.59 (GREEN) 0.61–0.68 (RED) 0.79–0.89 (NIR) 1.53–1.75 (SWIR) 0.43–0.47 (BLUE) 0.61–0.68 (RED) 0.78–0.89 (NIR) 1.58–1.75 (SWIR)	10 m 20 m 1.15 km to 1.7 km	60 (each HRVIR) 60 (each HRVIR) 2000
SPOT 5	HRG[1] HRG[1] HRS Vegetation instrument	0.51–0.73 (visible) 0.50–0.59 (GREEN) 0.61–0.68 (RED) 0.79–0.89 (NIR) 1.58–1.75 (SWIR) 0.50–0.59 (GREEN) 0.61–0.68 (RED) 0.79–0.89 (NIR) 1.58–1.75 (SWIR) 0.43–0.47 (BLUE) 0.61–0.68 (RED) 0.78–0.89 (NIR) 1.58–1.75 (SWIR)	2.5 m or 5 m 10 m 10 m 20 m 10 m 10 m all bands 1.15 to 1.7 km all bands	60 (each HRG) 60 (each HRG) 120 2000
Abbreviations: NIR: near-infrared, SWIR: shortwave infrared, TIR: thermal infrared, VIS: visible [1] Sensors pointable to either side of the ground track (±31°) [2] The ground sample distance (GSD) is the linear dimension across the ground area represented by a pixel, i.e., the pixel size on the ground.				

Table B.6   Sensors on the Indian Remote Sensing satellites and the ASTER sensor

Platform/ Operator	Sensor/ Altitude	Spectral resolution			Spatial resolution (m)	Swath width (km)	Revisit period (days)	Other characteristics
		Channel	Band	Waveband (μm)				
IRS-1A and B Indian National Remote Sensing Agency	LISS-I and II (each satellite carries both sensors)	1 2 3 4	Blue Green Red NIR	0.45–0.52 0.52–0.59 0.62–0.68 0.77–0.86	LISS-1 72.5 LISS-2 36.25	LISS-1 148 LISS-2 146	22	
IRS-1C and D/ Indian National Remote Sensing Agency	Panchromatic	Pan	Green-Red	0.50–0.75	5.8	70	24	Plus or minus 26ß off-nadir viewing
	LISS-III	2 3 4 5	Green Red NIR SWIR	0.52–0.59 0.62–0.68 0.77–0.86 1.55–1.70	23 23 23 70	142 142 142 148	24	The two satellites follow the same ground tracks offset by 12 days. Together they reduce the revisit period by half.
	WiFS	1 2	Red NIR	0.62–0.68 0.77–0.86	188 188	774	5 at equator	
Terra/ NASA	ASTER	1 2 3 4 5 6 7 8 9 10 11 12 13 14	Green Red NIR SWIR SWIR SWIR SWIR SWIR SWIR TIR TIR TIR TIR TIR	0.52–0.60 0.63–0.69 0.76–0.86 1.60–1.70 2.145–2.185 2.185–2.225 2.235–2.285 2.295–2.365 2.36–2.43 8.125–8.475 8.475–8.825 8.925–9.275 10.25–10.95 10.95–11.65	15 15 15 30 30 30 30 30 30 90 90 90 90 90	60 60 60 60 60 60 60 60 60 60 60 60 60 60	Acquisitions by request	15 m along-track stereo

Abbreviations: NIR: near-infrared, SWIR: shortwave infrared, TIR: thermal infrared, VIS: visible

Table B.7   Characteristics of the IKONOS, Orview-3, and Quickbird sensors

Platform/ Operator	Sensor	Spectral resolution			Spatial resolution at nadir (m)	Swath width (km)	Revisit period (days)
		Channel	Band	Waveband (μm)			
IKONOS/ Space Imaging	Panchromatic	Pan	VIS/NIR	0.45–0.90	1	11	1.5 to 3
	Multispectral	1 2 3 4	Blue Green Red NIR	0.45–0.52 0.51–0.60 0.63–0.70 0.76–0.85	4 4 4 4	11	
OrbView-3/ Orbital Imaging	Panchromatic	Pan	VIS/NIR	0.45–0.90	1	8	Less than 3
	Multispectral	1 2 3 4	Blue Green Red NIR	0.45–0.52 0.52–0.60 0.63–0.70 0.76–0.90	4 4 4 4	8	
QuickBird/ DigitalGlobe	Panchromatic	Pan	VIS/NIR	0.45–0.90	0.61	22	1 to 3.5 for selected locations
	Multispectral	1 2 3 4	Blue Green Red NIR	0.45–0.52 0.52–0.60 0.63–0.69 0.76–0.90	2.44 2.44 2.44 2.44	22	

Appendix **C** Remote sensing resources

Appendix C lists information and educational resources on remote sensing and related topics. Also included are sources for obtaining satellite and radar imagery. These lists are not comprehensive, and due to the nature of the Internet, some Web sources may have been moved or discontinued.

Aerial photography is widely available from local, state, provincial, and national government agencies. Imagery can also be obtained from firms that provide aerial survey services. Lidar, airborne digital imagery, and other specialized remote sensing services are available from a wide range of service providers that can be found on the Web sites of professional associations in the remote sensing, GIS, and geomatics fields.

Educational Web sites on remote sensing, geographic information systems (GIS), and earth sciences

Aerial Photography and Remote Sensing

Aerial Photography and Remote Sensing provides information on how to use remote sensing with GIS technology to answer environmental questions. Shannon Crum from the Geographer's Craft Project, Department of Geography, the University of Colorado at Boulder developed these materials. *www.colorado.edu/geography/gcraft/notes/remote/remote_f.html*

Columbia Earthscape

Columbia Earthscape provides a selection of links on a wide range of earth science resources available for teachers, students, scientists, and decision makers. *www.earthscape.org*

ESRI: Quick Intro to GIS, ArcView GIS, and ArcInfo

This Web site gives definitions and overviews of geographic information systems; ESRI® flagship GIS software, ArcInfo®; and ESRI view-and-query program, Arcview® GIS. *www.esri.com/industries/k-12/basicgis.html*

The Federation of American Scientists

The Federation of American Scientists Web site contains information on the benefits of using hyperspectral imagery, as well as links to resources on hyperspectral sensors and systems. *www.fas.org/irp/imint/hyper.htm*

Geodesy for the Layman

Geodesy for the Layman, maintained by the National Imagery and Mapping Agency (NIMA), provides the basic principles of geodesy, a branch of applied mathematics that deals with the earth's gravity, determining the earth's size and shape, and measuring and observing positions of points and areas of the earth's surface. *www.ngs.noaa.gov/PUBS_LIB/Geodesy4Layman/toc.htm*

The Geography Network

The Geography Network is a network of geographic information users and providers. Geography Network users can access many types of geographic content including dynamic maps, data, and more advanced Web services. The Geography Network is managed and maintained by ESRI. *www.geographynetwork.com*

Geomorphology from Space

Geomorphology from Space is an out-of-print 1986 NASA publication edited by Nicholas M. Short, Sr. and Robert W. Blair, Jr. This online version outlines the principles of geomorphology and the study of landforms and landscapes. *daac.gsfc.nasa.gov/DAAC_DOCS/geomorphology/GEO_HOME_PAGE.html*

Introduction to Remote Sensing for Agriculture

Introduction to Remote Sensing for Agriculture, provided by the U.S. Water Conservation Laboratory, outlines the principles of the remote sensing technology used in agricultural studies. *www.uswcl.ars.ag.gov/EPD/remsen/rsagintr.htm*

The NASA Distributed Active Archive Center Education Web page

The NASA Distributed Active Archive Center Education Web page is a compilation of educational sites for teachers and the general public. Topics include satellite imagery, the Terra spacecraft, and various earth sciences. *nasadaacs.eos.nasa.gov/education.html*

NASA's Jet Propulsion Laboratory Imaging Radar Web site

NASA's Jet Propulsion Laboratory Imaging Radar Web site offers information on imaging radar and in-depth looks at the publications, missions, and studies that use it. *southport.jpl.nasa.gov*

The NOAA National Geophysical Data Center Web site

The NOAA National Geophysical Data Center Web site offers an in-depth scientific discussion on digital terrain data. *www.ngdc.noaa.gov/seg/topo/topo2.shtml*

The Remote Sensing Applications Center (RSAC)

The Remote Sensing Applications Center (RSAC) assists the U.S. Forest Service in applying geospatial technology to the monitoring and mapping of natural resources. This Web site provides information on programs such as Burned Area Emergency Response (BAER) imagery support, the MODIS active fire mapping program, and other RSAC projects and remote sensing applications. *www.fs.fed.us/eng/rsac/index.html*

The University of Texas at El Paso Pan-American Center for Earth and Environmental Studies (PACES) Remote Sensing Portal
PACES is a NASA university research center that circulates data about geoscience, remote sensing, GIS technology, and other related topics. PACES maintains a large collection of satellite and aircraft imagery and related data covering most of the southwestern United States and northern Mexico. *paces. geo.utep.edu/research/remote_sensing/remote_sensing.shtml*

USGS: Geography Program Fact Sheets Web site
USGS: Geography Program Fact Sheets Web site provides information about USGS data products and services including aerial photography, digital elevation models, and various map products. *erg.usgs.gov/isb/pubs/pubslists/fctsht.html*

Visible Earth
Visible Earth is an image directory sponsored by NASA. Users are able to search the directory for images, visualizations, and animations of the earth. The site also provides links to imagery by subject. *visibleearth.nasa.gov*

Tutorials in remote sensing and related fields

The ARIA Educational Resources Web page
The ARIA Educational Resources Web page provides links to remote sensing tutorials and educational sites. This page is managed and maintained by the Arizona Regional Image Archive. *aria.arizona.edu/courses/courses.html*

The Basics of Remote Sensing From Satellites Tutorial Web site
The Basics of Remote Sensing From Satellites Tutorial Web site from NOAA shows how to interpret satellite imagery available from various sources on the Internet. *www.orbit. nesdis.noaa.gov/smcd/opdb/tutorial/intro.html*

Canada Centre for Remote Sensing (CCRS)
The Canada Centre for Remote Sensing Web site offers several online tutorials on remote sensing fundamentals and applications. *www.ccrs.nrcan.gc.ca/ccrs/homepg.pl?e*

Center for Airborne Remote Sensing and Technology and Applications Development
Operated by NASA, this Web site provides information and tutorials related to airborne remote sensing. *carstad.gsfc.nasa.gov*

The Chesapeake Bay and Mid-Atlantic from Space Remote Sensing Principles Tutorial
This tutorial is designed to present some of the basic concepts related to remote sensing and to give the reader enough information to understand what remote sensing is and how it can be useful. The tutorial also points the user to sources of Landsat 7 data products. *chesapeake.towson.edu/data/principles.asp*

NASA's Earth Observatory
NASA's Earth Observatory remote sensing tutorial provides an online text that covers subjects such as remote sensing history, methods, and use in NASA missions. The site features a glossary mode that allows the user to click on words and concepts in the text in order to obtain definitions. *earthobservatory.nasa.gov/Library/RemoteSensing*

NASA/Goddard Space Flight Center Remote Sensing Tutorial
This NASA sponsored tutorial educates students on the concepts of remote sensing, remote sensing technology, and interpretation of remotely sensed imagery. The tutorial is intended for teachers and college students, but may also be appropriate for students in grades 8 to 12. *rst.gsfc.nasa.gov*

The NASA Observatorium Web site
The NASA Observatorium Web site provides links to educational sites covering remote sensing basics, as well as links to a variety of online remote sensing tutorials. *observe.arc.nasa. gov/nasa/education/reference/main.html*

Remote Sensing: An On-Line Tutorial
This tutorial from Ohioview, a grassroots organization specializing in promoting the distribution of U.S. government satellite data for public use, steps users through remote sensing basics using a series of lecture slides and imagery examples. *dynamo.phy.ohiou.edu/tutorial/tutorial_files/frame.html*

Remote Sensing Core Curriculum
Remote Sensing Core Curriculum is managed by the American Society for Photogrammetry and Remote Sensing. It consists of a series of online educational modules covering

the principles and applications of remote sensing. With the sponsorship of multiple agencies and continuing contributions from educators around the world, the project is an evolving remote sensing educational resource. *www.r-s-c-c.org*

Remote Sensing Course On Line by Planetek Italia
This tutorial is available in Italian and English and requires online registration to complete the course. The course covers remote sensing history, basic techniques, system analysis, image processing, and practical application of remote sensing data. *www.planetek.it/corsotlr*

Remote Sensing On-line Courses from CTI Centre for Geography, Geology, and Meteorology, University of Leicester
Remote Sensing On-line Courses from CTI Centre for Geography, Geology and Meteorology, University of Leicester, is a list of links to remote sensing tutorials and classes. *www.geog.le.ac.uk/cti/online/rsonline.html*

A Remote Sensing Tutorial for Natural Resource Managers
This is a two-part tutorial that contains a lecture section and an interactive section. The lecture, or descriptive, section gives a step-by-step approach on how to incorporate remote sensing data into a natural resource project, including sections on what type of data to use, data preparation, and data interpretation. The interactive portion guides users through a real project involving mapping vegetation. *www.ag.unr.edu/serdp/tutorial/tutorial.htm*

Remote Sensing Using Satellites
Remote Sensing Using Satellites was developed under a grant from the National Science Foundation (NSF). This tutorial is designed to introduce remote sensing technology with the goal of making people better consumers of weather data. Users will learn about remote sensing technology, imagery interpretation, and have the opportunity to explore three recent hurricanes through a time sequence. *www.comet.ucar.edu/nsflab/web/index.htm*

R.S.A.T. Tutorials
R.S.A.T. Tutorials guide the user through a tour of satellite sensors, their spatial resolutions, spectral characteristics, processing, and thematic applications. Comparisons of Landsat Thematic Mapper, SPOT Panchromatic, India's IRS-1C Pan, and aerial photography are provided to help the user

visualize the information provided. Imagery examples from multiple U.S. cities, as well as British Columbia and Rio de Janeiro, are presented. *www.newc.com/rsat/tutorials.html*

The University of Arizona Remote Sensing Tutorials
The University of Arizona Remote Sensing Tutorials provides tutorials on radiometric calibration, image classification, and geometric correction of remote sensing imagery. *aria.arizona.edu/courses/tutorials/welcome.html*

The USGS Spectroscopy Lab Web site
The USGS Spectroscopy Lab Web site offers links to many online tutorials and information sources about imaging spectroscopy. *speclab.cr.usgs.gov*

Further reading

Books

Campbell, J. B. 2002. *Introduction to remote rensing.* 3rd ed. New York: The Guilford Press.

Drury, S. A. 1998. *A Guide to remote rensing: Interpreting images of the earth.* Oxford: Oxford University Press.

Henderson, F. M. and A. J. Lewis, eds. 1998. *Principles and application of imaging radar: Manual of remote sensing.* 3rd ed. Vol. 2. New York: John Wiley and Sons, Inc.

Jensen, J. R., 2000. *Remote sensing of the environment.* Upper Saddle River, N.J.: Prentice-Hall.

Kramer, H. 2002. *Observation of the earth and its environment: Survey of missions and sensors.* 4th ed. Berlin; New York: Springer-Verlag.

Lillesand, T. M., R. W. Kiefer, J. W. Chipman. 2004. *Remote sensing and image interpretation.* 5th ed. New York: John Wiley and Sons, Inc.

Rencz, A. N., and R. A. Ryerson eds. 1999. *Manual of remote sensing: Remote sensing for the earth science.* 3rd ed. New York: John Wiley and Sons, Inc.

Sabin, F. F. 1996. *Remote sensing: Principles and interpretations.* 3rd ed. New York: W. H. Freeman.

Periodicals

Canadian Journal of Remote Sensing
The official journal of the Canadian Remote Sensing Society.
www.ccrs.nrcan.gc.ca/ccrs/com/crss/cjrs/cjrs_e.html

Earth Observation Magazine
This publication discusses remote sensing, mapping, global positioning systems, geographic information systems, photogrammetry, and surveying.
www.eomonline.com

Earth Observation Quarterly
Publication of the European Space Agency.
esapub.esrin.esa.it/eoq/eoq.htm

Earth Observer
NASA newsletter published by Earth Observing System Project Science Office.
eospso.gsfc.nasa.gov/earth_observer.php

Geocarto International
A multidisciplinary journal of remote sensing and GIS published quarterly by the Geocarto International Centre.
www.geocarto.com/geocarto.html

GeoInformatica
An international journal on advances of computer science for geographic information systems.
www.kluweronline.com/issn/1384-6175

Geoworld
Geoworld reports the latest in spatial information.
www.geoplace.com/gw

IEEE Transactions on Geoscience and Remote Sensing
Remote sensing publication of the Institute of Electrical and Electronics Engineers (IEEE).
ieeexplore.ieee.org/Xplore/DynWel.jsp

Imagery Analysis Support Newsletter
A free, electronic newsletter offering articles on remote sensing imagery analysis.
imagery-analyst.com/iass.php

International Journal of Remote Sensing
This journal contains papers on the basic science, techniques, and applications of remote sensing technology.
www.tandf.co.uk/journals/tf/01431161.html

Journal of Photogrammetry and Remote Sensing
Official publication of the International Society for Photogrammetry and Remote Sensing (ISPRS).
www.elsevier.com/inca/publications/store/5/0/3/3/4/0

OE Magazine
Publication of the International Society for Optical Engineering.
oemagazine.com

Photogrammetric Engineering & Remote Sensing
Publication of the American Society for Photogrammetry and Remote Sensing.
www.asprs.org/publications.html

Photogrammetric Journal of Finland
Published by the Finnish Society of Photogrammetry and Remote Sensing and Institute of Photogrammetry and Remote Sensing at the Helsinki University of Technology, it contains articles, reports, and reviews in the field of photogrammetry, remote sensing, and spatial information sciences. *foto.hut.fi/seura/pjf.html*

Progress in Physical Geography
An international review of geographical work in the natural and environmental sciences. This publication contains a number of remote sensing review papers. *www.arnoldpublishers.com/journals/pages/pro_phy/03091333.htm*

Remote Sensing of the Environment
An interdisciplinary journal covering remote sensing topics in fields such as agriculture, meteorology, ecology, geology, and oceanography.
www.elsevier.com/wps/find/journaldescription.cwshome/505733/description#description

Spatial News
Geocomm's online newsletter for GIS, LBS, and geospatial industry professionals and students. *spatialnews.geocomm.com*

National level associations for remote sensing professionals in Canada and the United States

The American Congress on Surveying and Mapping

American Congress on Surveying and Mapping
6 Montgomery Village Avenue, Suite #403
Gaithersburg, MD 20879
Tel: 240-632-9716
Fax: 240-632-1321
E-mail: info@acsm.net
www.acsm.net

The American Society of Photogrammetry and Remote Sensing (ASPRS)

5410 Grosvenor Lane, Suite 210
Bethesda, MD 20814-2160
Tel: 301-493-0290
Fax: 301-493-0208
E-mail: asprs@asprs.org
www.asprs.org

Canadian Institute of Geomatics

1390 Prince of Wales Drive
Suite 400
Ottawa, Ontario K2C 3N6
Tel: 613-224-9851
Fax: 613-224-9577
www.cig-acsg.ca

Canadian Remote Sensing Society (CRSS) of the Canadian Aeronautics and Space Institute

130 Slater Street, Suite 618
Ottawa, Ontario K1P 6E2
Phone: 613-234-0191
Fax: 613-234-9039
E-mail: casi@casi.ca
www.casi.ca/index.php?pg=crss

Geomatics Industry Association of Canada (GIAC)

Suite 400, 1390 Prince of Wales Drive
Ottawa, Ontario, CANADA
K2C 3N6
Tel: 613-232-8770
Fax: 613-224-9577
E-mail: giac@giac.ca
www.giac.ca/site/index.cfm

Geoscience and Remote Sensing Society (GRSS) of the Institute of Electrical and Electronics Engineers (IEEE)

IEEE Corporate Office
3 Park Avenue, 17th Floor
New York, NY 10016-5997
Tel: +1 212 419 7900
Fax: +1 212 752 4929
ewh.ieee.org/soc/grss

Remote sensing and earth science glossaries

The Association for Geographic Information and the University of Edinburgh Department of Geography Online Dictionary of GIS Terms

www.geo.ed.ac.uk/agidict/welcome.html

The Canada Centre for Remote Sensing Glossary Database

www.ccrs.nrcan.gc.ca/ccrs/learn/terms/glossary/glossary_e.html

Index to Glossary Terms from the EROS Data Center

edcwww.cr.usgs.gov/glis/hyper/glossary/index

Remote Sensing Glossary from the Columbia University Remote Sensing Image Analysis Laboratory

www.ldeo.columbia.edu/rsvlab/glossary.html

Remote Sensing Glossary for Virtual Nebraska

www.casde.unl.edu/vn/glossary/intro.htm

Data resources

Government agencies

Canada Centre for Remote Sensing (CCRS)

The Canada Centre for Remote Sensing, Natural Resources Canada, coordinates a national research program that develops and applies remote sensing technology to sustainable development and environmental protection. CCRS also develops geospatial information applications and provides electronic access to spatial databases. The CCRS Web site offers catalogs of satellite images, maps, and free data, as well as the geospatial standards of Canada and information on CCRS ground stations and services. *www.ccrs.nrcan.gc.ca/ ccrs/homepg.pl?e*

Canada Centre for Remote Sensing
Natural Resources Canada
588 Booth Street
Ottawa, Ontario
Canada, K1A 0Y7
Tel: 613-947-1216
E-mail: info@ccrs.nrcan.gc.ca

European Space Agency (ESA)

The ESA is a conglomeration of European member states that have pooled intellectual and financial resources to run projects designed to find out more about the earth, the solar system, and the universe, as well as to develop satellite-based technologies. The ESA Web site offers image archives and information on past and present missions and satellites in orbit. *sci.esa.int/index.cfm*

ESA
8-10 rue Mario Nikis
75738 Paris
Cedex 15
France
Communication department:
Tel: 33-1-5369-7155
Fax: 33-1-5369-7690

Indian Space Research Organization (ISRO)

India's space program presently operates space systems including the Indian National Satellite (INSAT) for telecommunication, television broadcasting, meteorology and disaster warning, and Indian Remote Sensing Satellite (IRS) for resources monitoring and management. The ISRO Web site includes information on India's space program, current missions, and where to obtain ISRO remote sensing data. *www.isro.org*

For remote sensing data contact the National Remote Sensing Agency, an autonomous body under India's Department of Space.
NRSA Data Centre [NDC]
National Remote Sensing Agency
Department of Space
Government of India
Balanagar, Hyderabad 500 037

Tel: 00-91-040-387-8560 or 387-9572 Extn: 2327
Fax: 00-91-040-387-8664 or 387-8158
E-mail: sales@nrsa.gov.in
www.nrsa.gov.in

National Aeronautics and Space Administration (NASA)

NASA's Goddard Space Flight Center (GSFC) is dedicated to the pursuit of knowledge about the earth, solar system, and universe. It is home to the largest organization of scientists and engineers specializing in earth sciences in the United States. The GSFC Web site provides information about current NASA events, the status of space science missions, access to satellite photos, information on space flight technology under development, and the latest in space and earth science. The site also includes links to other imagery sources in the Photos section. *www.gsfc.nasa.gov*

Goddard Space Flight Center
Code 130, Office of Public Affairs
Greenbelt, MD 20771
Tel: 301-286-8955
E-mail: gsfcpao@pop100.gsfc.nasa.gov

National Space Development Agency of Japan (NASDA)

NASDA was founded in 1969 as a replacement for the National Space Development Center (NSDC) of the Science and Technology Agency (STA) of Japan. NASDA's goal is to forward exploration of the space environment through the development and launching of satellites and launch vehicles. The NASDA Web site includes information on current missions and satellites in orbit, a multimedia remote sensing imagery archive, and a history of Japan's space program. *www.nasda.go.jp/index_e.html*

National Oceanic and Atmospheric Administration (NOAA)

NOAA is an organization of the U.S. Department of Commerce. The agency provides information on weather, charts the seas and atmosphere, guides use and protection of ocean and coastal resources, and conducts research to improve understanding and stewardship of the environment. The NOAA Web site offers more information about the agency and access to archived GOES and POES imagery as well as aerial photography. Details about how to purchase imagery are also provided. *www.noaa.gov*

To purchase aerial photography
National Geodetic Survey - Information Services
National Ocean Service - NOAA
1315 East West Hwy.
Silver Spring, MD 20910
Tel: 301-713-2692
Fax: 301-713-4176
E-mail: info_center@ngs.noaa.gov
　　　　Joan.Rikon@noaa.gov
www.ngs.noaa.gov (Products and Services)
www.ngs.noaa.gov/PC_PROD/Catalog/aerial_photos.htm
(Current catalog of NOAA aerial photos)

To purchase satellite imagery:
E-mail: satinfo@nesdis.noaa.gov
www.goes.noaa.gov (For current satellite photos)
noaasis.noaa.gov (For information on satellite status)
www.saa.noaa.gov (For satellite picture archive)
lwf.ncdc.noaa.gov/oa/satellite/satelliteresources.html (For satellite education)
npoess.noaa.gov (For the NPOESS program)
dmsp.ngdc.noaa.gov/dmsp.html (For Defense Meteorological Satellite Program photos)

The United States Geological Survey (USGS)

The United States Geological Survey is the science agency for the U.S. Department of the Interior. The USGS provides for purchase index/mapping photography, survey photography, space photography, digital orthophoto quadrangles, as well as photography from the National Aerial Photography Program (NAPP), the National High Altitude Photography program (NHAP), and various multispectral aircraft scanners. The USGS also maintains an informational database of aerial photographic coverage of the United States and its territories that dates back to the 1940s and has a photographic library containing 300,000 photographs taken during geologic studies of the United States and its territories from 1869 to the present. The USGS Web site also provides links and information on how to obtain imagery from satellites including Landsat 1–7, the Terra satellite, and the NOAA satellite series. The Earth Resources Observation Systems (EROS) Data Center (EDC), a data management, systems development, and research field center for the USGS National Mapping Division, also offers imagery from these satellites.
www.usgs.gov

EROS Data Center
U.S. Geological Survey
47914 252nd St.
Sioux Falls, SD 57198
Tel: 605-594-6151
　　　800-252-4547
Fax: 605-594-6589
E-mail: custserv@usgs.gov

Commercial satellite and radar image sources

DigitalGlobe

DigitalGlobe offers imagery from its Quickbird satellite. Imagery is available with varying degrees of processing: Basic Imagery products are corrected only for radiometric distortions and adjustments for internal sensor geometry, optical and sensor distortions; Standard Imagery products are designed for users with knowledge of remote sensing applications and image processing tools that require data of modest absolute geometric accuracy or large area coverage or both; and Orthorectified Imagery products are designed for users who require an imagery product that is GIS-ready or for users who require a high degree of absolute geometric accuracy for analytical applications. Other products offered by DigitalGlobe include the AgroWatch™ program, which provides calibrated and quantifiable crop management maps for the production agriculture market; The Millenium Mosaic™, a natural color, 15-meter mosaic; SPOT products; and radar images and digital elevation models from Intermap Technologies. *www.digitalglobe.com*

DigitalGlobe
1900 Pike Road
Longmont, CO 80501-6700 USA
Tel: 800-655-7929 or 303-682-3800
For product inquiries and general questions:
Tel: 800-496-1225 or 303-702-5561
Fax: 303-702-5562
E-mail: info@digitalglobe.com

ImageSat International

ImageSat International is a commercial operator of the EROS A satellites. The digital imaging sensor produces panchromatic imagery with ground resolutions of 1.8 m GSD to 1m GSD. EROS A is designed to be highly maneuverable and quickly pointed and stabilized to image customer-specified sites at nadir (perpendicular to the surface) or at oblique angles up to 45 degrees. Oblique viewing enables the satellite to view virtually any site on the earth as often as two to three times per week. Future EROS B generation satellites are planned that will offer both panchromatic and multispectral imagery (visible and near-infrared bands). *www.imagesatintl.com*

ImageSat International N.V.
2 Kaufman Street, 17th Floor
Tel Aviv 61500 Israel
Tel: +972-3-796-0600
Fax: +972-3-516-3430

Intermap Technologies, Inc.

Intermap Technologies, Inc. specializes in photogrammetric mapping and airborne radar to produce low-cost, high-accuracy digital elevation models, radar image mosaics, and cartographic products for base mapping and visualization applications. Using its airborne STAR-3i® interferometric synthetic aperture radar (IfSAR), Intermap provides large-area mapping services for government agencies and private industry worldwide. *www.intermaptechnologies.com*

Denver office
Intermap Technologies, Inc.
E-mail: info@intermaptechnologies.com
400 Inverness Parkway, Suite 330
Englewood, CO 80112-5847
Tel: (303) 708-0955
Fax: (303) 708-0952

Calgary—Canadian Corporate Office
Intermap Technologies Corp.
E-mail: info@intermaptechnologies.com
#1000, 736-8th Ave. S.W.
Calgary, Alberta
Canada T2P 1H4
Tel: (403) 266-0900
Fax: (403) 265-0499

ORBIMAGE

ORBIMAGE processes images from a variety of remote sensing data sources, including its own OrbView satellites, in addition to Landsat, SPOT, and Radarsat. ORBIMAGE offers a variety of other products including the SeaStar Pro Fisheries Information Service, which provides fish-finding maps derived from OrbView-2 ocean imagery; imagery through its OrbView Cities program, a growing archive of high-resolution, one-meter panchromatic imagery of major U.S. and non-U.S. urban areas; and aerial imagery from its OrbView Cities aerial archive. Processing services include radiometric balancing, geopositioning, digital elevation data production, orthorectification and mosaicking, land-use and land-cover classification. *www.orbimage.com*

ORBIMAGE
21700 Atlantic Boulevard
Dulles, VA 20166
Tel: 703-480-7500 or 703-480-7539
Fax: 703-450-9570
E-mail: customer.support@orbimage.com

RADARSAT International (RSI)

Radarsat International's range of products is derived from the majority of today's commercially available Earth-observation satellites including Radarsat 1, Radarsat 2, Landsat 4/5, Landsat 7, IKONOS, IRS, ERS, QuickBird, and Envisat. Products include geocorrected imagery, digital elevation models, and application-specific products such as land classification maps and ocean oil seeps vectors. RSI's range of services includes near-real-time delivery for time-critical operations, as well as training, custom product generation, and GIS project implementation. *www.rsi.ca*

Client Services:
RADARSAT International
13800 Commerce Parkway
MacDonald Dettwiler Building
Richmond, British Columbia
Canada, V6V 2J3
Tel: (604) 244-0400
     Toll free (North America) 888-780-6444
Fax: (604) 244-0404
E-mail: info@rsi.ca

Space Imaging

Space Imaging collects, processes, and integrates data from its IKONOS satellite, as well as the Radarsat, Landsat, IRS satellite, and Digital Airborn Imagery services. Other services provided include developing maps and value-added applications for forestry, fire, and ecosystems; national and global security; airposts and harbors; and local, regional, and deferral governments. Space Imaging has a solutions organization on staff that specializes in cartography, remote sensing, image processing, database design and access, GIS application development, and data analysis. *www.spaceimaging.com*

Space Imaging
12076 Grant St.
Thornton, CO 80241
USA Tel: 800-232-9037
International Tel: 1-301-552-0537
E-mail: info@spaceimaging.com

SPOT Image

SPOT Image offers products derived from its SPOT satellite series images. SPOT Image offers imagery at levels of preprocessing that ranges from level 1A preprocessing (only detector equalization is performed) to level 3A preprocessing (orthorectified images). Other products include an ortho-imagery archive; digital elevation models; image maps; seamless, true-color, ten-meter resolution U.S. statewide coverage; coverages of major U.S. cities; and a line of precision agriculture image products. *www.spot.com*

SPOT Image
14595 Avion Parkway
Suite 500
Chantilly, VA 20151
Tel: 1-800-ASK-SPOT / (703) 715-3100
Fax: (703) 715-3120
General e-mail: info@spot.com
Technical/product support e-mail: techsupport@spot.com

Terraserver.com

Terraserver.com provides a subscription-based tool to satisfy a wide variety of mapping needs from environmental changes to mapping. Terraserver allows customers to select and print high-resolution images and view all imagery for free—down to eight meters of detail. Terraserver subscribers can view high-resolution images of one or two meters of detail and then purchase the images immediately.

Terraserver has imagery from more than fifty-six countries. Satellite images are generated from many sources including ORBIMAGE, SPIN 2, UK Perspectives, Air Photo USA, Aero-Data, ImageSat, SpaceShots, and GEOCOM/USGS. *www.terraserver.com*

Terraserver.com
5995 Chapel Rd.
Raleigh, NC 27607

Image galleries

All satellite image data providers include an image gallery on their Web sites. The following are additional sites that provide a wide range of useful remote sensing image examples.

Canada Centre for Remote Sensing
www.ccrs.nrcan.gc.ca/ccrs/data/data_e.html

European Space Agency Multimedia Galleries
www.esa.int/esaCP/index.html

Eurimage Image Gallery
www.eurimage.com/gallery/webfiles/gallery.html

NASA Image Galleries
Advanced Spaceborne Thermal Emission and Reflection Radiometer (ASTER)
Jet Propulsion Laboratory, NASA
asterweb.jpl.nasa.gov/gallery/default.htm

Airborne Sensor Facility Gallery
asapdata.arc.nasa.gov/Image.htm

Earth Observatory
earthobservatory.nasa.gov

GRIN Image Gallery
grin.hq.nasa.gov/BROWSE/earth_1.html

Johnson Space Centre Digital Image Collection
images.jsc.nasa.gov

Moderate Resolution Imaging Spectroradiometer (MODIS) sensor Image Gallery
modis.gsfc.nasa.gov/gallery/index.php

Planetary Photojournal
photojournal.jpl.nasa.gov/PIA.html

NOAA Operational Significant Events Imagery
www.osei.noaa.gov

USGS Landsat Image Gallery
landsat7.usgs.gov/gallery/detail/224

Index